Recent Advances in Emulsions and Applications

Recent Advances in Emulsions and Applications

Editors

César Burgos-Díaz
Mauricio Opazo-Navarrete
Eduardo Morales

MDPI

Basel • Beijing • Wuhan • Barcelona • Belgrade • Novi Sad • Cluj • Manchester

Editors
César Burgos-Díaz
Agriaquaculture Nutritional
Genomic Center (CGNA)
Temuco, Chile

Mauricio Opazo-Navarrete
Agriaquaculture Nutritional
Genomic Center (CGNA)
Temuco, Chile

Eduardo Morales
Universidad de La Frontera
Temuco, Chile

Editorial Office
MDPI
St. Alban-Anlage 66
4052 Basel, Switzerland

This is a reprint of articles from the Special Issue published online in the open access journal *Colloids and Interfaces* (ISSN 2504-5377) (available at: https://www.mdpi.com/journal/colloids/special_issues/Emulsion).

For citation purposes, cite each article independently as indicated on the article page online and as indicated below:

Lastname, A.A.; Lastname, B.B. Article Title. *Journal Name* **Year**, *Volume Number*, Page Range.

ISBN 978-3-0365-8950-3 (Hbk)
ISBN 978-3-0365-8951-0 (PDF)
doi.org/10.3390/books978-3-0365-8951-0

Contents

About the Editors vii

Preface ix

César Burgos-Díaz, Karla A. Garrido-Miranda, Daniel A. Palacio, Manuel Chacón-Fuentes, Mauricio Opazo-Navarrete and Mariela Bustamante
Food-Grade Oil-in-Water (O/W) Pickering Emulsions Stabilized by Agri-Food Byproduct Particles
Reprinted from: *Colloids* **2023**, *7*, 27, doi:10.3390/colloids7020027 . . . . **1**

Hualu Zhou, Dingkui Qin, Giang Vu and David Julian McClements
Impact of Operating Parameters on the Production of Nanoemulsions Using a High-Pressure Homogenizer with Flow Pattern and Back Pressure Control
Reprinted from: *Colloids* **2023**, *7*, 21, doi:10.3390/colloids7010021 . . . . **21**

Natalia Riquelme, Constanza Savignones, Ayelén López, Rommy N. Zúñiga and Carla Arancibia
Effect of Gelling Agent Type on the Physical Properties of Nanoemulsion-Based Gels
Reprinted from: *Colloids* **2023**, *7*, 49, doi:10.3390/colloids7030049 . . . . **37**

Giselle Vite, Samuel Lopez-Godoy, Pedro Díaz-Leyva and Anna Kozina
Improving the Size Distribution of Polymeric Oblates Fabricated by the Emulsion-in-Gel Deformation Method
Reprinted from: *Colloids* **2023**, *7*, 50, doi:10.3390/colloids7030050 . . . . **53**

Shashi Kiran Misra and Kamla Pathak
Nose-to-Brain Targeting via Nanoemulsion: Significance and Evidence
Reprinted from: *Colloids* **2023**, *7*, 23, doi:10.3390/colloids7010023 . . . . **65**

Rakesh Kumar Ameta, Kunjal Soni and Ajaya Bhattarai
Recent Advances in Improving the Bioavailability of Hydrophobic/Lipophilic Drugs and Their Delivery via Self-Emulsifying Formulations
Reprinted from: *Colloids* **2023**, *7*, 16, doi:10.3390/colloids7010016 . . . . **93**

Gina Kaysan, Theresa Hirsch, Konrad Dubil and Matthias Kind
A Microfluidic Approach to Investigate the Contact Force Needed for Successful Contact-Mediated Nucleation
Reprinted from: *Colloids* **2023**, *7*, 12, doi:10.3390/colloids7010012 . . . . **115**

Hayato Takase, Nozomi Watanabe, Koichiro Shiomori, Yukihiro Okamoto, Endang Ciptawati, Hideki Matsune and Hiroshi Umakoshi
Preparation of Hydrophobic Monolithic Supermacroporous Cryogel Particles for the Separation of Stabilized Oil-in-Water Emulsion
Reprinted from: *Colloids* **2023**, *7*, 9, doi:10.3390/colloids7010009 . . . . **135**

Pelin Salum, Çağla Ulubaş, Onur Güven, Levent Yurdaer Aydemir and Zafer Erbay
Casein-Hydrolysate-Loaded W/O Emulsion Preparation as the Primary Emulsion of Double Emulsions: Effects of Varied Phase Fractions, Emulsifier Types, and Concentrations
Reprinted from: *Colloids* **2022**, *7*, 1, doi:10.3390/colloids7010001 . . . . **145**

Mauricio Opazo-Navarrete, César Burgos-Díaz, Karla A. Garrido-Miranda and Sergio Acuña-Nelson
Effect of Enzymatic Hydrolysis on Solubility and Emulsifying Properties of Lupin Proteins (*Lupinus luteus*)
Reprinted from: *Colloids* **2022**, *6*, 82, doi:10.3390/colloids6040082 . . . . **159**

Sotirios Kiokias and Vassiliki Oreopoulou
Review on the Antioxidant Activity of Phenolics in o/w Emulsions along with the Impact of a Few Important Factors on Their Interfacial Behaviour
Reprinted from: *Colloids* **2022**, *6*, 79, doi:10.3390/colloids6040079 . . . . **173**

Rocío Díaz-Ruiz, Amanda Laca, Ismael Marcet, Lemuel Martínez-Rey, María Matos and Gemma Gutiérrez
Addition of Trans-Resveratrol-Loaded, Highly Concentrated Double Emulsion to Moisturizing Cream: Effect on Physicochemical Properties
Reprinted from: *Colloids* **2022**, *6*, 70, doi:10.3390/colloids6040070 . . . . **189**

Sirlene Adriana Kleinubing, Priscila Miyuki Outuki, Éverton da Silva Santos, Jaqueline Hoscheid, Getulio Capello Tominc, Mariana Dalmagro, et al.
Stability Studies and the In Vitro Leishmanicidal Activity of Hyaluronic Acid-Based Nanoemulsion Containing *Pterodon pubescens* Benth. Oil
Reprinted from: *Colloids* **2022**, *6*, 64, doi:10.3390/colloids6040064 . . . . **201**

Carolina Calderón-Chiu, Montserrat Calderón-Santoyo, Julio César Barros-Castillo, José Alfredo Díaz and Juan Arturo Ragazzo-Sánchez
Structural Modification of Jackfruit Leaf Protein Concentrate by Enzymatic Hydrolysis and Their Effect on the Emulsifier Properties
Reprinted from: *Colloids* **2022**, *6*, 52, doi:10.3390/colloids6040052 . . . . **217**

Julián Vera-Salgado, Carolina Calderón-Chiu, Montserrat Calderón-Santoyo, Julio César Barros-Castillo, Ulises Miguel López-García and Juan Arturo Ragazzo-Sánchez
Ultrasound-Assisted Extraction of *Artocarpus heterophyllus* L. Leaf Protein Concentrate: Solubility, Foaming, Emulsifying, and Antioxidant Properties of Protein Hydrolysates
Reprinted from: *Colloids* **2022**, *6*, 50, doi:10.3390/colloids6040050 . . . . **235**

Hadel A. Abo Enin, Ahad Fahd Alquthami, Ahad Mohammed Alwagdani, Lujain Mahmoud Yousef, Majd Safar Albuqami, Miad Abdulaziz Alharthi and Hashem O. Alsaab
Utilizing TPGS for Optimizing Quercetin Nanoemulsion for Colon Cancer Cells Inhibition
Reprinted from: *Colloids* **2022**, *6*, 49, doi:10.3390/colloids6030049 . . . . **247**

Daniela Sotomayor-Gerding, Eduardo Morales and Mónica Rubilar
Comparison between Quinoa and *Quillaja saponins* in the Formation, Stability and Digestibility of Astaxanthin-Canola Oil Emulsions
Reprinted from: *Colloids* **2022**, *6*, 43, doi:10.3390/colloids6030043 . . . . **261**

About the Editors

César Burgos-Díaz

César Burgos-Díaz holds a Master's degree in Molecular Biotechnology and completed his Ph.D. in Biotechnology at the University of Barcelona, Spain. Additionally, he has conducted postdoctoral research focused on the development of colloidal systems to encapsulate food-grade flavor compounds. Currently, he serves as a researcher in the Food Science and Technology unit at the Agriaquaculture Nutritional Genomic Center (CGNA) in Chile. Dr. César Burgos-Díaz is a recognized specialist in emulsion science, the microencapsulation of bioactive compounds, emulsion-based delivery systems, the development of plant-based foods, and the development of food products with improved properties. The results of his research have been published in various scientific papers and book chapters.

Mauricio Opazo Navarrete

Mauricio Opazo Navarrete is a Food Engineer and holds a Master's degree in Science and Food Engineering from the University of Bío-Bío, Chile. He completed his PhD studies in Food Process Engineering at Wageningen University in the Netherlands. He is currently a researcher at the Agriaquaculture Nutritional Genomic Center (CGNA) in Chile. He is a recognized specialist in the field of Food Science and Technology, with expertise encompassing proteins, food digestion, and the chemical and physical properties of food products. He has authored research publications in the domains of rheology, food digestion, emerging processing technologies, and colloidal systems within the realm of food science and technology.

Eduardo Morales

Eduardo Morales Antonio is a Food Engineer and holds a master's degree in Biotechnology Engineering Sciences from the Universidad de La Frontera (UFRO), Chile. He completed his PhD in Bioprocess Engineering Sciences at the UFRO, Chile. At the present, he is a researcher at the Núcleo Científico Tecnológico en Biorecursos (BIOREN-UFRO) in Chile. His lines of research include the Technology and processes of vegetable raw materials, encapsulation technologies for food development, and the structuring of vegetable oils. He is the author of research publications in the encapsulation of bioactive compounds, colloidal systems, and oil structures in food science and technology.

Preface

In recent years, there have been significant advances made in emulsion science in order to improve the quality and performance of different emulsion-based products using new techniques and structural designs. Emulsion systems have been employed for many years to develop a wide variety of commercial emulsified products, including food, pharmaceutical, and cosmetic products. Furthermore, this type of colloidal system has been utilized as a vehicle for the encapsulation and delivery of different bioactive compounds, such as antioxidants, vitamins, and fragrances. Therefore, a new generation of advanced emulsions may lead to products with enhanced quality and functionality.

The present reprint offers a comprehensive overview of recent advances in the development of novel emulsion systems and their potential applications in various industrial fields. Consequently, this Special Issue has established a collection of articles and reviews that present new studies addressing the development and application of advanced emulsion technologies. For instance, it includes chapters related to emulsions stabilized by particle-based emulsifiers (Pickering emulsions), emulsions-in-gels, nanoemulsion-based gels, emulsion-based delivery systems, the improvement of emulsifying properties through enzymatic treatment, double emulsions, and plant-based emulsifies, among others.

This Special Issue provides valuable information for researchers in the field of colloidal science, specifically in areas related to food, cosmetics, biomedicine, and pharmaceutical science.

Finally, the Editors would like to express their gratitude to all the authors who contributed to the success of this Special Issue, as well as the reviewers who evaluated the submissions in order to ensure the quality of the published manuscripts. The Editors would also like to extend their appreciation to the editorial staff of *Colloids and Interfaces* (MDPI) for their valuable, constant, and professional support in editing this reprint.

César Burgos-Díaz, Mauricio Opazo-Navarrete, and Eduardo Morales
Editors

colloids and interfaces

MDPI

Review

Food-Grade Oil-in-Water (O/W) Pickering Emulsions Stabilized by Agri-Food Byproduct Particles

César Burgos-Díaz [1,*,†], Karla A. Garrido-Miranda [1,*,†], Daniel A. Palacio [2,†], Manuel Chacón-Fuentes [1,†], Mauricio Opazo-Navarrete [1,†] and Mariela Bustamante [3,†]

1 Agriaquaculture Nutritional Genomic Center (CGNA), Temuco 4780000, Chile
2 Departamento de Polímeros, Facultad de Ciencias Químicas, Universidad de Concepción, Concepción 4030000, Chile
3 Department of Chemical Engineering, Scientific and Technological Bioresource Nucleus (BIOREN), Centre for Biotechnology and Bioengineering (CeBIB), Universidad de La Frontera, Temuco 4780000, Chile
* Correspondence: cesar.burgos@cgna.cl (C.B.-D.); karla.garrido@cgna.cl (K.A.G.-M.)
† These authors contributed equally to this work.

Abstract: In recent years, emulsions stabilized by solid particles (known as Pickering emulsions) have gained considerable attention due to their excellent stability and for being environmentally friendly compared to the emulsions stabilized by synthetic surfactants. In this context, edible Pickering stabilizers from agri-food byproducts have attracted much interest because of their noteworthy benefits, such as easy preparation, excellent biocompatibility, and unique interfacial properties. Consequently, different food-grade particles have been reported in recent publications with distinct raw materials and preparation methods. Moreover, emulsions stabilized by solid particles can be applied in a wide range of industrial fields, such as food, biomedicine, cosmetics, and fine chemical synthesis. Therefore, this review aims to provide a comprehensive overview of Pickering emulsions stabilized by a diverse range of edible solid particles, specifically agri-food byproducts, including legumes, oil seeds, and fruit byproducts. Moreover, this review summarizes some aspects related to the factors that influence the stabilization and physicochemical properties of Pickering emulsions. In addition, the current research trends in applications of edible Pickering emulsions are documented. Consequently, this review will detail the latest progress and new trends in the field of edible Pickering emulsions for readers.

Keywords: Pickering emulsions; agri-food byproducts; Pickering particles; emulsifying capacity

Citation: Burgos-Díaz, C.; Garrido-Miranda, K.A.; Palacio, D.A.; Chacón-Fuentes, M.; Opazo-Navarrete, M.; Bustamante, M. Food-Grade Oil-in-Water (O/W) Pickering Emulsions Stabilized by Agri-Food Byproduct Particles. *Colloids Interfaces* **2023**, 7, 27. https://doi.org/10.3390/colloids7020027

Academic Editor: Ramón G. Rubio

Received: 8 March 2023
Revised: 27 March 2023
Accepted: 29 March 2023
Published: 31 March 2023

1. Introduction

Many raw materials of plant origin such as soy, almond, rice, coconut, oat, and lupin, among others, are used to produce plant-based foods, which include meat and dairy analogues. However, the production of plant-based foods produces large amounts of valuable byproducts, which commonly are discarded by industry [1]. In this regard, there is a great opportunity to add value to these byproducts, thus contributing to the circular economy and environmental protection. Therefore, it is necessary to find specific and innovative applications of these food waste/byproducts. From this perspective, a new and relevant aspect is the use of agro-industrial byproducts as an alternative, renewable, and inexpensive source of "natural stabilizers" (solid particles) for various industrial applications [2]. Accordingly, it should be mentioned that development and research in this field are still very limited. Therefore, the utilization of novel particles obtained from "plant sources" could satisfy consumer demands to replace synthetic surfactants with ingredients with a lower environmental impact.

Emulsions are liquid–liquid (oil–water) colloidal systems that generally are obtained in the presence of surface-active molecules such as surfactants, amphiphilic polymers, or natural polymers (proteins) [3]. However, the use of surfactants in large-scale industrial applications and personal care products is not cost-effective, and in some cases, may cause adverse effects such as irritation and hemolytic behavior [4]. Therefore, the use of natural surface-active particles may be a promising alternative for use as surfactants. In this context, S. Pickering [5] observed that solid amphiphilic particles can also be used to stabilize emulsions, which were called Pickering emulsions. This type of system refers to emulsions that are not physically stabilized by conventional emulsifier molecules, but instead by solid colloidal particles [6]. The stabilization mechanism of this type of emulsion involves the partial adsorption of solid particles at the interface between oil and water (i.e., dual wettability) [4]. The adsorption of a particle at the oil–water interface is strongly influenced by its wettability, which depends on the oil–water interface contact angle [7].

In recent years, the study and use of Pickering emulsions have attracted a lot of interest in the field of food and pharmaceutical research [8]. Pickering emulsions have several advantages over traditional surfactant-stabilized emulsions, such as high stability against coalescence (even with large droplets), Ostwald ripening, and the emulsion does not contain surfactants, among others. Furthermore, in comparison to traditional emulsions, Pickering emulsions seem to be more appropriate in the development of encapsulation and delivery systems for bioactive compounds, for instance, antioxidants, probiotics, polyphenols, carotenoids, and tocotrienols, among others [8–10].

Many studies have been addressed to studying particle-stabilized Pickering emulsions. However, very few studies have been focused on food applications, mainly because many Pickering particles are of inorganic origin, which are not food-grade [6]. Therefore, to date, there is still a limited variety of food-grade inorganic particles due to their low biocompatibility and biodegradability. In addition, many of these particles require time-consuming, expensive, and environmentally unfriendly processes to be synthetized [11].

On the other hand, most emulsions are stabilized by using inorganic particles, restricting their used in food-grade formulations [12]. According to Jiang et al. [4], the first colloidal particles used to stabilize Pickering emulsions were inorganic particles, which have been extensively studied for this purpose. Among them, silica (as a colloidal particle) has been extensively used to stabilize Pickering emulsions owing to its ability to resist acidic and basic conditions, its easily modified surface, and the ability to control its size and structure [4]. Since most inorganic particles are not food-grade, it is necessary to explore new and effective Pickering stabilizers as an alternative to inorganic particles so that they are suitable for human consumption. In this context, the use of byproducts derived from natural sources, such as oil seeds, fruits, and legumes, has attracted great interest due to their techno-functional properties and low cost. However, studies on this area remain practically unexplored or are still very limited.

Based on the above considerations, this review provides a comprehensive overview of the recent advances in the stabilization of Pickering emulsions using diverse agri-food byproduct particles. Thus, several types of solid particles will be listed and discussed in detail. These novel approaches will provide relevant information on the behavior of solid plant-based particles in the development of new emulsion systems for food applications. In addition, this knowledge will help to valorize different byproducts/wastes as potential surface-active sources, thereby contributing to the circular economy.

2. Factors That Influence Pickering Emulsion Stabilization

2.1. Particle Wettability

Pickering emulsions have a stabilization mechanism different from emulsions stabilized by traditional emulsifiers and biopolymers with two distinct hydrophilic and hydrophobic regions. Pickering emulsions are stabilized through adsorption of solid particles at the oil–water interface and, therefore, they does not need to be amphiphilic [13]. Thus, the wettability of particles is a crucial parameter factor in particles adsorbing at the surface of the droplets and in the final stability of Pickering emulsions [7]. In particle-stabilized emulsions, whether the emulsion is oil/water (O/W) or water/oil (W/O) type is dependent on the wettability of the particle, which is determined by the contact angle in water at the oil–particle–water interface [14]. Likewise, the adsorption of particles at the oil–water interface is strongly influenced by its hydrophobicity. According to the contact angles of solid particles (Figure 1), hydrophilic particles (i.e., with a contact angle of <90° measured through the water phase) should stabilize O/W emulsions. Conversely, hydrophobic particles (i.e., with a contact angle of >90°) should better stabilize W/O emulsions [7]. However, particles that are completely wetted by water or oil remain dispersed in that phase and cannot form an emulsion.

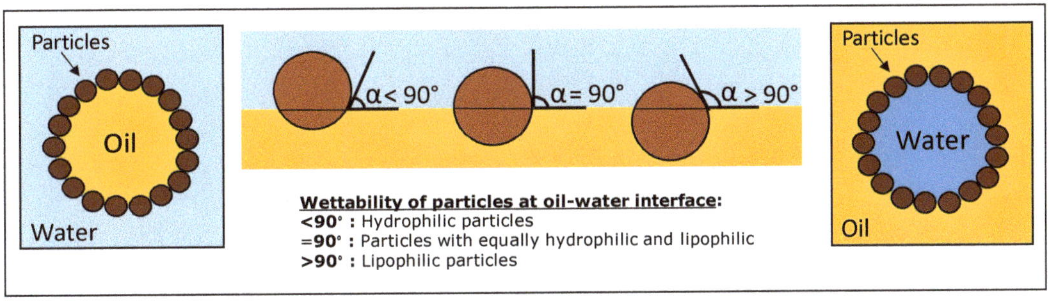

Figure 1. Schematic representation of wettability of Pickering stabilizers. The solid particles adsorbed at the oil–water interface to stabilize the oil droplets of the emulsion.

In the majority of cases, the effectiveness of Pickering stabilizers is associated with their wettability, which is highly influenced by the particles' hydrophobic nature and consequently can directly impact the type of Pickering emulsion formed (O/W or W/O). Thus, a strong adsorption at the oil–water interface occurs due to the partial wettability of spherical solid particles, resulting in robust steric hindrance. This can prevent droplet coalescence and flocculation in the emulsion by the steric mechanism [13]. Apart from the wettability of particles, other external factors can also play a crucial role in emulsion stability, including the particle size, pH, particle concentration, ionic strength, the droplet size of emulsion, particle type, and proportion of the oil phase [7]. It is important to mention that although a contact angle of around 90° is theoretically considered optimal for stabilizing Pickering emulsions [1], various other factors, such as particle size, electrical potential, and particle shape, among others [1], can also impact the formation of Pickering emulsions.

2.2. Particle Concentration

Another factor that can influence the stability of the Pickering emulsion is the concentration of solid particles. The stability of the emulsion and droplet size are significantly influenced by the concentration of particles [2,15]. This is because solid particles need to be adsorbed at the oil–water interface of droplets to perform as emulsifiers, and thus the emulsion stability tends to increase proportionally with particle concentration [7]. Burgos-Díaz et al. [6,8] observed that when the concentration of food-grade particles increased, the emulsion was stable against creaming for 45 days. This behavior was attributed to the greater amount of particles that could cover the oil–water interface and thus improve the emulsion stability (Figure 2). Likewise, the particle concentration had an impact on the emulsion droplet size.

According to Li et al. [13], if the particle concentration is insufficient to cover the newly formed droplets during emulsion preparation, these droplets will only be partly covered by particles (Figure 3). Thus, emulsion droplets can progressively merge, leading to fast coalescence and large droplets. On the contrary, with the increase in the concentration, more particles can adsorb at the interface (oil–water) to form a single or multi-layer structure, which avoids the coalescence of emulsion droplets and thus stabilizes the emulsion (Figure 3).

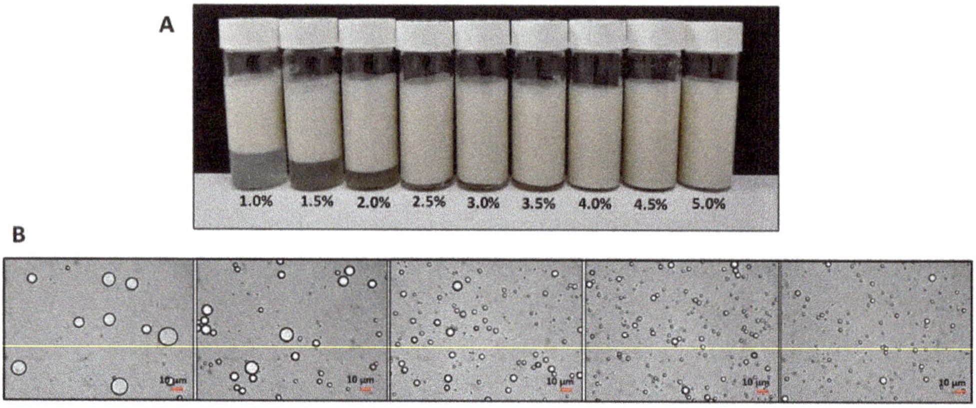

Figure 2. (**A**) Images of O/W emulsions stabilized by lupin hull at different concentrations (1–5%, *w*/*w*) after storage for 45 days. (**B**) Optical micrographs of the O/W Pickering emulsions stabilized at different byproduct concentrations (1.0–5.0%, *w*/*w*). The images were acquired at 40× magnification. The emulsions were prepared at a fixed oil concentration (20%, *w*/*w*) (image adapted from Burgos-Díaz et al. [1]).

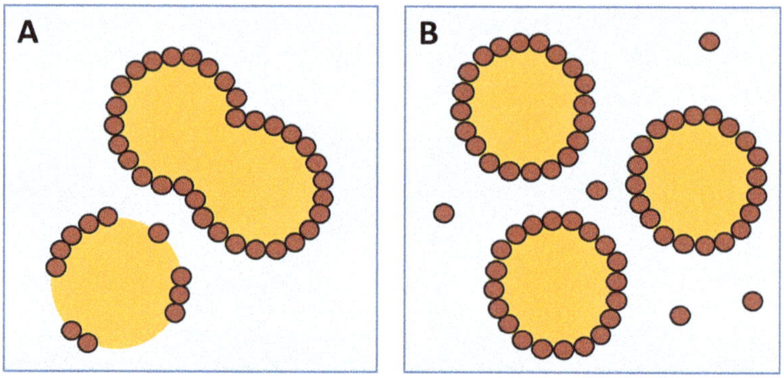

Figure 3. Schematic representation of the oil droplet distribution. (**A**) Droplets partly covered by solid particles; (**B**) droplets totally covered by solid particles.

2.3. Morphology of Solid Particles

Food-grade particles of different shapes have attracted great interest from the scientific community because particle shape plays a vital role in emulsion stability. Particle shapes can be classified as regular spherical or irregular shapes with anisotropic morphology [16]. The factors that influence the distinct shapes of solid particles are associated with their origin or source, their structure, their properties, and the methods used for their preparation [13]. In this regard, non-spherical particles such as fibers, rods, and cubes have been utilized to stabilize Pickering emulsions [17]. Li et al. [13] reported that the properties of Pickering emulsions can be influenced by the type and shape of solid particles. In particular, the shape of the particles governs their behavior at the interface (oil–water) and their ability to stabilize the emulsion. Particles (as a Pickering stabilizer) with different shapes may have different densities, desorption energy, and capillary forces between adjacent particles, which can significantly impact the stabilization principles and applications of the emulsions. Additionally, the particle shape can play a critical role in the stabilization of emulsions by altering the wettability behavior of particles and the interactions between adjacent particles. According to Burgos-Díaz et al. [1], the stabilization of a Pickering emulsion can be influenced not only by the particle wettability but also by the particle shape, size, and electrical potential. Figure 4 illustrates the use of Pickering stabilizers obtained from different agri-food byproducts, which have different shapes and sizes, to stabilize O/W emulsions.

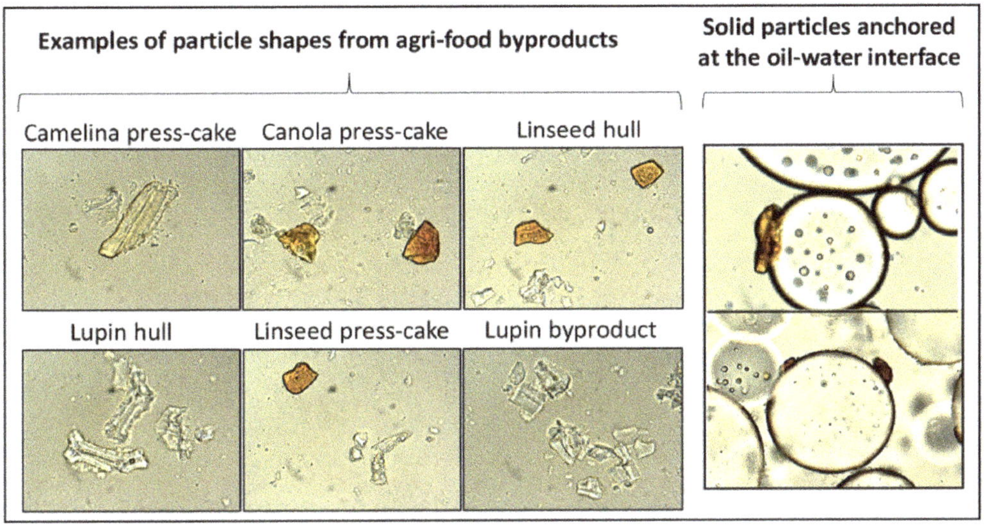

Figure 4. Micrographs corresponding to the morphology of particles obtained from different byproducts and used to stabilize Pickering emulsions (image adapted from Burgos-Díaz et al. [1]).

2.4. Oil Volume Fraction

The emulsion stability and type (O/W or W/O) can be also influenced by the dispersed phase volume. In this regard, the stability and type of emulsion can be influenced by the oil fraction present between the disperse and continuous phases. Burgos-Díaz et al. [6] showed that when the oil volume fraction increased from 5% to 20%, the emulsion droplet size varied in all emulsions stabilized with Pickering particles at different concentrations. The authors observed that at a fixed Pickering stabilizer concentration, the size of the oil droplets increased as the oil fraction increased. This behavior is expected, since when the particle concentration is kept relatively low and constant, the number of particles present may not be enough to provide complete stabilization for oil droplets as the oil volume fraction is raised. As a result, the number of particles may not be enough to completely cover the surface of freshly formed oil droplets during the emulsification process, leading to droplet coalescence and oiling off. Figure 5 shows a schematic representation of the effect of particle and oil concentration on the droplet size of the emulsion.

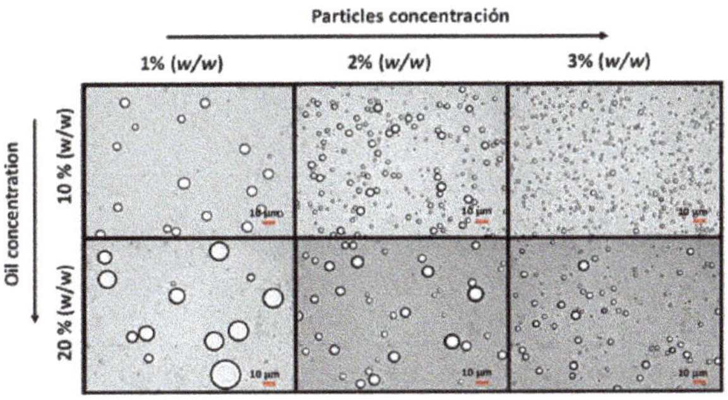

Figure 5. Micrographs of the effect of particle and oil concentration on droplet size of an O/W Pickering emulsion.

3. Preparation of Pickering Emulsions

Various methods or techniques have been used for the preparation of both O/W and W/O Pickering emulsions. Among them, rotor-stator homogenization, high-pressure homogenization, and sonication are the most commonly employed to formulate Pickering emulsions [18,19]. Nevertheless, monodisperse Pickering emulsions can also be produced through membrane emulsification and microfluidics [19]. In the majority of research, the common method for preparing Pickering emulsions is the "rotor–stator homogenization method" (UltraTurrax), which consists of a homogenizer with a rotor and a stator with openings. In this technique, the rotation speed and the homogenization time are the primary parameters that affect the emulsion droplet size with a rotor–stator homogenizer [18]. On the other hand, the "high-pressure homogenization method" is a continuous emulsification process in which a pre-emulsification is required to obtain a coarse emulsion (primary emulsion), which is then passed through the slits of a high-pressure homogenizer, and its cavitation, turbulence, and shear are utilized to form a fine emulsion [20]. Conversely, the "ultrasonic method" also uses cavitation, turbulence, and shear stress to prepare emulsions, promoting adoption of the Pickering stabilizer on a two-phase interface. During emulsion preparation, cavitation produces localized high temperatures, high pressures, and stress, which is beneficial to the formation of Pickering emulsions [20].

Recently, techniques such as membrane emulsification and microfluidic emulsification have also been applied to prepare Pickering emulsions. Regarding the "membrane emulsification method", this method presses a pure dispersed phase or primary emulsion into a microporous membrane and controls the injection rate and shearing conditions to prepare a Pickering emulsion [20]. The two main types of membrane emulsification techniques are "direct membrane emulsification" and "premix membrane emulsification". In direct membrane emulsification, the dispersed phase is pressed or injected through a microporous membrane into the continuous phase. The same principle is applied in "premix membrane emulsification", except that it is a pre-emulsified mixture that is pressed through the membrane [18]. Apparently, the pore size of the membrane, the viscosities of the continuous phase and the dispersed phase, and the magnitude of the surface tension are important factors affecting the droplet size of the Pickering emulsion [18]. Finally, "microfluidic technology" is a drop-by-drop technology used for preparing Pickering emulsions. In this method, the dispersed phase flows horizontally and the continuous phase flows vertically, and when they intersect, the dispersed phase forms spherical droplets under the influence of the continuous phase drag [20].

4. Agri-Food Byproducts as a Source of Pickering Stabilizers

Different functional ingredients, such as proteins, polysaccharides, fibers, flavor molecules, and phytochemicals, can be generally found in agri-food byproducts [3,21]. In addition, this type of byproduct is an excellent source of natural emulsifier molecules and therefore this makes them good candidates to develop new solid amphiphilic particles or Pickering stabilizers. The great potential of Pickering particles is that these natural stabilizers can be an effective alternative to synthetic surfactants commonly used for stabilizing emulsions. In general, Pickering stabilizers can be classified as non-food-grade Pickering particles, e.g., calcium carbonate, barium sulfate, silica, clays (montmorillonite and laponite), and particles based on synthetic polymers (polystyrene, poly(N-isopropylacrylamide), and PS-polybutadiene block (PB)-block-PMMA) [22], and food-grade Pickering particles, e.g., polysaccharide particles (starch, chitosan, and cellulose), fat crystals, complex particles, flavonoid particles, food-grade wax, protein-based particles proteins (zein, whey, soy, and lupin), and byproducts (apple peel pomace, pomace, cocoa, and rapeseed press-cake) [1]. Most food-grade particles such as starch, cellulose, chitin, and proteins must be physically or chemically modified to improve their interfacial functionality and stability in Pickering emulsions [20].

In recent years, various byproducts have been investigated as potential natural Pickering stabilizers (Figure 6), including apple pomace [23], citrus fibers [24], cocoa, rapeseed press-cake [25], protein from moringa seed residues [26], tea residues [27], perilla protein isolate extracted from oilseed residues [28], polysaccharide extract from peanut oil residues [29], lupine hull, lupine byproducts, camelina press-cake, linseed hull, and linseed press-cake [1], among others. Hence, it is necessary to give immediate attention to managing agri-food byproducts and valorize them as edible techno-functional ingredients for the food industry.

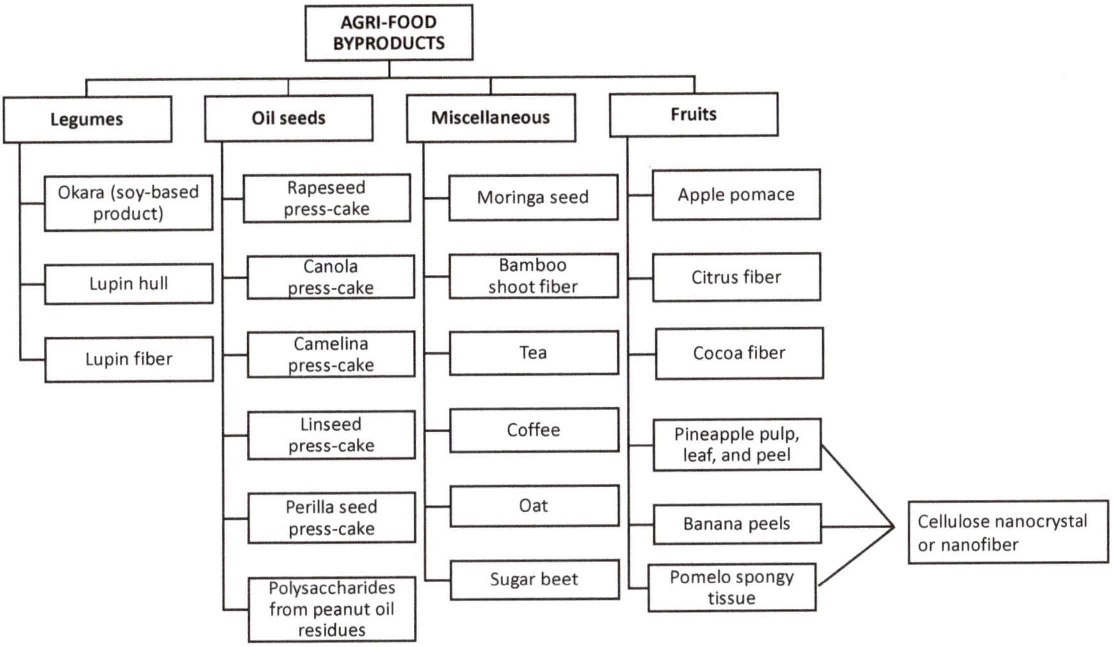

Figure 6. Classification of agri-food byproducts as a source of Pickering stabilizers.

4.1. Legume Byproducts

Legume byproducts have a high content of insoluble proteins and fibers, which make them good candidates to obtain novel solid particles to stabilize Pickering emulsions. In addition, these byproducts possess other bioactive molecules of interest to the food industry (phenols, carotenoids, phytosterols, and fibers), which also can provide benefits to emulsion stability and human health [30–32].

For instance, okara is a byproduct of the processing of soy-based products (milk, tofu, or protein isolate) composed of insoluble dietary fiber (IDF), protein [33], and polysaccharides [34,35]. As a result of its nutritional composition (mainly proteins, fiber, and carbohydrates), the effectiveness of the okara solid particles as Pickering stabilizers has been explored. For example, Yang et al. [36] reported the extraction of nanoparticles from insoluble okara soybean polysaccharides and their utilization as Pickering stabilizers. Bao et al. [37] modified okara insoluble dietary fiber (OIDF) by fermentation using *Kluyveromyces marxianus*. The modified OIDF exhibited a porous structure with a honeycomb shape, emulsification properties, and a significant increase in water and oil holding capacity.

Burgos-Díaz et al. [1] reported that lupin byproducts (a byproduct based on insoluble fiber–protein compounds and lupin hull) can be used effectively as particle-based emulsifiers to prepare stable O/W Pickering emulsions. Their study showed that all samples contained surface-active agents, including proteins (30.80–17.90 g/100 g) and dietary fiber (60.59–67.10 g/100 g). In addition, the authors also determined that the presence of proteins, both soluble and insoluble, is relevant because they are the main surface-active compounds found in raw plant materials.

4.2. Oil Seed Byproducts

Oilseed cakes (e.g., from canola seed, camelina seed, rapeseed, and linseed) are byproducts obtained during the pressing process to obtain oils [38]. After the pressing process, a 60% byproduct (press-cake) is obtained, which contains a high protein content (25–35%) [39], 8–10% residual oil (triglycerides), and different amounts of cellulose, lignin [40], glucosinolates, phenolic compounds, phytic acid, and other compounds [41,42].

The use of press-cakes as Pickering stabilizers has been previously reported in a study performed by Joseph and coworkers [3,25], who evaluated the utilization of rapeseed press-cake powders as Pickering stabilizers, determining that this byproduct can effectively stabilize oil-in-water Pickering emulsions. The authors attributed this ability to the higher amount of amphiphilic species such as insoluble protein (34–38 g/100 g), polyphenols (0.8 eq gallic acid/100 g), insoluble polysaccharide–polyphenol, and protein–polyphenol complexes. Burgos-Díaz and coworkers [1] evaluated the emulsifying properties of different oilseed byproducts, such as canola, camelina, and linseed press-cake. In this study, the authors reported that camelina press-cakes showed the best results as Pickering stabilizers, which could be attributed to the presence of macronutrients mainly associated with the content of protein (46.71 g/100 g) and fibers (38.58 g/100 g) since these macronutrients can perform as natural emulsifiers.

Another type of byproduct used for stabilizing emulsions is perilla seed press-cakes, which are obtained during the processing of perilla oilseeds to obtain oil. This obtained residue has a high content of proteins (35–45%), fibers (55–65%), phytic acid, polysaccharides, and phenolic compounds [28]. Zhao et al. [43] determined that perilla seed protein has a high surface hydrophobicity (119.29), solubility (78.71%), and a small particle size (318 nm), and therefore it can be considered as a natural emulsifier.

On the other hand, the peanut oil extraction process produces three fractions: (i) an oil-rich cream fraction, (ii) a protein-rich water fraction, and (iii) a sediment fraction (neutral and acidic polysaccharides) [44]. Acidic polysaccharides have been found to have a high particle aggregation ability, while alkaline-extracted polysaccharides reduce the oil–water surface tension [29,45]. Based on the antecedents described previously, Ye et al. [29] obtained polysaccharides from peanut oil processing residues by the alkaline method to be used as a Pickering stabilizer. The obtained complexes (PEC) comprised polysaccharides (56–68%) and proteins (13–18%). The presence of protein improved the hydrophobicity of the PECs. Furthermore, as the extraction pH value increased to 10.0, the protein in the PECs covalently bound to the polysaccharide and the polysaccharide conformation unfolded simultaneously, leading to the particle size increasing from 264 ± 5 nm to 360 ± 13 nm, which in turn resulted in the higher emulsifying capacity of the PECs [29]. Therefore, oilseed byproducts have great potential to develop edible Pickering stabilizers.

4.3. Fruit Byproducts

Fruit and vegetable industries generate 14.8 Mt of waste and byproducts in the European Union [46]. These types of byproducts can be considered hazardous from an environmental point of view, since they can affect the deterioration of drinking water quality, contaminate aqueous media, inhibit seed germination, and cause intestinal disorders in animals. Nevertheless, with appropriate treatment, they can represent a cost-effective raw material which is abundant in valuable functional compounds [47], which make them good ingredients as Pickering stabilizers. For this reason, various studies have shown the potential of apple pomace to be used as a natural stabilizer due to its good emulsifying properties. The emulsifying properties of apple pomace powders were attributed to their insoluble fraction content (90.5% wt over dry matter) and they are responsible for the stability and formation of a network in Pickering emulsions [23,48,49]. Soluble components alone may not be enough to stabilize oil-in-water emulsions, since they could provide a synergistic or antagonistic effect at the interface with insoluble particles [50]. Another byproduct is the fiber obtained from citrus peel, which is a residue from the pectin industry and orange juice production. Citrus fibers are mainly composed of proteins (~8%), pectins

(~35–42%), cellulose (>45%), hemicellulose, and lignin, which can vary in concentration depending on the origin. The fibers have a high water retention capacity and apparent viscosity, which is dependent on their structure [51–53]. For instance, Qi et al. [24] developed a Pickering emulsion using dietary fibers fragmented by ultra-high-pressure homogenization. However, in contrast to other particles, the fibers form a network, which swells and can bind oil droplets to the surface and absorb water into the network.

The chocolate industry produces mainly two byproducts: cocoa fiber resulting from the grinding of the husk and cocoa powder obtained from the grinding of the cocoa press-cake. Typically, the cocoa powder contains different functional components, such as phenolic compounds, hydrocolloids, sugars, proteins, fibers, theobromine, and lipids, which account for approximately 10–25% of its content [54]. Gould et al. [55] reported the use of cocoa mass, cocoa fibers, cocoa particles, and cocoa with different fat contents to obtain Pickering emulsions. In addition, Joseph et al. [25,54] produced Pickering emulsions using cocoa powder defatting with hexane as a stabilizer. They reported that the cocoa powder had a soluble fraction of 23.6 g/100 g and a protein content of 25 g/100 g, which could explain its ability to stabilize emulsions.

Fruit byproducts can be used to obtain particles with improved characteristics (nanocrystals or cellulose nanofibers) using different techniques such as enzymatic hydrolysis, hydrothermal treatment, acid or basic hydrolysis, high-pressure homogenization, and ultrasound, among others [56–58].

In this area, cellulose nanocrystals (CNC) stand out as Pickering stabilizers since they have a low density, high surface area, high crystallinity, amphiphilicity, low toxicity, and excellent functionality [59]. Tang et al. [60] compared the efficiency of Pickering stabilizers of cellulose nanocrystals (CNCs) obtained by acid hydrolysis from pineapple pulp (PPu), pineapple leaf (PL), and pineapple peel (PPe) residues. PPu nanocrystals showed the best emulsifying properties, attributed to their high cellulose content (96.10%), a CNC yield of 23.26%, and an average length of 141.5 nm. Foo et al. [61] obtained CNCs by acid hydrolysis of empty fruit bunches (EFB) available from the palm oil industry, in which they obtained a yield of 53.23%, a crystallinity of 76.55%, and a hydrodynamic diameter of 69.22 nm, while the width and length of the CNCs were 16 nm and 257 nm, respectively.

Another Pickering stabilizer is cellulose nanofiber (CNF), used for its low crystallinity and high compatibility with proteins and lipids. In addition, its hydrophilic nature, together with its large size (long), causes the CNF to overlap and join disordered networks and macroscopic gels, acting as a viscosity modifier. CNFs from banana peels, which are produced by chemical and enzymatic treatments, show a potential to be used as natural stabilizers [62]. Costa et al. [63] evaluated the influence of the processes of ultrasound and high-pressure homogenization on the CNF from banana peels. The study determined that the ultrasonication power and pressure led to shortening of the length, modifications in the crystallinity index, expansion of hydrophobic domains, and enhancement in zeta potential values (from −16.1 to −44.1 mV) of CNFs, which favored the stabilization of emulsions. Pomelo spongy tissue has been used to obtain cellulose nanofibers (CNFs) by its high dietary fiber content. Wen et al. [64] obtained CNFs by a chemical method (acidic and basic) and using a high-pressure homogenizer. They obtained low lignin values (0.70%), high cellulose contents (79.68%), and particle sizes of >3 μm in length and 33–64 nm in width.

4.4. Other Agri-Food Byproducts

Moringa seed residue protein is one of the main byproducts of Moringa oil extraction. Huang et al. [26] extracted the protein to be used as a Pickering stabilizer, achieving a Moringa seed residue protein (MSRP) yield of 14.24%. The particle diameter of the positively charged MSRP was greater than 233 nm. MSRP was observed to be soluble at a pH of 5 (206.89 mg/g) and in the presence of 0.2 M NaCl (202.55 mg/g). Tea residues are another resource used to extract plant proteins. This byproduct is obtained from the production of tea beverages and the extraction of bioactive components [65]. Tea residues are 90% insoluble protein, which can be extracted by alkaline or enzymatic methods [66].

Ren et al. [67] fabricated water-insoluble tea protein nanoparticles (TWIPNs). TWIPNs were obtained by the nanoprecipitation method and determined to be irregular colloidal particles with a hydrodynamic diameter greater than 300 nm and a zeta potential greater than −30 mV at ionic strengths of 0–400 mM and a fixed concentration of TWIPNs (2.0%); these properties indicate the potential of these nanoparticles to stabilize emulsions [67].

He et al. [68] studied the use of bamboo shoot fibers as plant food particle stabilizers for O/W Pickering emulsions. In this study, the bamboo shoot fibers were treated by high-pressure homogenization, which produced a filamentous morphology of fibers. A rheological analysis showed that the fibers present a shear thinning behavior. The findings suggest that the dietary fibers derived from bamboo shoots possess a soft nature and appropriate shape for producing stable edible Pickering emulsions with potential applications in the food industry.

Huc-Mathis et al. [48] analyzed the properties of oat and sugar beet residues for potential application as Pickering stabilizers. They obtained a particle size of 6.60 ± 0.01 µm for oats and 5.80 ± 0.02 µm for sugar beet using only a micronizing process. The insoluble contents of sugar beet and oats were 93.7 ± 0.8% and 94.0 ± 0.1%, respectively, and the Zeta potentials were −31 ± 5 mV and −22.6 ± 1.2 mV, respectively. The results obtained, in addition the fact that sugar beets present emulsifying properties associated with their pectins and fibers [69] and the saponins present in oat extract possess emulsifying properties [70], indicated propertied that made it possible to use these byproducts as Pickering stabilizers.

Coffee residues have a lignin content of between 20 and 27% (*wt/wt*) [71], which is indicative of a hydrophobic molecule. However, it has been demonstrated that lignin may exhibit hydrophilic, hydrophobic, and amphiphilic characteristics depending on the botanical source and the methods used for extraction. Gould et al. [72] performed a hydrothermal treatment to remove lignin from coffee particles. This treatment produced an increase in the hydrophobicity of the particle surface, which improved their emulsifying properties.

5. O/W Pickering Emulsions Stabilized by Different Agri-Food Byproduct Particles

This section provides an overview of the recent studies on the stabilization of Pickering emulsions (O/W) using edible natural particles. Thus, different works on Pickering emulsions stabilized by different agri-food byproducts are described below. For instance, Yang et al. [36] showed that it was possible to stabilize O/W Pickering emulsions by using nanoparticles from insoluble soybean polysaccharides of okara. The results showed that nanoparticles have high emulsifying and gelling properties, which favored the stabilization of Pickering emulsions. It should be noted that the concentration of particles and the oil content in the formation of the emulsion favored their stability [36]. Bao et al. [37] showed that fermentation-modified okara insoluble dietary fiber (OIDF) exhibits excellent properties to be considered as a Pickering stabilizer. These particles had a strong electrostatic interaction, smaller droplets, and a higher encapsulation efficiency (95%) and yield than unmodified OIDF.

Huc-Mathis et al. [48] evaluated the ability of apple pomace as a Pickering stabilizer. Their first investigation showed that apple powder had better emulsifying properties than oat bran. The insoluble fibers contributed to emulsion stability through the Pickering mechanism for 15 days, which was favored by the stabilization provided by proteins and pectin in the soluble fraction [50]. Then, the same authors designed an experiment to analyze the influence of apple powder, microcrystalline cellulose, and oil content on the properties of Pickering emulsions. They determined that the higher the apple powder content, the lower the oil droplet size. Likewise, the highest concentrations of fat and polymer resulted in an increase in the values of the elastic modulus, G′, and viscosity, with a simultaneous decrease in the tan(δ) value [23]. In their latest research, they compared apple powder to sugar beet powder and oat bran. These byproducts were able to produce stable emulsions that were resistant to coalescence. Of these, apple powder provided supplementary stabilization by preventing drainage of the continuous phase. The smallest emulsion oil droplets were obtained with sugar beet, followed by apple then oat, which was associated with the emulsifying properties of the byproducts [48]. On the other hand, Lu et al. [49] studied micronized apple pomace as a novel food-grade emulsifier for stabilizing O/W Pickering emulsions. These emulsions exhibited a smaller droplet size and improved gelation and antioxidant properties when the particle size was reduced.

Qi et al. [24] studied the use of citrus fibers for the stabilization of O/W emulsions, evaluating the Pickering mechanism and the fiber-based network effect. This study showed that the O/W emulsions were more stable using a fiber concentration of 2% (*wt*/*v*) and 25% (*v*/*v*) oil. These emulsions had a storage stability of $\geq$60 days and they were not obviously influenced by changes in the pH, ionic strength of NaCl, or temperature. He et al. [68] used water-insoluble dietary fibers from bamboo shoots to stabilize O/W emulsions. The emulsions remained stable against coalescence for at least 4 weeks, and their stability was not affected by changes in pH, ionic strength, or pasteurization conditions.

Gould et al. [55] showed the ability to form coalescence-stable O/W emulsions using cocoa particles. They also determined that increasing the cocoa particle concentration reduced the average droplet size. On the other hand, Joseph et al. [54] prepared O/W Pickering emulsions stabilized by defatted cocoa powder. For this, the emulsions were studied using three techniques, i.e., rotor–stator, sonication, and microfluidization, where the microfluidization technique was the most efficient emulsification technique. The emulsion obtained with this technique showed the highest anchorage rate, the smallest droplets (4.2 µm), and the emulsion was stable after 90 days. In addition, Joseph et al. [3] investigated the production of powdered re-dispersible Pickering stabilizers based on cocoa powder and rapeseed press-cakes. The rehydrated O/W Pickering emulsions showed an average droplet size of <100 µm, which was larger compared to their parent emulsions. The authors stated that this behavior could be attributed to the coalescence phenomena occurring during the drying process. The formulations based on rapeseed press-cakes provided excellent results, since after spray drying, the re-dispersed emulsion exhibited almost the same characteristics as the original emulsion in terms of size distribution. Finally, Joseph et al. [25] obtained O/W Pickering emulsions stabilized by cocoa, rapeseed, and lupin solid particles. The authors also tested three emulsification techniques and found that the thinnest emulsions and the highest anchorage ratios were obtained using microfluidization, independently of the nature of the particles. The emulsions showed narrow droplet size distributions, especially in the presence of rapeseed and cocoa powder. Finally, the authors concluded that rapeseed powder derived from defatted press-cakes was the most effective in terms of interfacial coverage.

Gould et al. [72] showed that ground coffee residue particles can be used as a Pickering stabilizer to prepare O/W emulsions. These emulsions exhibited a droplet size of around 100 µm and no change in microstructure during a 12 week storage period. The droplet size was also unaffected by pH changes. They also determined their stability against coalescence under shear and their stability under pasteurization conditions (10 min at 80 °C) [72].

Burgos-Díaz et al. [1] characterized and compared six food-grade Pickering stabilizers obtained from various sources of agri-food byproducts (canola press-cake, camelina press-cake, linseed hull, linseed press-cake, lupin byproduct, and lupin hull). The study showed that emulsions stabilized with camelina press-cake, lupin hull, and lupin byproduct at concentrations of $\geq$3.5% (*wt*/*wt*) exhibited remarkable stability against creaming for at least 45 days of storage. The size of the droplets was significantly influenced by the concentration of particles and the type of raw material utilized. Microscopy studies showed that the solid particles were anchored to the surfaces of the oil droplets, which is clear evidence of the formation of a Pickering emulsion stabilized by solid particles [1].

Foo et al. [73] obtained Pickering nanoemulsions with nanocrystalline cellulose from an empty oil palm fruit bunch. The main results found in this study were based on the role of lignocellulosic residues in the formation and stability of Pickering nanoemulsions, in which nanocrystalline cellulose significantly contributed to the stabilization of the nanoemulsion, as well as to the size of the particles obtained. The nanoemulsion exhibited a droplet size of approximately 400 nm, and a high stability for 6 months against the formation of creaming and coalescence. These results are promising, since the stabilization obtained by nanocrystalline cellulose could help to obtain improved products in the food and personal care industries [73]. In addition, pomelo spongy tissue cellulose nanofibers (PCNFs) have been used to stabilize Pickering emulsions [64]. The emulsions obtained showed an excellent stability and a strong concentration-dependent effect of PCNFs; at higher concentrations of PCNFs, increasingly stable emulsions were observed. Tang et al. [60] reported that the emulsions stabilized with pineapple cellulose nanocrystals (PCNCs) obtained from pineapple peel were stable after storage for 50 days, which was attributed to the microstructure of the PCNCs. On the other hand, Costas et al. [63] investigated the impact of emulsification conditions using ultrasound and a high-pressure homogenizer on the physicochemical characteristics of emulsions stabilized by cellulose nanofibers (CNFs) sourced from banana peel. The authors determined that both emulsification processes presented oil droplet creaming. The ultrasonic emulsification process resulted in a reduction in the length and aspect ratio of the CNFs, thus forming smaller emulsion droplets due to the obtained particles achieved to cover the droplet–emulsion interface, which prevented coalescence of the emulsion.

Huang et al. [26] evaluated the stability of Pickering emulsions using moringa seed residue proteins through the effect of pH and ionic strength. The results showed that the emulsion presented high zeta potential values and excellent morphological characteristics. Concerning the emulsion stability, a cream layer appeared rapidly within 2 h; however, after 30 days of storage, the emulsions showed no apparent changes, indicating a good stability against creaming. On the other hand, at pH 5 and 0.2 M NaCl, elastic networks can be obtained in Pickering emulsions stabilized with moringa seed residues, so stabilization with this type of residue could play an important role in emulsifying and stabilizing processes in the food and beverage industry [26]. In addition, Ren et al. [67] obtained Pickering emulsions stabilized with nanoparticles of tea-water-insoluble proteins (TWIPs). The main results showed that the Pickering emulsions generated firm and thick surface layers as the concentration of nanoparticles increased, which helped to reduce the size of the emulsion droplets at the amounts of oil and used water (4:6). According to an analysis of creaming stability, a cream layer was visible on top of the emulsions after 6 h of emulsion preparation. However, during 3–40 days of storage, no evident change was observed in the creaming behavior. This indicated the excellent anti-creaming ability of the TWIPs [67].

According to Zhao et al. [43], perilla protein isolate extracted from cold pressing residues displayed excellent functional properties, including a high foaming ability (90.67%), emulsification capacity (3390.09 m^2/g), and water holding capacity (2.17 g/g). Therefore, this byproduct exhibited an excellent capacity to stabilize Pickering emulsions based on its small droplet size, high ζ-potential, and low creaming index of emulsion droplets. Liu et al. [28] showed that increasing the perilla seed protein (PSP) concentration (0.25 at 1.0 wt %) was favorable to avoid the aggregation/flocculation of emulsions. They found that PSP-stabilized emulsions were stable at NaCl concentrations of less than 150 mmol/L and at pHs between 3.0 and 9.0. Moreover, the stability of the emulsion was enhanced against aggregation and creaming when it was subjected to a temperature of 70 °C. Finally, polysaccharide–protein complexes (PECs) of peanut sediment from aqueous extraction processes were used as Pickering stabilizers by Ye et al. [29]. They determined that a 4% wt concentration of PEC10.0 (pH = 10) stabilized Pickering emulsions with a superior creaming stability, which was mainly attributed to the fact that the relatively high viscosity limited droplet movement. These emulsions were stable for 20 days, with an average particle size of 16.96 μm for an oil fraction of 0.6.

Recent studies using agri-food byproducts as Pickering stabilizers in emulsions are summarized in Table 1.

Table 1. Pickering emulsions based on agri-food byproducts.

Type of Particle	Particle Load	Emulsification Method	Droplet Size (μm)	Storage	Reference
Nanoparticles from insoluble soybean polysaccharides of okara	1% wt	UltraTurrax (8000 rpm for 4 min) and microfluidization (40 Mpa)	~20	14 days	[36]
Modified okara insoluble dietary fiber	0.8% wt	UltraTurrax (13,000 rpm for 2 min) and ultrasonicator (500 W for 6 min)	~1	28 days	[37]
Apple pomace	100 mg powder/g oil	UltraTurrax (10,000 rpm for 3 min)	45	15 days	[50]
Apple pomace	123 mg/g of oil	UltraTurrax (10,000 rpm for 6 min)	28.7	57 days	[48]
Sugar beet			17.8	16 days	
Oat bran			–		
Apple pomace	3.2% wt	UltraTurrax (20,000 rpm for 1 min)	9.86	30 days	[49]
Citrus fiber	2% *wt/v*	UltraTurrax (10,000 rpm for 6 min) and microfluidization (300 bar for 3 min)	~100	15 days	[24]
Cocoa Powder	6% *wt/wt*	UltraTurrax (8000 rpm for 2 min)	5	100 days	[55]
Cocoa Powder	2.5% wt	Microfluidization (800 bar, and 6 passes)	4.2	90 days	[54]
Rapessed press-cake	2.5% wt	Microfluidization (8×10^7 Pa, and 6 passes)	4.1	90 days	[25]
Coffee residue particles	8% wt	UltraTurrax (9000 rpm for 2 min)	100	84 days	[72]
Nanocrystalline cellulose from empty oil palm fruit bunches	1% wt	Ultrasonicator (70% amplitude for 90 s) and microfluidization (15,000 psi with 1–20 passes)	0.389	6 months	[73]
Water-insoluble bamboo shoot dietary fiber	0.3% wt	UltraTurrax (2 min and 12,000 rpm)	10.9	30 days	[68]
Moringa seed residue protein	0.02 g/mL	UltraTurrax (15,000 rpm for 4 min)	1.97	30 days	[26]
Tea-water-insoluble protein nanoparticles	4% wt	UltraTurrax (20,000 rpm for 2 min)	18.7	40 days	[67]
Pineapple cellulose nanocrystals	0.1% *wt/v*	Ultrasonicator (70% amplitude for 5 min)	6.8	50 days	[60]
Perilla protein isolate (cold pressing residues)	2% *wt/v*	UltraTurrax (15,000 rpm for 2 min)	27.55	7 days	[43]
Polysaccharides and proteins from peanuts (pH = 10)	4% wt	UltraTurrax (8800 rpm for 1 min)	16.96	20 days	[29]

6. Applications and Future Trends

In recent years, the superior performance of Pickering emulsions has been reported in various fields such as cosmetics, pharmacy, biomedicine, and food [13]. Compared with conventional emulsions, food-grade Pickering emulsions have several advantages, such as increased safety, good stability, environmental friendliness, and excellent biocompatibility [20]. According to Xia et al. [74], this type of emulsion has been studied and characterized and has a wide spectrum of applications in the food industry. For instance, Pickering emulsions have been used as fat substitutes [75,76], delivery systems for nutraceuticals [77,78], and even in the manufacturing of food-grade cleaning agents [79], as is detailed in Table 2. Regarding fat substitutes, Pickering emulsions have been demonstrated as a butter substitute. A study showed that the use of Pickering emulsions (stabilized by ethyl cellulose and camellia seed oil) could replace cream in the production of frozen yoghurt and ice cream [74]. In terms of nutraceutical delivery, Pickering emulsions are suitable delivery systems to enhance the physical stability, the compatibility with food matrices, the oxidative stability, and the protection of labile bioactive compounds. Thus, Pickering solid particles adsorbed on the oil–water interface can form a physical barrier and prevent the degradation of nutrients [74].

On the other hand, Pickering emulsions have been also applied in other fields indirectly related to food science; they have been applied in biomedicine in order to improve the solubility of poorly soluble drugs [80]. In biomedicine, the utilization of Pickering emulsions as a drug carrier or delivery system is due to their excellent biocompatibility, since these emulsions are prepared using non-toxic raw materials [81]. Another feature that has enhanced the applications of these types of emulsions is their greater stability compared with conventional emulsions. For instance, Pickering emulsions are not easily affected by environmental stresses, such as temperature, ionic strength, and pH.

Table 2. Current applications of food-grade Pickering emulsions.

Application	Main Products	Purpose	References
Fat substitutes	Butter Yoghurt Ice cream	- More satisfactory food quality regarding sensory evaluation and physicochemical characterization. - Reduction in calorie intake. - Extension in shelf life without changing color and texture. - More advantages in the oil digestion process. - Preparation of oleogels to replace fat.	[75,76,82,83]
Delivery systems for bioactive compounds	Curcumin Hesperidin β-carotene	- Enhance physical stability, compatibility with several food matrixes and protection of labile bioactive compounds. - Improve the resistance against thermal degradation. - Improve bioavailability in the simulated gastrointestinal tract.	[10,77,78]
Cleaning agents	Green detergent from corncob	Cleaning oil stains in an eco-friendly and safe way.	[79]

Food-grade Pickering emulsions have evolved with the new trends that have emerged in recent years. These evolutions include Pickering double emulsions, nutraceutical co-delivery, multilayer Pickering emulsions, Pickering emulsions fixed in gels, preparation of porous materials, and responsive Pickering emulsions [74]. Briefly, "Pickering double emulsions" can be classified as water-in-oil-in-water (W/O/W) or oil-in-water-in-oil (O/W/O) emulsions. Regarding O/W/O, this emulsion consists of a continuous oil system containing water droplets with smaller oil droplets inside them, whereas W/O/W emulsions are water-continuous systems containing oil droplets within which smaller water droplets are dispersed [74]. One potential application of this type of colloidal system is in the production of a reduced-fat emulsion product that has a lower oil content but with a similar texture perceived in the mouth. Moreover, Pickering double emulsions can encapsulate and protect bioactive compounds in the inner phase of Pickering double emulsions, which are then subsequently released during the digestive process [74].

Nutraceutical co-delivery: Other emerging trends within the functional food industry include products with bioactive compounds or nutraceutical micro/nano encapsulated products. In this context, Pickering emulsions as a delivery system have been widely reported. However, most of these studies describe the encapsulation of a single molecule in the disperse phase emulsion and there have been few reports on the co-delivery of bioactive compounds in Pickering emulsions at the same time [74]. Therefore, the current applications in this area are focused on the development of encapsulation systems that include two or more bioactive compounds to develop new food products. Some studies have shown that the combination of some compounds in the preparation of nutraceuticals, such as curcumin and resveratrol, increases antioxidant effects, and even synergistic effects have been reported [84].

Multilayer Pickering emulsion: This colloidal system refers to emulsions which are stabilized by multiple layers of insoluble particles. Multilayer emulsions have been applied to protect different labile lipophilic compounds and have shown better physical stability against environmental stresses, such as heat treatment, freeze-drying, and ionic strength, among others [85]. Despite the potential applications of multilayer Pickering emulsions, this type of multilayer emulsion has not been investigated before [74].

Pickering emulsions fixed in gels: "Pickering emulsion gel" refers to colloidal systems based on oleogels and hydrogels. Oleogel-based Pickering emulsions have shown greater stability in comparison to other emulsion systems. For example, Pickering emulsions stabilized by ovotransferrin fibrils and based on oleogels demonstrated a remarkable stability during storage and a high level of stability during freeze–thaw cycles [78]. In addition, oleogels have been used to replace saturated fats and trans fats in food.

Preparation of porous materials: This new trend has great advantages because it presents particles that can facilitate the fabrication of porous materials. This could allow the introduction of some functional groups within particles or even as an absorbent of heavy metals due to the porous characteristics they present [74,86].

Responsive Pickering emulsions: This type of emulsion, also known as "stimuli-response Pickering emulsions", has attracted attention because of their potential applications for emulsion polymerization and target delivery of bioactive compounds. In this context, pH-responsive emulsions are the most common stimuli-responsive Pickering emulsions. These emulsion systems (pH-responsive) are characterized by their simplicity and diversity of materials available for use because variations in the pH of these systems result in changes in the surface behavior of the materials [74,87].

Dairy products: The physicochemical characteristics of dairy products could be improved by the use of Pickering emulsions, as they transfer their stability to these products. As a consequence, the shelf-life of milk and its derivatives could be prolonged. Therefore, there is a wide range of potential applications in the dairy industry [74,75,88–90].

On the other hand, edible Pickering emulsions are commonly liquids. However, in some cases, the liquid formulation presents complications when it is incorporated into food matrix systems, which complicates food processing [75]. One of the strategies used to avoid this problem is the use of hydrogels, which immobilizes Pickering emulsions. For instance, it has been observed that hydrogels stabilized with alginates improve oil retention, inhibit lipid oxidation, and also help the controlled release of compounds of interest [91]. Finally, it is important to mention that Pickering emulsions are still considered a novel type of colloidal system, which is an indicator of the need for new perspectives and future trends in the application of these emulsion systems [81].

7. Conclusions

Pickering emulsions have not only gained ground in the food industry but have also attracted the interest of the pharmaceutical industry in recent years. However, new food trends are pushing scientists to pay more attention to the sustainability of the bio-based particles they are using to stabilize emulsions. Thereby, the use of agri-food byproducts represents an excellent opportunity to explore new particle sources to be used as natural stabilizers. Until now, the advantageous stability of Pickering emulsions has been demonstrated, which is largely attributed to the diverse range of stabilization mechanisms in comparison to traditional emulsifiers. Therefore, exploring new, more natural, food-grade stabilizers seems to be a logical step in this field. Thus, this review briefly described a variety of natural particles from different agri-food byproducts used as Pickering stabilizers. Owing to the recent information exposed, the use of agri-food byproducts from legumes, oil seeds, and fruits is proven to be an alternative, renewable, and inexpensive source of solid amphiphilic particles to be used as "natural stabilizers" in different industrial applications. In addition, the exposed information shows that Pickering emulsions have great potential in fields other than food, such as interfacial catalysis, biomedicine, drug delivery, functional materials, and others.

Author Contributions: Conceptualization, C.B.-D. and K.A.G.-M.; resources, C.B.-D. and K.A.G.-M.; writing—original draft preparation, C.B.-D., K.A.G.-M., D.A.P., M.O.-N., M.C.-F. and M.B.; review and editing, supervision, C.B.-D. and K.A.G.-M. The authors have contributed substantially to the work reported. All authors have read and agreed to the published version of the manuscript.

Funding: This research was funded by ANID through FONDECYT-REGULAR project N° 1210136 and FONDECYT-POSTDOCTORADO N° 3220459.

Institutional Review Board Statement: Not applicable.

Informed Consent Statement: Not applicable.

Data Availability Statement: Not applicable.

Acknowledgments: We acknowledge the Chilean Agency for Research and Development (ANID), FONDECYT-REGULAR project N° 1210136 and FONDECYT-POSTDOCTORADO N° 3220459.

Conflicts of Interest: The authors declare no conflict of interest.

References

1. Burgos-Díaz, C.; Mosi-Roa, Y.; Opazo-Navarrete, M.; Bustamante, M.; Garrido-Miranda, K. Comparative Study of Food-Grade Pickering Stabilizers Obtained from Agri-Food Byproducts: Chemical Characterization and Emulsifying Capacity. *Foods* **2022**, *11*, 2514. [CrossRef] [PubMed]
2. Yang, Y.; Fang, Z.; Chen, X.; Zhang, W.; Xie, Y.; Chen, Y.; Liu, Z.; Yuan, W. An Overview of Pickering Emulsions: Solid-Particle Materials, Classification, Morphology, and Applications. *Front. Pharmacol.* **2017**, *8*, 287. [CrossRef] [PubMed]
3. Joseph, C.; Savoire, R.; Harscoat-Schiavo, C.; Pintori, D.; Monteil, J.; Faure, C.; Leal-Calderon, F. Redispersible Dry Emulsions Stabilized by Plant Material: Rapeseed Press-Cake or Cocoa Powder. *LWT-Food Sci. Technol.* **2019**, *113*, 108311. [CrossRef]
4. Jiang, H.; Sheng, Y.; Ngai, T. Pickering Emulsions: Versatility of Colloidal Particles and Recent Applications. *Curr. Opin. Colloid Interface Sci.* **2020**, *49*, 1–15. [CrossRef]
5. Pickering, S.U. CXCVI—Emulsions. *J. Chem. Soc. Trans.* **1907**, *91*, 2001–2021. [CrossRef]
6. Burgos-Díaz, C.; Wandersleben, T.; Olivos, M.; Lichtin, N.; Bustamante, M.; Solans, C. Food-Grade Pickering Stabilizers Obtained from a Protein-Rich Lupin Cultivar (AluProt-CGNA®): Chemical Characterization and Emulsifying Properties. *Food Hydrocoll.* **2019**, *87*, 847–857. [CrossRef]
7. Gonzalez Ortiz, D.; Pochat-Bohatier, C.; Cambedouzou, J.; Bechelany, M.; Miele, P. Current Trends in Pickering Emulsions: Particle Morphology and Applications. *Engineering* **2020**, *6*, 468–482. [CrossRef]
8. Burgos-Díaz, C.; Opazo-Navarrete, M.; Soto-Añual, M.; Leal-Calderón, F.; Bustamante, M. Food-Grade Pickering Emulsion as a Novel Astaxanthin Encapsulation System for Making Powder-Based Products: Evaluation of Astaxanthin Stability during Processing, Storage, and Its Bioaccessibility. *Food Res. Int.* **2020**, *134*, 109244. [CrossRef] [PubMed]
9. Haji, F.; Cheon, J.; Baek, J.; Wang, Q.; Tam, K.C. Application of Pickering Emulsions in Probiotic Encapsulation—A Review. *Curr. Res. Food Sci.* **2022**, *5*, 1603–1615. [CrossRef] [PubMed]
10. Mwangi, W.W.; Lim, H.P.; Low, L.E.; Tey, B.T.; Chan, E.S. Food-Grade Pickering Emulsions for Encapsulation and Delivery of Bioactives. *Trends Food Sci. Technol.* **2020**, *100*, 320–332. [CrossRef]
11. Mwangi, W.W.; Ho, K.W.; Ooi, C.W.; Tey, B.T.; Chan, E.S. Facile Method for Forming Ionically Cross-Linked Chitosan Microcapsules from Pickering Emulsion Templates. *Food Hydrocoll.* **2016**, *55*, 26–33. [CrossRef]
12. Araiza-Calahorra, A.; Wang, Y.; Boesch, C.; Zhao, Y.; Sarkar, A. Pickering Emulsions Stabilized by Colloidal Gel Particles Complexed or Conjugated with Biopolymers to Enhance Bioaccessibility and Cellular Uptake of Curcuminal. *Curr. Res. Food Sci.* **2020**, *3*, 178–188. [CrossRef]
13. Li, W.; Jiao, B.; Li, S.; Faisal, S.; Shi, A.; Fu, W.; Chen, Y.; Wang, Q. Recent Advances on Pickering Emulsions Stabilized by Diverse Edible Particles: Stability Mechanism and Applications. *Front. Nutr.* **2022**, *9*, 738. [CrossRef] [PubMed]
14. Perrin, L.; Gillet, G.; Gressin, L.; Desobry, S. Interest of Pickering Emulsions for Sustainable Micro/Nanocellulose in Food and Cosmetic Applications. *Polymers.* **2020**, *12*, 2385. [CrossRef] [PubMed]
15. Levine, S.; Bowen, B.D.; Partridge, S.J. Stabilization of Emulsions by Fine Particles II. Capillary and van Der Waals Forces between Particles. *Colloids Surf.* **1989**, *38*, 345–364. [CrossRef]
16. Low, L.E.; Siva, S.P.; Ho, Y.K.; Chan, E.S.; Tey, B.T. Recent Advances of Characterization Techniques for the Formation, Physical Properties and Stability of Pickering Emulsion. *Adv. Colloid Interface Sci.* **2020**, *277*, 102117. [CrossRef]

17. Ming, Y.; Xia, Y.; Ma, G. Aggregating Particles on the O/W Interface: Tuning Pickering Emulsion for the Enhanced Drug Delivery Systems. *Aggregate* **2022**, *3*, e162. [CrossRef]
18. Albert, C.; Beladjine, M.; Tsapis, N.; Fattal, E.; Agnely, F.; Huang, N. Pickering Emulsions: Preparation Processes, Key Parameters Governing Their Properties and Potential for Pharmaceutical Applications. *J. Control. Release* **2019**, *309*, 302–332. [CrossRef]
19. Kempin, M.V.; Kraume, M.; Drews, A. W/O Pickering Emulsion Preparation Using a Batch Rotor-Stator Mixer—Influence on Rheology, Drop Size Distribution and Filtration Behavior. *J. Colloid Interface Sci.* **2020**, *573*, 135–149. [CrossRef]
20. Chen, L.; Ao, F.; Ge, X.; Shen, W. Food-Grade Pickering Emulsions: Preparation, Stabilization and Applications. *Molecules* **2020**, *25*, 3202. [CrossRef]
21. Baiano, A. Recovery of Biomolecules from Food Wastes—A Review. *Molecules* **2014**, *19*, 14821–14842. [CrossRef] [PubMed]
22. Schrade, A.; Landfester, K.; Ziener, U. Pickering-Type Stabilized Nanoparticles by Heterophase Polymerization. *Chem. Soc. Rev.* **2013**, *42*, 6823–6839. [CrossRef] [PubMed]
23. Huc-Mathis, D.; Guilbaud, A.; Fayolle, N.; Bosc, V.; Blumenthal, D. Valorizing Apple By-Products as Emulsion Stabilizers: Experimental Design for Modeling the Structure-Texture Relationships. *J. Food Eng.* **2020**, *287*, 110115. [CrossRef]
24. Qi, J.R.; Song, L.W.; Zeng, W.Q.; Liao, J.S. Citrus Fiber for the Stabilization of O/W Emulsion through Combination of Pickering Effect and Fiber-Based Network. *Food Chem.* **2021**, *343*, 128523. [CrossRef]
25. Joseph, C.; Savoire, R.; Harscoat-Schiavo, C.; Pintori, D.; Monteil, J.; Faure, C.; Leal-Calderon, F. Pickering Emulsions Stabilized by Various Plant Materials: Cocoa, Rapeseed Press Cake and Lupin Hulls. *LWT-Food Sci. Technol.* **2020**, *130*, 109621. [CrossRef]
26. Huang, Z.; Huang, X.; Zhou, W.; Zhang, L.; Liu, F.; Li, J.; Peng, S.; Cao, Y.; Li, Y.; Li, R.; et al. Fabrication and Stability of Pickering Emulsions Using Moringa Seed Residue Protein: Effect of PH and Ionic Strength. *Int. J. Food Sci. Technol.* **2021**, *56*, 3484–3494. [CrossRef]
27. Ren, Z.; Chen, Z.; Zhang, Y.; Lin, X.; Weng, W.; Li, B. Pickering Emulsions Stabilized by Tea Water-Insoluble Protein Nanoparticles From Tea Residues: Responsiveness to Ionic Strength. *Front. Nutr.* **2022**, *9*, 840. [CrossRef]
28. Liu, N.; Chen, Q.; Li, G.; Zhu, Z.; Yi, J.; Li, C.; Chen, X.; Wang, Y. Properties and Stability of Perilla Seed Protein-Stabilized Oil-in-Water Emulsions: Influence of Protein Concentration, PH, NaCl Concentration and Thermal Treatment. *Molecules* **2018**, *23*, 1533. [CrossRef]
29. Ye, J.; Hua, X.; Zhao, Q.; Dong, Z.; Li, Z.; Zhang, W.; Yang, R. Characteristics of Alkali-Extracted Peanut Polysaccharide-Protein Complexes and Their Ability as Pickering Emulsifiers. *Int. J. Biol. Macromol.* **2020**, *162*, 1178–1186. [CrossRef]
30. Tassoni, A.; Tedeschi, T.; Zurlini, C.; Cigognini, I.M.; Petrusan, J.I.; Rodríguez, Ó.; Neri, S.; Celli, A.; Sisti, L.; Cinelli, P.; et al. State-of-the-Art Production Chains for Peas, Beans and Chickpeas—Valorization of Agro-Industrial Residues and Applications of Derived Extracts. *Molecules* **2020**, *25*, 1383. [CrossRef]
31. Mateos-Aparicio, I.; Redondo-Cuenca, A.; Villanueva-Suárez, M.J.; Zapata-Revilla, M.A.; Tenorio-Sanz, M.D. Pea Pod, Broad Bean Pod and Okara, Potential Sources of Functional Compounds. *LWT-Food Sci. Technol.* **2010**, *43*, 1467–1470. [CrossRef]
32. Mateos-Aparicio, I.; Redondo-Cuenca, A.; Villanueva-Suárez, M.J. Broad Bean and Pea By-Products as Sources of Fibre-Rich Ingredients: Potential Antioxidant Activity Measured in Vitro. *J. Sci. Food Agric.* **2012**, *92*, 697–703. [CrossRef] [PubMed]
33. Stanojevic, S.P.; Barac, M.B.; Pesic, M.B.; Vucelic-Radovic, B.V. Composition of Proteins in Okara as a Byproduct in Hydrothermal Processing of Soy Milk. *J. Agric. Food Chem.* **2012**, *60*, 9221–9228. [CrossRef] [PubMed]
34. Ullah, I.; Yin, T.; Xiong, S.; Zhang, J.; Din, Z.-U.; Zhang, M. Structural Characteristics and Physicochemical Properties of Okara (Soybean residue) Insoluble Dietary Fiber Modified by High-Energy Wet Media Milling. *LWT-Food Sci. Technol.* **2017**, *82*, 15–22. [CrossRef]
35. Porfiri, M.C.; Vaccaro, J.; Stortz, C.A.; Navarro, D.A.; Wagner, J.R.; Cabezas, D.M. Insoluble Soybean Polysaccharides: Obtaining and Evaluation of Their O/W Emulsifying Properties. *Food Hydrocoll.* **2017**, *73*, 262–273. [CrossRef]
36. Yang, T.; Liu, T.X.; Li, X.T.; Tang, C.H. Novel Nanoparticles from Insoluble Soybean Polysaccharides of Okara as Unique Pickering Stabilizers for Oil-in-Water Emulsions. *Food Hydrocoll.* **2019**, *94*, 255–267. [CrossRef]
37. Bao, Y.; Xue, H.; Yue, Y.; Wang, X.; Yu, H.; Piao, C. Preparation and Characterization of Pickering Emulsions with Modified Okara Insoluble Dietary Fiber. *Foods* **2021**, *10*, 2982. [CrossRef]
38. Moreno-González, M.; Girish, V.; Keulen, D.; Wijngaard, H.; Lauteslager, X.; Ferreira, G.; Ottens, M. Recovery of Sinapic Acid from Canola/Rapeseed Meal Extracts by Adsorption. *Food Bioprod. Process.* **2020**, *120*, 69–79. [CrossRef]
39. Arntfield, S.D. Proteins from Oil-Producing Plants. In *Proteins in Food Processing*, 2nd ed.; Woodhead Publishing: Sawston, UK, 2018; pp. 187–221.
40. Parodi, E.; La Nasa, J.; Ribechini, E.; Petri, A.; Piccolo, O. Extraction of Proteins and Residual Oil from Flax (*Linum usitatissimum*), Camelina (*Camelina sativa*), and Sunflower (*Helianthus annuus*) Oilseed Press Cakes. *Biomass Convers. Biorefinery* **2021**, *13*, 1915–1926. [CrossRef]
41. Fetzer, A.; Herfellner, T.; Stäbler, A.; Menner, M.; Eisner, P. Influence of Process Conditions during Aqueous Protein Extraction upon Yield from Pre-Pressed and Cold-Pressed Rapeseed Press Cake. *Ind. Crops Prod.* **2018**, *112*, 236–246. [CrossRef]
42. Li, T.; Dai, T.; Ahlström, C.; Thuvander, J.; Rayner, M.; Matos, M.; Gutiérrez, G.; Östbring, K. The Effect of Precipitation PH on Protein Recovery Yield and Emulsifying Properties in the Extraction of Protein from Cold-Pressed Rapeseed Press Cake. *Molecules* **2022**, *27*, 2957.
43. Zhao, Q.; Wang, L.; Hong, X.; Liu, Y.; Li, J. Structural and Functional Properties of Perilla Protein Isolate Extracted from Oilseed Residues and Its Utilization in Pickering Emulsions. *Food Hydrocoll.* **2021**, *113*, 106412. [CrossRef]

44. Li, P.; Zhang, W.; Han, X.; Liu, J.; Liu, Y.; Gasmalla, M.A.A.; Yang, R. Demulsification of Oil-Rich Emulsion and Characterization of Protein Hydrolysates from Peanut Cream Emulsion of Aqueous Extraction Processing. *J. Food Eng.* **2017**, *204*, 64–72. [CrossRef]
45. Ye, J.; Hua, X.; Zhao, Q.; Zhao, W.; Chu, G.; Zhang, W.; Yang, R. Chain Conformation and Rheological Properties of an Acid-Extracted Polysaccharide from Peanut Sediment of Aqueous Extraction Process. *Carbohydr. Polym.* **2020**, *228*, 115410. [CrossRef] [PubMed]
46. Marić, M.; Grassino, A.N.; Zhu, Z.; Barba, F.J.; Brnčić, M.; Rimac Brnčić, S. An Overview of the Traditional and Innovative Approaches for Pectin Extraction from Plant Food Wastes and By-Products: Ultrasound-, Microwaves-, and Enzyme-Assisted Extraction. *Trends Food Sci. Technol.* **2018**, *76*, 28–37. [CrossRef]
47. Fierascu, R.C.; Sieniawska, E.; Ortan, A.; Fierascu, I.; Xiao, J. Fruits By-Products—A Source of Valuable Active Principles. A Short Review. *Front. Bioeng. Biotechnol.* **2020**, *8*, 319. [CrossRef]
48. Huc-Mathis, D.; Almeida, G.; Michon, C. Pickering Emulsions Based on Food Byproducts: A Comprehensive Study of Soluble and Insoluble Contents. *J. Colloid Interface Sci.* **2021**, *581*, 226–237. [CrossRef]
49. Lu, Z.; Ye, F.; Zhou, G.; Gao, R.; Qin, D.; Zhao, G. Micronized Apple Pomace as a Novel Emulsifier for Food O/W Pickering Emulsion. *Food Chem.* **2020**, *330*, 127325. [CrossRef]
50. Huc-Mathis, D.; Journet, C.; Fayolle, N.; Bosc, V. Emulsifying Properties of Food By-Products: Valorizing Apple Pomace and Oat Bran. *Colloids Surf. A Physicochem. Eng. Asp.* **2019**, *568*, 84–91. [CrossRef]
51. Chatsisvili, N.T.; Amvrosiadis, I.; Kiosseoglou, V. Physicochemical Properties of a Dressing-Type o/w Emulsion as Influenced by Orange Pulp Fiber Incorporation. *LWT-Food Sci. Technol.* **2012**, *46*, 335–340. [CrossRef]
52. Lundberg, B.; Pan, X.; White, A.; Chau, H.; Hotchkiss, A. Rheology and Composition of Citrus Fiber. *J. Food Eng.* **2014**, *125*, 97–104. [CrossRef]
53. Wallecan, J.; McCrae, C.; Debon, S.J.J.; Dong, J.; Mazoyer, J. Emulsifying and Stabilizing Properties of Functionalized Orange Pulp Fibers. *Food Hydrocoll.* **2015**, *47*, 115–123. [CrossRef]
54. Joseph, C.; Savoire, R.; Harscoat-Schiavo, C.; Pintori, D.; Monteil, J.; Leal-Calderon, F.; Faure, C. O/W Pickering Emulsions Stabilized by Cocoa Powder: Role of the Emulsification Process and of Composition Parameters. *Food Res. Int.* **2019**, *116*, 755–766. [CrossRef] [PubMed]
55. Gould, J.; Vieira, J.; Wolf, B. Cocoa Particles for Food Emulsion Stabilisation. *Food Funct.* **2013**, *4*, 1369–1375. [CrossRef] [PubMed]
56. Picot-Allain, M.C.N.; Emmambux, M.N. Isolation, Characterization, and Application of Nanocellulose from Agro-Industrial By-Products: A Review. *Food Rev. Int.* **2021**, 1–29. [CrossRef]
57. Kazmi, M.Z.H.; Karmakar, A.; Michaelis, V.K.; Williams, F.J. Separation of Cellulose/Hemicellulose from Lignin in White Pine Sawdust Using Boron Trihalide Reagents. *Tetrahedron* **2019**, *75*, 1465–1470. [CrossRef]
58. Baksi, S.; Saha, S.; Birgen, C.; Sarkar, U.; Preisig, H.A.; Markussen, S.; Wittgens, B.; Wentzel, A. Valorization of Lignocellulosic Waste (Crotalaria Juncea) Using Alkaline Peroxide Pretreatment under Different Process Conditions: An Optimization Study on Separation of Lignin, Cellulose, and Hemicellulose. *J. Nat. Fibers* **2019**, *16*, 662–676. [CrossRef]
59. Dai, H.; Wu, J.; Zhang, H.; Chen, Y.; Ma, L.; Huang, H.; Huang, Y.; Zhang, Y. Recent Advances on Cellulose Nanocrystals for Pickering Emulsions: Development and Challenge. *Trends Food Sci. Technol.* **2020**, *102*, 16–29. [CrossRef]
60. Tang, L.; Liao, J.; Dai, H.; Liu, Y.; Huang, H. Comparison of Cellulose Nanocrystals from Pineapple Residues and Its Preliminary Application for Pickering Emulsions. *Nanotechnology* **2021**, *32*, 495708. [CrossRef]
61. Foo, M.L.; Ooi, C.W.; Tan, K.W.; Chew, I.M.L. A Step Closer to Sustainable Industrial Production: Tailor the Properties of Nanocrystalline Cellulose from Oil Palm Empty Fruit Bunch. *J. Environ. Chem. Eng.* **2020**, *8*, 104058. [CrossRef]
62. Tibolla, H.; Pelissari, F.M.; Rodrigues, M.I.; Menegalli, F.C. Cellulose Nanofibers Produced from Banana Peel by Enzymatic Treatment: Study of Process Conditions. *Ind. Crops Prod.* **2017**, *95*, 664–674. [CrossRef]
63. Costa, A.L.R.; Gomes, A.; Tibolla, H.; Menegalli, F.C.; Cunha, R.L. Cellulose Nanofibers from Banana Peels as a Pickering Emulsifier: High-Energy Emulsification Processes. *Carbohydr. Polym.* **2018**, *194*, 122–131. [CrossRef] [PubMed]
64. Wen, J.; Zhang, W.; Xu, Y.; Yu, Y.; Lin, X.; Fu, M.; Liu, H.; Peng, J.; Zhao, Z. Cellulose Nanofiber from Pomelo Spongy Tissue as a Novel Particle Stabilizer for Pickering Emulsion. *Int. J. Biol. Macromol.* **2022**, *224*, 1439–1449. [CrossRef]
65. Morikawa, C.K.; Saigusa, M. Recycling Coffee Grounds and Tea Leaf Wastes to Improve the Yield and Mineral Content of Grains of Paddy Rice. *J. Sci. Food Agric.* **2011**, *91*, 2108–2111. [CrossRef]
66. Ren, Z.; Chen, Z.; Zhang, Y.; Zhao, T.; Ye, X.; Gao, X.; Lin, X.; Li, B. Functional Properties and Structural Profiles of Water-Insoluble Proteins from Three Types of Tea Residues. *LWT-Food Sci. Technol.* **2019**, *110*, 324–331. [CrossRef]
67. Ren, Z.; Chen, Z.; Zhang, Y.; Lin, X.; Li, B. Novel Food-Grade Pickering Emulsions Stabilized by Tea Water-Insoluble Protein Nanoparticles from Tea Residues. *Food Hydrocoll.* **2019**, *96*, 322–330. [CrossRef]
68. He, K.; Li, Q.; Li, Y.; Li, B.; Liu, S. Water-Insoluble Dietary Fibers from Bamboo Shoot Used as Plant Food Particles for the Stabilization of O/W Pickering Emulsion. *Food Chem.* **2020**, *310*, 125925. [CrossRef] [PubMed]
69. Maravić, N.; Šereš, Z.; Nikolić, I.; Dokić, P.; Kertész, S.; Dokić, L. Emulsion Stabilizing Capacity of Sugar Beet Fibers Compared to Sugar Beet Pectin and Octenyl Succinate Modified Maltodextrin in the Production of O/W Emulsions: Individual and Combined Impact. *LWT-Food Sci. Technol.* **2019**, *108*, 392–399. [CrossRef]
70. Ralla, T.; Salminen, H.; Edelmann, M.; Dawid, C.; Hofmann, T.; Weiss, J. Oat Bran Extract (Avena Sativa L.) from Food by-Product Streams as New Natural Emulsifier. *Food Hydrocoll.* **2018**, *81*, 253–262. [CrossRef]

71. Pujol, D.; Liu, C.; Gominho, J.; Olivella, M.À.; Fiol, N.; Villaescusa, I.; Pereira, H. The Chemical Composition of Exhausted Coffee Waste. *Ind. Crops Prod.* **2013**, *50*, 423–429. [CrossRef]
72. Gould, J.; Garcia-Garcia, G.; Wolf, B. Pickering Particles Prepared from Food Waste. *Materials* **2016**, *9*, 791. [CrossRef] [PubMed]
73. Foo, M.L.; Ooi, C.W.; Tan, K.W.; Chew, I.M.L. Preparation of Black Cumin Seed Oil Pickering Nanoemulsion with Enhanced Stability and Antioxidant Potential Using Nanocrystalline Cellulose from Oil Palm Empty Fruit Bunch. *Chemosphere* **2022**, *287*, 132108. [CrossRef]
74. Xia, T.; Xue, C.; Wei, Z. Physicochemical Characteristics, Applications and Research Trends of Edible Pickering Emulsions. *Trends Food Sci. Technol.* **2021**, *107*, 1–15. [CrossRef]
75. Wei, Z.; Huang, Q. Edible Pickering Emulsions Stabilized by Ovotransferrin–Gum Arabic Particles. *Food Hydrocoll.* **2019**, *89*, 590–601. [CrossRef]
76. Kargar, M.; Fayazmanesh, K.; Alavi, M.; Spyropoulos, F.; Norton, I.T. Investigation into the Potential Ability of Pickering Emulsions (Food-Grade Particles) to Enhance the Oxidative Stability of Oil-in-Water Emulsions. *J. Colloid Interface Sci.* **2012**, *366*, 209–215. [CrossRef] [PubMed]
77. Fu, D.; Deng, S.; McClements, D.J.; Zhou, L.; Zou, L.; Yi, J.; Liu, C.; Liu, W. Encapsulation of β-Carotene in Wheat Gluten Nanoparticle-Xanthan Gum-Stabilized Pickering Emulsions: Enhancement of Carotenoid Stability and Bioaccessibility. *Food Hydrocoll.* **2019**, *89*, 80–89. [CrossRef]
78. Wei, Z.; Cheng, J.; Huang, Q. Food-Grade Pickering Emulsions Stabilized by Ovotransferrin Fibrils. *Food Hydrocoll.* **2019**, *94*, 592–602. [CrossRef]
79. Liu, B.; Li, T.; Wang, W.; Sagis, L.M.C.; Yuan, Q.; Lei, X.; Cohen Stuart, M.A.; Li, D.; Bao, C.; Bai, J.; et al. Corncob Cellulose Nanosphere as an Eco-Friendly Detergent. *Nat. Sustain.* **2020**, *3*, 448–458. [CrossRef]
80. Wei, Z.; Wang, C.; Zou, S.; Liu, H.; Tong, Z. Chitosan Nanoparticles as Particular Emulsifier for Preparation of Novel PH-Responsive Pickering Emulsions and PLGA Microcapsules. *Polymer* **2012**, *53*, 1229–1235. [CrossRef]
81. Guo, Q. Progress in the Preparation, Stability and Functional Applications of Pickering Emulsion. *IOP Conf. Ser. Earth Environ. Sci.* **2021**, *639*, 012028. [CrossRef]
82. Feng, X.; Sun, Y.; Yang, Y.; Zhou, X.; Cen, K.; Yu, C.; Xu, T.; Tang, X. Zein Nanoparticle Stabilized Pickering Emulsion Enriched with Cinnamon Oil and Its Effects on Pound Cakes. *LWT-Food Sci. Technol.* **2020**, *122*, 109025. [CrossRef]
83. Jiang, Y.; Zhang, C.; Yuan, J.; Wu, Y.; Li, F.; Li, D.; Huang, Q. Effects of Pectin Polydispersity on Zein/Pectin Composite Nanoparticles (ZAPs)as High Internal-Phase Pickering Emulsion Stabilizers. *Carbohydr. Polym.* **2019**, *219*, 77–86. [CrossRef]
84. Guo, C.; Yin, J.; Chen, D. Co-Encapsulation of Curcumin and Resveratrol into Novel Nutraceutical Hyalurosomes Nano-Food Delivery System Based on Oligo-Hyaluronic Acid-Curcumin Polymer. *Carbohydr. Polym.* **2018**, *181*, 1033–1037. [CrossRef] [PubMed]
85. Wei, Z.; Gao, Y. Physicochemical Properties of β-Carotene Bilayer Emulsions Coated by Milk Proteins and Chitosan-EGCG Conjugates. *Food Hydrocoll.* **2016**, *52*, 590–599. [CrossRef]
86. Zhou, F.Z.; Yu, X.H.; Zeng, T.; Yin, S.W.; Tang, C.H.; Yang, X.Q. Fabrication and Characterization of Novel Water-Insoluble Protein Porous Materials Derived from Pickering High Internal-Phase Emulsions Stabilized by Gliadin-Chitosan-Complex Particles. *J. Agric. Food Chem.* **2019**, *67*, 3423–3431. [CrossRef] [PubMed]
87. Ruan, Q.; Guo, J.; Wan, Z.; Ren, J.; Yang, X. PH Switchable Pickering Emulsion Based on Soy Peptides Functionalized Calcium Phosphate Particles. *Food Hydrocoll.* **2017**, *70*, 219–228. [CrossRef]
88. Zhu, Y.; McClements, D.J.; Zhou, W.; Peng, S.; Zhou, L.; Zou, L.; Liu, W. Influence of Ionic Strength and Thermal Pretreatment on the Freeze-Thaw Stability of Pickering Emulsion Gels. *Food Chem.* **2020**, *303*, 125401. [CrossRef]
89. Fasihi, H.; Fazilati, M.; Hashemi, M.; Noshirvani, N. Novel Carboxymethyl Cellulose-Polyvinyl Alcohol Blend Films Stabilized by Pickering Emulsion Incorporation Method. *Carbohydr. Polym.* **2017**, *167*, 79–89. [CrossRef]
90. Fasihi, H.; Noshirvani, N.; Hashemi, M.; Fazilati, M.; Salavati, H.; Coma, V. Antioxidant and Antimicrobial Properties of Carbohydrate-Based Films Enriched with Cinnamon Essential Oil by Pickering Emulsion Method. *Food Packag. Shelf Life* **2019**, *19*, 147–154. [CrossRef]
91. Lim, H.P.; Ho, K.W.; Surjit Singh, C.K.; Ooi, C.W.; Tey, B.T.; Chan, E.S. Pickering Emulsion Hydrogel as a Promising Food Delivery System: Synergistic Effects of Chitosan Pickering Emulsifier and Alginate Matrix on Hydrogel Stability and Emulsion Delivery. *Food Hydrocoll.* **2020**, *103*, 105659. [CrossRef]

colloids and interfaces

MDPI

Article

Impact of Operating Parameters on the Production of Nanoemulsions Using a High-Pressure Homogenizer with Flow Pattern and Back Pressure Control

Hualu Zhou [1], Dingkui Qin [1], Giang Vu [1] and David Julian McClements [1,2,*]

[1] Biopolymers and Colloids Laboratory, Department of Food Science, University of Massachusetts, Amherst, MA 01003, USA
[2] Department of Food Science & Bioengineering, Zhejiang Gongshang University, 18 Xuezheng Street, Hangzhou 310018, China
* Correspondence: mcclements@foodsci.umass.edu; Tel.: +1-(413)-545-2275; Fax: +1-(413)-545-1262

Abstract: The main objective of this study was to establish the relative importance of the main operating parameters impacting the formation of food-grade oil-in-water nanoemulsions by high-pressure homogenization. The goal of this unit operation was to create uniform and stable emulsified products with small mean particle diameters and narrow polydispersity indices. In this study, we examined the performance of a new commercial high-pressure valve homogenizer, which has several features that provide good control over the particle size distribution of nanoemulsions, including variable homogenization pressures (up to 45,000 psi), nozzle dimensions (0.13/0.22 mm), flow patterns (parallel/reverse), and back pressures. The impact of homogenization pressure, number of passes, flow pattern, nozzle dimensions, back pressure, oil concentration, emulsifier concentration, and emulsifier type on the particle size distribution of corn oil-in-water emulsions was systematically examined. The droplet size decreased with increasing homogenization pressure, number of passes, back pressure, and emulsifier-to-oil ratio. Moreover, it was slightly smaller when a reverse rather than parallel flow profile was used. The emulsifying performance of plant, animal, and synthetic emulsifiers was compared because there is increasing interest in replacing animal and synthetic emulsifiers with plant-based ones in the food industry. Under fixed homogenization conditions, the mean particle diameter decreased in the following order: gum arabic (0.66 μm) > soy protein (0.18 μm) > whey protein (0.14 μm) ≈ Tween 20 (0.14 μm). The information reported in this study is useful for the optimization of the production of food-grade nanoemulsions using high-pressure homogenization.

Keywords: homogenizer; nanoemulsions; particle size; nozzles; back pressures; plant-based

Citation: Zhou, H.; Qin, D.; Vu, G.; McClements, D.J. Impact of Operating Parameters on the Production of Nanoemulsions Using a High-Pressure Homogenizer with Flow Pattern and Back Pressure Control. *Colloids Interfaces* **2023**, *7*, 21. https://doi.org/10.3390/colloids7010021

Academic Editors: César Burgos-Díaz, Mauricio Opazo-Navarrete and Eduardo Morales

Received: 24 January 2023
Revised: 9 March 2023
Accepted: 13 March 2023
Published: 16 March 2023

1. Introduction

Oil-in-water nanoemulsions are colloidal dispersions containing small oil droplets dispersed in water, with the droplets being coated by emulsifying agents, such as surfactants, phospholipids, proteins, and/or polysaccharides [1–3]. The mean particle diameter in nanoemulsions typically ranges from around 20 to 200 nm, depending on the formulation and preparation method used [2,4]. Nanoemulsions have several potential advantages for certain commercial applications due to their small droplet dimensions [5]. First, the relatively small droplet size increases their resistance to gravitational separation and aggregation, which increases the shelf life of emulsified products [6]. Second, nanoemulsions with sufficiently small droplets (<50 nm) are optically transparent because they only scatter light very weakly, which is useful for the development of optically clear foods and beverages [7]. Third, the small dimensions of the oil droplets in nanoemulsions means they are rapidly and completely digested within the gastrointestinal tract, which increases the bioavailability of any hydrophobic bioactive agents encapsulated inside them [8,9].

Indeed, nanoemulsions can be used in many different commercial products, including foods, beverages, cosmetics, drugs, biomedicines, and agrochemicals [10–12]. For example, they can be used in the medical industry as drug delivery systems or for the development of diagnostic tools, such as contrast agents [12]. They can be used in the agrochemical industry as delivery systems to improve the effectiveness and reduce the environmental impact of pesticides [11]. The ability to create nanoemulsions with well-defined and tunable droplet size distributions is important in many of these commercial applications because the stability and efficacy of these colloidal delivery systems depend on this particle characteristic [13]. In particular, it is usually desirable to produce nanoemulsions with narrow particle size distributions and adjustable mean particle diameters to improve stability and obtain more reliable release characteristics. Consequently, there is great interest in controlling the particle size characteristics of nanoemulsions to improve their functional performance [1,14–16].

Food-grade nanoemulsions are typically produced using two distinct approaches: low-energy physicochemical and high-energy mechanical methods [17]. The low-energy methods are based on controlled alterations in composition and/or temperature, which cause the system to move to a different part of the phase diagram, thereby leading to the spontaneous assembly of nanoscale oil droplets [18,19]. Representative examples of this homogenization approach include spontaneous emulsification and phase inversion temperature methods [18]. The main advantage of low-energy methods is that no expensive equipment is required, but the main disadvantage is that they can typically only be carried out using high concentrations of synthetic surfactants and specific kinds of oil. In contrast, high-energy methods use mechanical forces to create nanoemulsions from mixtures of oil, water, and emulsifiers [20]. Specialized mechanical devices have been developed to generate the high level of disruptive forces required to create nanoemulsions, including high-pressure valve homogenizers [21], microfluidizers [22,23], and sonicators [24,25]. The main advantage of this method is that a wide range of oils and emulsifiers can be utilized, but the main disadvantage is that specialized equipment is required.

High-pressure valve homogenizers are one of the most common mechanical devices used to produce nanoemulsions in the food industry, especially for nanoemulsions produced from low- or intermediate-viscosity fluids [21,26–28]. In most cases, a coarse emulsion premix is produced first using a high-shear mixer that is then fed into the homogenizer, which increases the overall efficiency of the particle size reduction process [29,30]. The coarse emulsions pass through the inlet of the homogenizer and then are pulled into a chamber that forces them through a narrow nozzle using a piston [31]. As the coarse emulsion passes through the nozzle, the large droplets are broken down into smaller ones by the intense, disruptive forces generated inside the device, which are usually a combination of shear, turbulence, and cavitation forces. The nature of the disruptive forces depends on the homogenization device and operating conditions used and can be manipulated to alter the particle size distribution produced. The nozzle dimensions also have an impact on the efficiency of droplet disruption, with smaller nozzles normally producing smaller droplets [31]. However, larger nozzle dimensions are often more suitable for homogenizing highly viscous fluids so as to ensure a good flow rate and avoid clogs. The number of times to pass through the nozzle also determines the size of final nanoemulsions, with the droplet size decreasing with an increasing number of passes until a plateau is reached [31–33]. The physicochemical properties of the materials used to prepare nanoemulsions also influence the size of the droplets that can be produced during homogenization, such as oil type, emulsifier type, oil concentration, emulsifier concentration, and emulsifier-to-oil ratio [34–38]. To produce small droplets, it is critical that there is enough emulsifier present to cover all of the new oil–water interface generated during homogenization. Otherwise, some droplet coalescence occurs inside the device, leading to a larger droplet size.

In this study, we systematically examined the performance of a recently introduced commercial homogenizer at producing oil-in-water nanoemulsions using food-grade ingredients. This device is a type of high-pressure homogenizer, and so the results obtained can be related to other homogenization devices within this category, but it also has some additional features that provide flexibility when producing nanoemulsions with specific particle size distributions. These additional features include the ability to vary the homogenization pressure over a wide range, the number of passes, the nozzle dimensions, the flow pattern, and the back pressure. We examined the impact of these parameters, as well as emulsifier-to-oil concentration ratio, oil concentration, emulsifier concentration, and emulsifier type, on the particle size distributions. Moreover, we compared the emulsifying properties of plant-derived, animal-derived, and synthetic emulsifiers in producing nanoemulsions because there is growing interest in the food industry in creating plant-based foods to replace animal ones for ethical, health, and environmental reasons. The information obtained in this study should therefore be useful to promote the use of nanoemulsions in a wide range of applications within the agrochemical, food, supplement, cosmetics, personal care, biomedical, and pharmaceutical industries.

2. Materials and Methods

2.1. Materials

Corn oil (Mazola, ACH Food Company, Memphis, TN, USA) was purchased from a local supermarket and stored in a refrigerator (4 °C) prior to use. Tween 20 were purchased from Sigma-Aldrich (Sigma Chemical Co., St. Louis, MO, USA). Soy protein isolate was kindly provided by ADM (Decatur, IL, USA). Whey protein isolate was kindly provided by Davisco Foods International Inc. (Le Sueur, MN, USA). Gum Arabic was kindly provided by TIC Gums (Belcamp, MD, USA). Double distilled water was used to prepare all samples.

2.2. Preparation of the Coarse Emulsions

A hand-held shearing device (M133/1281-0, Biospec Products, Inc., Bartlesville, OK, USA) was used to blend oil, water, and emulsifier together in a glass container for 4 min. This led to the formation of coarse oil-in-water emulsions containing relatively large oil droplets ($d > 1000$ nm). These large droplets were relatively unstable to creaming and phase separation, so it was important to transfer the coarse emulsions to the high-pressure homogenizer used to prepare the nanoemulsions relatively quickly (within a few minutes).

2.3. Impact of Operating Conditions and Formulation on Nanoemulsion Formation

Nanoemulsions were prepared using a high-pressure homogenizer (Nano DeBEE Gen II, BEE International Inc, South Easton, MA, USA). This device can be operated at operating pressures up to 45,000 psi. It is also possible to adjust the instrument to use parallel or reverse flow patterns, different nozzle dimensions, and different back pressures. Experiments were carried out to establish the impact of these machine settings on the droplet size distributions of oil-in-water nanoemulsions. Experiments were therefore carried out using parallel or reverse flow profiles, different nozzle dimensions (Z5/Z8, 0.13/0.22 mm), and with or without back pressure. To provide comparable results, the following parameters were fixed for these experiments: emulsifier type (Tween 20); emulsifier concentration (2 wt%); oil concentration (10 wt%); operating pressure (12,000 psi); the number of passes (three). For the other experiments, the following default machine settings were used: Z5 nozzle, reverse flow pattern, and no back pressure. The impact of homogenization pressure was assessed from 12,000 to 45,000 psi. The impact of emulsifier concentration was assessed from 0.2 to 5 wt%. The number of passes was assessed from 0 to 8. The impact of formulation parameters was also assessed using different oil concentrations, emulsifier concentrations, emulsifier-to-oil concentration ratios, and emulsifier types (whey protein, soy protein, gum Arabic, and Tween 20).

2.4. Characterization of the Particle Size Parameters of Nanoemulsions

Laser diffraction (Mastersizer 2000, Malvern Instruments, Worcestershire, United Kingdom) was used to measure the mean particle diameter (d_{32} and d_{43}) and particle size distribution of the nanoemulsions. Here, d_{32} and d_{43} are the surface-weighted and volume-weighted mean particle diameters, respectively [39]: $d_{32} = (\Sigma n_i d_i^3)/(\Sigma n_i d_i^2)$ and $d_{43} = (\Sigma n_i d_i^4)/(\Sigma n_i d_i^3)$, where n_i and d_i are the number and diameter of the droplets in the *i*th size category, and Σ represents the sum over all the categories (i = 1 to N). For laser diffraction, the particle size distribution of a colloidal dispersion is determined by measuring the variation of the intensity of light as a function of the scattering angle when a laser beam is directed through the sample. The particle size distribution that gives the best fit between the measured scattering pattern and the predictions made using Mie theory is then determined. The mean particle diameters are then calculated from the particle size distribution. The "Mean Particle Diameter" in all plots refers to the d_{32} value if not clearly specified. A measure of the width of the distribution was also calculated from the particle size distribution for some samples: $W = (\Sigma\phi_i(d_{i.} - d_{43})^2/N)^{1/2}$, where d_i and ϕ_i are the particle size and the volume fraction of droplets in the *i*th size category, respectively, while N is the total number of size categories [39].

2.5. Statistical Analysis

The average values and standard errors of mean particle diameters were calculated from at least three measurements using Microsoft Excel. ANOVA (post hoc Tukey HSD test) software was used to ascertain the significant difference between differences between samples ($p < 0.05$).

3. Results and Discussion

3.1. Impact of Homogenizer Operating Parameters on Particle Size

Initially, we examined the impact of operating parameters on the particle size characteristics of 10 wt% corn oil-in-water nanoemulsions stabilized by 2 wt% Tween 20 produced using the high-pressure homogenizer. Several parameters were examined, including the flow pattern, nozzle, and effects of back pressure.

3.1.1. Flow Pattern

The nature of the flow pattern within a homogenizer impacts the efficiency of droplet disruption by altering the relative magnitudes of cavitation, shear, and turbulent forces, as well as the residence time of the droplets in the disruption zone. For this reason, we examined the impact of two different flow patterns (parallel and reverse) that could be created within the homogenizer on the size of the oil droplets produced during homogenization (Figure 1). For the parallel flow pattern, the coarse emulsion is placed in the inlet reservoir, and then pressure from an intensifier pump forces it through a small nozzle. The large droplets in the coarse emulsion are broken down into smaller ones, leading to the formation of nanoemulsions in the emulsifying cell. These nanoemulsions then flow through a cooling system before being collected. For this flow pattern, back pressure is used to reduce the output flow of the nanoemulsions.

For the reverse flow pattern, there are two different channels that the nanoemulsions can pass through; one is similar to the parallel flow channel, while the other is a confined reverse flow channel (Figure 1). The reverse flow channel creates a reverse stream of nanoemulsion that impinges on the input stream, thereby creating additional disruptive forces that reduce the size of the oil droplets. As a result, the reverse flow pattern should be more effective at reducing the proportion of large droplets in the nanoemulsions than the parallel flow pattern.

As expected, the experimental measurements showed that more nanoemulsions with smaller mean droplet diameters and a narrower particle size distribution were produced using the reverse flow pattern compared to the parallel one (Figure 2). Even so, this effect was relatively small. For instance, the mean particle diameter (d_{32}) was 0.137 $\pm$ 0.001 µm for reverse flow and 0.144 $\pm$ 0.001 µm for parallel flow (less than 5% difference).

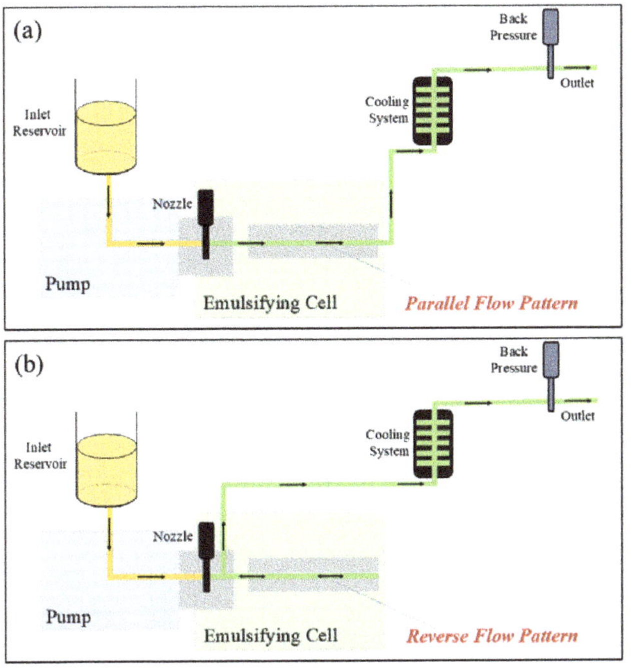

Figure 1. Schematic illustration of (**a**) parallel and (**b**) reverse flow type within the homogenizer used to produce the nanoemulsions.

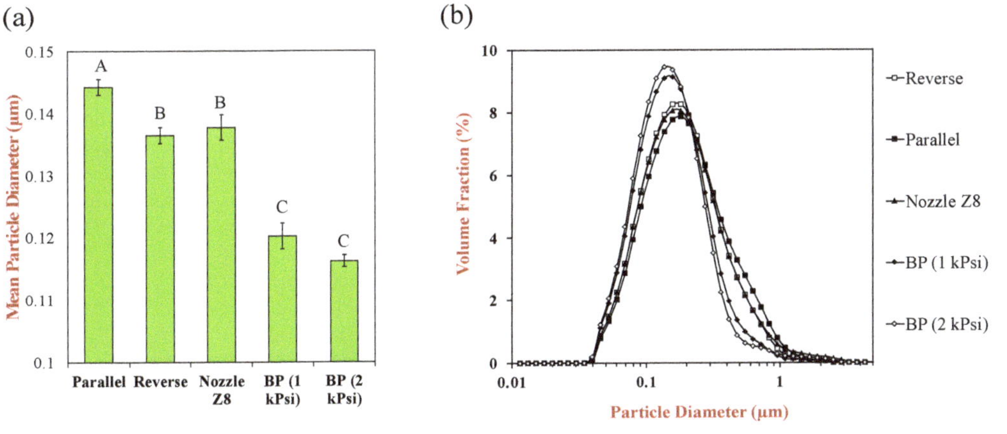

Figure 2. The impact of different homogenization parameters on (**a**) the mean particle diameter and (**b**) particle size distribution of 10% corn oil-in-water nanoemulsions. The default parameters were: 12,000 psi operating pressure, 3 passes, Z5 nozzle, no back pressure, and reverse flow pattern. The letters (A, B, C) represent the significance of the samples ($p < 0.05$). In (**a**), the error bars represent the standard deviations of multiple measurements. For (**b**), the widths of the distributions calculated from the particle size distribution data were: 0.18, 0.21, 0.26, 0.17, and 0.20 µm from the "Reverse" to "BP (2 KPsi)" sample, respectively.

3.1.2. Impact of Nozzle Dimensions

The impact of nozzle dimensions was also examined for nanoemulsions that were all produced using the reverse flow profile. Nanoemulsions were produced using a small nozzle (Z5: 0.13 mm) and a large nozzle (Z8: 0.22 mm). The samples produced using the small nozzle are labelled "Reverse", while those produced using the large nozzle are labeled as "Nozzle Z8" in Figure 2. We hypothesized that the smaller nozzle would produce nanoemulsions containing smaller droplets, but the difference was not significant. For instance, the mean particle diameter (d_{32}) was 0.137 ± 0.001 μm for the small nozzle and 0.138 ± 0.002 μm for the large nozzle (less than 1% difference). Thus, the nozzle dimensions had practically no impact on the formation of nanoemulsions for this particular formulation.

3.1.3. Impact of Back Pressure

The impact of applying back pressure to the nanoemulsions during homogenization was then examined. Previous studies have shown that applying back pressure can alter the cavitation zones around a nozzle, which can alter the efficiency of droplet disruption inside a homogenizer [40,41]. We found that applying back pressure led to a significant reduction in particle size. In this case, the samples were labelled as "Reverse", "BP 1 KPsi", and "BP 2 KPsi" for nanoemulsions prepared using 0, 1000 psi and 2000 psi of back pressure, respectively. The mean particle diameters (d_{32}) were 0.137 ± 0.001, 0.122 ± 0.002, and 0.116 ± 0.001 μm for these three nanoemulsions, respectively (Figure 2a). Moreover, the width of the particle size distribution became narrower at the higher back pressures (Figure 2b). These results suggest that the use of back pressure is beneficial for producing nanoemulsions containing smaller droplets. This effect can be attributed to the alteration in the cavitation pattern around the nozzle in the presence of back pressure, as well as the increased retention time of the nanoemulsions in the disruptive zone.

3.1.4. Number of Passes

Another effective mean of controlling the size of the droplets in nanoemulsions is to alter the number of times they are passed through the homogenizer nozzle. The influence of the number of passes through the homogenizer on the mean droplet diameter and particle size distribution was therefore measured (Figure 3). Prior to passing through the homogenizer, the mean particle diameter of the coarse emulsion was around 14.2 μm. These emulsions were relatively unstable to creaming and oiling-off, which can be attributed to the increase in gravitational forces and coalescence rate with increasing droplet size [39]. A single pass through the homogenizer caused a pronounced decrease in the mean particle diameter, which can be attributed to the intense, disruptive forces (cavitation, shear, and turbulence) the emulsions experience as they pass through the homogenizer nozzle. The mean particle diameter then continued to decrease as the number of passes was increased, but there was only a fairly modest reduction in mean particle size with an increasing number of passes. For example, the mean particle diameter decreased from 0.217 ± 0.006 μm after one pass to 0.134 ± 0.003 μm after eight passes. The number of passes had an important impact on the particle size distributions of the nanoemulsions (Figure 3b). The width of the particle size distribution decreased with an increasing number of passes from one to six passes and then remained relatively constant. For instance, the calculated widths of the distributions were 10.4, 0.46, 0.41, 0.33, 0.25, 0.23, 0.21, 0.21, and 0.19 μm from zero to eight passes, respectively. This effect can mainly be attributed to the fact that a larger fraction of the droplets in the nanoemulsions has a chance to pass through the intense, disruptive zone inside the homogenizer as the number of passes increases.

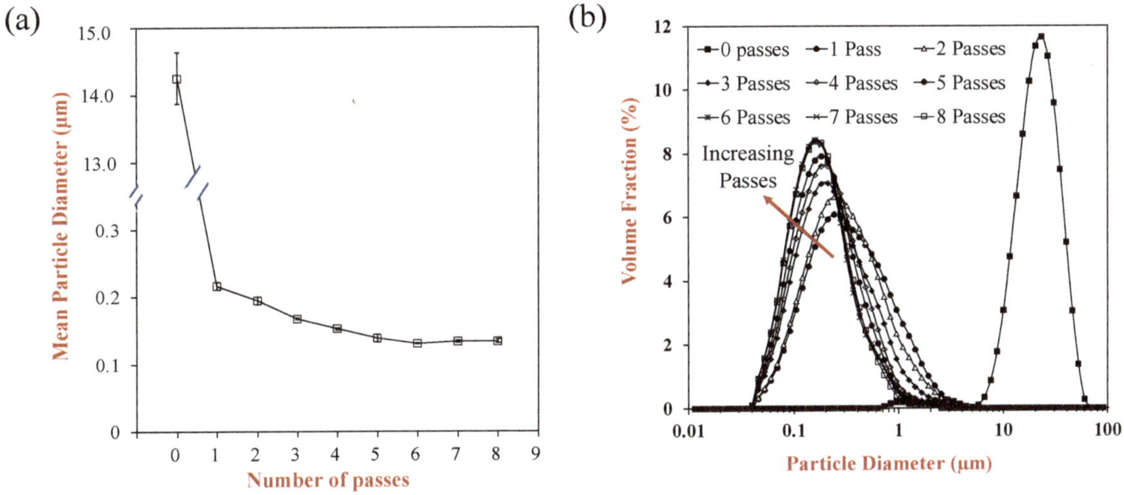

Figure 3. Impact of the number of passes on the particle size characteristics of nanoemulsions (10 wt% corn oil, 2 wt% Tween 20 surfactant): (**a**) the mean particle diameter and (**b**) particle size distribution. For (**a**), the standard deviations calculated from repeated measurements were 0.385, 0.006, 0.006, 0.001, 0.001, 0.005, 0.001, 0.001, and 0.003 μm from 0 to 8 passes, respectively. For (**b**), the widths of the distributions calculated from the particle size distribution data were: 10.4, 0.46, 0.41, 0.33, 0.25, 0.23, 0.21, 0.21, and 0.19 μm from 0 to 8 passes, respectively.

These results show that nanoemulsions with smaller droplets and narrower particle size distributions can be produced by increasing the number of passes through the homogenizer. However, the time and energy requirements increase for each additional pass, which would increase energy costs and reduce the sustainability of the process. Consequently, a limited number of passes would be preferred for most industrial applications. For this reason, we used three passes as a default number of passes in this study. Other researchers have also reported a decrease in droplet size with the number of passes for nanoemulsions produced using high-pressure homogenizers [32,42].

3.1.5. Homogenization Pressure

Controlling the operating pressure of a homogenizer can also be utilized to alter the particle size of nanoemulsions since the magnitude of the disruptive energy generated inside the device increases as the pressure used to force the emulsion through the nozzle is raised [43]. As expected, the mean droplet diameter decreased with increasing operating pressure, which agrees with previous studies [32,44]. Previous studies with high-pressure homogenizers have shown there is often a linear log–log relationship between the mean particle diameter and operating pressure when there is sufficient emulsifier present to cover all the droplets formed [43]. As expected, there was a linear log–log relationship between the mean particle diameter and homogenization pressure with a slope of around −0.084 (see Figure 4 caption). However, the reduction in mean particle diameter with increasing pressure was relatively small. For example, the mean particle diameter decreased from around 0.137 ± 0.001 at 12,000 psi to 0.123 ± 0.001 μm at 45,000 psi, which indicates that an approximately three-fold increase in pressure only caused a 10% reduction in particle size. Moreover, there was only a small change in the droplet size distribution with increasing pressure (Figure 4b).

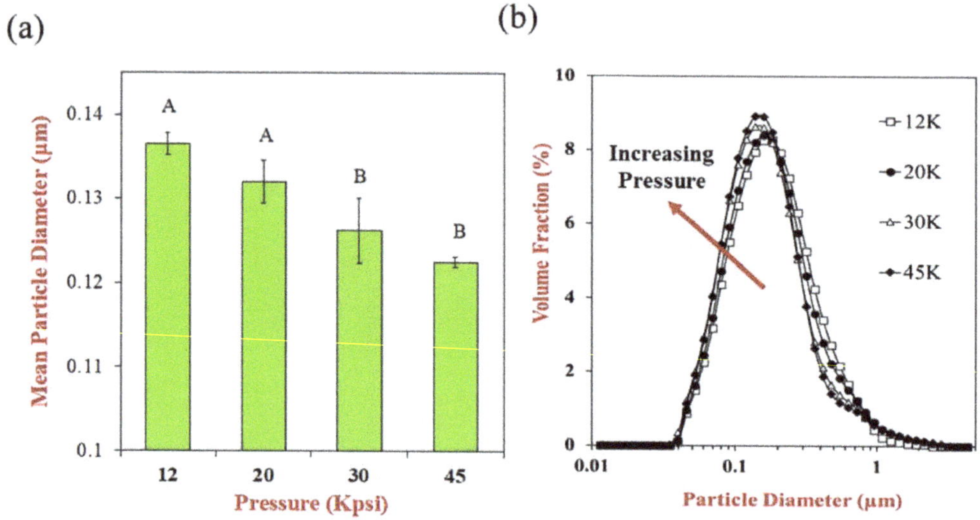

Figure 4. The impact of homogenizer pressure parameter on (**a**) mean particle diameter and (**b**) particle size distributions of 10 wt% corn oil-in-water nanoemulsions stabilized by Tween 20. The letters (A, B) represent significant differences between samples ($p < 0.05$). For (**a**), the error bars represent the standard deviations of repeated measurements. A linear relation between the Log (Mean Particle Diameter/µm) and Log (Pressure/KPsi) was found: $\log(D) = -0.084\log(P) - 0.773$, $R^2 = 0.990$. For (**b**), the widths of the distributions calculated from the particle size distribution data were: 0.18, 0.24, 0.24, and 0.24 µm from the 12 k, 20 k, 30 k, and 45 k samples, respectively.

As the operating pressure is raised, the energy costs of the homogenization process increase, which is undesirable from a cost and sustainability perspective. Consequently, it may be better to use as low an operating pressure as possible to create nanoemulsions with sufficient stability and functional performance. For this reason, we used an operating pressure of 12,000 psi as a default value to produce nanoemulsions with small droplets without incurring excessively high energy costs.

3.2. Impact of Nanoemulsion Properties

In addition to the operating conditions of the homogenizer, the nature of the components used to prepare a nanoemulsion also impacts the mean droplet diameter and particle size distribution [43]. For this reason, we examined the impact of emulsifier concentration (at a fixed oil concentration and at a fixed emulsifier-to-oil ratio) and emulsifier type on the efficiency of homogenization.

3.2.1. Emulsifier Concentration at a Fixed Oil Concentration

Initially, we examined the impact of emulsifier concentration on the mean particle diameter and particle size distribution of nanoemulsions with a fixed oil concentration, i.e., 10 wt% (Figure 5). As the emulsifier concentration was increased, the mean particle diameter of the nanoemulsions decreased significantly from 0.2 to 2 wt% Tween 20 and then remained relatively constant from 2 to 5 wt% Tween 20. This reduction in particle size with increasing emulsifier concentration can be attributed to two main physicochemical phenomena [39]. First, the time required for the oil droplet surfaces to become saturated with emulsifier molecules is reduced at higher emulsifier concentrations, which inhibits the recoalescence of the oil droplets inside the homogenizer. Second, smaller droplets have a larger specific surface area than larger ones and so require more emulsifiers to stabilize them. The minimum concentration of emulsifier required to completely cover all of the oil droplet surfaces in a nanoemulsion can be calculated using the following equation [45]:

$$C_s = \frac{6 \times \Gamma \times \Phi}{d_{32}} \tag{1}$$

here C_s is the minimum concentration of emulsifier required to reach saturation (kg/m^3), Γ is the surface load of the emulsifier at saturation, Φ is the dispersed phase volume fraction, and d_{32} is the surface-weighted mean droplet diameter [39,46]. To a first approximation, the saturation concentration of Tween 20 was calculated to be about 1 wt% (d_{32} = 130 nm, Φ = 0.1, $\Gamma = 2.0 \times 10^{-6}$ kg/m^2) [47], which is fairly close to the value of 2 wt% observed experimentally. The fact that a higher concentration was required than predicted may have been because not all of the surfactant molecules adsorbed to the oil–water interfaces in the nanoemulsions. For instance, some of the surfactant molecules may have been present as monomers or micelles in the aqueous phase.

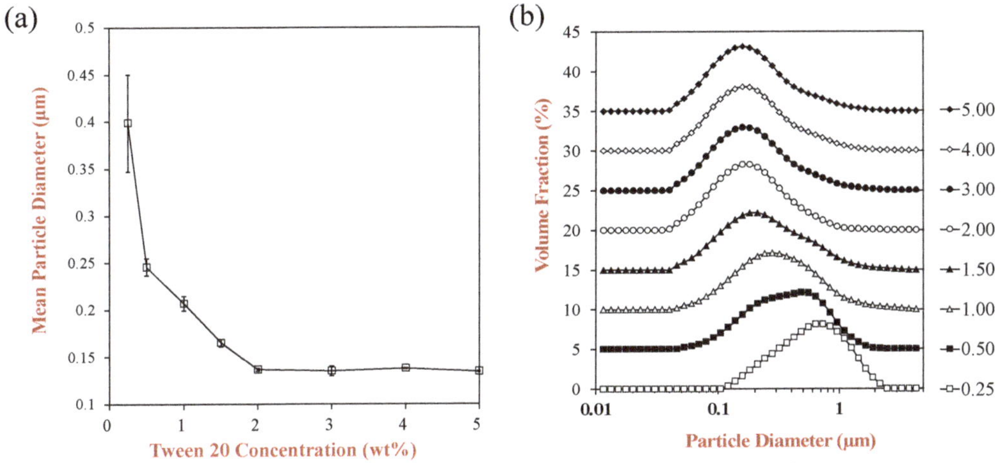

Figure 5. The influence of Tween 20 concentration on (**a**) the mean particle diameter (d_{32}) and (**b**) particle size distribution of nanoemulsions containing 10 wt% corn oil. Default processing conditions were used for all samples, as described in the text. In (**a**), the error bars represent the standard deviations of repeated measurements. For (**b**), the widths of the distributions calculated from the particle size distribution data were: 0.41, 0.30, 0.41, 0.32, 0.18, 0.26, 0.24, and 0.23 μm from the "0.25" to "5.00" samples, respectively.

Practically, it is important to use the minimum amount of emulsifier that can produce stable nanoemulsions containing small droplets; otherwise, ingredient costs will be increased. For this reason, 2 wt% Tween 20 was selected as the default concentration to produce the nanoemulsions.

3.2.2. Oil Concentration

For many commercial applications, it is necessary to create nanoemulsions containing a relatively high amount of oil, as this impacts their appearance, rheology, sensory attributes, and delivery properties. For this reason, we examined the impact of oil concentration on the formation of nanoemulsions. In these experiments, the emulsifier-to-oil concentration was fixed at 2:10, as this ratio was found to create small droplets in the previous section. The impact of increasing the oil concentration on the mean particle diameter (d_{32} and d_{43}) and particle size distributions of the nanoemulsions was then examined (Figure 6). Both d_{32} and d_{43} decreased with increasing oil concentration, with the effect being particularly noticeable for the d_{43} values, which are more sensitive to the larger droplets in an emulsion. The widths of the particle size distribution also decreased with increasing oil concentration (Figure 6b). These results show that using a higher oil concentration is beneficial for producing nanoemulsions with smaller droplets and narrower size distributions. Moreover, this would be beneficial from a cost and time perspective since the total operating time of the homogenizer could be reduced. The origin of the decrease in droplet size with increasing droplet concentration may have been due to the increase in viscosity of the nanoemulsions at higher droplet concentrations. A higher viscosity leads to greater shear stresses being generated during homogenization, thereby leading to smaller droplet sizes. Conversely, a higher droplet concentration may also lead to more frequent droplet–droplet collisions, which could promote coalescence within the homogenizer. It should be noted that we could not form nanoemulsions at 60 wt% oil or higher, which was because they were too viscous to pump through the homogenizer.

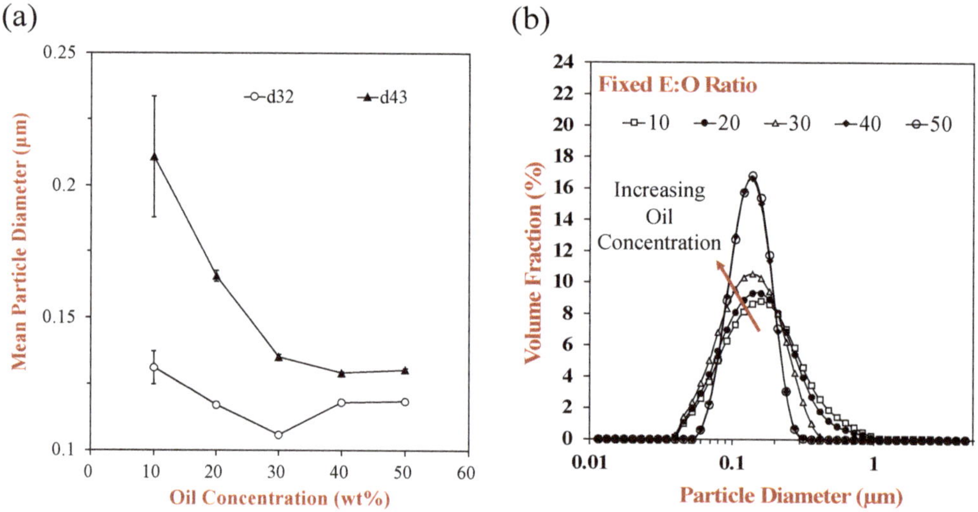

Figure 6. The influence of oil concentration on the (**a**) mean particle diameters (d_{32} and d_{43}) and (**b**) particle size distributions of corn oil-in-water nanoemulsions stabilized by Tween 20 (emulsifier-to-oil ratio = 0.2). For (**a**), the error bars represent the standard deviations of repeated measurements. For (**b**), the widths of the distributions calculated from the particle size distribution data were: 0.18, 0.12, 0.07, 0.04, and 0.04 μm from the "10" to "50" samples, respectively.

3.2.3. Emulsifier-To-Oil Concentration

When formulating an oil-in-water nanoemulsion, it is possible to increase the oil concentration while keeping the emulsifier concentration constant or while keeping the emulsifier-to-oil concentration constant, which impacts the size of the droplets formed (Figure 7). At a fixed emulsifier concentration, which is 2 wt%, the mean particle diameter increased with increasing oil concentration, which can be attributed to the fact that there was insufficient emulsifier present to cover all of the droplet surfaces in the concentrated nanoemulsions. As a result, droplet coalescence occurred during homogenization, leading to an increase in droplet size [48]. In contrast, at a fixed emulsifier-to-oil ratio (= 0.2), the mean particle diameter decreased with increasing oil concentration, which can be attributed to the reasons discussed in the previous section, i.e., increased disruptive forces for more viscous nanoemulsions. These experiments highlight the importance of keeping the emulsifier-to-oil ratio constant when increasing the oil concentration.

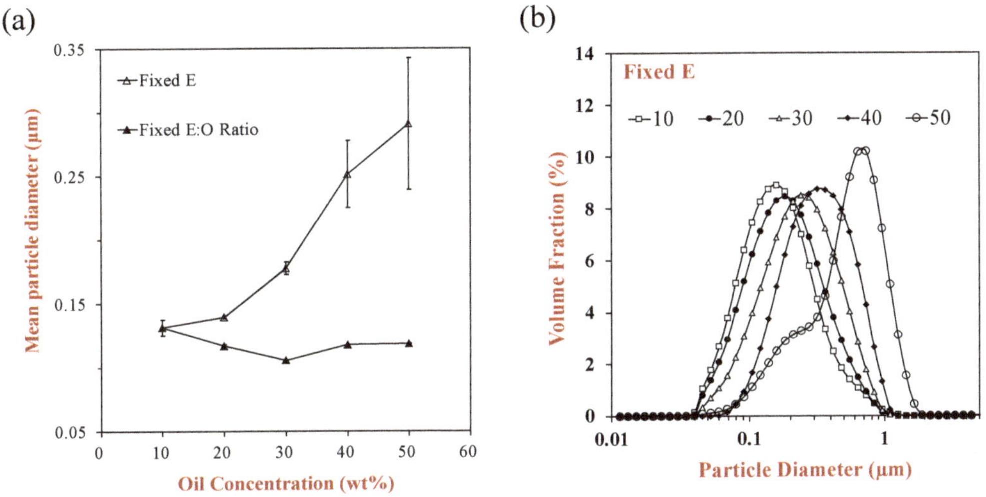

Figure 7. The influence of oil concentration on the mean particle diameter (d_{32}) and (**b**) particle size distributions of corn oil-in-water nanoemulsions at a fixed emulsifier concentration or fixed emulsifier-to-oil ratio (=0.2). For (**a**), the error bars represent the standard deviations of repeated measurements. For (**b**), the calculated widths of the distributions were: 0.14, 0.14, 0.16, 0.20, and 0.32 mm for 10, 20, 30, 40, and 50% oil concentration, respectively. For (**b**), the widths of the distributions calculated from the particle size distribution data were: 0.14, 0.14, 0.16, 0.20, and 0.32 μm from the "10" to "50" samples, respectively.

3.2.4. Emulsifier Type

In principle, a wide variety of different types of emulsifiers can be used to formulate edible nanoemulsions. Each of these emulsifiers has a different functional performance because of its different molecular characteristics. In this study, we compared the performance of a synthetic non-ionic surfactant (Tween 20) with that of three natural emulsifiers: whey protein (animal), soy protein (plant), and gum Arabic polysaccharide (plant). These emulsifiers were selected because there is a growing emphasis in the food industry on the creation of plant-based foods because of ethical, environmental, and animal welfare concerns [5,49]. The type of emulsifier used had a significant impact on the formation of the nanoemulsions (Figure 8). The droplets in the nanoemulsions produced using gum Arabic were considerably larger than those in the nanoemulsions produced using surfactants or proteins. This effect can mainly be attributed to the fact that gum Arabic is a mixture of relatively large hydrophilic polysaccharide molecules that adsorb slowly to droplet surfaces and form thick interfaces. Indeed, a previous study reported that the surface load (Γ) of gum Arabic (24.8 mg/m^2) was over 10-fold higher than that of whey protein (2.2 mg/m^2) in oil-in-water nanoemulsions [45]. The whey and soy proteins were able to form nanoemulsions containing considerably smaller droplets than the gum Arabic. This effect can be attributed to the fact that these are relatively small amphiphilic molecules that rapidly move to the oil–water interfaces during homogenization and form thin interfaces around the oil droplets. The relatively low surface load of these globular proteins means that less emulsifier is required to cover a given droplet surface area, thereby leading to smaller droplet sizes [50]. The particle size distributions of the nanoemulsions produced by the soy and whey proteins had a fairly similar shape, but the mean particle diameter was slightly lower for the whey proteins than the soy proteins, which may be because of the smaller molecular dimensions of the whey proteins [51]. The non-ionic surfactant also gave small droplets during homogenization, which can also be attributed to its small size, rapid adsorption, and low surface load.

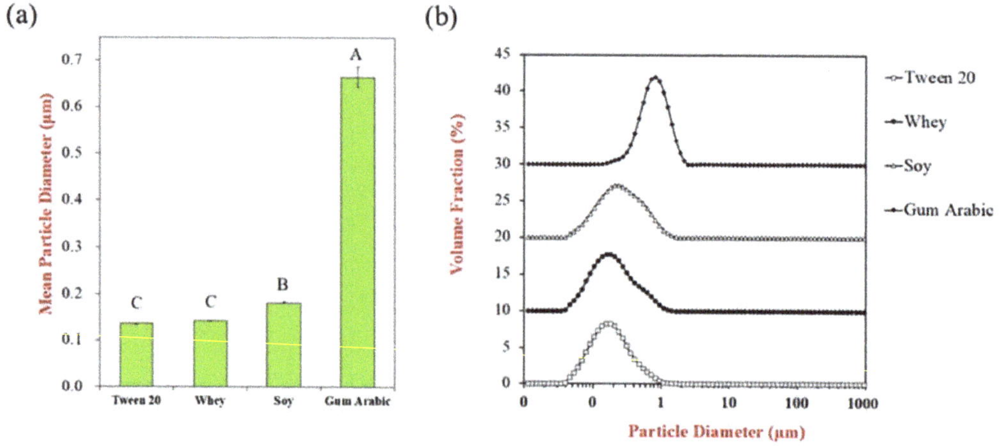

Figure 8. The influence of four emulsifier types on (**a**) the mean particle diameter and (**b**) the particle size distributions. The letters (A, B, C) represent the significance of the samples ($p < 0.05$). For (**a**), the error bars represent standard deviations. For (**b**), the widths of the distributions calculated from the particle size distribution data were: 0.18, 0.19, 0.24, and 0.37 μm from the "Tween 20" to "Gum Arabic" samples, respectively.

Several previous studies have also shown that emulsifier type plays a critical role in the formation of nanoemulsions containing small droplets, which was also related to differences in the molecular and physicochemical attributes of the emulsifiers [52,53]. Overall, our results highlight the importance of selecting an appropriate emulsifier for forming nanoemulsions.

4. Conclusions

In this study, we systematically investigated the major factors influencing the ability of a high-pressure homogenizer to form nanoemulsions. The results of our study are useful for understanding the parameters impacting the formation of nanoemulsions using high-pressure homogenizers in general. However, the specific homogenizer used in our study also had some additional features that provide increased flexibility in controlling the droplet size distributions of nanoemulsions due to its ability to vary the back pressure, flow, profile, and nozzle size. The impact of homogenizer operating conditions (e.g., number of passes, homogenization pressure, flow profile, back pressure, and nozzle size) and formulation parameters (e.g., emulsifier concentration, oil concentration, and emulsifier type) on the mean particle diameter and particle size distribution of the nanoemulsions was examined. The droplet size could be reduced by increasing the homogenization pressure, number of passes, back pressure, and reducing the nozzle size. It could also be reduced by increasing the emulsifier concentration or by increasing the oil concentration (at a fixed emulsifier-to-oil ratio). Nanoemulsions containing relatively small droplets (d_{32} < 150 nm) could be produced using a non-ionic surfactant (Tween 20), as well as a plant protein (soy protein) or animal protein (whey protein). However, a plant polysaccharide (gum Arabic) was not able to produce nanoemulsions containing small droplets, which was mainly attributed to its relatively large molecular dimensions. The information obtained in this study may facilitate the more efficient formation of nanoemulsions containing small droplet sizes. In particular, it shows that the number of passes, homogenization pressure, and emulsifier concentration should be optimized to produce small droplets without increasing energy use.

Author Contributions: Conceptualization, H.Z. and D.J.M.; methodology, H.Z., G.V. and D.Q.; software, H.Z. and D.J.M.; validation, H.Z. and D.J.M.; formal analysis, H.Z. and D.J.M.; investigation, H.Z., G.V. and D.Q.; resources, D.J.M.; data curation, H.Z. and G.V.; writing—original draft preparation, H.Z.; writing—review and editing, H.Z. and D.J.M.; visualization, H.Z. and D.J.M.; supervision, D.J.M.; project administration, D.J.M.; funding acquisition, D.J.M. All authors have read and agreed to the published version of the manuscript.

Funding: This work was partly based upon work supported by funding from the Good Food Institute.

Institutional Review Board Statement: Not applicable.

Informed Consent Statement: Not applicable.

Data Availability Statement: The data that support the findings of this study are available from the corresponding author upon reasonable request.

Acknowledgments: We thank BEE International Inc. Company for kindly lending us the high-pressure homogenizer used in this study.

Conflicts of Interest: The authors declare no conflict of interest.

References

1. Aswathanarayan, J.B.; Vittal, R.R. Nanoemulsions and Their Potential Applications in Food Industry. *Front. Sustain. Food Syst.* **2019**, *3*, 95. [CrossRef]
2. Gupta, A.; Eral, H.B.; Hatton, T.A.; Doyle, P.S. Nanoemulsions: Formation, properties and applications. *Soft Matter* **2016**, *12*, 2826–2841. [CrossRef]
3. Silva, H.D.; Cerqueira, M.A.; Vicente, A.A. Nanoemulsions for Food Applications: Development and Characterization. *Food Bioprocess Technol.* **2012**, *5*, 854–867. [CrossRef]
4. McClements, D.J.; Rao, J. Food-Grade Nanoemulsions: Formulation, Fabrication, Properties, Performance, Biological Fate, and Potential Toxicity. *Crit. Rev. Food Sci. Nutr.* **2011**, *51*, 285–330. [CrossRef]
5. McClements, D.J. Advances in edible nanoemulsions: Digestion, bioavailability, and potential toxicity. *Prog. Lipid Res.* **2021**, *81*, 101081. [CrossRef]
6. Kumar, N.; Verma, A.; Mandal, A. Formation, characteristics and oil industry applications of nanoemulsions: A review. *J. Pet. Sci. Eng.* **2021**, *206*, 109042. [CrossRef]
7. Dasgupta, N.; Ranjan, S.; Gandhi, M. Nanoemulsions in food: Market demand. *Environ. Chem. Lett.* **2019**, *17*, 1003–1009. [CrossRef]
8. Li, G.; Zhang, Z.; Liu, H.; Hu, L. Nanoemulsion-based delivery approaches for nutraceuticals: Fabrication, application, characterization, biological fate, potential toxicity and future trends. *Food Funct.* **2021**, *12*, 1933–1953. [CrossRef]
9. Walia, N.; Dasgupta, N.; Ranjan, S.; Ramalingam, C.; Gandhi, M. Food-grade nanoencapsulation of vitamins. *Environ. Chem. Lett.* **2019**, *17*, 991–1002. [CrossRef]
10. Che Marzuki, N.H.; Wahab, R.A.; Abdul Hamid, M. An overview of nanoemulsion: Concepts of development and cosmeceutical applications. *Biotechnol. Biotechnol. Equip.* **2019**, *33*, 779–797. [CrossRef]
11. Mustafa, I.F.; Hussein, M.Z. Synthesis and Technology of Nanoemulsion-Based Pesticide Formulation. *Nanomaterials* **2020**, *10*, 1608. [CrossRef] [PubMed]
12. Roy, A.; Nishchaya, K.; Rai, V.K. Nanoemulsion-based dosage forms for the transdermal drug delivery applications: A review of recent advances. *Expert Opin. Drug Deliv.* **2022**, *19*, 303–319. [CrossRef] [PubMed]
13. Palmieri, D.; Brasili, F.; Capocefalo, A.; Bizien, T.; Angelini, I.; Oddo, L.; Toumia, Y.; Paradossi, G.; Domenici, F. Improved hybrid-shelled perfluorocarbon microdroplets as ultrasound- and laser-activated phase-change platform. *Colloids Surf. A Physicochem. Eng. Asp.* **2022**, *641*, 128522. [CrossRef]
14. Ezhilarasi, P.N.; Karthik, P.; Chhanwal, N.; Anandharamakrishnan, C. Nanoencapsulation Techniques for Food Bioactive Components: A Review. *Food Bioprocess Technol.* **2013**, *6*, 628–647. [CrossRef]
15. Fathi, M.; Mozafari, M.R.; Mohebbi, M. Nanoencapsulation of food ingredients using lipid based delivery systems. *Trends Food Sci. Technol.* **2012**, *23*, 13–27. [CrossRef]
16. Naseema, A.; Kovooru, L.; Behera, A.K.; Kumar, K.P.P.; Srivastava, P. A critical review of synthesis procedures, applications and future potential of nanoemulsions. *Adv. Colloid Interface Sci.* **2021**, *287*, 102318. [CrossRef]
17. Rehman, R.; Younas, A.; Ullah, S.; Ano, A.B.; Muzaffar, R.; Altaf, A.A.; Altaf, A.A. Recent Developments in Nano-Emulsions? Preparatory Methods and their Applications: A Concise Review. *Pak. J. Anal. Environ. Chem.* **2022**, *23*, 175–193. [CrossRef]
18. Safaya, M.; Rotliwala, Y.C. Nanoemulsions: A review on low energy formulation methods, characterization, applications and optimization technique. *Mater. Today Proc.* **2020**, *27*, 454–459. [CrossRef]
19. Komaiko, J.S.; McClements, D.J. Formation of Food-Grade Nanoemulsions Using Low-Energy Preparation Methods: A Review of Available Methods. *Compr. Rev. Food Sci. Food Saf.* **2016**, *15*, 331–352. [CrossRef]

20. Sneha, K.; Kumar, A. Nanoemulsions: Techniques for the preparation and the recent advances in their food applications. *Innov. Food Sci. Emerg. Technol.* **2022**, *76*, 102914. [CrossRef]
21. Floury, J.; Desrumaux, A.; Legrand, J. Effect of Ultra-high-pressure Homogenization on Structure and on Rheological Properties of Soy Protein-stabilized Emulsions. *J. Food Sci.* **2002**, *67*, 3388–3395. [CrossRef]
22. Villalobos-Castillejos, F.; Granillo-Guerrero, V.G.; Leyva-Daniel, D.E.; Alamilla-Beltrán, L.; Gutiérrez-López, G.F.; Monroy-Villagrana, A.; Jafari, S.M. Chapter 8—Fabrication of Nanoemulsions by Microfluidization. In *Nanoemulsions*; Jafari, S.M., McClements, D.J., Eds.; Academic Press: Cambridge, MA, USA, 2018; pp. 207–232.
23. Li, Y.; Deng, L.; Dai, T.; Li, Y.; Chen, J.; Liu, W.; Liu, C. Microfluidization: A promising food processing technology and its challenges in industrial application. *Food Control* **2022**, *137*, 108794. [CrossRef]
24. Jafari, S.M.; He, Y.; Bhandari, B. Nano-Emulsion Production by Sonication and Microfluidization—A Comparison. *Int. J. Food Prop.* **2006**, *9*, 475–485. [CrossRef]
25. Taha, A.; Ahmed, E.; Ismaiel, A.; Ashokkumar, M.; Xu, X.; Pan, S.; Hu, H. Ultrasonic emulsification: An overview on the preparation of different emulsifiers-stabilized emulsions. *Trends Food Sci. Technol.* **2020**, *105*, 363–377. [CrossRef]
26. Jo, Y.-J.; Kwon, Y.-J. Characterization of β-carotene nanoemulsions prepared by microfluidization technique. *Food Sci. Biotechnol.* **2014**, *23*, 107–113. [CrossRef]
27. Håkansson, A.; Trägårdh, C.; Bergenståhl, B. Dynamic simulation of emulsion formation in a high pressure homogenizer. *Chem. Eng. Sci.* **2009**, *64*, 2915–2925. [CrossRef]
28. Floury, J.; Desrumaux, A.; Lardières, J. Effect of high-pressure homogenization on droplet size distributions and rheological properties of model oil-in-water emulsions. *Innov. Food Sci. Emerg. Technol.* **2000**, *1*, 127–134. [CrossRef]
29. Gallassi, M.; Gonçalves, G.F.N.; Botti, T.C.; Moura, M.J.B.; Carneiro, J.N.E.; Carvalho, M.S. Numerical and experimental evaluation of droplet breakage of O/W emulsions in rotor-stator mixers. *Chem. Eng. Sci.* **2019**, *204*, 270–286. [CrossRef]
30. Paximada, P.; Tsouko, E.; Kopsahelis, N.; Koutinas, A.A.; Mandala, I. Bacterial cellulose as stabilizer of o/w emulsions. *Food Hydrocoll.* **2016**, *53*, 225–232. [CrossRef]
31. Hidajat, M.J.; Jo, W.; Kim, H.; Noh, J. Effective Droplet Size Reduction and Excellent Stability of Limonene Nanoemulsion Formed by High-Pressure Homogenizer. *Colloids Interfaces* **2020**, *4*, 5. [CrossRef]
32. Qian, C.; McClements, D.J. Formation of nanoemulsions stabilized by model food-grade emulsifiers using high-pressure homogenization: Factors affecting particle size. *Food Hydrocoll.* **2011**, *25*, 1000–1008. [CrossRef]
33. Lee, L.; Norton, I.T. Comparing droplet breakup for a high-pressure valve homogeniser and a Microfluidizer for the potential production of food-grade nanoemulsions. *J. Food Eng.* **2013**, *114*, 158–163. [CrossRef]
34. Tan, Y.; Zhang, Z.; Liu, J.; Xiao, H.; McClements, D.J. Factors impacting lipid digestion and nutraceutical bioaccessibility assessed by standardized gastrointestinal model (INFOGEST): Oil droplet size. *Food Funct.* **2020**, *11*, 9936–9946. [CrossRef]
35. Wooster, T.J.; Golding, M.; Sanguansri, P. Impact of Oil Type on Nanoemulsion Formation and Ostwald Ripening Stability. *Langmuir* **2008**, *24*, 12758–12765. [CrossRef] [PubMed]
36. Norton, I.T.; Frith, W.J. Microstructure design in mixed biopolymer composites. *Food Hydrocoll.* **2001**, *15*, 543–553. [CrossRef]
37. Chung, C.; Smith, G.; Degner, B.; McClements, D.J. Reduced Fat Food Emulsions: Physicochemical, Sensory, and Biological Aspects. *Crit. Rev. Food Sci. Nutr.* **2016**, *56*, 650–685. [CrossRef] [PubMed]
38. Karthik, P.; Ezhilarasi, P.N.; Anandharamakrishnan, C. Challenges associated in stability of food grade nanoemulsions. *Crit. Rev. Food Sci. Nutr.* **2017**, *57*, 1435–1450. [CrossRef]
39. McClements, D.J. *Food Emulsions: Principles, Practices, and Techniques*; CRC Press: Boca Raton, FL, USA, 2015.
40. Schlender, M.; Minke, K.; Spiegel, B.; Schuchmann, H.P. High-pressure double stage homogenization processes: Influences of plant setup on oil droplet size. *Chem. Eng. Sci.* **2015**, *131*, 162–171. [CrossRef]
41. Schlender, M.; Minke, K.; Schuchmann, H.P. Sono-chemiluminescence (SCL) in a high-pressure double stage homogenization processes. *Chem. Eng. Sci.* **2016**, *142*, 1–11. [CrossRef]
42. Kuhn, K.R.; Cunha, R.L. Flaxseed oil—Whey protein isolate emulsions: Effect of high pressure homogenization. *J. Food Eng.* **2012**, *111*, 449–457. [CrossRef]
43. Hakansson, A. Emulsion Formation by Homogenization: Current Understanding and Future Perspectives. In *Annual Review of Food Science and Technology*; Doyle, M.P., McClements, D.J., Eds.; Annual Reviews: San Mateo, CA, USA, 2019; Volume 10, pp. 239–258.
44. Tan, C.P.; Nakajima, M. β-Carotene nanodispersions: Preparation, characterization and stability evaluation. *Food Chem.* **2005**, *92*, 661–671. [CrossRef]
45. Bai, L.; Huan, S.; Gu, J.; McClements, D.J. Fabrication of oil-in-water nanoemulsions by dual-channel microfluidization using natural emulsifiers: Saponins, phospholipids, proteins, and polysaccharides. *Food Hydrocoll.* **2016**, *61*, 703–711. [CrossRef]
46. Tcholakova, S.; Denkov, N.D.; Sidzhakova, D.; Ivanov, I.B.; Campbell, B. Interrelation between Drop Size and Protein Adsorption at Various Emulsification Conditions. *Langmuir* **2003**, *19*, 5640–5649. [CrossRef]
47. Courthaudon, J.-L.; Dickinson, E.; Dalgleish, D.G. Competitive adsorption of β-casein and nonionic surfactants in oil-in-water emulsions. *J. Colloid Interface Sci.* **1991**, *145*, 390–395. [CrossRef]
48. Hakansson, A.; Tragardh, C.; Bergenstahl, B. Studying the effects of adsorption, recoalescence and fragmentation in a high pressure homogenizer using a dynamic simulation model. *Food Hydrocoll.* **2009**, *23*, 1177–1183. [CrossRef]

49. Kumar, M.; Tomar, M.; Punia, S.; Dhakane-Lad, J.; Dhumal, S.; Changan, S.; Senapathy, M.; Berwal, M.K.; Sampathrajan, V.; Sayed, A.A.S.; et al. Plant-based proteins and their multifaceted industrial applications. *Lwt-Food Sci. Technol.* **2022**, *154*, 112620. [CrossRef]
50. Tan, Y.; Lee, P.W.; Martens, T.D.; McClements, D.J. Comparison of Emulsifying Properties of Plant and Animal Proteins in Oil-in-Water Emulsions: Whey, Soy, and RuBisCo Proteins. *Food Biophys.* **2022**, *17*, 409–421. [CrossRef]
51. McClements, D.J.; Lu, J.K.; Grossmann, L. Proposed Methods for Testing and Comparing the Emulsifying Properties of Proteins from Animal, Plant, and Alternative Sources. *Colloids Interfaces* **2022**, *6*, 19. [CrossRef]
52. Wanyi, W.; Lu, L.; Zehan, H.; Xinan, X. Comparison of emulsifying characteristics of different macromolecule emulsifiers and their effects on the physical properties of lycopene nanoemulsions. *J. Dispers. Sci. Technol.* **2020**, *41*, 618–627. [CrossRef]
53. Flores-Andrade, E.; Allende-Baltazar, Z.; Sandoval-González, P.E.; Jiménez-Fernández, M.; Beristain, C.I.; Pascual-Pineda, L.A. Carotenoid nanoemulsions stabilized by natural emulsifiers: Whey protein, gum Arabic, and soy lecithin. *J. Food Eng.* **2021**, *290*, 110208. [CrossRef]

colloids and interfaces

MDPI

Article

Effect of Gelling Agent Type on the Physical Properties of Nanoemulsion-Based Gels

Natalia Riquelme [1], Constanza Savignones [1], Ayelén López [1], Rommy N. Zúñiga [2] and Carla Arancibia [1,*]

[1] Department of Food Science and Technology, Technological Faculty, Universidad de Santiago de Chile, Obispo Umaña 050, Estación Central 9170201, Chile; natalia.riquelme.h@usach.cl (N.R.)
[2] Department of Biotechnology, Universidad Tecnológica Metropolitana, Las Palmeras, 3360, Ñuñoa 7800003, Chile; rommy.zuniga@utem.cl
* Correspondence: carla.arancibia@usach.cl; Tel.: +56-(2)-27184518

Abstract: Senior populations may experience nutritional deficiencies due to physiological changes that occur during aging, such as swallowing disorders, where easy-to-swallow foods are required to increase comfort during food consumption. In this context, the design of nanoemulsion-based gels (NBGs) can be an alternative for satisfying the textural requirements of seniors. This article aimed to develop NBGs with different gelling agents, evaluating their physical properties. NBGs were prepared with a base nanoemulsion (d = 188 nm) and carrageenan (CA) or agar (AG) at two concentrations (0.5–1.5% *w*/*w*). The color, rheology, texture, water-holding capacity (WHC) and FT-IR spectra were determined. The results showed that the CA-based gels were more yellow than the AG ones, with the highest hydrocolloid concentration. All gels showed a non-Newtonian flow behavior, where the gels' consistency and shear-thinning behavior increased with the hydrocolloid concentration. Furthermore, elastic behavior predominated over viscous behavior in all the gels, being more pronounced in those with AG. Similarly, all the gels presented low values of textural parameters, indicating an adequate texture for seniors. The FT-IR spectra revealed non-covalent interactions between nanoemulsions and hydrocolloids, independent of their type and concentration. Finally, the CA-based gels presented a higher WHC than the AG ones. Therefore, NBG physical properties can be modulated according to gelling agent type in order to design foods adapted for seniors.

Keywords: nanoemulsion-based gels; agar; carrageenan; physical properties

Citation: Riquelme, N.; Savignones, C.; López, A.; Zúñiga, R.N.; Arancibia, C. Effect of Gelling Agent Type on the Physical Properties of Nanoemulsion-Based Gels. *Colloids Interfaces* **2023**, 7, 49. https://doi.org/10.3390/colloids7030049

Academic Editors: Georgi G. Gochev and Reinhard Miller

Received: 25 January 2023
Revised: 16 June 2023
Accepted: 29 June 2023
Published: 4 July 2023

1. Introduction

The increase in the size of the senior population (>60 years old) has become one of the most significant challenges for the food industry, which is required to design foods focused on this population group that respond to their sensory, biological, and nutritional requirements due to the physiological changes caused by aging [1]. For example, these individuals often have bad teeth or dental prostheses that decrease their chewing performance [2]. They also present changes in salivary discharge and food swallowing disorders, such as dysphagia, which can lead to dehydration, malnutrition, and aspiration pneumonia, reducing their quality of life [3,4]. For this reason, food textures for senior adults should be easy-to-swallow and moist, and the food should be easily disintegrated and mixed in the mouth, avoiding mastication with the teeth [5]. A promising alternative to texture-modified food focused on this population is developing gels, where a biopolymer-based network is built to retain water or colloidal dispersions [6]. Consequently, these gels are suitable for seniors, since they are ingested through compression between the tongue and the palate until their complete disintegration without any chewing processing, facilitating their swallowing [7]. Therefore, these matrices can be an excellent sensory alternative for older people, because these gels could be designed to provide pleasant consumption experiences [8].

Alternatively, the food industry has innovated itself through developing ingredients and functional foods using novel technologies, such as nanoemulsions [9]. Nanoemulsions are dispersions of two immiscible liquids, where the dispersed phase has a droplet size between 20 and 200 nm [10]. These systems allow for the carriage of bioactive lipid compounds and their incorporation into aqueous products [9]. In addition, they have several advantages related to conventional emulsions due to their smaller droplet size and increased interfacial area. For example, nanoemulsions can protect lipid compounds from interaction with other food components or unfavorable conditions [11], improve lipid compounds' bioaccessibility during digestion [12,13], and extend their kinetic stability [14], among others. For this reason, new nanoemulsified structures, such as nanoemulsion-based gels, can be used to develop products adapted to the oral requirements of older people, where different hydrocolloids and their mixtures can be applied to obtain easy-to-swallow foods for people with chewing or swallowing dysfunctions.

Nanoemulsion-based gels' structure can usually be formed in two steps, through preparing the nanoemulsions and then turning the nanoemulsions into gels [15]. This last stage can be performed through nanoemulsion lipid droplets, aggregation, forming a network structure, or through their dispersion into hydrocolloid dispersions [16]. In both cases, the food matrix corresponds to a combination of these different structures (the emulsion and gel), which have good physical stability and mechanical properties [17]. However, nanoemulsion-based gel properties depend on the interactions between oil droplets and hydrocolloids [15,18], which can make it difficult to elaborate on stable gel-based foods, since the interactions between ingredients affect their physical properties, especially the texture characteristics [19]. For this reason, selecting the hydrocolloid type is crucial for obtaining stable nanoemulsion-based gels with controlled rheological and textural properties focused on the needs of seniors, such as soft, moist, and easy-to-chew and -swallow characteristics [20]. In this sense, food-grade biopolymers, such as proteins and polysaccharides, are considered for nanoemulsion-based gels' preparation [21], where polysaccharides form a more stable structure in food products while proteins are prone to denaturation, limiting their applications.

Carrageenan and agar correspond to two biopolymers obtained from red marine algae, which are widely used in the food industry due to their numerous applications as thickening, gelling, and emulsifying agents [22]. In addition, they are considered a food additive and generally recognized as safe by the Food and Drug Administration and European Food and Safety Agency [23]. In the case of carrageenan, its structure is composed of a linear polysaccharide with alternating units of D-galactose and 3,6-anhydrous-galactose linked by α-1,3 and β-1,4 glycosidic bonds, which form a high-molecular-weight biopolymer [24]. In addition, among the different forms of carrageenan, κ-carrageenan is the most important one and can form gels in aqueous dispersions [25]. In the case of agar, it is a natural polysaccharide also extracted from various species of red algae. This polysaccharide consists of two fractions (agarose and agaropectin) in variable proportions, where agarose is responsible for the gelling properties [26], since it can form aggregates from compact and ordered helical structures [27], allowing for the formation of a firm or weak gel. The chemical structure of agar is composed of D-galactopyranosyl residues and 3,6-anhydrous-L-galactopyranosyl monomers linked together and alternating α-1,3 and β-1,4 bonds [23].

A few studies have been conducted on nanoemulsion-based gels, mainly related to their use as fat replacers [28–30] and their application as encapsulation systems of bioactive compounds [31–33]. However, the development of food products based on nanoemulsion-filled gels has not yet been reported. These food matrices may be ideal for the elaboration of easy-to-swallow desserts since these products are one of the most appreciated by seniors for their sensory properties and palatability [34]. Therefore, it is necessary to study how nanoemulsion-based gels can lead to a wide range of rheological and mechanical properties, which depend on the nature of the components (lipid phase, emulsifier, and hydrocolloid type and concentration) and their interactions [19]. In this context, this work aimed to evaluate the effects of hydrocolloid type and concentration on the physical properties of nanoemulsion-based gels in order to obtain easy-to-swallow gels that can be used as potential foods for seniors with swallowing problems. In this sense, it is possible to hypothesize that using carrageenan and agar as gelling agents would allow one to obtain nanoemulsion-based gels with targeted physical properties, which could be an alternative for the development of food products such as puddings or custard desserts.

2. Materials and Methods

2.1. Materials

Nanoemulsions were prepared with canola oil (Belmont, Watt's S.A., San Bernardo, Chile) as a lipid phase; soy lecithin (Metarin P-Cargill, Blumos S.A., Santiago, Chile) and pea protein isolate (Nutralys® F85M, Roquette, Lestron, France) as emulsifiers; and purified water obtained from a reverse-osmosis system (Vigaflow S.A., Colina, Chile) as the aqueous phase. In addition, carrageenan (κ-carrageenan, Sabores.cl, Santiago, Chile) or agar (Tractor Bean, Santiago, Chile) was added into the nanoemulsions as gelling agents.

2.2. Preparation of Nanoemulsion-Based Gels

The base nanoemulsion preparation consisted of the following steps: (i) First, the aqueous phase was prepared by dispersing the pea protein (1% w/w) in purified water using a magnetic stirrer (Arex, Velp Scientifica, Usmate Velate, Italy) at 350 rpm for 40 min. Then, soy lecithin (3% w/w) was added to the aqueous phase and stirred for 40 min at 800 rpm until its complete dispersion. The aqueous phase was stored at 4 ± 1 °C in a glass beaker for 24 h. (ii) Subsequently, a coarse emulsion was prepared by dispersing the lipid phase (5% w/w of canola oil) in the aqueous phase using a high-speed homogenizer (IKA T25, Ultra Turrax, Germany) at 10,000 rpm for 10 min. (iii) Finally, the pre-emulsion was subjected to a high-energy homogenization process using an ultrasound device (VCX500, Sonics, Orlando, FL, USA) to reduce the particle size on the nanometric scale. The process conditions were as follows: 20 kHz, 90% amplitude, and 19.5 min with 15 and 10 s work and rest intervals, respectively. The base nanoemulsion presented narrow monomodal droplet size distributions, and its mean droplet size was relatively small (Figure 1), with values equal to 188 ± 1 nm and a polydispersity index of 0.14, which were determined using Zetasizer (NanoS90, Malvern Instruments, Malvern, UK).

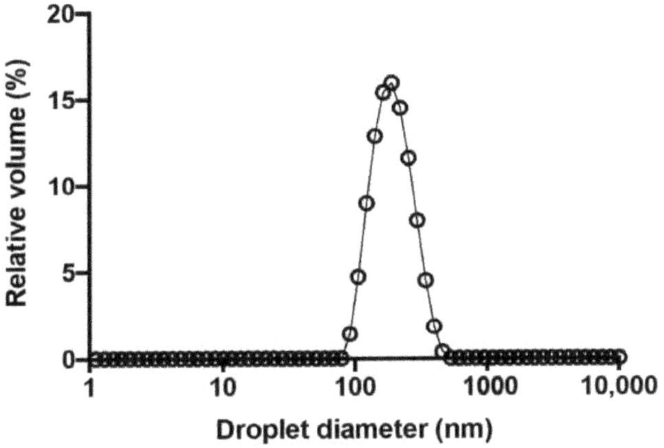

Figure 1. Particle size distribution of base nanoemulsion prepared with pea protein and soy lecithin as emulsifiers (1% and 3% w/w, respectively).

Finally, the nanoemulsion-based gels (NBG) were prepared by mixing the base nanoemulsion with different hydrocolloids (as gelling agent) at two concentrations: κ-carrageenan (0.5 and 1.5% w/w) and agar (1.0 and 1.5% w/w). These concentrations were chosen according to preliminary work (data not show) to obtain NBGs with similar apparent viscosity at a shear rate of 10 s^{-1}. This consideration was defined to obtain negligible distortion in the physical characterization results based on the effects of differences in the initial viscosity of the gels. The dispersion was carried out with a propeller stirrer (F20100151, Velp Scientifica, Usmate Velate, Italy) at 50 rpm in a thermoregulated bath (Heating bath B-100, Buchi, Flawil, Switzerland) at 80 ± 1 °C for 40 min. After that, the dispersions obtained were cooled at room temperature (25 ± 1 °C) and stored at 4 ± 1 °C for 24 h until further characterization. Each formulation was prepared in duplicate.

2.3. *Optical Properties of Nanoemulsion-Based Gels*

The optical properties of the nanoemulsion-based gels were determined using a colorimeter (ChromaMeter CR-410, Konica Minolta, Osaka, Japan), which was calibrated using a standard calibrated plate (L*: 93.46, a*: 0.42 and b*: 4.08). A 20 g piece of each sample was deposited in a Petri dish (60 mm internal diameter) to homogenize the surface, so that the CIELab parameters could be obtained: L* (brightness), a* (redness–greenness), and b* (blueness–yellowness). Also, the whiteness index (WI) of the nanoemulsion-based gels was calculated using Equation (1) [35]:

$$\mathrm{WI}(\%) = 100 - \sqrt{(100 - \mathrm{L}^*)^2 + \mathrm{a}^{*2} + \mathrm{b}^{*2}} \tag{1}$$

2.4. *Rheological Properties of Nanoemulsion-Based Gels*

2.4.1. Flow Properties

The flow behavior of all the NBGs was characterized using a rotational rheometer (Rheolab QC, Anton Paar, Austria) equipped with a concentric cylinder geometry (CC27, Anton Paar, Austria). The flow curves were determined by recording the shear stress values of the samples from 1 to 100 s^{-1} and from 100 to 1 s^{-1} for 120 s. The measurements were carried out at 37 ± 1 °C, controlled with a Peltier system. The samples were placed in the geometry and allowed to stand for 10 min before measurement to recover their structure and reach the test temperature [36]. The experimental values of the flow curves were fitted to the Ostwald–de Waele model (Equation (2)):

$$\sigma = K\dot{\gamma}^n \tag{2}$$

where σ is shear stress (Pa), K is the consistency index (Pa s), $\dot{\gamma}$ is the shear rate (s^{-1}), and n is the flow index (dimensionless). Apparent viscosity at a shear rate of 10 s^{-1} ($\eta_{10\ s^{-1}}$) was used as a parameter for comparing the samples, since this shear rate represents the effort that is spent during the swallowing process of semisolid foods [37,38].

2.4.2. Viscoelastic Properties

Viscoelastic assays of the NBGs were carried out at 37 ± 1 °C with a controlled stress rheometer (MCR 72, Anton-Paar, Austria) equipped with a parallel-plate geometry (50 mm diameter, 2 mm gap). After loading the sample into the rheometer, it was allowed to stand for 10 min to stabilize and reach the test temperature. Small-amplitude oscillation sweeps (SAOS) were performed to analyze the viscoelastic properties of the samples. First, the linear viscoelasticity zone (LVR) was determined, where stress sweeps were conducted between 0.1 and 2000 Pa at 1 Hz. Then, a frequency sweep from 0.1 to 10 Hz was performed at 3 Pa, which was selected because it was within the LVR. Finally, the oscillatory rheological parameters (G', G'', tan δ and complex dynamic viscosity-η^*) at 1 Hz were calculated to compare the viscoelastic properties of the different NBGs. Measurements were performed in duplicate.

2.5. *Textural Properties*

Cylindrical samples of the NBGs (4 cm diameter and 2 cm height) were subjected to texture profile analysis (TPA) using a texture analysis device (Zwick/Roell Z0.5, Zwick

GmbH & CO, Ulm, Germany). For this purpose, the samples were subjected to a two-cycle compression, where in each compression cycle, the NBG sample was compressed to 20% of its original height using an aluminum cylinder probe (5 cm diameter) operated at 0.1 mm/s. After the assays, the hardness (maximum force during the first compression); cohesiveness (ratio between the areas obtained from the second and first compressions); adhesiveness (negative maximum force after the first compression); springiness (ratio between the height of maximum force during the second and first compressions); and chewiness (value obtained from hardness × cohesiveness × springiness) parameters were calculated to compare the samples.

2.6. Fourier Transform Infrared Spectroscopy (FT-IR)

Fresh NBGs were characterized using Fourier transform infrared spectroscopy (FT-IR) to identify possible molecular interactions between the components of the gels. The measurements were conducted using an FT-IR spectrometer with an attenuated total reflectance unit (ATR) (Diamond Two, Perkin Elmer, England). All samples were scanned in the wavenumber range of 4000–500 cm^{-1} at a resolution of 1 cm^{-1}. Calibration was performed using a background spectrum recorded from the clean and empty cell at room temperature (25 °C). The FT-IR spectra were smoothed and baseline-corrected using the SpectrumTM 10 software.

2.7. Water-Holding Capacity of Nanoemulsion-Based Gels

The NBGs' water-holding capacity (*WHC*) was determined according to the methodology proposed by Liu et al. [39]. For this purpose, 20 g of each sample was placed in a 50 mL centrifuge tube and centrifuged (Universal 32R, Hettich, Tuttlingen, Germany) at 8000 rpm for 30 min at 4 ± 1 °C. Later, the water released from the sample was weighed on an analytical balance (AUX120, Shimadzu, Japan). The *WHC* of the nanoemulsion-based gels was calculated using Equation (3):

$$WHC\ (\%) = \frac{W_T - W_F}{W_T} \times 100\% \quad (3)$$

where W_T is the mass of the total water of each sample and W_F is the mass of the water released after the centrifugation process.

2.8. Statistical Analysis

The statistical analyses were performed using two-way ANOVA (Analysis of Variance) and Tukey's test with a significance level of a = 0.05 using the XLStat© 2019 software (Addinsoft, Paris, France). The experiments were performed in duplicate, and each replica was measured at least twice (minimum of four measurements). The results were reported as the average of all the measurements and their corresponding standard deviation.

3. Results and Discussion

3.1. Visual Appearance of Nanoemulsion-Based Gels

The visual appearance of the nanoemulsion-based gels (NBGs) prepared with different types and concentrations of hydrocolloids is shown in Figure 2. All NBGs presented a homogeneous morphology, indicating that all the samples formed a stable nanoemulsion gel with no phase separation. Also, all the samples could retain their shape after gel formation (Figure 2) because of the formation of strong interactions between the different components of the NBGs.

Figure 2. Photographs of the nanoemulsion-based gels with different types and concentrations of hydrocolloids.

3.2. Color Properties

Food color is traditionally represented using the CIELab color space, where L* denotes brightness, a* denotes the variation from green to red, and b* represents the variation from blue to yellow [40]. Differences between samples in the CIELab parameters were found, depending on the hydrocolloid type and concentration (Table 1). In the case of agar-based gels, no significant differences ($p > 0.05$) in the color parameters were obtained, regardless of the agar concentration. In addition, these gels showed a high whiteness index percentage (over 83%), which slightly varied with the increase in the hydrocolloid concentration (Table 1). On the contrary, the color of the κ-carrageenan-based gels differed depending on their concentration. At the lowest κ-carrageenan concentration, no differences were obtained with respect to the agar-based gels (Table 1 and Figure 2). However, after increasing the κ-carrageenan concentration from 0.5 to 1.5% *w*/*w*, the color of the gels changed to a more yellow color, and the whiteness index decreased significantly. These differences could be due to the intrinsic color of these hydrocolloids. According to Martín del Campo et al. [41], most types of commercial carrageenan retain a cream-yellow color after extraction and purification, with some pigments remaining in the product. For this reason, the gels with κ-carrageenan showed a more yellow color, especially at the highest concentrations.

Table 1. Color parameters of nanoemulsion-based gels with different hydrocolloid types and concentrations.

Hydrocolloid		L*	a*	b*	Whiteness Index (%)
Type	Concentration (% *w*/*w*)				
Carrageenan	0.5	85.2 ± 0.8 [b]	−0.52 ± 0.06 [b]	8.19 ± 0.16 [b]	83.05 ± 0.63 [b]
	1.5	83.3 ± 0.3 [c]	0.28 ± 0.05 [a]	18.41 ± 0.45 [a]	76.03 ± 0.10 [c]
Agar	1.0	86.7 ± 0.2 [a]	−0.77 ± 0.04 [c]	8.05 ± 0.16 [b]	84.42 ± 0.24 [a]
	1.5	86.5 ± 0.1 [ab]	−0.83 ± 0.01 [c]	8.71 ± 0.01 [b]	83.89 ± 0.07 [a]

Note. Mean values with a common superscript letter do not differ significantly ($p > 0.05$, Tukey's test).

3.3. Rheological Characterization of Nanoemulsion-Based Gels

Rheology plays an important role in understanding structural breakdown during the oral processing of food [42], and rheological characterization could offer guidance to improve safe food consumption for people with swallowing difficulties, according to the International Dysphagia Diet Standardization Initiatives (IDDSI) [43]. For this reason, nanoemulsified gels were characterized based on their flow and viscoelastic properties to obtain an adequate food rheology, which could be used for developing safe food products for seniors.

3.3.1. Flow Properties

In general, all NBGs presented a non-Newtonian and shear-thinning flow behavior (Figure 3A), showing progressive structure disruption under the application of the shear rate (decreased viscosity values with an increasing shear rate). This behavior is ideal for developing easy-to-swallow foods. since it demonstrates that a non-Newtonian behavior can render products safer to swallow than Newtonian fluids, posing a lower aspiration risk [44]. The hysteresis area between the upward and downward curves was observed in all the samples, indicating a thixotropic behavior, which was more evident at the highest hydrocolloid concentration. This phenomenon may be due to the formation of strong intermolecular forces among the hydrocolloid chains, since there are a more significant number of junction points among them, allowing for gel structure shaping. In this sense, these intermolecular interactions tend to break during the upward shear rate, and there is not enough time for a partial or total recovery of the gel structure, giving rise to the observed hysteresis [45].

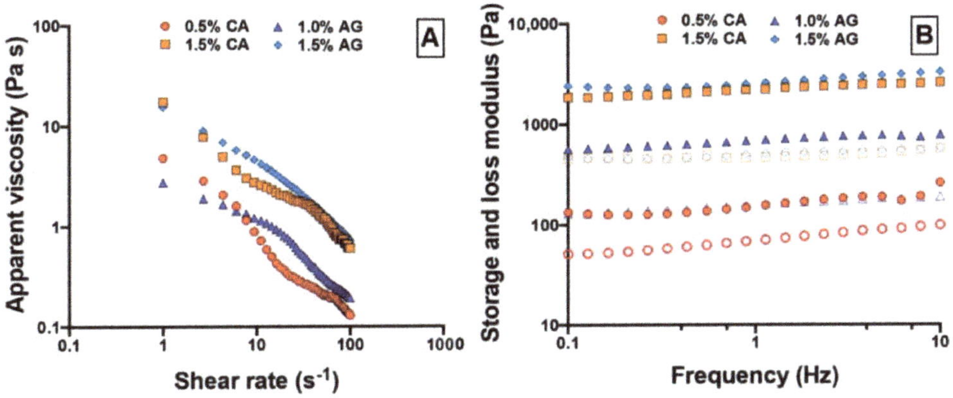

Figure 3. (**A**) Viscosity curves (upward and downward curves) and (**B**) mechanical spectra (G′: filled symbols, and G″: empty symbols) of the nanoemulsion-based gels with different hydrocolloid types and concentrations. CA: carrageenin, AG: agar.

On the one hand, the experimental data for the descendent flow curves of all the gels were fitted to the Ostwald–de Waele model (power law), with R^2 values between 0.96–0.98 and 0.81–0.92 for the agar- and κ-carrageenan-based gels, respectively. Table 2 shows the flow parameters obtained from the fitted data for all the NBGs, where significant differences (p-value < 0.05) were observed between samples, depending on the type and concentration of hydrocolloid. Regarding the consistency index (K), the κ-carrageenan-based gels showed the highest values, regardless of the hydrocolloid concentration. These differences could be due to the anionic sulfate content in this hydrocolloid. Carrageenan has a higher sulfate content than agar; hence, it can form firmer gels through disulfide bonding between its chains [46], which increases the gels' consistency.

Table 2. Rheological properties of the nanoemulsion-based gels with different hydrocolloid types and concentrations.

Hydrocolloid		K	n	η_{10s-1}	G′	G″	tan δ	η*
Type	Concentration (% w/w)	(Pa s)	(-)	(Pa s)	(Pa)	(Pa)	(-)	(Pa s)
Carrageenan	0.5	6.2 ± 0.2 [c]	0.17 ± 0.01 [c]	0.97 ± 0.02 [b]	155 ± 9 [d]	66 ± 5 [d]	0.40 ± 0.05 [a]	25 ± 1 [d]
	1.5	16.8 ± 0.8 [a]	0.28 ± 0.01 [b]	3.26 ± 0.26 [a]	2269 ± 45 [b]	537 ± 28 [a]	0.21 ± 0.01 [b]	328 ± 6 [b]
Agar	1.0	2.8 ± 0.2 [d]	0.45 ± 0.04 [a]	0.77 ± 0.05 [b]	682 ± 58 [c]	163 ± 3 [c]	0.24 ± 0.02 [b]	103 ± 11 [c]
	1.5	10.0 ± 0.8 [b]	0.43 ± 0.01 [a]	3.02 ± 0.33 [a]	2597 ± 38 [a]	484 ± 24 [b]	0.20 ± 0.01 [b]	378 ± 11 [a]

Note. K: consistency index, n: flow index, η_{10s-1}: apparent viscosity at a shear rate of 10 s^{-1}. G′: storage modulus, G″: loss modulus, tan δ: loss tangent angle, η*: complex dynamic viscosity at 1 Hz. Mean values with a common superscript letter do not differ significantly ($p > 0.05$, Tukey's test).

On the other hand, the NBGs showed flow index values < 1 (Table 2), confirming their shear-thinning flow behavior. However, significant differences (p-value < 0.05) in the flow index values were observed between samples, depending on the hydrocolloid type. The κ-carrageenan-based gels were more pseudoplastic than the agar-based gels (lower n values), which could be due to the structural differences between these hydrocolloids, which can be linked directly to their rheological properties [47]. In this sense, κ-carrageenan can form a helix structure, given the reduction in conformational flexibility. The latter is caused by binding between the cations of the sulfate group and anhydrous-galactosyl residue [48], where the polymeric chains can be aligned more easily with the shear rate. In turn, agar-based gels have a more complex structure due to inter- and intramolecular hydrogen bonding in the gel network, forming a tetrahedral ice-like structure [23] and decreasing their ability to align with the flow.

Finally, the apparent viscosity values, at a shear rate of 10 s^{-1} (η_{10s-1}), presented differences due to the hydrocolloid concentration (Table 2). NBGs with the lower concentration (0.5% κ-carrageenan and 1.0% agar) did not show significant differences in the η_{10s-1} values (p-value > 0.05). However, the η_{10s-1} values increased (p-value < 0.05) with the increase in the hydrocolloid concentration, regardless of the hydrocolloid type (Table 2). This result was expected, because the concentrations of each gelling agent were selected to obtain gels with a similar apparent viscosity. In addition, according to the International Dysphagia Diet Standardization Initiative [43], all NBG textures can be classified as level 4 (as a puree), as they do not require biting, chewing, or oral preparation. This gel texture can be chewed more easily, which is a desired characteristic for the development of foods for seniors, since it makes the swallowing process more effortless and safer.

3.3.2. Viscoelastic Properties

Viscoelastic properties were assessed within the linear viscoelastic region (LVR) to obtain more information about the NBG structure. The mechanical spectra (viscoelastic moduli as a function of frequency) are shown in Figure 3B. The mechanical spectra were generally similar for both hydrocolloids, where the storage modulus was always higher than the loss modulus (G′ > G″) in the range of frequencies studied. This result indicates that all the samples presented a gel-like behavior due to the formation of a well-developed network in all the NBGs [49], confirming that both hydrocolloids can form an internal three-dimensional network between oil droplets [50]. Additionally, G′ and G″ were significantly affected by the hydrocolloid concentration (Figure 3B), where the NBGs with the highest hydrocolloid concentration showed higher G′ and G″ values. An increase in hydrocolloid concentration causes polymer chains to be closer, obtaining more bonds between structural elements and forming a more compact and stronger gel network [19]. However, some differences were detected between the samples. The loss modulus of the κ-carrageenan-based gels showed a slight dependency as a function of frequency (0.1–10 Hz) (Figure 3B), which suggests a weaker internal structure and, therefore, a characteristic behavior of soft gels [51]. This result suggested that the κ-carrageenan-based gels could be deformed more easily during their consumption than the agar ones, which is a positive result, since gels with low hardness are adequate for older people [52].

Regarding the viscoelastic parameters at 1 Hz (Table 2), the agar-based NBGs were observed to show higher values (p-value < 0.05) for both moduli (G′ and G″) than the κ-carrageenan-based ones, indicating the formation of a more structured gel network [53]. In addition, the phase angle (tan δ, G″/G′), which provides information about the viscous modulus and the internal structure of different NBGs [54], confirmed that the κ-carrageenan-based gels corresponded to weaker-structured systems. Despite this, Suebsaen et al. [52] and Ishihara et al. [55] mentioned that tan δ values between 0.1 and 1.0 correspond to rheological criteria for the safe swallowing of foods. In this sense, all the NBGs might be suitable for safe swallowing for elderly populations (0.2 < tan δ < 0.4). On the other hand, the complex viscosity (η*) values also showed differences between hydrocolloid types (Table 2), where the agar-based gels showed the highest values compared to the κ-carrageenan-gels, confirming the results obtained for flow behavior (Table 2).

Finally, these results suggest that both hydrocolloids (κ-carrageenan and agar) can form weak nanoemulsion-filled gels, where little effort is needed to bite and chew them, easily destroying their structures during oral processing. Therefore, these gels can be used to develop food products with adequate textural properties for seniors.

3.4. Texture Properties

The mechanical properties of food matrices designed for people with dysphagia problems are important for a safe swallowing process, where hardness, cohesiveness, and adhesiveness textural parameters are relevant for physiological behaviors and bolus flow patterns [42].

NBG texture was evaluated through texture profile analysis (TPA), which is frequently used to mimic the chewing process by compressing the sample twice on a flat surface, simulating the first two bites during food consumption at a constant speed [56]. Figure 4 shows the TPA curves of the NBGs with 20% compression, where differences in the TPA profiles of the samples were observed according to the hydrocolloid type and concentration. The gels with κ-carrageenan required greater force for their compression than the agar gels, which also increased with a higher hydrocolloid concentration. Moreover, a low fracturability during the first compression was observed in the agar gels, since the curve showed two peaks during the first compression (Figure 4), while the κ-carrageenan gels did not show this behavior, given that only one peak was observed during the first compression (Figure 4).

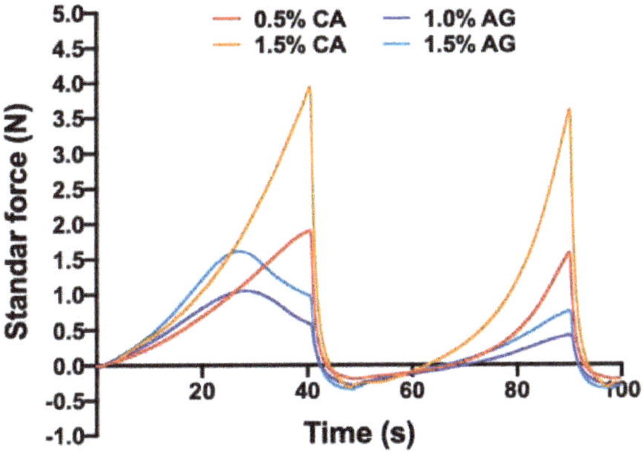

Figure 4. Texture profile analysis (TPA) curves of the nanoemulsion-based gels with different hydrocolloid types and concentrations. CA: carrageenin and AG: agar.

The textural parameters determined from the TPA curve were hardness, adhesiveness, cohesiveness, and springiness (Table 3), where differences between samples were observed due to the hydrocolloid type and concentration. The NBGs based on κ-carrageenan showed higher values ($p < 0.05$) of hardness than the agar-based ones (Table 3). According to Wada et al. [57], easy-to-swallow foods should have hardness values under 15,000 N/m^2; hence, all NBGs suit seniors with dysphagia disorders. Despite this, κ-carrageenan gels can form a more integrated gel network that increases the strength of compression because of its high molecular weight (788.7 g/mol) in comparison to agar (336.3 g/mol) [58]. Finally, this result agrees with the higher consistency index and apparent viscosity obtained for these samples (Table 2).

Table 3. TPA parameters of the nanoemulsion-based gels with different hydrocolloid types and concentrations.

Hydrocolloid		Hardness (N/m^2)	Adhesiveness (J/m^2)	Cohesiveness (-)	Springiness (-)
Type	Concentration (% *w/w*)				
Carrageenin	0.5	251.8 ± 24.5 ^{b}	0.02 ± 0.002 ^{c}	0.24 ± 0.02 ^{b}	0.56 ± 0.01 ^{b}
	1.5	438.7 ± 26.6 ^{a}	0.05 ± 0.002 ^{b}	0.46 ± 0.02 ^{a}	0.83 ± 0.01 ^{a}
Agar	1.0	152.1 ± 3.9 ^{b}	0.05 ± 0.005 ^{b}	0.14 ± 0.01 ^{c}	0.53 ± 0.02 ^{b}
	1.5	216.2 ± 16.9 ^{b}	0.09 ± 0.007 ^{a}	0.21 ± 0.03 ^{b}	0.82 ± 0.04 ^{a}

Note. Mean values with a common superscript letter do not differ significantly ($p > 0.05$, Tukey's test).

All the NBGs showed a slight adherence because the area between the two compressions was small (Figure 4). Despite this, significant differences ($p < 0.05$) between samples were observed depending on the hydrocolloid type and concentration (Table 3). In general, the agar-based gels presented higher adhesiveness than the κ-carrageenan ones. Adhesive foods are associated with an increased choking risk and require increased lingual effort to propel them into and through the pharynx [59]. Thus, low-adhesive textures could facilitate the swallowing process of older people, since they do not stick to the oral surface [60]. Also, Hadde et al. [61] mentioned that level 4 of the IDDSI for different foods with modified textures with adhesiveness values <0.5 would be safe for senior people.

Cohesiveness is a parameter related to the disintegration of gels in fragments during the chewing process, which was affected by both the hydrocolloid type and its concentration in this study. As shown in Table 3, the cohesiveness values of NBGs based on agar were lower than those of κ-carrageenan gels (Table 3), suggesting that agar gels are more brittle and can easily break during swallowing [32]. κ-Carrageenan gels, on the contrary, can maintain their gel structure during the two deformation processes. These differences could be related to the gels' skeletal structure and hardness, where the presence of disulfide bond groups in the κ-carrageenan gels made them stronger and more cohesive [62]. Despite this, all samples showed lower cohesiveness (0.14–0.46) values, consistent with parameters of the Japanese dysphagia-modified diet (values < 0.9) [63], being optimal for older people with dysphagia problems.

Finally, the springiness values were affected by the hydrocolloid concentration alone. As the hydrocolloid concentration increased, so did the gels' springiness (Figure 4 and Table 3). This behavior may be attributed to the increased number of hydrogen-bonding groups with the increasing hydrocolloid concentration, which improves the resistance of gels to deformation [64]. Springiness can reflect gel elasticity, where higher values suggest a homogeneous and well-connected gel structure [65]. In this sense, all gels presented low springiness values, especially at a lower hydrocolloid concentration, indicating that gels can be immediately deformed after the first compression into many small pieces, improving the swallowing process for seniors.

3.5. Fourier Transform Infrared (FTIR) Analysis

The FTIR spectra of the gels were obtained in order to understand the molecular interactions between the components of the nanoemulsion-based gels. The FTIR spectra of the canola oil, base nanoemulsion, and NBGs are presented in Figure 5. First, the canola oil spectra showed characteristics peaks (Figure 5A) that corresponded to the absorption bands of stretching (3008 cm^{-1}) and bending (rocking) (1417 cm^{-1}) vibrations of *cis* =C-H group; stretching vibrations (asymmetrical and symmetrical) of a -CH_2 group at 2923 and 2853 cm^{-1}, respectively; stretching vibrations of a -C=O group (ester) at 1743 cm^{-1}; bending (scissor) vibrations of -CH_2 and CH_3 (-C-H) groups at 1465 cm^{-1}; bending symmetrical vibrations of a -CH_3 group at 1377 cm^{-1}; stretching and bending vibrations of -C-O and -CH_2- groups at 1160 cm^{-1}; and bending (rocking) vibrations of a -$(CH_2)n$- group at 722 cm^{-1}, which was also observed by Jamwal et al. [66]. In the case of nanoemulsion, the absorption peaks of canola oil were also observed in the spectrum (Figure 5A). However, a new peak at 1621 cm^{-1} was obtained, corresponding to the stretching vibrations of pea protein's amide I region (C=O and C=N), which was used as an emulsifier [67]. Also, the stretching vibrations of soy lecithin's phosphate group (PO_2- and *p*-O-C) were found in the absorption region between 1090 and 850 cm^{-1} [68].

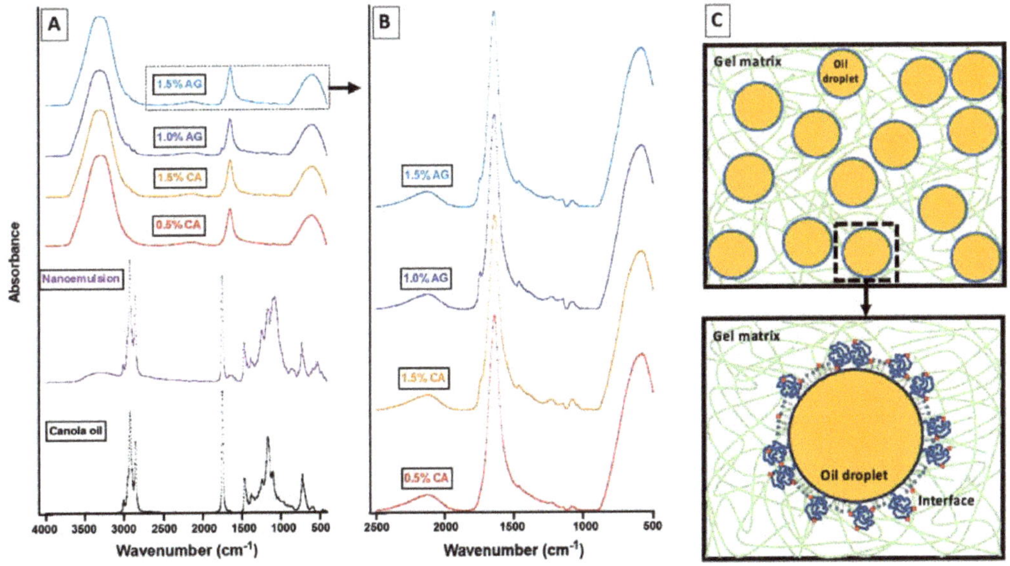

Figure 5. FTIR spectra from the nanoemulsion-based gels with different hydrocolloid types and concentrations (**A**,**B**) and schematic representation of nanoemulsion-based gels' structure (**C**). CA: carrageenin and AG: agar.

Different spectra were observed when all the NBGs were compared (Figure 5A), suggesting that new interactions were established between the nanoemulsion ingredients and hydrocolloids (κ-carrageenan and agar). All spectra of the NBGs revealed similar absorption peaks, independent of the hydrocolloid type and concentration. For example, the predominate absorption region appeared at around 3600–3200 cm^{-1} with a peak of ~3330 cm^{-1} (Figure 5A), representing the stretching vibration of O-H groups from the water (due to the high hydration level in the sample). However, the NBGs also presented a bending vibration of O-H groups at 1638 cm^{-1} [69], indicating inter and intramolecular hydrogen bonds formed during the formation of nanoemulsion-based gels with both hydrocolloids [50]. Additionally, a new absorption region between 450 and 650 cm^{-1} with a peak of ~590 cm^{-1} was observed in all the samples (Figure 5A), which can be associated with the sulfate groups of κ-carrageenan and agar [70]. Figure 5B shows a zoomed spectrum in a wavenumber range between 2500 and 500 cm^{-1} for each NBG, where new absorption peaks at 2131 and 1081 cm^{-1} are observed in the NBG. These peaks corresponded to the stretching vibrations of C≡C and C–O bonds, respectively, because of the molecular structure of the hydrocolloids [71]. On the other hand, the absorption peaks observed in the canola oil and nanoemulsion spectra were not distinguished in the NBG spectra, indicating the incorporation of both hydrocolloids into oil droplets through new molecular interactions. The changes in the absorption peaks of the FTIR spectra also revealed a good incorporation of both hydrocolloids into the base nanoemulsion, since non-covalent interactions such as O-H stretching were founded. Therefore, these results suggest that oil droplets present in the NBGs could act as active fillers (Figure 5C), because the interface could be connected with the gel network through emulsifiers (pea protein and soy lecithin) with non-covalent bonds [15].

3.6. Water-Holding Capacity

The water-holding capacity (WHC) was studied to evaluate the physical stability of the NBGs. In general, the NBGs with κ-carrageenan presented higher WHC values (75–87%) than the ones with agar (43–80%), especially at lower concentrations (Figure 6). These results can be related to NBG hardness (Table 3), since the κ-carrageenan-based gels presented the highest values, which positively affected the WHC values of these gels (Table 3). Based on this, the water-holding capacity of emulsion-based gels is related to their microstructural properties, especially the pore size and strength of the gel network [19]. Accordingly, κ-carrageenan can form a porous structure that binds water during gelation [72]. This fact means that κ-carrageenan can bind the free water molecules using hydrogen bonds due to the presence of anionic sulfate groups that improve its water-holding capacity [73]. In turn, agar-based gels are characterized by a poor WHC, since their structure is weaker and more brittle (Table 3), reducing water retention [74,75]. In addition, the concentration also has a significant effect on WHC, since the NBGs with the highest hydrocolloid concentration (1.5% w/w) presented a %WHC >80% (Figure 6). This behavior may be due to the gel structure becoming stronger and denser at the highest hydrocolloid concentration, retaining the water in the gel network and reducing the amount of free water after centrifugation. Also, it should be noted that the effect of concentration was more pronounced in the agar-based gels, where there was a significant increase in the %WHC (42.8% and 79.9% WHC for 1.0% and 1.5% w/w agar, respectively) (Figure 6). In this sense, we can hypothesize that agar-based gels might be more elastic (Figure 3 and Table 2) when their WHC is lower, because polysaccharide–polysaccharide interactions predominated in the gel network of this sample. For this reason, the polymeric chains formed by this hydrocolloid make them prone to syneresis, because they cannot hold water securely through capillary forces [19]. Instead, in the κ-carrageenan gels, the polysaccharide–water interactions could predominate due to the capability of sulfate groups to bind water molecules, as previously mentioned. These gels show lower cohesivity and a more brittle gel network than κ-carrageenan gels, promoting syneresis.

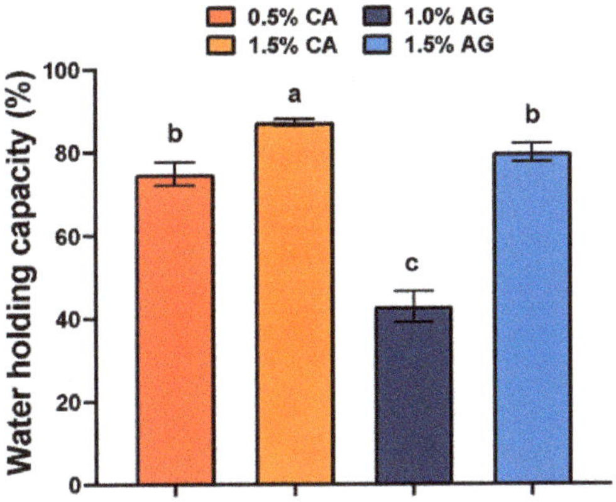

Figure 6. Water-holding capacity of the nanoemulsion-based gels with different hydrocolloid types and concentrations. CA: carrageenin and AG: agar. Mean values with a common letter do not differ significantly ($p > 0.05$, Tukey's test).

4. Conclusions

The results of this study showed that most of the physical properties of NBGs depend on the type and concentration of hydrocolloids. All NBGs presented a shear-thinning flow behavior with different levels of consistency and pseudo-plasticity. Similarly, the viscoelasticity of all the NBGs showed a predominantly elastic behavior over viscous behavior, making this effect more pronounced in the agar-based gels. Therefore, because they can be easily swallowed, it is possible to form gels with different rheological properties using agar and carrageenan as gelling agents. Regarding textural properties, the obtained NBGs were characterized by a low adhesiveness and cohesiveness, which means they could be suitable for consumption by seniors. FTIR spectra analysis revealed interactions between the components of the NBGs, suggesting that the oil droplets of nanoemulsions act as active fillers in the gel network. In addition, the water-holding capacity also depended on the type and concentration of hydrocolloid, since the NBGs with κ-carrageenan retained more water than the NBGs with agar due to the different network structure that each hydrocolloid forms. Finally, it is possible to obtain gels based on nanoemulsions with different physical properties, which could be useful for developing foods with varied textures for the senior population with swallowing problems.

Author Contributions: Conceptualization C.A.; methodology C.S., A.L. and N.R.; formal analysis and writing—original draft preparation N.R.; writing—review and editing C.A. and R.N.Z.; funding acquisition and project administration C.A. All authors have read and agreed to the published version of the manuscript.

Funding: This research was funded by Vicerrectoria de Investigación, Innovación y Creación (VRIIC, USACH, Chile) by project Dicyt-Regular No. 082171AA (Carla Arancibia) and research contract No. USA2155_Dicyt (Natalia Riquelme).

Institutional Review Board Statement: Not applicable.

Informed Consent Statement: Not applicable.

Data Availability Statement: Not applicable.

Acknowledgments: The authors thank Blumos S.A. for providing free soy lecithin and pea protein samples.

Conflicts of Interest: The authors declare no conflict of interest.

References

1. Calligaris, S.; Moretton, M.; Melchior, S.; Mosca, A.C.; Pellegrini, N.; Anese, M. Designing food for the elderly: The critical impact of food structure. *Food Funct.* **2022**, *13*, 6467–6483. [CrossRef]
2. Song, X.; Perez-Cueto, F.J.; Bredie, W.L. Sensory-driven development of protein-enriched rye bread and cream cheese for the nutritional demands of older adults. *Nutrients* **2018**, *10*, 1006. [CrossRef]
3. Cichero, J. Age-related changes to eating and swallowing impact frailty: Aspiration, choking risk, modified food texture and autonomy of choice. *Geriatrics* **2018**, *3*, 69. [CrossRef]
4. Alsanei, W.A.; Chen, J. Food structure development for specific population groups. In *Handbook of Food Structure Development*; Spyropoulos, F., Lazidis, A., Norton, I., Eds.; Royal Society of Chemistry: London, UK, 2019; pp. 459–479.
5. Aguilera, J.M.; Park, D. Texture-modified foods for the elderly: Status, technology, and opportunities. *Trends Food Sci. Technol.* **2016**, *57*, 156–164. [CrossRef]
6. Hadde, E.K.; Chen, J. Texture and texture assessment of thickened fluids and texture-modified food for dysphagia management. *J. Texture Stud.* **2021**, *52*, 4–15. [CrossRef] [PubMed]
7. Kohyama, K.; Ishihara, S.; Nakauma, M.; Funami, T. Fracture phenomena of soft gellan gum gels during compression with artificial tongues. *Food Hydrocoll.* **2021**, *112*, 106283. [CrossRef]
8. Munialo, C.D.; Kontogiorgos, V.; Euston, S.R.; Nyambayo, I. Rheological, tribological and sensory attributes of texture-modified food for dysphagia patients and the elderly: A review. *Int. J. Food Sci. Technol.* **2019**, *55*, 1862–1871. [CrossRef]
9. Tan, C.; McClements, D.J. Application of advanced emulsion technology in the food industry: A review and critical evaluation. *Foods* **2021**, *10*, 812. [CrossRef]
10. Jena, G.K.; Parhi, R.; Sahoo, S.K. Nanoemulsions in Food Industry. In *Application of Nanotechnology in Food Science, Processing and Packaging*; Springer: Cham, Switzerland, 2022; pp. 73–91.
11. Li, G.; Zhang, Z.; Liu, H.; Hu, L. Nanoemulsion-based delivery approaches for nutraceuticals: Fabrication, application, characterization, biological fate, potential toxicity and future trends. *Food Funct.* **2021**, *12*, 1933–1953. [CrossRef] [PubMed]
12. Salvia-Trujillo, L.; Soliva-Fortuny, R.; Rojas-Graü, M.A.; McClements, D.J.; Martín-Belloso, O. Edible nanoemulsions as carriers of active ingredients: A review. *Annu. Rev. Food Sci. Technol.* **2017**, *8*, 439–466. [CrossRef]
13. McClements, D.J. Advances in edible nanoemulsions: Digestion, bioavailability, and potential toxicity. *Prog. Lipid Res.* **2021**, *81*, 101081. [CrossRef] [PubMed]
14. Islam, F.; Saeed, F.; Afzaal, M.; Hussain, M.; Ikram, A.; Khalid, M.A. Food grade nanoemulsions: Promising delivery systems for functional ingredients. *J. Food Sci. Technol.* **2023**, *60*, 1461–1471. [CrossRef] [PubMed]
15. Lin, D.; Kelly, A.L.; Miao, S. Preparation, structure-property relationships and applications of different emulsion gels: Bulk emulsion gels, emulsion gel particles, and fluid emulsion gels. *Trends Food Sci. Technol.* **2020**, *102*, 123–137. [CrossRef]
16. Lu, Y.; Mao, L.; Hou, Z.; Miao, S.; Gao, Y. Development of emulsion gels for the delivery of functional food ingredients: From structure to functionality. *Food Eng. Rev.* **2019**, *11*, 245–258. [CrossRef]
17. Torres, O.; Murray, B.; Sarkar, A. Emulsion microgel particles: Novel encapsulation strategy for lipophilic molecules. *Trends Food Sci. Technol.* **2016**, *55*, 98–108. [CrossRef]
18. Lu, Y.; Zhang, Y.; Yuan, F.; Gao, Y.; Mao, L. Emulsion gels with different proteins at the interface: Structures and delivery functionality. *Food Hydrocoll.* **2021**, *116*, 106637. [CrossRef]
19. Farjami, T.; Madadlou, A. An overview on preparation of emulsion-filled gels and emulsion particulate gels. *Trends Food Sci. Technol.* **2019**, *86*, 85–94. [CrossRef]
20. Xu, G.; Kang, J.; You, W.; Li, R.; Zheng, H.; Lv, L.; Zhang, Q. Pea protein isolates affected by ultrasound and NaCl used for dysphagia's texture-modified food: Rheological, gel, and structural properties. *Food Hydrocoll.* **2023**, *139*, 108566. [CrossRef]
21. Montes de Oca-Ávalos, J.M.; Borroni, V.; Huck-Iriart, C.; Navarro, A.S.; Candal, R.J.; Herrera, M.L. Relationship between formulation, gelation kinetics, micro/nanostructure and rheological properties of sodium caseinate nanoemulsion-based acid gels for food applications. *Food Bioproc. Tech.* **2020**, *13*, 288–299. [CrossRef]
22. Liao, Y.; Sun, Y.; Wang, Z.; Zhong, M.; Li, R.; Yan, S.; Li, Y. Structure, rheology, and functionality of emulsion-filled gels: Effect of various oil body concentrations and interfacial compositions. *Food Chem. X* **2022**, *16*, 100509. [CrossRef]
23. Rhein-Knudsen, N.; Meyer, A.S. Chemistry, gelation, and enzymatic modification of seaweed food hydrocolloids. *Trends Food Sci. Technol.* **2021**, *109*, 608–621. [CrossRef]
24. Zia, K.M.; Tabasum, S.; Nasif, M.; Sultan, N.; Aslam, N.; Noreen, A.; Zuber, M. A review on synthesis, properties and applications of natural polymer-based carrageenan blends and composites. *Int. J. Biol. Macromol.* **2017**, *96*, 282–301. [CrossRef] [PubMed]
25. Bui, V.T.; Nguyen, B.T.; Nicolai, T.; Renou, F. Mixed iota and kappa carrageenan gels in the presence of both calcium and potassium ions. *Carbohydr. Polym.* **2019**, *223*, 115107. [CrossRef] [PubMed]
26. Bertasa, M.; Dodero, A.; Alloisio, M.; Vicini, S.; Riedo, C.; Sansonetti, A.; Scalarone, D.; Castellano, M. Agar gel strength: A correlation study between chemical composition and rheological properties. *Eur. Polym. J.* **2020**, *123*, 109442. [CrossRef]
27. Lee, W.K.; Lim, Y.Y.; Leow, A.T.C.; Namasivayam, P.; Abdullah, J.O.; Ho, C.L. Factors affecting yield and gelling properties of agar. *J. Appl. Phycol.* **2017**, *29*, 1527–1540. [CrossRef]
28. de Souza Paglarini, C.; de Figueiredo Furtado, G.; Biachi, J.P.; Vidal, V.A.S.; Martini, S.; Forte, M.B.S.; Lopes Cunha, R.; Pollonio, M.A.R. Functional emulsion gels with potential application in meat products. *J. Food Eng.* **2018**, *222*, 29–37. [CrossRef]

29. Lucas-González, R.; Roldán-Verdu, A.; Sayas-Barberá, E.; Fernández-López, J.; Pérez-Álvarez, J.A.; Viuda-Martos, M. Assessment of emulsion gels formulated with chestnut (*Castanea sativa* M.) flour and chia (*Salvia hispanica* L) oil as partial fat replacers in pork burger formulation. *J. Sci. Food Agric.* **2020**, *100*, 1265–1273. [CrossRef]
30. Nacak, B.; Öztürk-Kerimoğlu, B.; Yıldız, D.; Çağındı, Ö.; Serdaroğlu, M. Peanut and linseed oil emulsion gels as potential fat replacer in emulsified sausages. *Meat Sci.* **2021**, *176*, 108464. [CrossRef]
31. Chen, X.; McClements, D.J.; Wang, J.; Zou, L.; Deng, S.; Liu, W.; Liu, C. Coencapsulation of (−)-Epigallocatechin-3-gallate and quercetin in particle-stabilized W/O/W emulsion gels: Controlled release and bioaccessibility. *J. Agric. Food Chem.* **2018**, *66*, 3691–3699. [CrossRef]
32. Li, J.; Xu, L.; Su, Y.; Chang, C.; Yang, Y.; Gu, L. Flocculation behavior and gel properties of egg yolk/κ-carrageenan composite aqueous and emulsion systems: Effect of NaCl. *Food Res. Int.* **2020**, *132*, 108990. [CrossRef]
33. Pan, Y.; Li, X.M.; Meng, R.; Xu, B.C.; Zhang, B. Investigation of the formation mechanism and curcumin bioaccessibility of emulsion gels based on sugar beet pectin and laccase catalysis. *J. Agri Food Chem.* **2021**, *69*, 2557–2563. [CrossRef] [PubMed]
34. Riquelme, N.; Robert, P.; Arancibia, C. Understanding older people perceptions about desserts using word association and sorting task methodologies. *Food Qual. Prefer.* **2022**, *96*, 104423. [CrossRef]
35. Chen, H.; Mao, L.; Hou, Z.; Yuan, F.; Gao, Y. Roles of additional emulsifiers in the structures of emulsion gels and stability of vitamin E. *Food Hydrocoll.* **2020**, *99*, 105372. [CrossRef]
36. Riquelme, N.; Robert, P.; Troncoso, E.; Arancibia, C. Influence of the particle size and hydrocolloid type on lipid digestion of thickened emulsions. *Food Funct.* **2020**, *11*, 5955–5964. [CrossRef] [PubMed]
37. Chen, J.; Stokes, J.R. Rheology and tribology: Two distinctive regimes of food texture sensation. *Trends Food Sci. Technol.* **2012**, *25*, 4–12. [CrossRef]
38. Sharma, M.; Pico, J.; Martinez, M.M.; Duizer, L. The dynamics of starch hydrolysis and thickness perception during oral processing. *Food Res. Int.* **2020**, *134*, 109275. [CrossRef]
39. Liu, W.; Gao, H.; McClements, D.J.; Zhou, L.; Wu, J.; Zou, L. Stability, rheology, and β-carotene bioaccessibility of high internal phase emulsion gels. *Food Hydrocoll.* **2019**, *88*, 210–217. [CrossRef]
40. Ghirro, L.C.; Rezende, S.; Ribeiro, A.S.; Rodrigues, N.; Carocho, M.; Pereira, J.A.; Santamaria-Echart, A. Pickering emulsions stabilized with curcumin-based solid dispersion particles as mayonnaise-like food sauce alternatives. *Molecules* **2022**, *27*, 1250. [CrossRef]
41. Martín del Campo, A.; Fermín-Jiménez, J.A.; Fernández-Escamilla, V.V.; Escalante-García, Z.Y.; Macías-Rodríguez, M.E.; Estrada-Girón, Y. Improved extraction of carrageenan from red seaweed (*Chondracantus canaliculatus*) using ultrasound-assisted methods and evaluation of the yield, physicochemical properties and functional groups. *Food Sci. Biotech.* **2021**, *30*, 901–910. [CrossRef]
42. Giura, L.; Urtasun, L.; Belarra, A.; Ansorena, D.; Astiasarán, I. Exploring tools for designing dysphagia-friendly foods: A review. *Foods* **2021**, *10*, 1334. [CrossRef]
43. IDDSI. The IDDSI Framework. International Dysphagia Diet Standardization Initiative Retriever. Available online: https://iddsi.org/IDDSI/media/images/Complete_IDDSI_Framework_Final_31July2019.pdf (accessed on 16 December 2022).
44. Nakauma, M.; Funami, T. Structuring for Elderly Foods. In *Food Hydrocolloids*; Fang, Y., Zhang, H., Nishinari, K., Eds.; Springer: Singapore, 2021.
45. Bazunova, M.V.; Shurshina, A.S.; Lazdin, R.Y.; Kulish, E.I. Thixotropic properties of solutions of some polysaccharides. *Russ. J. Phy. Chem. B* **2020**, *14*, 685–690. [CrossRef]
46. Thiviya, P.; Gamage, A.; Liyanapathiranage, A.; Makehelwala, M.; Dassanayake, R.S.; Manamperi, A.; Merah, O.; Mani, S.; Koduru, J.R.; Madhujith, T. Algal polysaccharides: Structure, preparation and applications in food packaging. *Food Chem.* **2022**, *405*, 134903. [CrossRef]
47. Goff, H.D.; Guo, Q. The role of hydrocolloids in the development of food structure. In *Handbook of Food Structure Development*; Spyropoulos, F., Lazidis, A., Norton, I., Eds.; Royal Society of Chemistry: London, UK, 2019; pp. 1–28.
48. Makshakova, O.N.; Faizullin, D.A.; Zuev, Y.F. Interplay between secondary structure and ion binding upon thermoreversible gelation of κ-carrageenan. *Carbohydr. Polym.* **2020**, *227*, 115342. [CrossRef]
49. Zou, Y.; Yang, X.; Scholten, E. Rheological behaviour of emulsion gels stabilized by zein/tannic acid complex particles. *Food Hydrocoll.* **2018**, *77*, 363–371. [CrossRef]
50. Liu, C.; Li, Y.; Liang, R.; Sun, H.; Wu, L.; Yang, C.; Liu, Y. Development and characterization of ultrastable emulsion gels based on synergistic interactions of xanthan and sodium stearoyl lactylate. *Food Chem.* **2023**, *400*, 133957. [CrossRef]
51. Shi, Z.; Shi, Z.; Wu, M.; Shen, Y.; Li, G.; Ma, T. Fabrication of emulsion gel based on polymer sanxan and its potential as a sustained-release delivery system for β-carotene. *Int. J. Biol. Macromol.* **2020**, *164*, 597–605. [CrossRef]
52. Suebsaen, K.; Suksatit, B.; Kanha, N.; Laokuldilok, T. Instrumental characterization of banana dessert gels for the elderly with dysphagia. *Food Biosci.* **2019**, *32*, 100477. [CrossRef]
53. Liang, X.; Ma, C.; Yan, X.; Zeng, H.; McClements, D.J.; Liu, X.; Liu, F. Structure, rheology and functionality of whey protein emulsion gels: Effects of double cross-linking with transglutaminase and calcium ions. *Food Hydrocoll.* **2020**, *102*, 105569. [CrossRef]
54. Moret-Tatay, A.; Rodríguez-García, J.; Martí-Bonmatí, E.; Hernando, I.; Hernández, M.J. Commercial thickeners used by patients with dysphagia: Rheological and structural behaviour in different food matrices. *Food Hydrocoll.* **2015**, *51*, 318–326. [CrossRef]

55. Ishihara, S.; Nakauma, M.; Funami, T.; Odake, S.; Nishinari, K. Swallowing profiles of food polysaccharide gels in relation to bolus rheology. *Food Hydrocoll.* **2011**, *25*, 1016–1024. [CrossRef]
56. Funami, T.; Nakauma, M. Instrumental food texture evaluation in relation to human perception. *Food Hydrocoll.* **2022**, *124*, 107253. [CrossRef]
57. Wada, S.; Kawate, N.; Mizuma, M. What type of food can older adults masticate? Evaluation of mastication performance using color-changeable chewing gum. *Dysphagia* **2017**, *32*, 636–643. [CrossRef] [PubMed]
58. PubChem, National Library of Medicine: National Center for Biotechnology Information. Available online: https://pubchem.ncbi.nlm.nih.gov/compound/ (accessed on 18 January 2023).
59. Cichero, J.A.Y. Adjustment of food textural properties for elderly patients. *J. Texture Stud.* **2016**, *47*, 277–283. [CrossRef]
60. Guo, Q. Understanding the oral processing of solid foods: Insights from food structure. *Compr. Rev. Food Sci. Food Saf.* **2021**, *20*, 2941–2967. [CrossRef]
61. Hadde, E.K.; Prakash, S.; Chen, W.; Chen, J. Instrumental texture assessment of IDDSI texture levels for dysphagia management. Part 2: Texture modified foods. *J. Texture Stud.* **2022**, *53*, 617–628. [CrossRef]
62. Jiang, S.; Ma, Y.; Wang, Y.; Wang, R.; Zeng, M. Effect of κ-carrageenan on the gelation properties of oyster protein. *Food Chem.* **2022**, *382*, 132329. [CrossRef]
63. Gallego, M.; Barat, J.M.; Grau, R.; Talens, P. Compositional, structural design and nutritional aspects of texture-modified foods for the elderly. *Trends Food Sci. Technol.* **2022**, *119*, 152–163. [CrossRef]
64. Ren, Y.; Jiang, L.; Wang, W.; Xiao, Y.; Liu, S.; Luo, Y.; Xie, J. Effects of Mesona chinensis Benth polysaccharide on physicochemical and rheological properties of sweet potato starch and its interactions. *Food Hydrocoll.* **2020**, *99*, 105371. [CrossRef]
65. Moreno, H.M.; Domínguez-Timón, F.; Díaz, M.T.; Pedrosa, M.M.; Borderías, A.J.; Tovar, C.A. Evaluation of gels made with different commercial pea protein isolate: Rheological, structural and functional properties. *Food Hydrocoll.* **2020**, *99*, 105375. [CrossRef]
66. Jamwal, R.; Kumari, S.A.; Sharma, S.; Kelly, S.; Cannavan, A.; Singh, D.K. Recent trends in the use of FTIR spectroscopy integrated with chemometrics for the detection of edible oil adulteration. *Vib. Spectrosc.* **2021**, *113*, 103222. [CrossRef]
67. Lu, Y.; Ma, Y.; Zhang, Y.; Gao, Y.; Mao, L. Facile synthesis of zein-based emulsion gels with adjustable texture, rheology and stability by adding β-carotene in different phases. *Food Hydrocoll.* **2022**, *124*, 107178. [CrossRef]
68. Kuligowski, J.; Quintás, G.; Garrigues, S.; de la Guardia, M. Determination of lecithin and soybean oil in dietary supplements using partial least squares-Fourier transform infrared spectroscopy. *Talanta* **2008**, *77*, 229–234. [CrossRef] [PubMed]
69. Duman, O.; Polat, T.G.; Diker, C.Ö.; Tunç, S. Agar/κ-carrageenan composite hydrogel adsorbent for the removal of Methylene Blue from water. *Int. J. Biol. Macromol.* **2020**, *160*, 823–835. [CrossRef]
70. Gómez-Ordóñez, E.; Rupérez, P. FTIR-ATR spectroscopy as a tool for polysaccharide identification in edible brown and red seaweeds. *Food Hydrocoll.* **2011**, *25*, 1514–1520. [CrossRef]
71. Yang, Y.; Zhang, M.; Alalawy, A.I.; Amutairi, F.M.; Al-Duais, M.A.; Wang, J.; Salama, E.S. Identification and characterization of marine seaweeds for biocompounds production. *Environ. Technol. Innov.* **2021**, *24*, 101848. [CrossRef]
72. Tang, H.; Tan, L.; Chen, Y.; Zhang, J.; Li, H.; Chen, L. Effect of κ-carrageenan addition on protein structure and gel properties of salted duck egg white. *J. Sci. Food Agric.* **2021**, *101*, 1389–1395. [CrossRef]
73. Yu, W.; Wang, Z.; Pan, Y.; Jiang, P.; Pan, J.; Yu, C.; Dong, X. Effect of κ-carrageenan on quality improvement of 3D printed Hypophthalmichthys molitrix-sea cucumber compound surimi product. *LWT* **2022**, *154*, 112279. [CrossRef]
74. Mao, R.; Tang, J.; Swanson, B.G. Water holding capacity and microstructure of gallan gels. *Carbohydr. Polym.* **2001**, *46*, 365–371. [CrossRef]
75. Ryu, J.N.; Jung, J.H.; Lee, S.Y.; Ko, S.H. Comparison of physicochemical properties of agar and gelatin gel with uniform hardness. *Food Eng. Prog.* **2012**, *16*, 14–19.

colloids and interfaces

MDPI

Article

Improving the Size Distribution of Polymeric Oblates Fabricated by the Emulsion-in-Gel Deformation Method

Giselle Vite [1], Samuel Lopez-Godoy [1], Pedro Díaz-Leyva [2] and Anna Kozina [1,*]

[1] Instituto de Química, Universidad Nacional Autónoma de México, Mexico City 04510, Mexico; giselle.vite.q@gmail.com (G.V.); lopez.godoy.samuel@gmail.com (S.L.-G.)

[2] Departamento de Física, Universidad Autónoma Metropolitana Iztapalapa, Mexico City 09340, Mexico; pdleyva@xanum.uam.mx

* Correspondence: akozina@unam.mx; Tel.: +52-5556224437

Abstract: The optimization of fabrication conditions for colloidal micron-sized oblates obtained by the deformation of an oil-in-hydrogel emulsion is reported. The influence of the type of emulsion stabilizer, ultrasonication parameters, and emulsion and gel mixing conditions was explored. The best conditions with which to obtain more uniform particles were using polyvinyl alcohol as an emulsion stabilizer mixed with the gelatine solution at 35 °C and slowly cooling to room temperature. Four fractionation methods were applied to oblates to improve their size uniformity. The iterative differential centrifugation method produced the best size polydispersity reduction.

Keywords: colloids; oblates; emulsion-in-gel; fractionation

Citation: Vite, G.; Lopez-Godoy, S.; Díaz-Leyva, P.; Kozina, A. Improving the Size Distribution of Polymeric Oblates Fabricated by the Emulsion-in-Gel Deformation Method. *Colloids Interfaces* **2023**, *7*, 50. https://doi.org/10.3390/colloids7030050

Academic Editors: César Burgos-Díaz, Mauricio Opazo-Navarrete and Eduardo Morales

Received: 3 April 2023
Revised: 17 May 2023
Accepted: 25 May 2023
Published: 12 July 2023

1. Introduction

Auto-organization of colloidal particles is an active field of research due to the possibilities of application of this knowledge in the fabrication of new functional materials [1–5]. For spherical submicrometer particles with different types of isotropic interactions, a wide variety of structures was observed, both ordered (such as crystals) and disordered (such as fluids, gels, glasses, etc.) [6–10]. When a binary mixture of spherical colloids of different sizes is used, the structure complexity increases and the types of order become much more diverse [11]. Another way of increasing the structural divergence is by using anisotropic colloids as more complex building blocks for self-assembly. Therefore, recent studies have demonstrated much interest in anisotropically shaped particles such as colloidal rods, dumbbells, ellipsoids, spherocylinders, cubes, etc. [12–15]. Since inter-particle interactions between such colloids depend on the particle orientation, new self-assembly routes were discovered [16,17]. While there are a considerable number of computer simulations and theoretical predictions concerning the self-assembly of such anisotropic colloids, the experimental verification of predicted structures is still in its initial stage. This is mainly due to the fact that syntheses of such building blocks in the large amount required for the self-assembly studies are not well established. While many scientific reports have described the fabrication of colloids of many different shapes, usually only a small amount of these may be obtained as a batch, which complicates the conduction of the assembly studies [14,18].

Colloidal oblates have gained considerable attention due to the high degree of directionality in their inter-particle interactions. The preferable orientation of oblates in their self-organization is along their minor axes, forming stacks and columnar structures [19–23]. Such an organization is an example of a more complex phase behavior. Therefore, it is expected that the directionality would affect both equilibrium and non-equilibrium suspension properties in the bulk and at fluid interfaces. In addition, the particle dynamics are expected to be more complex due to the directionality of oblate translation and rotation. Besides the formation of novel structures, colloidal oblates may be potentially applied as a model for blood cells [24], drug carriers [25,26], emulsion stabilizers [27], and to the fabrication of photonic materials [28,29]. For the majority of studies and applications, the main requirements for oblates are to be fabricated in a good amount and with high uniformity in size and shape. Therefore, there is a need for a versatile synthesis of these colloids.

One of the most common synthetic methods of colloidal oblates is based on the mechanical deformation of pre-synthesized spherical polymeric particles (polystyrene, PS, or poly(methyl methacrylate), PMMA) embedded in an elastic matrix [30,31]. The advantages of this method include the ability to obtain a larger number of particles from a single synthesis and their relatively uniform size and shape. The main drawback is the need for precise control of deformation at a rather high temperature (above 135 °C). In addition, pre-formed, almost monodisperse, solid spherical particles are used, which reduces the variety of the ellipsoid materials to mainly PS or PMMA. Courbaron et al. suggested a similar method of fabrication, but with an oil-in-hydrogel emulsion to be deformed at room temperature [32]. The oil could be solidified by photo-cross-linking while the droplets are deformed. A large yield of oblates could be obtained, but with a rather large particle polydispersity stemming mainly from the original emulsion. Although the particle size was not as uniform as in the case of the deformation of solid particles, considerable advantages of this method include the possibility of using different materials as an oil phase and the deformation that takes place at room temperature. These advantages could potentially be used in the fabrication of drug carriers. The importance of the particle shape has been underlined for drug delivery systems [33]. For instance, ellipsoid particles were shown to have different adhesion in blood vessels and during endocytosis [34,35]. Therefore, an emulsion-in-gel deformation method might open up the possibility of using biocompatible materials for drug encapsulation in-situ, which is not accessible by the deformation of pre-fabricated solid spherical particles.

In the present work, we explored the conditions for producing colloidal oblates by deforming an oil-in-gel emulsion. We clarified the influence of each fabrication step on the oblate size and polydispersity. We demonstrated that the initial emulsion droplet polydispersity has a strong effect on the ellipsoid size distribution. Thus, we explored different methods of oblate fractionation to optimize the yield of more uniform particles. This work represents a significant advancement in that a large amount of ellipsoids (approximately 600 mg) were produced from a single batch. After particle fractionation, about 100 mg of particles with considerably reduced polydispersity were obtained. Unlike previously reported works, this is the first optimized method suggested for the fabrication of uniform ellipsoids with a large yield starting from an emulsion. Our method may be useful for the large-scale production of ellipsoid carriers for bioactive compounds since it does not require extreme fabrication conditions.

2. Materials and Methods

The following steps can be used to describe the overall process of particle fabrication: (1) emulsification of the polymerizable oil; (2) emulsion-in-gel preparation; (3) emulsion-in-gel deformation followed by deformed droplet solidification; and (4) particle recovery and fractionation.

2.1. Emulsion Preparation

An oil phase and an aqueous phase were prepared to make an oil-in-water emulsion. The oil phase consisted of 0.6 g of the mixture of 1,6-hexanediol diacrylate (polymerizable oil monomer) and 1-hydroxycyclohexylphenyl ketone (photoinitiator) at 1% wt of the oil phase. The aqueous phase was 5 mL of deionized water (resistivity $\rho = 18.1$ MΩ·cm) with a surfactant. Two types of surfactants were tested: (1) sodium dodecyl sulfate at a concentration of 10 mM and (2) polyvinyl alcohol (PVA, 87–89% hydrolyzed, M_w = 13,000–23,000) at 1% wt. All the reactants were obtained from Sigma-Aldrich, and deionized water was used throughout the study.

In a 10 mL vial, an oil phase was deposited on top of an aqueous phase, and the sample was homogenized by using an ultrasonic processor (500 W, Cole-Parmer, Vernon Hills, IL, USA) for 10 min at a constant power of 100 W and an amplitude of 20%.

2.2. Emulsion-in-Gel Preparation

To form a gel, 9 g of an aqueous solution of gelatin (G1890, Sigma-Aldrich, St. Louis, MO, USA) at 4% wt was prepared at 80 °C to dissolve gelatin completely at constant stirring. Then, the emulsion-in-gel was prepared using two methods. In the first method, the emulsion was added directly to the gelatin solution at 80 °C with moderate stirring and then the mixture was cooled down to room temperature, with gelatin forming a solid gel. In the second method, a gelatin solution was first cooled down to about 35–40 °C before its complete solidification. Then an emulsion was added, and the mixture was left to solidify, trapping the oil droplets during its cooling down to room temperature. The gel solidification after mixing with the emulsion was performed in a custom-made PTFE rectangular mold with dimensions $L \times W \times H$ of $56 \times 24 \times 7$ mm, which allowed us to avoid further gel cutting by using a well-shaped emulsion-in-gel brick ready for deformation.

2.3. Gel Deformation and Droplet Solidification

A rectangular gel brick was then mechanically deformed in a custom-made press (see Supplementary Materials). Prior to deformation, the press was placed inside a custom-made wooden box on a Dewar flask filled with ice and illuminated with a UV lamp (100 W, B100AP, Analytik Jena, Jena, Germany) at a distance of 25 cm. Such a collocation allowed for precise temperature control of the gel at 22 ± 0.5 °C during the entire 2.5-h droplet solidification. The sample brick was covered with a rectangular glass, and the deformation was applied along the vertical axis by moving down the screw sliding cylinder. The deformation degree (the ratio between the final and initial brick heights) was kept constant in all the cases and was equal to 57%. After the deformation and droplet solidification, the brick was dissolved in warm water, and the particles were recovered by the sedimentation-redispersion method.

2.4. Characterization

Particle size and size distribution were characterized by scanning electron microscopy (SEM, JSM-7800F Jeol, Tokyo, Japan). The equatorial radius of the oblates was measured with the Digimizer software using only the particles oriented in such a way that their images were circular. Over 300 particles were analyzed. A Gaussian distribution function was then fitted to the data to obtain the average particle equatorial radius R and standard deviation SD. The polydispersity index (PDI) was then calculated as $PDI = SD/R$. The minor axis (pole-to-pole) oblate dimensions and the corresponding PDI were ignored due to the difficulties in the location of particles oriented with their profiles perpendicular to the observation direction. Fractionated particles were observed under an optical microscope (AxioImager, Zeiss, Jena, Germany) in bright and dark fields. For selected samples, particle size distribution was measured by dynamic light scattering (DLS) using the CONTIN method. For this, a dilute dispersion of oblates was prepared in water and measured at a scattering angle of 90° during 300 s at 20 °C in a 3DLS spectrometer (LS Instruments, Fribourg, Switzerland) equipped with a 632.8 nm laser, an index-matching bath, and a temperature control.

2.5. Fractionation

Several methods of particle fractionation were applied. The first one was differential centrifugation. For this, 500 mg of ellipsoids were dispersed in 50 mL of a 10 mM SDS solution and centrifuged for 3 min at the following rpm values: 400 ($11\times g$), 500 ($18\times g$), 650 ($30\times g$), and 800 ($45\times g$). After each centrifugation cycle, the supernatant was decanted, dispersed again, and sedimented at a higher speed. Therefore, four fractions were obtained using this method.

The second method used was the Bibette method [36], where 500 mg of particles were dispersed in 100 mL of an 8.2 mM SDS solution, which corresponds to one critical micelle concentration (CMC) of SDS [37]. The suspension was poured into a tall glass cylinder (diameter of about 10 mm) and left to settle under gravity for 24 h. Then, the sediment was collected, and the particles from the supernatant were recovered, washed with water and ethanol, and dispersed in a 2 CMC SDS solution. The procedure was repeated for 3, 4, and 5 CMC of SDS solution. Thus, five fractions of particles were obtained.

The third method consisted of particle fractionation along a sucrose density gradient. A 15-mL falcon tube was filled with five layers (2 mL each) of sucrose solutions from 40 to 20% wt in steps of 5% wt. Then, 2 mL of the sample was deposited on top of the gradient at a concentration of 20 mg/mL in a 10 mM SDS solution, which was followed by centrifugation at 4000 rpm (1115× *g*) for 5 min. Five fractions of particles were recovered and analyzed.

Iterative differential centrifugation, an extension of the first method of differential centrifugation, was used as the fourth method. For this method, 40 mg of particles were dispersed in 15 mL of a 10 mM SDS solution and centrifuged at 4000 rpm (1115× *g*) for 5 min. The supernatant was withdrawn and observed under an optical microscope (AxioImager, Zeiss). Only very small particles were observed. The supernatant was then stored in a separate tube, while the sediment was dispersed and centrifuged again twice under the same conditions. The supernatant of the three runs was collected as the first fraction, F1. The second fraction F2 was formed in a similar way by the particles left in the supernatant after three-run sedimentation at 4000 rpm (1115× *g*) for 1 min. The third fraction F3 was obtained by collecting the supernatants during ten-run sedimentation at 3500 rpm (865× *g*) for 1 min. The particles were observed under the microscope after each run to rule out the presence of larger particles. In a similar way, the fourth fraction F4 was obtained by collecting the supernatant after the five-run sedimentation at 1000 rpm (70× *g*) for 1 min. The sediment was recovered, forming the fifth fraction, F5.

3. Results and Discussion

3.1. Emulsion-in-Gel Preparation

Figure 1 shows SEM images of spherical particles obtained after emulsion preparation and droplet solidification without gel for emulsions stabilized by two surfactants: SDS and PVA.

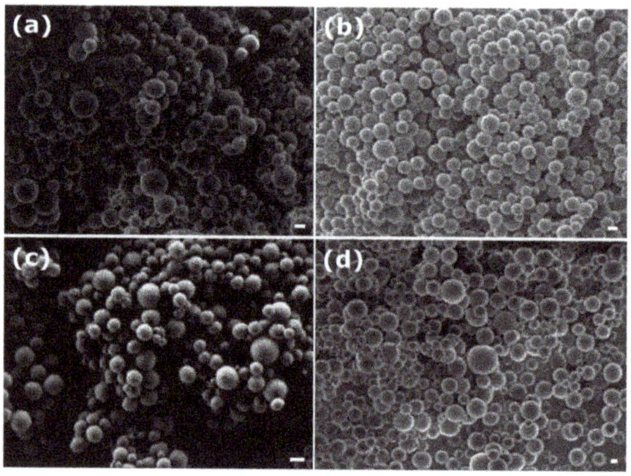

Figure 1. SEM images of spherical polymeric particles solidified in an emulsion prepared by ultrasonication using (**a**) 10 mM aqueous SDS solution during 10 min of ultrasonication at the ultrasonic processor amplitude of 20%, (**b**) 1% wt PVA solution as a continuous phase during 10 min of ultrasonication at the ultrasonic processor amplitude of 20%, (**c**) 1% wt PVA solution during 10 min of ultrasonication at the ultrasonic processor amplitude of 50%, and (**d**) 10 mM aqueous SDS solution during 3 min of ultrasonication at the ultrasonic processor amplitude of 20%. Scale bars = 1 μm.

As is seen from the figure, in all the cases, spherical particles were successfully formed. The following particle characteristics were obtained after 10 min of ultrasonication at 20% amplitude: average radius R of 710 nm and $PDI = 0.35$ using 10 mM SDS solution (Figure 1a) and 720 nm and $PDI = 0.14$ for 1% PVA solution (Figure 1b). As one may notice, although the average particle size did not change significantly, the polydispersity decreased more than twice with the use of PVA. This is connected with the higher viscosity of PVA solution as compared to 10 mM SDS, which reduces cavitation during the ultrasound treatment [38]. In addition, a larger medium viscosity hinders droplet coalescence, which results in a more uniform particle size distribution. Curiously, an increase of the ultrasonic processor amplitude for PVA-stabilized emulsion up to 50% resulted in a decrease of the average particle radius to 470 nm with a drastic increase of PDI up to 0.53 (Figure 1c). A similar result was observed previously on the fabrication of oil-in-water nanoemulsions by ultrasonication, where a large applied amplitude resulted in promoted droplet coalescence [39]. Therefore, the low amplitude is more favorable for PDI reduction. A decrease in sonication time to 3 min at the same amplitude of 20% did not have a significant effect on PDI (0.36) as shown for SDS-stabilized emulsion in Figure 1d, while the average radius increased to 900 nm. Previous studies revealed the absence of a significant impact of processing time on particle size and polydispersity, while the physicochemical properties of the dispersed and continuous phases as well as the type and concentration of the surfactant had a much larger contribution to the droplet rupture [39–41].

To check the reproducibility of the method, various samples were prepared. Table 1 summarizes particle radii and PDI.

Table 1. Average radius and PDI for spherical particles prepared by ultrasonication at 20% amplitude for 10 min.

Sample	Surfactant	R, nm	PDI
E1	SDS	710	0.32
E2	SDS	750	0.26
E3	SDS	550	0.27
E4	SDS	710	0.35
E5	PVA	720	0.14
E6	PVA	610	0.20
E7	PVA	850	0.22
E8	PVA	620	0.19

As can be seen, the ultrasonication method produces spheres with a relatively high polydispersity for SDS-stabilized emulsions with a radius of about 700 nm. Similar radii but lower PDI are observed for PVA-stabilized emulsions. While PVA helps to reduce PDI, it does not completely solve the problem since the polydispersity varies in the range of 0.14–0.22, which is still considered large. Nevertheless, further PDI reduction may be laborious and, as will be shown later, not a necessary process since the droplet trapping in the gel and their further deformation contribute to an increase in particle PDI.

The next step was the emulsion-in-gel preparation, which involved mixing the emulsion with the gelatin solution. Two ways of mixing were evaluated: at high (80 °C) and low (35 °C) temperatures. Figure 2 shows SEM images of spherical particles E8 and E5 that were UV-cured before their mixing with gelatin solution (in the original emulsion) and after their emulsion-in-gel solidification, curing, and recovery without deformation (EG8 and EG5). Sample EG8 was recovered after mixing with the gelatin solution at 80 °C followed by gel solidification and oil curing. Sample EG5 was recovered after mixing with the gelatin solution at 35 °C, followed by gel solidification and oil curing.

As can be seen, the particles solidified in the gel (EG8 and EG5) are less uniform in size as compared to the original emulsions. Sample EG8 (Figure 2b) resulted in a radius of 820 nm with a PDI of 0.47, while sample EG5 had an average radius of 970 nm and a PDI of 0.24. Thus, on mixing emulsions with gels, both average particle size and polydispersity increase, indicating a possible emulsion droplet coalescence during the gelation. High temperature promotes emulsion destabilization, which results in a drastic increase in polydispersity. Lowering the mixing temperature helps to hinder droplet coalescence, which nevertheless results in a PDI increase of about 10%. Despite an increase in PDI at the low temperature, this way of mixing was used in further ellipsoids preparation. At even lower temperatures, satisfactory emulsion mixing is limited by the gelatin's extremely high viscosity.

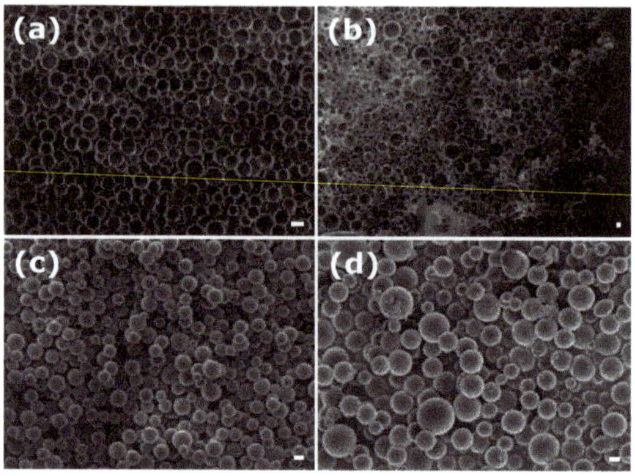

Figure 2. SEM images of solidified spherical polymeric particles in emulsion: (**a**,**c**) without gel; (**b**,**d**) in the gel. The upper row corresponds to samples E8 and EG8, and the lower row to samples E5 and EG5. Scale bars = 1 μm.

3.2. Emulsion-in-Gel Droplet Deformation

Figure 3 shows SEM images of spherical particles and their corresponding deformed particles.

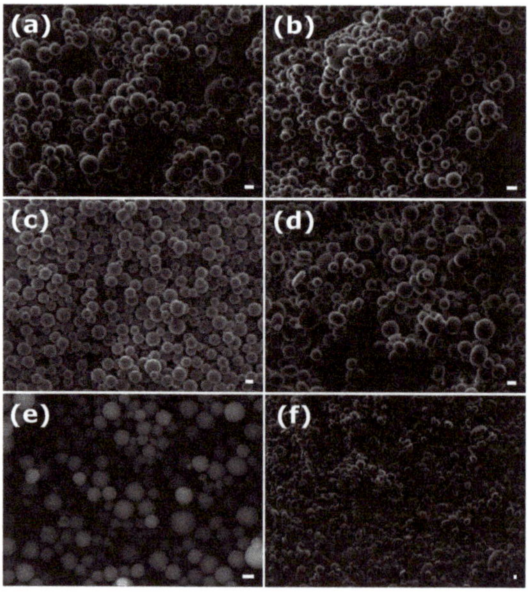

Figure 3. SEM images of spherical polymeric particles solidified in emulsion (**left column**) and their corresponding deformed particles (**right column**): (**a**) E1, (**b**) EGD1, (**c**) E5, (**d**) EGD5, (**e**) E6, and (**f**) EGD6. Scale bars = 1 μm.

In all the cases, deformation took place, which resulted in ellipsoidal particles of oblate shape. Similarly, Figure 4 shows samples obtained by the deformation of emulsions E2, E4, E7, and E8.

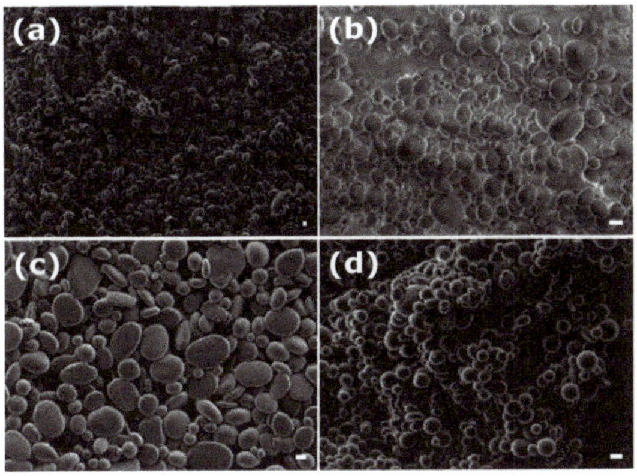

Figure 4. SEM images of deformed particles obtained from the corresponding emulsions: (**a**) EGD2 from E2, (**b**) EGD4 from E4, (**c**) EGD7 from E7, and (**d**) EGD8 from E8. Scale bars = 1 μm.

Table 2 summarizes the deformed particles' equatorial radii and PDI (see Supplementary Materials for particle size distributions).

Table 2. Average radius and PDI for spherical (R and PDI) and deformed (R_{ob} and PDI_{ob}) particles.

Sample	Surfactant	R, nm	PDI	R_{ob}, nm	PDI_{ob}
EGD1	SDS	710	0.32	990	0.53
EGD2	SDS	750	0.26	930	0.46
EGD4	SDS	710	0.35	1030	0.41
EGD5	PVA	720	0.14	1100	0.37
EGD6	PVA	610	0.20	900	0.40
EGD7	PVA	850	0.22	960	0.35
EGD8	PVA	620	0.19	750	0.38

The analysis shows that particle deformation causes a significant increase in particle size and polydispersity. While the increase in the radius is expected since a spherical droplet is pressed, such a drastic increase in PDI indicates emulsion destabilization during both gelation and deformation. Another reason for such an increase in PDI may be a difference in particle deformation as a function of particle size. At a constant external deformation, smaller particles are expected to deform less than the large ones due to the larger Laplace pressure. Different degrees of deformation result in differences in the radii of ellipsoids, which significantly widen the particle size distribution.

3.3. Particle Fractionation

The first fractionation method used was differential centrifugation. Figure 5 shows SEM images of the four fractions of sample EGD4.

The following values for average radii and PDI were obtained (see Supplementary Materials for particle size distributions): fraction 1 (400 rpm, 11× g) R_{ob} = 1090 nm, PDI_{ob} = 0.41; fraction 2 (500 rpm, 18× g) R_{ob} = 1110 nm, PDI_{ob} = 0.35; fraction 3 (650 rpm, 30× g) R_{ob} = 1220 nm, PDI_{ob} = 0.44; and fraction 4 (800 rpm, 45× g) R_{ob} = 1170 nm, PDI_{ob} = 0.37. It is clear that neither the average particle size nor the PDI changed significantly. These results could not be considered satisfactory since the polydispersity was too large. The possible reasons for such a poor separation could be a large particle concentration during the fractionation and a too-narrow interval of centrifugal force. Such conditions could favor particle sedimentation in clusters or aggregates, hindering differentiation in particle settling velocities. Therefore, other methods were implemented to improve particle fractionation.

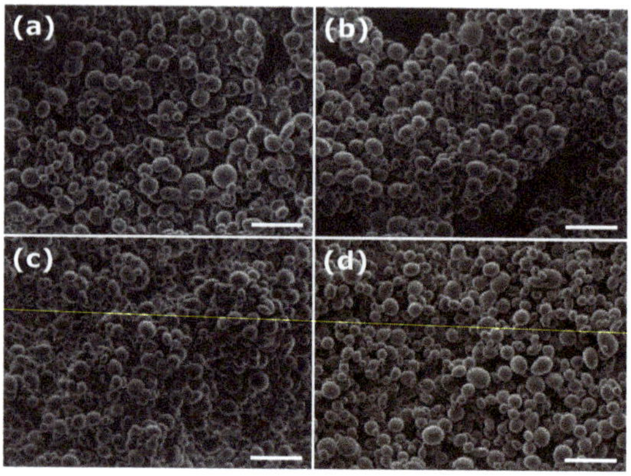

Figure 5. SEM images of fractions of deformed particles obtained by centrifugation for 3 min at: (**a**) 400 (11× *g*), (**b**) 500 (18× *g*), (**c**) 650 (30× *g*), and (**d**) 800 (45× *g*) rpm. Scale bars = 10 μm.

Table 3 summarizes particle sizes and PDI for five fractions obtained by the Bibette method from sample EGD5 (see Supplementary Materials for the corresponding particle size distributions).

Table 3. Average particle radius and PDI for deformed particles obtained by fractionation with the Bibette method.

Sample	CMC	R_{ob}, nm	PDI_{ob}
EGD5	0	1100	0.37
EGD5F1	1	1130	0.24
EGD5F2	2	1100	0.21
EGD5F3	3	950	0.14
EGD5F4	4	650	0.23
EGD5F5	5	980	0.23

As can be seen from the table, fractionation was able to reduce the particle polydispersity by at least 0.13 (EGD5F1). The intermediate fraction resulted in being the least polydisperse, with $PDI_{ob} = 0.14$. We connected it with the intermediate depletion attraction strength, which allows more selective particle aggregation. This fractionation method appears to have the potential to reduce the polydispersity of intermediate fractions. However, the rest of the fractions would require additional fractionation. Thus, this method can be recommended for ellipsoid fractionation but implies additional fractionation steps. The achievement of this method is the fraction EGD5F3, with a PDI value similar to that reported earlier on the deformation of almost monodisperse polymeric particles [30].

The next method applied was fractionation in a density gradient. Table 4 summarizes particle sizes and PDI for five fractions obtained by this method from sample EGD5 (see Supplementary Materials for particle size distributions).

The uppermost and smallest in its average size sample, EGD5F1, improved its PDI, reducing it to 0.25. Nevertheless, the rest of the fractions did not improve significantly, either in their radius or in PDI, probably due to the presence of small particles in all the fractions. The heaviest fraction, EGD5F5, contained extremely large particles with an average size of 24 μm. Such large particles were not observed in the original emulsion, E5. Therefore, the appearance of these gigantic ellipsoids is connected with the deformation process in gel, probably due to the droplet coalescence during the gelation and gel deformation.

These gigantic particles were not detected in the deformed EGD5 sample by SEM, probably due to the sample preparation method in which the dry sample was withdrawn from the uppermost part of the tube, whereas such large particles are expected to settle on the bottom due to their extremely fast sedimentation. Considering the above results, we can conclude that the method of fractionation in a density gradient is not efficient for the studied dispersion.

Table 4. Average particle radius and *PDI* for deformed particles obtained by fractionation in a sucrose density gradient. The fraction numbering starts from the uppermost fraction.

Sample	Sucrose, % wt	R_{ob}, nm	PDI_{ob}
EGD5	0	1100	0.37
EGD5F1	20	515	0.25
EGD5F2	25	780	0.34
EGD5F3	30	740	0.33
EGD5F4	35	730	0.30
EGD5F5	40	24,000	0.36

The last method tested was iterative differential centrifugation. Figure 6 shows light microscopy images for the four fractions.

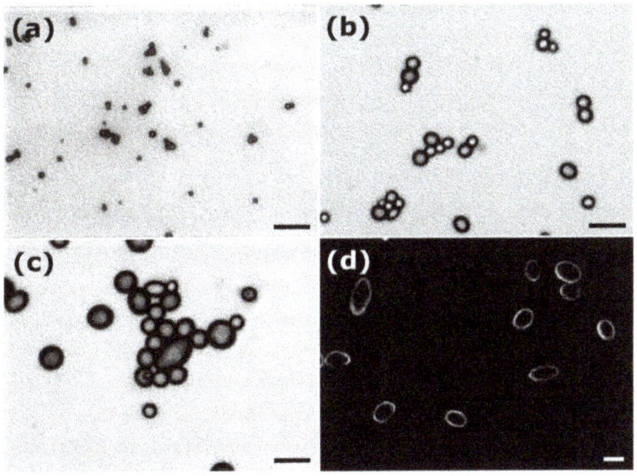

Figure 6. Light microscopy images of fractions of deformed particles obtained by iterative differential centrifugation: (**a**) EGD5F1 (1115× *g*; 5 min), (**b**) EGD5F2 (1115× *g*; 1 min), (**c**) EGD5F3 (865× *g*; 1 min), and (**d**) EGD5F4 (70× *g*; 1 min). Scale bars = 10 μm.

The average particle size and *PDI* of the corresponding fractions obtained from sample EGD5 are summarized in Table 5 (see Supplementary Materials for particle size distributions).

Table 5. Average particle radius and *PDI* for deformed particles obtained by iterative differential centrifugation. The fraction numbering starts from the fraction of the smallest particle size.

Sample	g-Force, g	R_{ob}, nm	PDI_{ob}
EGD5	0	1100	0.37
EGD5F1	1115 (5 min)	490	0.14
EGD5F2	1115 (1 min)	1450	0.19
EGD5F3	865 (1 min)	2740	0.28
EGD5F4	70 (1 min)	16,600	0.33
EGD5F5	sediment	24,600	0.25

The first two fractions containing the smallest average particles have considerably improved their PDI. In fact, these PDI values are comparable with the one for fraction EGD5F3 of the Bibette method and with those reported earlier [30]. Thus, this result may be considered good. The larger fractions remained quite polydisperse. Figure 7 shows the size distributions obtained by DLS for initial spherical particles E5, ellipsoids EGD5 before fractionation, and the particle fractions EGD5F1, EGD5F2, and EGD5F3.

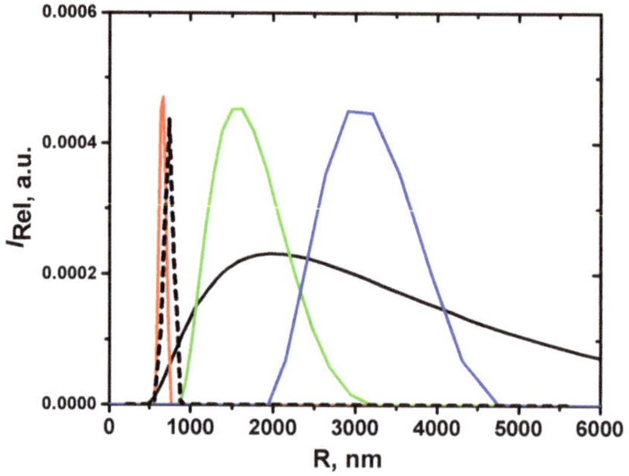

Figure 7. Size distributions of particles obtained from DLS using the CONTIN method: spherical particles E5 (black dashed line), a deformed initial EGD5 sample (black solid line), and the fractions EGD5F1 (red line), EGD5F2 (green line), and EGD5F3 (blue line).

As can be seen, the initial size distribution is significantly narrowed after the first fractionation. As a result, we can conclude that fractionation by iterative centrifugation has the potential to significantly reduce the size distribution of oblates.

4. Conclusions

We demonstrated the versatility of producing colloidal oblates by deforming an oil-in-gel emulsion. Controlled deformation at a constant temperature yields a large amount of oblates (about 500 mg from one deformation cycle). Therefore, this method of oblate fabrication may be potentially extended to the encapsulation of biologically active substances. Even with proper deformation control, the size of the droplet has a significant impact on the shape of the resulting oblates. Therefore, a broad size distribution of oblate sizes is inevitable. Nevertheless, it is possible to improve the size uniformity by further sample fractionation. The most efficient method of fractionation was found to be iterative differential fractionation.

Supplementary Materials: The following supporting information can be downloaded at: https://www.mdpi.com/article/10.3390/colloids7030050/s1, Figure S1: Schematic representation of the custom-made press for the gel sample deformation: (1) metallic lid; (2) metallic base; (3) PTFE center groove; (4) glass sample cell; (5) PTFE press sliding cylinder; (6) sliding rails; (7) screw sliding cylinder.; Figure S2: Particle size distributions obtained from the analysis of SEM images of spherical polymeric particles solidified in emulsion before droplet deformation (upper row) and deformed particles obtained from the corresponding emulsions (lower row): (a) E1, (b) E5, (c) E6, (d) EGD1, (e) EGD5, and (f) EGD5; Figure S3: Size distributions of particles obtained from DLS using the CONTIN method: spherical particles E6 (black dashed line) and deformed initial EGD6 oblates (black solid line); Figure S4: Particle size distributions obtained from the analysis of SEM images of the four fractions of sample EGD4 (as given in the legend) after differential centrifugation; Figure S5: Particle size distributions obtained from the analysis of SEM images of the effective fractions of sample EGD5 (as given in the legend) after the Bibette fractionation method; Figure S6: Particle size distributions obtained from the analysis of SEM images of the eefective fractions of sample EGD5 (as given in the legend) after fractionation in a sucrose density gradient; Figure S7: Particle size distributions obtained from the analysis of SEM images of the effective fractions of sample EGD5 (as given in the legend) after iterative differential centrifugation; Figure S8: Light microscopy images taken 1 day (a,b), and 14 days (c,d) after preparation of sample EGD5F2 (obtained by iterative centrifugation) by dispersion in water. Scale bars = 10 μm.

Author Contributions: Conceptualization, A.K.; methodology, S.L.-G. and P.D.-L.; formal analysis, S.L.-G.; investigation, G.V. and S.L.-G.; resources, P.D.-L. and A.K.; writing—original draft, A.K.; writing—review and editing, S.L.-G., P.D.-L. and A.K.; supervision, A.K.; funding acquisition, A.K. All authors have read and agreed to the published version of the manuscript.

Funding: This research was funded by CONACyT (grant CB-A1S-21124) and DGAPA-UNAM (grant IN100322).

Data Availability Statement: Data is contained within the article. Supporting data and figures are provided in Supplementary Materials.

Acknowledgments: We thank Samuel Tehuacanero-Cuapa for SEM image acquisition.

Conflicts of Interest: The authors declare no conflict of interest.

References

1. Whitesides, G.M.; Grzybowski, B. Self-assembly at all scales. *Science* **2002**, *295*, 2418–2421. [CrossRef] [PubMed]
2. Iyer, A.S.; Paul, K. Self-assembly: A review of scope and applications. *IET Nanobiotechnol.* **2015**, *9*, 122–135. [CrossRef] [PubMed]
3. Vogel, N.; Retsch, M.; Fustin, C.-A.; del Campo, A.; Jonas, U. Advances in colloidal assembly: The design of structure and hierarchy in two and three dimensions. *Chem. Rev.* **2015**, *115*, 6265–6311. [CrossRef] [PubMed]
4. Zhang, H.; Bu, X.; Yip, S.; Liang, X.; Ho, J.C. Self-assembly of colloidal particles for fabrication of structural color materials toward advanced intelligent systems. *Adv. Intell. Syst.* **2020**, *2*, 1900085. [CrossRef]
5. Li, Z.; Fan, Q.; Yin, Y. Colloidal self-assembly approaches to smart nanostructured materials. *Chem. Rev.* **2022**, *122*, 4976–5067. [CrossRef]
6. Poon, W.C.K.; Pusey, P.N. Phase transition of spherical colloids. In *Observation, Prediction, and Simulation of Phase Transitions in Complex Fluids*; Baus, M., Rull, L.F., Ryckaert, J.-P., Eds.; Kluwer Academic Publishers: Dordrecht, The Netherlands, 1995.
7. Poon, W.C.K. *Colloidal Suspensions*; Clarendon Press: Oxford, UK, 2012.
8. Zaccarelli, E. Colloidal gels: Equilibrium and non-equilibrium routes. *J. Phys. Condens. Matter* **2007**, *19*, 323101. [CrossRef]
9. Pusey, P.N.; Zaccarelli, E.; Valeriani, C.; Sanz, E.; Poon, W.C.K.; Cates, M.E. Hard spheres: Crystallization and glass formation. *Phil. Trans. R. Soc. A* **2009**, *367*, 4993–5011. [CrossRef]
10. Lu, P.J.; Weitz, D.A. Colloidal particles: Crystals, glasses, and gels. Annu. Rev. Condens. *Matter Phys.* **2013**, *4*, 217–233.
11. Lopez-Godoy, S.; Díaz-Leyva, P.; Kozina, A. Self-assembly in binary mixtures of spherical colloids. *Adv. Coll. Inter. Sci.* **2022**, *308*, 102748. [CrossRef]
12. Sacanna, S.; Pine, D.J. Shape-anisotropic colloids: Building blocks for complex assemblies. *Curr. Opin. Coll. Inter. Sci.* **2011**, *16*, 96–105. [CrossRef]
13. Dugyala, V.R.; Daware, S.V.; Basavaraj, G.M. Shape anisotropic colloids: Synthesis, packing behavior, evaporation driven assembly, and their application in emulsion stabilization. *Soft Matter* **2013**, *9*, 6711. [CrossRef]
14. Lee, K.J.; Yoon, J.; Lahann, J. Recent advances with anisotropic particles. *Curr. Opin. Coll. Inter. Sci.* **2011**, *16*, 195–202. [CrossRef]
15. Li, T.; Lilja, K.; Morris, R.J.; Brandani, G.B. Langmuir–Blodgett technique for anisotropic colloids: Young investigator perspective. *J. Coll. Inter. Sci.* **2019**, *540*, 420–438. [CrossRef]
16. Krishnamurthy, S.; Kalapurakal, R.A.M.; Mani, E. Computer simulations of self-assembly of anisotropic colloids. *J. Phys. Condens. Matter* **2022**, *34*, 273001. [CrossRef]
17. Furst, E.M. Directed self-assembly. *Soft Matter* **2013**, *9*, 9039–9045. [CrossRef]
18. Yang, S.-M.; Kim, S.-H.; Lima, J.-M.; Yi, G.-R. Synthesis and assembly of structured colloidal particles. *J. Mater. Chem.* **2008**, *18*, 2177–2190. [CrossRef]
19. Cuetos, A.; Martínez-Haya, B. Columnar phases of discotic spherocylinders. *J. Chem. Phys.* **2008**, *129*, 214706. [CrossRef]
20. Fejer, S.N.; Chakrabarti, D.; Wales, D.J. Self-assembly of anisotropic particles. *Soft Matter* **2011**, *7*, 3553–3564. [CrossRef]
21. Kao, P.-K.; Solomon, M.J.; Ganesan, M. Microstructure and elasticity of dilute gels of colloidal discoids. *Soft Matter* **2022**, *18*, 1350–1363. [CrossRef]
22. Lu, J.; Bu, X.; Zhang, X.; Liu, B. Self-assembly of shape-tunable oblate colloidal particles into orientationally ordered crystals, glassy crystals and plastic crystals. *Soft Matter* **2021**, *17*, 6486–6494. [CrossRef]
23. Hsiao, L.C.; Schultz, B.A.; Glaser, J.; Engel, M.; Szakasits, M.E.; Glotzer, S.C.; Solomon, M.J. Metastable orientational order of colloidal discoids. *Nat. Commun.* **2015**, *6*, 8507. [CrossRef] [PubMed]
24. Doshia, N.; Zahr, A.S.; Bhaskar, S.; Lahann, J.; Mitragotri, S. Red blood cell-mimicking synthetic biomaterial particles. *Proc. Natl. Acad. Sci. USA* **2009**, *106*, 21495–21499. [CrossRef] [PubMed]

25. Chen, J.; Clay, N.; Kong, H. Non-spherical particles for targeted drug delivery. *Chem. Eng. Sci.* **2015**, *125*, 20–24. [CrossRef] [PubMed]
26. Kapate, N.; Clegg, J.R.; Mitragotri, S. Non-spherical micro- and nanoparticles for drug delivery: Progress over 15 years. *Adv. Drug Delivery Rev.* **2021**, *177*, 113807. [CrossRef]
27. de Folter, J.W.J.; Hutter, E.M.; Castillo, S.I.R.; Klop, K.E.; Philipse, A.P.; Kegel, W.K. Particle shape anisotropy in Pickering emulsions: Cubes and peanuts. *Langmuir* **2014**, *30*, 955–964. [CrossRef]
28. Ding, T.; Liu, Z.-F.; Song, K.; Clays, K.; Tung, C.-H. Photonic crystals of oblate spheroids by blown film extrusion of prefabricated colloidal crystals. *Langmuir* **2009**, *25*, 10218–10222. [CrossRef]
29. Cho, Y.-S.; Kim, Y.K.; Chung, K.C.; Choi, C.J. Deformation of colloidal crystals for photonic band gap tuning. *J. Dispers. Sci. Technol.* **2011**, *32*, 1408–1415. [CrossRef]
30. Florea, D.; Wyss, H.M. Towards the self-assembly of anisotropic colloids: Monodisperse oblate ellipsoids. *J. Coll. Int. Sci.* **2014**, *416*, 30–37. [CrossRef]
31. Ahn, S.J.; Ahn, K.H.; Lee, S.J. Film squeezing process for generating oblate spheroidal particles with high yield and uniform sizes. *Colloid Polym. Sci.* **2016**, *294*, 859–867. [CrossRef]
32. Courbaron, A.-C.; Cayre, O.J.; Paunov, V.N. A novel gel deformation technique for fabrication of ellipsoidal and discoidal polymeric microparticles. *Chem. Commun.* **2007**, *2007*, 628–630. [CrossRef]
33. Mathaes, R.; Winter, G.; Besheer, A.; Engert, J. Non-spherical micro- and nanoparticles: Fabrication, characterization and drug delivery applications. *Expert Opin. Drug Deliv.* **2015**, *12*, 481–492. [CrossRef] [PubMed]
34. Doshi, N.; Prabhakarpandian, B.; Rea-Ramsey, A.; Pant, K.; Sundaram, S.; Mitragotri, S. Flow and adhesion of drug carriers in blood vessels depend on their shape: A study using model synthetic microvascular networks. *J. Control. Rel.* **2010**, *146*, 196–200. [CrossRef] [PubMed]
35. Yoo, J.-W.; Doshi, N.; Mitragotri, S. Endocytosis and intracellular distribution of PLGA particles in endothelial cells: Effect of particle geometry. *Macromol. Rapid Commun.* **2010**, *31*, 142–148. [CrossRef] [PubMed]
36. Bibette, J. Depletion interactions and fractionated crystallization for polydisperse emulsion purification. *J. Coll. Int. Sci.* **1991**, *147*, 474–478. [CrossRef]
37. Moroi, Y.; Motomura, K.; Matuura, R. The critical micelle concentration of sodium dodecyl sulfate-bivalent metal dodecyl sulfate mixtures in aqueous solutions. *J. Coll. Int. Sci.* **1974**, *46*, 111–117. [CrossRef]
38. Peters, D. Ultrasound in materials chemistry. *J. Mater. Chem.* **1996**, *6*, 1605–1618. [CrossRef]
39. Pratap-Singh, A.; Guo, Y.; Ochoa, S.L.; Fathordoobady, F.; Singh, A. Optimal ultrasonication process time remains constant for a specific nanoemulsion size reduction system. *Sci. Rep.* **2021**, *11*, 9241. [CrossRef]
40. Gupta, A.; Eral, H.B.; Hatton, T.A.; Doyle, P.S. Controlling and predicting droplet size of nanoemulsions: Scaling relations with experimental validation. *Soft Matter* **2016**, *12*, 1452–1458. [CrossRef]
41. Belgheisi, S.; Motamedzadegan, A.; Milani, J.M.; Rashidi, L.; Rafe, A. Impact of ultrasound processing parameters on physical characteristics of lycopene emulsion. *J. Food Sci. Technol.* **2021**, *58*, 484–493. [CrossRef]

colloids and interfaces

MDPI

Review

Nose-to-Brain Targeting via Nanoemulsion: Significance and Evidence

Shashi Kiran Misra [1] and Kamla Pathak [2,*]

1 School of Pharmaceutical Sciences, CSJM University, Kanpur 208024, Uttar Pradesh, India
2 Faculty of Pharmacy, Uttar Pradesh University of Medical Sciences Saifai, Etawah 206130, Uttar Pradesh, India
* Correspondence: deanpharmacy2015@gmail.com

Abstract: Background: Non-invasive and patient-friendly nose-to-brain pathway is the best-suited route for brain delivery of therapeutics as it bypasses the blood–brain barrier. The intranasal pathway (olfactory and trigeminal nerves) allows the entry of various bioactive agents, delivers a wide array of hydrophilic and hydrophobic drugs, and circumvents the hepatic first-pass effect, thus targeting neurological diseases in both humans and animals. The olfactory and trigeminal nerves make a bridge between the highly vascularised nasal cavity and brain tissues for the permeation and distribution, thus presenting a direct pathway for the entry of therapeutics into the brain. **Materials:** This review portrays insight into recent research reports (spanning the last five years) on the nanoemulsions developed for nose-to-brain delivery of actives for the management of a myriad of neurological disorders, namely, Parkinson's disease, Alzheimer's, epilepsy, depression, schizophrenia, cerebral ischemia and brain tumours. The information and data are collected and compiled from more than one hundred Scopus- and PubMed-indexed articles. **Conclusions:** The olfactory and trigeminal pathways facilitate better biodistribution and bypass BBB issues and, thus, pose as a possible alternative route for the delivery of hydrophobic, poor absorption and enzyme degradative therapeutics. Exploring these virtues, intranasal nanoemulsions have proven to be active, non-invasiveand safe brain-targeting cargos for the alleviation of the brain and other neurodegenerative disorders.

Keywords: non-invasive; blood–brain barrier; intranasal; nanoemulsions; neurological disorders

Citation: Misra, S.K.; Pathak, K. Nose-to-Brain Targeting via Nanoemulsion: Significance and Evidence. *Colloids Interfaces* **2023**, *7*, 23. https://doi.org/10.3390/colloids7010023

Academic Editors: César Burgos-Díaz, Mauricio Opazo-Navarrete and Eduardo Morales

Received: 17 January 2023
Revised: 12 March 2023
Accepted: 14 March 2023
Published: 17 March 2023

1. Introduction

Brain diseases, including dementia, epilepsy, migraine, autoimmune disorders (Parkinson's, Alzheimer's and prion's disease), brain tumours and acute ischemic brain haemorrhages, require extensive clinical care due to a significantly high rate of morbidity and mortality worldwide [1]. Most of the developed brain-targeting medications relieve symptomatic brain deregulatory functions and are inefficient in providing satisfactory therapeutic responses. The major issues overlaid are (i) lipophilic blood–brain barrier (BBB), (ii) complexity of the microenvironment of the brain and (iii) abnormal protein status. Central nervous system (CNS)vessels containing arterioles and venules are regular, continuous and non-apertured. These vessels are involved in the regulation and exchange of ions and molecules throughout the braincells. The exclusive morphology of CNS vessels and the barrier function of the BBB guard the brain from the entry of antigens, toxins and pathogens. The BBB diverts blood from the interstitial fluid and serves as an efficient barricade for the diffusion of most of the actives to reach the brain receptors of the CNS. Functionally, it is a dynamic regulator that transports nutrients and checks the entry of heavy and undesirable (lipophilic) molecules across the extracellular fluid of the brain. Lipophilic molecules with an optimum Log P (approximately 1.5–2.7) and molecularweight of 600 Daltons can freely permeate the BBB [2].

The brain endothelial cells (BECs) form the walls of brain blood vessels and are highly polarized in contrast to the endothelial cells of other tissues. BECs are jointed together

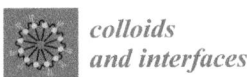

by tight junctions that again confine the paracellular flux of ions and limit transcellular exchange between blood and brain cells, i.e., transcytosis and pinocytosis. BECs coordinate a series of metabolic, transportation and physiological functions with the interaction of several neural, vascular and neural components involved in the management of health and diseased conditions of the body. BBB-adorned p-glycoprotein efflux transporters also restrict the movement of bioactive/macromolecules and their receptor binding for a desired pharmacological response. Expression of both efflux and influx transporters between blood and brain cells ismediated via BECs. Efflux transporters (such as p-gp) that removetoxins tend to diffuse cell membranes passively, while the influx transporters are a kind of carrier that engage to deliver nutrients and ions to the brain cells [3]. Apart from being an active cellular self-defence barrier, the BBB strictly monitors the CNS microenvironment and communicates and acclimatises with the conduct of CNS cells in the progression of brain disorders [4].

These attributes of BBB create hurdles for the permeation of therapeutic or bioactive agents to reach the brain tissues, thus exhibiting obstructions to combat CNS diseases. A report of WHO healthcare statistics states that approximately 1.5 million people are sufferers of CNS disorders, including autoimmune Parkinson's, Alzheimer's diseases and schizophrenia [5].

Although the current drug delivery approaches have displayed a vibrant picture of effective CNS treatments with survival rates surging, still, there are unsolved issues for the complete cure/therapy of most CNS disorders. There is definitely a need for an advanced therapeutic system that enables the potential crossing of the BBB at an adequate level to attain the desired pharmacological action. The BBB limits the entry of 98% of the low molecular weight molecules and hence, drastically reduces bioavailability. A report retrieved from the comprehensive medicinal chemistry database signifies that a mere 5% of therapeutics out of 7000 with 357 Dalton molecular weight and 2.57 Log P (partition coefficient) have the potential to penetrate the BBB and exhibit potent action for mitigation of insomnia, depression and schizophrenia [6]. Invasive strategies, including intra parenchymal, intracranial and intra-cerebro ventricular injections, are administered for direct drug delivery to the brain. These local strategies are interlinked and emphasize the use of electromagnetic field and ultrasound approaches. These are preferred to treat psychological and neurological disorders but are painful and risky as well [7]. Hence, a non-invasive approach such as the nose-to-brain delivery pathway is suggested that bypasses the BBB, reduces toxicity and delivers the therapeutics at the target site. The intranasal delivery route facilitates the direct delivery of the drug into the cerebrospinal fluid following the olfactory path [8].

Complexity of Nose-to-Brain Drug Delivery Path

The nose is an integral human olfactory/respiratory part and contains a 60 μm thick area of 1.25–2 m^2. The nasal septum divides the nasal cavity into the nasal vestibule, olfactory and respiratory regions. The nose entrance or nasal vestibule is covered with squamous epithelium that contains vibrissae (nasal hair) and oil glands. The respiratory region comprises the maximum surface area and holds the ciliated respiratory epithelium and vascularised nasal turbinate made up of erectile and sinusoid tissues [9]. The olfactory part is positioned on the rooftop of the nasal cavity and lined with olfactory epithelium (pseudostratified). Adhered olfactory nerves directly connect with CNS by circumventing the BBB. The olfactory region has a surface area of 2–10 cm^2. After crossing the olfactory epithelium, a therapeutic agent transports intracellularly along the associated olfactory nerve. Lipophilic molecules follow paracellular passive diffusion, while hydrophilic drugs move through a carrier-mediated transport way [10].

The nasal region contains two nerve terminations, i.e., olfactory and trigeminal. Both neuropathways originate from the nasal cavity at the olfactory neuroepithelium and terminate in the brain.

Interestingly, these routes are the single way that links the brain to the exterior environment and transports countless neurotherapeutic agents. A diverse range of neurotherapeutics (macro and low molecular weight) is generously distributed to the CNS via this route [11]. Figure 1 portrays the typical route, including diverse transport systems and nerves for the effective delivery of therapeutic agents following the nose-to-brain pathway. Moreover, the associated blood–nerve barrier comprising the end oneurial microvessels and perineurium is involved in the regulation of the permeability of ions and molecules from both these pathways into the CNS. Hence, the blood–nerve barrier is equally important, and proper attention must be specified while designing a nose-to-brain drug delivery system. The non-invasive intranasal pathway bypasses both the BBB and first-pass metabolism. Hence, it is assumed to be superior by virtue of myriad merits, including lower dose strength, ease of administration, safety, instant therapeutic action and reduced systemic toxicity with improved patient compliance [12]. The focus of intranasal delivery is to transport brain-targeted therapeutics in the desired concentration at a predetermined rate. Drug metabolism, degradation and clearance can influence the efficiency of the drug at the site [13].

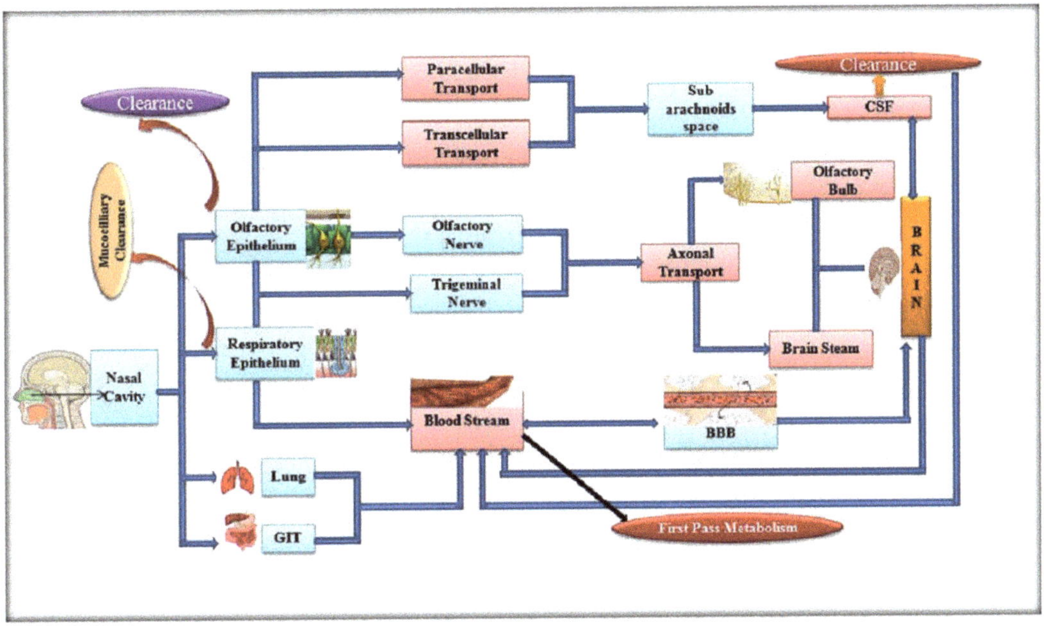

Figure 1. Complex pathway involved in the transport of therapeutics through nose-to-brainroute.

The enhanced surface area of the stratified squamous naso-epithelium is extremely permeable, porous and vascularised, hence permitting fast drug absorption to occur at a smaller dose. Nasal drug delivery systems do not need any drug modification or coupling with a carrier for the distribution and effectiveness of therapeutic agents [14].

2. Nanoemulsions

Researchers are consistently engaged in the development of promising and potential non-invasive brain drug delivery systems. To achieve the targeted CNS response, various scientific investigations have been reported that define the olfactory and trigeminal routes and successfully deliver CNS medicaments following the nose-to-brain pathway. Emulsions are colloidal systems with two distinct phases, in which one is dispersed and another is dispersion media. Their droplet or globular dimension typically varies from 200 nm to 100 μm. Countless bioactive, nutraceutical and therapeutic agents can be feasibly incorporated or encapsulated in this biphasic system for conventional and controlled effects as well [15]. However, conventional emulsions exhibit issues such asphysical instability (aggregation and gravitational separation) and a poor ability to control the release of the entrapped active pharmaceutical agent (API) at the target site.

Nanoemulsions are formulated via two broad methodologies, low-energy (phase inversion) and high-energy methods (ultrasonication, microfluidic, high-pressure homogenization). The low-energy method is based on phase inversion temperature and produces a smaller globular dimension with the involvement of less energy, whereas the high-energy methods are reliant on energetic mechanical strategies and disruptive forces for the generation of smaller oil globules.

A few advanced emulsions, including multilayer emulsion, multiple emulsion, nanoemulsion, pickering emulsion and high internal phase emulsion, have been introduced for the delivery of bioactive agents. Table 1 compiles salient features of advanced types of emulsion as drug delivery systems. However, these advanced emulsions are not cost-effective and are tedious to formulate.

Table 1. Salient features and drug delivery applications of advanced emulsions.

Advanced Emulsion	Structure	Salient Features	Applications	Ref.
Nanoemulsion	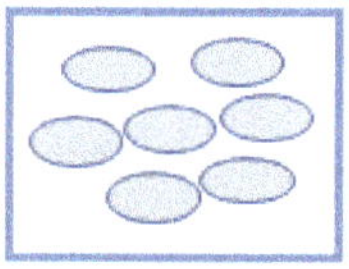	Nanoscaled droplets provide thermodynamically (kinetically stable) colloidal dispersions. Controlled and effective lipophilic bioactive agents can be transported.	Widely explored in food industry for its superb stability and shelf life of encapsulated colouring/flavouring components/nutraceutical and bioactive agents	[16]
Multilayer emulsion	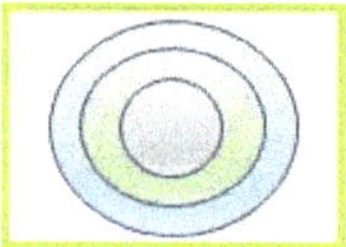	It contains hydrated lipophilic multi-molecular layers around oil (dispersed phase). Formed multilayer film provides adequate viscosity; hence, stable formulation is obtained.	Delivery and bio-accessibility of fat-soluble curcumin and other pH-sensitive bio-active agents	[17]
Multiple emulsion	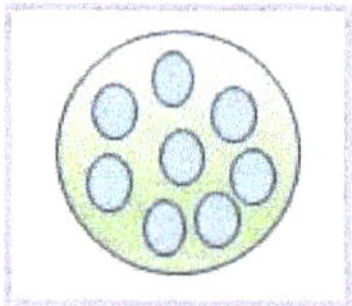	These are emulsions within emulsion: i.e., o/w/o and w/o/w. Multiple emulsion limits bioactive degradation and is suitable for drug delivery via diverse routes such as ocular, oral, parenteral, nasal and pulmonary.	Delivery of low caloric food products, designing of easily spreadable creams, controlled and targeted dosages	[18]
Pickering emulsion	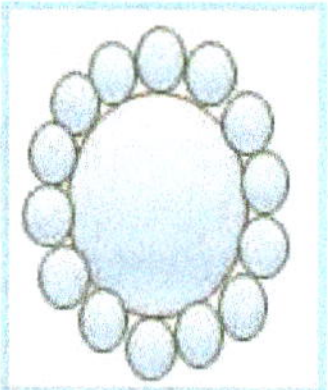	Instead of surface-active agents, solid particles confine at the interface of both phases and provide stability. These emulsions are more stable, tunable, feasible, stimuli-responsive and less toxic compared to the traditional ones.	Photocatalysis, protein recognition, purification of water, cosmeceutical products	[19]
High internal phase emulsion	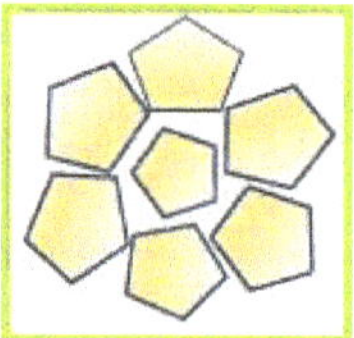	HIPEs have high internal phase volume and are used to swap hydrogenated fats added in food stuffs. Hence, they improve stability and functionality and stabilize the system. The volume fraction is more than 0.74, ensuing formation of solid dispersed phase or polyhedral in the form of thin films.	Preparation of food stuffs, fuel and cosmetics	[20]

These lipophilic systems have a fine globular size, preferably in nanodimensions. These structures are well absorbed through the mucosal layer and are hence suitable for site-specific drug delivery. Among all, the nanoemulsions are the most successful among other advanced systems due to their fine size, leveraged surface area, the onset of action, biodistribution and stability. Both the o/w and w/o types of emulsions are modulated to target naso-mucosa for drug delivery across the brain tissues. Exceptionally high encapsulation of lipophilic drugs in o/w type nanoemulsion offers enhanced solubility, better absorption and improved bioavailability with minimum enzymatic degradation [21]. Nanoemulsions are also preferred for the formulation of mucoadhesive systems as these facilitate a long residence time and enhanced nasal absorption with limited clearance. The o/w type of nanoemulsions is better suited for brain-targeting delivery systems owing to their small size, lipophilicity and efficiency in permeating the BBB. Despite the merits offered, nose-to-brain delivery faces significant challenges and limits its frequent use. Lower bioavailability, susceptibility to enzymatic degradation and higher nasal clearance present the grey side of this route.

Moreover, mucociliary clearance, a smaller volume of the nasal cavity and its limited surface area further limit the absorption of sensitive drugs aimed to target brain tissues. Therefore, different nanoengineered mucoadhesive and in situ nanoemulgels (thermo-responsive and pH-responsive) have been designed to provide prolonged and effective drug delivery.

2.1. Oil and Globule Size

Newly discovered drug molecules display poor aqueous solubility, which hence, severely affects the pharmacokinetic and pharmacodynamic characters. Oil, an integral part of nanoemulsion, resolves this issue and favours the maximum solubility of the drug entity. One can choose vegetable oil, triglycerides/diglycerides and essential polyunsaturated fatty acids depending upon the lipophilicity of the therapeutic agent. Higher oil concentration increases the globule scale of nanoemulsion and causes poor permeation across nasal mucosa [22]. Reports demonstrated that linolenic acid with two *cis*-double bonds and eighteen carbon monocarboxylic fatty acids couldefficiently cross the BBB, suggesting that entry of these oils in brain cells is carried out in a selective and distinct manner. Hence, it is suitable for the formulation of nanoemulsion [23]. For brain targeting, the lipidic component (5–20%) is selected for the formulation of the o/w type of nanoemulsion. Solubility of the oil highly matters in the selection of oil, as it must solubilize the drug molecule for efficient action at the brain target site. Usually, lipids, including cottonseed oil, coconut oil, sesame oil and soybean oil, are preferred alone to their mixture [24]. Further, added materials (surfactant and co-surfactant), process parameters (viscosity ratio, mixing time), rheology (shearing stress and shear rate) and type of methodology (microfluidization, phaseinversion, ultrasonication and high-pressure homogenization) decisively control the globule size distribution in nanoemulsion [25]. As discussed earlier, the nasal route comprises the olfactory and trigeminal pathways that are the foremost channels for brain-targeting drug delivery systems. Anatomically, the typical diameter of the olfactory axon in a human goes up to 700 nm [26]. Thus, the oil globular size should be kept within this limit to maintain adequate transportation and retention as well. Reports state that the smaller the oil globular dimension, i.e., 80, 100 and 200 nm, the higher the mucosal retention property for 16, 14 and12 h, respectively, in the arena of the nostril, exhibiting sluggish mucociliary clearance compared to the larger-scaled particles. Moreover, larger particles (greater than 900 nm) are also unable to reach the olfactory bulb and havepoor bioavailability inthe region of the brain [27].

Further, the type of oils and their properties are also significant while aiming brain targeting through nanoemulsion. The literature envisages that linolenic acid, omega-6 fatty acids, oleic acid and linolenic acids are frequently employed for the development of nanoemulsion to be delivered through the intranasal route. Ahmad et al. have developed an intranasal nanoemulsion containing amiloride (drug), oleic acid (oil) and Tween 20 and Carbitol as surfactant and cosurfactant. The selection of oil, its concentration and developed globular size played a prominent role in enhancing bioavailability in brain tissues. Outcomes suggested that an average globular size of 100 nm with a polydispersity index of 0.231 and negative zeta potential (−9.83 mV) assisted hurdle-free transportation and validated the biodistribution. Formulated nanoemulsion exhibited high drug content. i.e., ~98.38% at pH 6.4 ± 0.18. Both nose-to-brain transport (approximately 586%) and brain targeting efficacy (1992%) in rodents explained higher bioavailability compared to the intravenous route, thus suggesting the efficacy of the drug in epileptic sufferers [28].

2.2. Surfactant and Stability

Surfactants not only reduce the interfacial tension between two distinct phases but also manage the stability of the biphasic colloidal system by preventing phase separation [29]. Further, the surfactants improve the solubility of therapeutics; alter the fluidity of the epithelium cell junctions, thus enhancing the permeation of these drugs through the nasal–mucosal pathway.

Additionally, surfactants may change the structural integrity of mucosal lining, so the selection of a concentration of surfactant is crucial while preparation of nanoemulsion aimed atbrain targeting [30].Reports suggest that surfactants such as phosphatidylcholine, bilesalt, chitosan, starch, Tweens (tween 60, 80), sodium dodecyl sulphate and Span (Span 60, 80),poloxamers, casein, polyethylene glycol containing block polymers and few proteins and lipids are frequently utilized for the development of nanoemulsions [31].

Kinetically stable nanoemulsion takes a long time to separate out in its different phases. However, diverse establishment steps, including creaming, cracking, coalescence, flocculation and Ostwald ripening, are often befall with emulsions. The globules come close due to electrostatic attraction force and form a single big entity that mediates flocculation-type destabilization. At the same time, coalescence occurs due to the merging of oil globules or water droplets. In the case of nanoemulsion, steric stabilization is stronger by virtue of the adsorbed coating of surfactant on the droplet, resulting in an enhanced repulsive maximum that checks instabilities. Formulation factors such as surfactant concentration, ionic strength, chemical configuration, solubility and temperature distinctly affect the stability of nanoemulsion. The rate of destabilization is highly concerned with increased concentration of surfactant and more diffusion of oil that reduce Gibbs elasticity. Hence, surfactant plays a critical role in a dual manner; i.e., the formulation and stability of nanoemulsion. Customized polymeric emulsifiers are suggested while preparing nanoemulsions owing to their flexibility, tailored size and hydrophobicity [32].

Cosurfactants assist surfactants in the reduction of boundary tension or interfacial tension up to the desired level. The mechanism or role-play of cosurfactants is similar to the surfactant; hence, an optimized concentration and desired HLB (hydrophilic–lipophilic balance) offering compound should be selected by constructing a ternary phase diagram (concentration of oil, surfactant and cosurfactant). Chiral alcohols, tween 80, polyethylene glycols 600, butan-1-ol and sorbitol are preferred as cosurfactant while formulating nanoemulsions [33]. Zeta potential poses a prominent role in maintaining the thermodynamic stability of nanoemulsions. Reports demonstrate that optimum zeta potential value. i.e., ±30 mV offers an electrostatically stabilized drug delivery system. Further, the retention time in the nasal cavity was also controlled by this parameter. The nasal mucosa is adorned with negatively charged mucin at its surface; hence, positive-charge-containingnanoemulsionsaresuitableforgoodadherence in such cases [34]. Several of the literature reports depict most of the brain-targeted nanoemulsions with a negative charge and zeta potentialof −10 mV [35].

3. Intranasal Nanoemulsions for the Management of Brain Diseases

The intranasal pathway (olfactory and trigeminal nerves) lets the entry of various bioactive agents circumvent the hepatic first-pass effect, thus targetingneurological diseases [36]. Several investigations and research have suggested the vital role ofdrug-loaded nanoemulsions in the mitigation of countless brain disorders proves to be a potential alternative to oral drug delivery systems. Further, mucoadhesive polymer enriched nanoemulsion prolongs residence time and therapeutic effect by weakening rapid clearance from the nasal mucosa. Diedrich et al. (2022) developed a chitosan-coated luteolin nanoemulsion for effective brain targeting after intranasal administration in a neuroblastoma sufferer child. Obtained nanoemulsion possessed ~68 nm average particle size and positive zeta potential (~13 mV). The formulation exhibited 85.5% encapsulation efficiency and 72 h prolonged in vitro release of luteolin. Ex vivo performed Baker–Lonsdale kinetic model depicted approximately six times higher permeation across the nasal mucosa. Moreover, pharmacokinetic studies of the single dose administered intranasal nanoemulsion revealed a tentimes higher drug half-life and nearly a four times higher luteolin biodistribution in brain tissues that further suggested potential usage of developed chitosan-coated luteolin nanoemulsion for the management of brain neuroblastoma. A performed in vivo study exhibited complete inhibition of the growth of blastoma cells at a concentration of 2 μM [37,38].

3.1. Nanoemulsions for Neurological Disorders

Neurological disorders include a broad range of disabilities that progressively impair the nervous system associated with vital functions of the body. i.e., mobility, sensation, coordination, reasoning, and learning [39]. Epilepsy, autism, Parkinson's, Alzheimer's and other neuromuscular disorders are just a few to be named in this category. These specific ailments require precise delivery of therapeutic agents inside the brain for effective results. BBB presents a formidable hurdle that limits the efficacy of conventional formulations [40]. For instance, acetylcholinesterase inhibitors such as rivastigmine, memantine and galantamine have the potential to manage Alzheimer's disease but possess poor brain targeting owing to theirerratic pharmacokinetic and pharmacodynamic profiles [41]. Nanosized nanoemulsion and its structural architecture facilitate site-specific targeted drug release with minimum adverse effects. Figure 2 summarizes the role of nanoemulsions in the management of brain disorders.

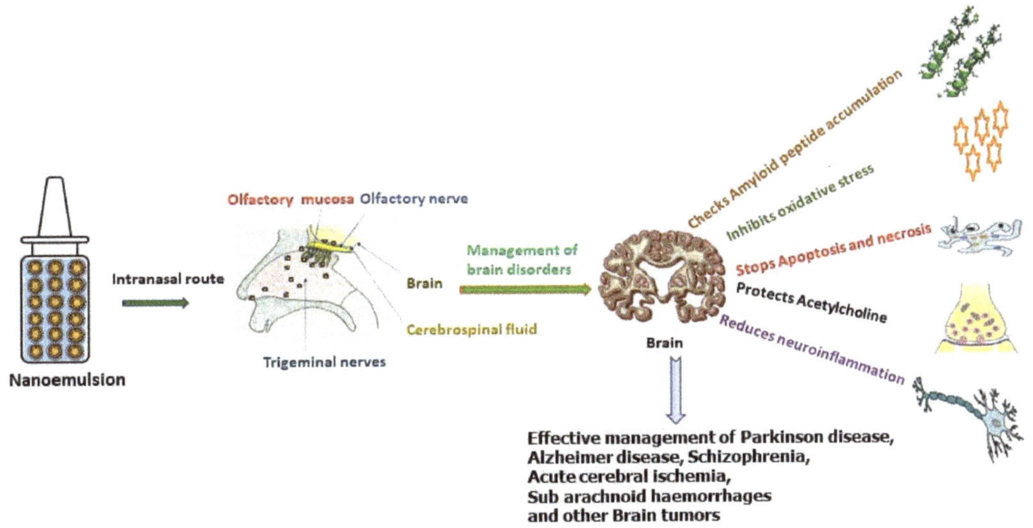

Figure 2. Intra nasal delivered nanoemulsion and management of brain disorders.

Jiang et al. (2022) explored herbal huperzine for the management of the neurological issue. i.e., Alzheimer's disorder. Huperzine is obtained from *Huperzia serrata*, a Chinese club moss, and has poor brain transportation. To upgrade its properties, we formulated huperzine nanoemulsion (hup-NE) modified with lactoferrin (Lf-hup-NE) that could transport the drug to the affected sites of the brain via olfactory–trigeminal nerves. Performed pharmacokinetic studies displayed greater mean residence time of modified Lf-hup-NE (4.07 h) compared to non-modified nanoemulsion (3.03 h). Similarly, Cmax and $AUC_{0\text{–}t}$ of Lf-hup-NE were also higher. i.e., ~52.29 ng/mL and ~245.09 ng/mL X h compared to developed non-modified nanoemulsions ~50.54 ng/mL and 123.63 ng/mL Xh, respectively. A rat model olfactory nerve transaction was designed to observe nanoemulsion transportation via the nose-to-brain route. Of the nanoemulsion, 50µLwas intranasal administered both in olfactory nerve transacted rats and normal rats, and in vivo fluorescence images of treated rats were collected that depicted retention of the drug in brain regions. The results confirmed successful drug absorption and transportation of hup-NE and modified Lf-hup-NE via blood circulation that implicated both direct and indirect drug delivery. P2 signals intensity of modified Lf-hup-NE was high owing to its greater efficiency of translocation in brain areas through intranasal delivery. Further, the developed modified Lf-hup-NE had the potential to inhibit P-gp efflux protein and enhanced drug concentration there. This novel formulation can be exploited for better transportation and accumulation of herbal huperzine in brain cells and can achieve better therapeutic responses [42].

One of the progressive neurodegenerative motor neuron disorders is amyotrophic lateral sclerosis. It is characterized by symptoms such as the depletion of upper and lower motor neurons. A BCS class II drug, riluzole, has limited bioavailability (60%) due to poor penetration across BBB. This drug is indicated for the management of amyotrophic lateral sclerosis (ALS). Parikh et al. (2016) formulated riluzole containing o/w nanoemulsion by phase titration method. Sefsol 218 and tween 80/carbitol (1:1) were employed as oil substitutes and surfactants, respectively. The developed nanoemulsion was thermodynamically stable with a drop size of ~23 nm. Further, the formulation was free of nasal ciliotoxicity and significantly increased brain uptake of riluzole ($p < 4.1 \times 10^{-6}$) via intranasal delivery on comparing with oral nanoemulsion to the Wistar albino rats. This novel nanoemulsion has displayed a promising alternative approach for the treatment of people living with ALS [43].

Recently, nose-to-brain delivery of bromocriptine mesylate- and glutathione-embedded nanodimension emulsion was designed by Ashhar et al. (2022) for the effective treatment of another distinguished neurodegenerative Parkinson's disease. This fatal disorder is provoked by the generation of free radicals in neurons (dopaminergic); consequently, oxidative stress-induced neuron degradation occurs. The researchers designed intranasal dosage for containing bromocriptine mesylate and glutathione-embedded nanoemulsion to relieve oxidative stress. Performed DPPH radical scavenging analysis revealed enhanced antioxidant action due to the combined effect of both bromocriptine mesylate and glutathione present in nanoemulsion. The formulated nanoemulsion was estimated for depth of permeation with confocal laser scanning microscopy after intranasal administration, which resulted in superior penetration to the brain cells. Further, the pharmacokinetic study estimated $AUC_{0\text{–}8}$ of nanoemulsion that revealed a higher concentration of both compounds in brain regions of the Wistar rat model after intranasal administration. The nasal ciliotoxicity study in Wistar rats explained the biocompatibility of the formulated nanoemulsion. Thereafter, the biochemical study displayed a reduced level of interleukin-6, alpha tumour necrosis factor and thiobarbituric acid, which are reactive substances, and concluded that bromocriptine mesylate- and glutathione-loaded nanoemulsion has the potential to overcome oxidative stress level in persons living with Parkinson's disease [44].

Another chronic CNS complaint: 'epilepsy' is frequently characterized by instant senseless seizures due to neurons' electric instabilities in the brain region. BCS class II topiramate(anticonvulsant) has been employed for the clinical management of partial and generalized seizures. Its poor bioavailability on oral administration is due to poor entry across BBB. Being a substrate of P-gp transporter, we reported the limited efficacy against epilepsy. An o/w nanoemulsion was prepared utilizing Capmul MCM C8 (2% *w*/*w*) by

Patel and coworkers (2020) for the improvement of the brain delivery of topiramate. A blend of 32% surfactant and co-surfactant Tween 20 and carbitol at a ratio of 2:1 was included to customize the globular size, PDI, zeta potential and viscosity. The resulting nanoemulsion possessed a ~4.73 nm mean particle size with a stability of six months. The brain uptake efficiency of the intranasally delivered piramate contained a nanoemulsion that was quite higher ($p < 1.8 \times 10^{-8}$) compared to the orally delivered nanoemulsion. The pharmacokinetic studies in Wistar albino rats depicted enhanced bioavailability with minimum adverse effects after delivering the nanoemulsion through the intranasal route [45]. Table 2 gathers recent studies on developed nanoemulsions and their applications for the treatment of various brain diseases.

Table 2. Recently developed nanoemulsions for the management of diverse neurologicaldisorders.

A. Alzheimer Disease				
Therapeutic Agent	**Objective**	**Study Model**	**Outcome**	**Ref.**
Memantine	To bypass BBB for effective management of Alzheimer's disease	Particle size, in vitro release and radiolabelling with technetiumpertechnetate (^{99m}Tc).	Average globular size~11 nm. 80% drug release in nasal simulated fluid. ~98% cell viability. Higher uptake of drug (3.6% radioactivity) within 1.5 h in brain of rat.	[46]
Donepezil	To enhance drug delivery in brain	Particle size In vitro release radio labelling with Technetium pertechnetate	~65 nm average globular size, 0.084 poly dispersity index (PDI) and −10.7 mV Zeta potential 99% drug release in phosphatebuffer and 98% in simulated cerebrospinal mediawithin 4 h and 2 h, respectively. Dose-dependent radical scavenging and toxicity	[47]
Huperzine A	Selectively Acetylcholinesterase inhibition via huperzine loaded nanoemulsion modified with lactoferrin	Particle size and zeta potential, in vitromodel (hCMEC/D3cells), drug-targeting index	15.24 average particle size, 0.12 PDI and negative 4.48 mV zetapotential Drug-targetingindexwas 3.2 ± 0.75 Nanoemulsion was uptake by transcytosis and transported via specific multidrug resistance associated proteins transporters	[48]
Resveratrol	Coconut oil-based resveratrol nanoemulsion was targeted for the alleviation of Alzheimer's disease	Effect of process variables, in vitro and in vivo permeation study in goat	Average particle size ~110 nm, −21.13 mV and 88.54% drug release in 8 h. Added coconut oil enhanced permeation efficiency. Optimum drugconcentrationwas 2.64 mg/mL. Higher resveratrol concentration in blood and brain (~1234.87 ng/mL and~5762.30 ng/mL in the brain, respectively, at 2 mg/kg dose. Superior permeation of prepared nanoemulsion compared to suspension in Goat nasal mucosa.	[49]

Table 2. *Cont.*

Therapeutic Agent	Objective	Study Model	Outcome	Ref.
		A. Alzheimer Disease		
		B. Parkinson disease		
Quercetin	Nanoemulsion comprising food-grade quercetin and oil (Capmul MCM NF) was aimed at neuroprotective action	HR-TEM, area electron diffraction studies and in vivo studies (cytotoxic study) and oxidative stress in wild-type *Caenorhabditis elegans* N2 strain	Spherical and ~50 nm particle size nanoemulsion potentially enhanced mitochondrial (3.080 ±1.1) and fat content (2.64± 0.1) and reduced aggregation. Downregulated reactive Oxygen species content in *C elegans* N2 strain. Dose and time-dependent toxicity. i.e., for A549 cells 300 µg/mL in 48 h. No toxicity against lymphocytes.	[50]
Naringenin	Vitamin E adorned naringenin nanoemulsion was aimed at combating Parkinson's disease	Structural property, refractive index and transmittance. Multiple behavioural analysis and oxidative study	Average smaller droplet size (~38.7 nm), 0.14 PDI with optimum zetapotential (−27.4 mV) and 19.6 Pas dynamic viscosity. 1.43 Refractive index and 98.1% transmittance indicated isotropic nature and stability of prepared nanoemulsion Three times high permeation coefficient and flux of nanoemulsion compared to suspension AUC0–48 hfor NE in brain 5345.1 and blood 3777.6 ng/mL X h Intranasally administered NE reversed Parkinson's in 6-OHDA-persuaded rats while given with levodopa	[51]
Selegiline	Intranasal delivery monoamine oxidase B inhibitor (selegiline) nanoemulsion for better treatment of neuro degenerative disease	Quality by design approach for optimization of variables and behavioural study in rats	Optimized nanoemulsion (~61 nm) was quite stable (−34 mV). PDI (~0.203) Transmittance (~99.8%) and refractive index (1.30 ± 0.01) Almost 3.7-fold selegiline drug permeation observed compared to its oral suspension. Significantly improved locomotor activity and muscle coordination in experimental rat model on comparing with oral route.	[52]

Table 2. *Cont.*

A. Alzheimer Disease				
Therapeutic Agent	**Objective**	**Study Model**	**Outcome**	**Ref.**
Hyaluronic acid coloaded with resveratrol and curcumin)	Mucoadhesive polyphenols containing nanoemulsion were targeted for brain delivery	Antioxidant potential, in vitro and ex vivo and in vivo assessment of polyphenols in rat brain	Average particle size ~115 nm with negative zeta potential (23.9 mV). Higher nasal mucosa adhesive potential of the developed nanoemulsion Preserved polyphenolic antioxidant efficacy and protected from their degradation. Controlled diffusion was achieved for 6 h and showed ex vivo permeation efflux for resveratrol (2.86 $\mu g/cm^2$) and curcumin (2.09 $\mu g/cm^2$) across sheep nasal mucosa. Increased amount of two polyphenols in brain	[53]
C. Epilepsy				
Amiloride	To enhance brain bioavailability of nanoemulsion containing diuretic amiloride	3^3 factorial central composite design, Physicochemical parameters, in vitro drug release and antiepileptic effect in mice	Concentration of olei cacid (2.5%), Tween 20 and Carbitol (10%) and sonication time (45 s) were optimised, which resulted in ~89 nm hydrodynamic diameter. ~1.38 refractive index, 6.4± 0.2 pH and ~41 cp viscosity of nanoemulsion ~99% percentage transmittance and ~80% cumulative drug release. Drastically increased nose-to-brain transport (~586%) and efficiency (~1992%) in mice model Improved seizure threshold in epileptic rodent model and induced seizure in mice.	[54]
Letrozole	Nanoemulsion contained aromatase inhibitor letrozole designed for brain delivery while avoiding peripheral responses	Structural analysis, in vitro and ex vivo drug diffusion study and behavioural study	Spherical and smaller particles of mean diameter. i.e., 96 nm and 0.162 PDI. Negative −7.12 zetapotential Prolonged drug release during intranasal administration that results in high concentration letrozole in brain region. Enhanced neuroprotective and antiepileptic effects. Inhibition and diversion of aromatomization and metabolic pathways of testosterone into 17-beta estradiol observed.	[55]

Table 2. *Cont.*

A. Alzheimer Disease				
Therapeutic Agent	**Objective**	**Study Model**	**Outcome**	**Ref.**
Oxcarbazepine	PLGA tri-block polymer admixed oxcarbazepine emulsome (emulsion +liposome) and its thermogel were developed to amplify drug concentration in brain	In vitro study, histopathological analysis, and pharmacokinetic parameters	Concentration dependent rheological behaviour. The emulsome exhibited sustained effect of 81.1% for 24 h. Higher drug transport in brain tissue uptake in rat model compared to drug solution and suspension (Trileptal®). 3818.8 ng/mL and 5699.9 ng/mL C_{max} of nanoemul gels in plasma and brain, respectively. Histopathology study of nasal tissues revealed mild anti-inflammatory response and vascular congestion without toxicity.	[56]
D. Neuroprotection				
Naringenin	To explore bioflavonoid naringenin for its neuroprotective effect owing to its antioxidant and anti-inflammatory virtues	Particle size and zeta potential, neuroprotective assay (Aβ triggered ROS, total tau, amyloid precursor protein) against neuroblastoma cells, i.e., SH-SY5Y	Average droplet size 113.8 nm with narrow PDI~0.312. Percentage transmittance ~97.01% and +12.4 mV zeta potential. Enhanced neurotoxic effects SH-SY5Y cells due to down-regulation of amyloid precursor protein. Nanoemulsion checked amyloid-genesis and reduced level of phosphorylated tau at low concentration of 0.125 μM in comparison with bare drug 25 μM.	[57]
Curcumin	Improvement of solubility and bioavailability issues of curcumin added in nanoemulsion for the novel neuroprotective action	Oxidative stress and mitochondrial complex I activity	Curcumin-loaded nanoemulsion significantly enhanced motor activity at 25 mg/kg dose Lessened lipoperoxidation and improved antioxidant activity in mitochondria while compared to free curcumin. Curcumin-loaded nanoemulsion inhibited Mitochondrial complex 1 activity	[58]
Chia seed oil	Developed nanoemulsion containing Chia seed oil had potential to overcome issues of Parkinson's disease and neuroprotective actions	Motor and behavioural evaluation. i.e., rotarod and locomotor tests and biochemical evaluation	Solubility, bioavailability, and stability of Chia seed oil contained nanoemulsion was escalated Potential application of developed nanoemulsion is suggested for neuroprotective effect in neurodegenerative disorders	[59]

Table 2. *Cont.*

A. Alzheimer Disease				
Therapeutic Agent	**Objective**	**Study Model**	**Outcome**	**Ref.**
E. Depression and Schizophrenia				
Paroxetine	Selective serotonin reuptake inhibitor 'paroxetine 'embedded intranasal nanoemulsion was explored for treatment of depression that also avoided first-pass metabolism	Average particle size, PDI, zetapotential, permeation, behavioural studies, Forced swimming test in Wistar rat and antidepressant studies	Spherical droplets of size ~58 nm with 0.339 PDI and −33 mV zeta potential. Good% transmittance and refractive index. i.e., 100.6% and 1.41, respectively 2.57 times higher permeation of nanoemulsion compared to paroxetine suspension. Improved anti-depression action due to enhancement of reduced level of glutathione on intranasal delivery. Decreased level of Thiobarbituric acid reactive Substance (TBARS).	[60]
Aripiprazole	Quinolinone derivative antidepressant aripiprazole nanoemulsion was designed for treatment of schizophrenia and bipolar disorder	Response surface method and central composite rotatable design for optimization	Optimized overhead stirring time (120 min), high shear homogenization (15 min) and rpm (4400) resulted in nanosized (~62 nm) droplets and 3.72 mPa viscosity. At pH 7.4, osmolality of nanoemulsion was ~297 mOsm/kg Stable for three months at variable temperatures. i.e., 4 °C, 25 °C and 45 °C.	[61]
Asenapine maleate	Mucoadhesive nanoemulsion of antipsychotic asenapine maleate prepared for improved nasomucosal adhesion, efficient brain-targeting and safety	Sigle-dose pharmacokinetic study, animal behaviour study, ex vivo ciliotoxicity in sheep nasal mucosa	Spherical particles with average diameter of 21.2 nm with narrow PDI (0.355). Increased drug concentration in Wistar rat brain (within 1 h) after intranasal delivery compared to intravenous route (3 h). Enhanced brain targeting capacity (284.33 ng/mL). Improved locomotor activity with no extrapyramidal symptoms in sheep nasal mucosa.	[62]

Mucoadhesive buspirone-loaded nanoemulsion was formulated for direct delivery to the brain and modification of bioavailability after administering through an intranasal route. A total of 5% of *w*/*v* hydroxypropyl beta-cyclodextrin and chitosan hydrochloride (1% *w*/*v*) were selected for the preparation of mucoadhesive nanoemulsion. The assay resulted in a 61% improvement in bioavailability, which exhibited peak plasma concentration in rats' brains at 30 min lesser compared to bare buspirone nanoemulsion (60 min). Further, after nasal administration, buspirone–chitosan nanoemulsion exhibited 2.5 times higher $AUC_{0–480}$ in the brain (~711 ng/g) compared to I/V administration (~282 ng/g) and bare buspirone nasal formulation (~354 ng/g). Mucoadhesive buspirone–chitosan nanoemulsion revealed a high percentage of drug transport (75.77%) and targeting efficiency in the brain region [63]. Recently, the BCS class II drug 'melatonin' was aimed to develop a mucoadhesive nanoemulsion to enhance brain bioavailability and alleviatedepression. The formulated nanoemulsion improved the poor aqueous solubility of melatonin, and added chitosan provided a mucoadhesive property that exhibited prolonged retention time (0.641 min) in the brain region. The locomotor activity of model rats exhibited improved behavioural responses [64].

3.2. Nanoemulsions for Brain Tumour

Managing and treating brain tumours has always remained the most crucial and challenging task, as approximately 0.2 million clinical diagnoses of the brain and other associated CNS malignancies are reported worldwide. One of the WHO surveys stated that nearly 80%of primary brain tumours are concerned with the origin of gliomas (glial cells) [65]. Glioblastoma multiforme, or GBM, is the most prominent and aggressive grade IV glioma that affects nearly8%of individuals in a population of 100,000. This malignant brain tumour is highly fatal, has a very low survival rate (1 year), and requires immediate surgery and other treatments, including neuroimaging and photodynamic/radiotherapy [66]. Scientific reports confirmed that CD73 is accountable for the development of adenosine that is overexpressed in glioblastoma cells and hence targeted to treat this tumour in the brain region. Originally, surface enzyme CD73 was a biomarker and described as a lymphocyte differentiation antigen, and its inhibition checks glioblastoma pathogenesis [67]. Failure in chemotherapy may be due to impaired therapeutic action via poor permeation across BBB. i.e., microvasculature environment of the CNS.

FDA-approved first-line antineoplastic therapeutic 'temozolomide or temodar®' is widely employed for the management of glioblastoma multiforme or GBM owing to permeation efficacy across BBB. It is a DNA alkylating drug administered orally and intravenously due to its very short half-life. However, administering high doses leads to severe adverse toxicities, including cardiomyopathy, oral ulceration, myelo-suppression and hematological issues [68]. To avoid these critical issues at present novel strategies such as topical implants, convection-enhanced delivery and nanotechnological-based formulations are clinically required for the management of brain gliomas. The last decade was the era of successful brain-targeted drug delivery approaches following a peculiar nasal path where trigeminal and olfactory nerves facilitate precise drug transportation due to the sole connection between the central nervous system and brain region. This non-invasive and safe route represents the quick onset of action with minimum systemic toxicity and adverse risks that comply with brain tumour sufferers. Moreover, low therapeutic doses and bypassing hepatic first-pass metabolism offer extramerits to the intranasal antineoplastic drug-delivery system.

However, half-life clearance (15–20 min) and limited volume of the formulation (25–200 μL) in a single dose pose strains and must be considered while formulating nose-to-brain delivery [69].

Biocompatible lipidic nanoemulsions are a highly preferred delivery system for intranasal delivery and brain targeting. The extended residing time in the nasal cavity releases the adequate therapeutic agent from the nanostructured emulsion. Bayanati et al. (2021) designed in situ emulsion-based gel of antineoplastic drug temozolomide through low energy technique for the chemotherapy of glioblastoma. The formulation bypassed BBB after delivering through the intranasal route. A suitable pseudoternary phase diagram with various quantities of triacetin, labrasol (surfactant) and transcutol®P (permeation enhancer) was established and was added for the preparation of the nanoemulsion. The nanoemulsion contained a range of particle sizes from 19–23 nm with PDI from 0.18–0.25. A slight positive zeta potential. i.e., ~1.6 might be due to the non-ionic nature of the selected surfactant/co-surfactant. By contrast, the developed in situ gel with a blend of amphiphilic poloxamer 407 and 188 exhibited a slightly increased mean particle size (16.25 nm) and PDI (~0.35) with an acceptable pH (~6.5) that showed a compatibility with nasal mucosa without irritation. The augmented viscosity (113.57 cp) of the formulation also advocated prolonged nasal retention due to reduced mucociliary clearance. Further, in vitro release efficiencies of both nanoemulsion and its in situ gel were nearly 90% and 87%, respectively, and appeared to be sustained compared to the nasal solution after 6 h. Noticeable risen in percentage mucoadhesion was observed in situ gel (37.03%) compared to nanoemulsion (20.35%) owing to the gel-forming amphiphilic poloxamer that might have formed a noncovalent entanglement with nasal mucosa. Further, the developed in situ nanoemulsion exhibited a 1.52 times higher permeation compared to the control solution. Added labrasol and transcutol®P efficiently enhanced the solubility of temozolomide and worked as permeation enhancers across nasal olfactory–trigeminal pathways. Labrasol inhibited P-gp and acted as an efflux transporter. Both temozolomide nanoemulsion and gel with poloxamer 407 were stable during freeze-thaw cycles and performed centrifugation. Gamma scintigraphy revealed an accumulation of radio-labelled temozolomide in the brain after intranasal delivery and concluded efficient uptake to the affected site of the brain [70]. Table 3 outlines the newly developed nanoemulsions for the amelioration of brain tumours.

Table 3. Recently developed nanoemulsion for the treatment of brain glioblastoma (data collected for last 5 years).

Therapeutic Agent	Objective	Evaluated Parameters	Result	Application	Ref.
Kaempferol	Investigation of glioma cells inhibition efficacy with and without chitosan-loaded kaempferol nanoemulsion and compared mucoadhesive properties after intranasal delivery	Average droplet size (180 nm) pH (5.56 ± 0.02) Viscosity (11.48 ± 0.13 cp25 °C) Drug content (96.37 ± 2.67%) Association efficiency (99.35 ± 0.10%) Permeation Efficiency across nasal mucosa (13.04 μg/cm^2)	Chitosan improved residing properties in the nasal mucosa owing to its mucoadhesive property, hence enhancing permeation and reducing nasal clearance of drug. Further, chitosan-KPF nanoemulsion inhibited C6 glioma cells viability via rapidly triggering apoptosis compared to without chitosan-loaded kaempferol nanoemulsion.	Treatment of glioma	[71]
Curcumin(CUR) and quercetin(QUE)	Two phytoconstituents containing nanoemulsions were designed for synergistically inhibiting growth of glioblastoma U373MG cells	Average droplet size 93 nm PDI 0.149 Drug content 42.4% for CUR and 55.1% for QUE Zetapotential −14.8 mV Drug release in pH6.4 For CUR, 95.84% For QUE, 94.0%	Optimized nanoemulsion displayed significantly high percentage of brain targeting efficiency, i.e., ~178% for curcumin and ~170% for quercetin. Successful nose-to-brain delivery of both curcumin and quercetin (~44% and ~38%, respectively) indicated potential CNS targeting through intranasal pathway. Further, the nanoemulsion exhibited synergistically site-specific targeting of human glioblastoma cells compared to doxorubicin drug.	Treatment of human glioblastoma	[72]

Table 3. *Cont.*

Therapeutic Agent	Objective	Evaluated Parameters	Result	Application	Ref.
CD73SiRNA	Cationic nanoemulsion composed of si-RNACD73R was successfully developed and targeted to inhibit brain tumour growth	Cell viability 30–50% Glioma cells inhibition 60–80% Tumour growth reduction 60% decreased adenosine level 95%	Developed si-RNA-CD73R Nanoemulsion enabled silencing of surface enzyme CD73 and overexpression of adenosine in the brain of rats after nasal delivery. Administered nanoemulsion reduced growth of glioblastoma cells and adenosine level in cancerous cells.	Treatment of brain tumour	[73]
Disulfiram inclusion complex with copper ion	Ion-sensitive disulfiram nanoemulsion was explored for the management of glioblastoma	Average particlesize ~63.4 nm. Zeta potential (−23.5 mV) Prolonged drug release 50% at 4 h and and 75% at 12 h	Performed in vitro studies indicated effective inhibition of proliferation of glioblastoma cells. i.e., C6 andU87. Further, formulated in situ gel nanoemulsion showed excellent uptake and brain target capability by producing highest signal fluorescence in the brain of rats.	Glioblastoma targeting therapy	[74]
Polyphenolic flavonoid quercetin	Enhancement of bioavailability and permeability of quercetin-loaded nanoemulsion across blood–brain barrier	Mean droplet size 125.5 nm PDI 0.251 Entrapment efficiency ~87.0%. Cmax 5962.7 ng/mL after 4 h Mean resident time 46.13 h. AUC is 5.32 times higher than pure drug	Intranasal-administered quercetin nanoemulsion improved solubility, therapeutic index and permeability in the brain region.	Brain cancer	[75]

An amalgamation of antineoplastic paclitaxel and C (6)-ceramide-loaded oil in water nanoemulsion was developed, including a high concentration of polyunsaturated fatty acid (pinenut oil) for enhanced therapeutic action against human U-118 glioblastoma cells. The average particle size of the nanoemulsion was observed at 200 nm. Epi-fluorescent microscopy was employed for the uptake and biodistribution of rhodamine-labelled paclitaxel and nitro-benzofurazone-labelled C (6)-ceramide nanoemulsion in glioblastoma cells. On intranasal administration, the developed nanoemulsion exhibited prominently high cytotoxicity and apoptosis within the malignant cells, thus suggesting a promising approach for the therapy of aggressive glioblastoma tumours [76]. Disulfiram, an alcohol withdrawal drug, is an FDA-approved antineoplastic drug that has been clinically proven in various cancerous cases, including glioblastoma. Its inclusion complex with copper ions has already gained success in the management of adenosine over-expressive human brain tumour. Qu et al. (2021) designed a novel inclusion complex of disulfiram and hydroxypropyl −β-cyclodextrin with a copper ion that augmented drug solubility and antitumour activity with increased safety in vitro.

Disulfiram is considered an active antitumour agent while complexed with copper ions. Here, disufiram is entrapped within the structure of hydroxyl propyl β-cyclodextrin; thus, the solubility and safety profile of the drug were amplified. An intense fluorescence signal was observed in the Wistar male rat brain model that indicated extreme brain targeting via the intranasal delivery of the drug. Developed inclusion complex embedded nanoemulsion promoted apoptosis after intranasal administration, thus inhibiting cell proliferation and tumour growth. Furthermore, histopathological outcomes displayed nonobvious damage to normal cells [77].

3.3. Nanoemulsions in Cerebral Ischemia

Cerebral ischemia is a condition aroused by acute brain injury that results in inadequate blood flow in the brain region. It is also a medical emergency that, if ignored, may lead to cerebral infarction and, ultimately, permanent brain disability. Broad cerebral ischemia is categorized as global and focal. The former happens due to shock and systemic hypotension.

Structural and functional cardiac issues, including arrhythmia, mediate global ischemia. At the same time, the obstruction of arterial blood flow (thrombosis or embolism) to the brain and the irreversible neuronal loss leads to focal ischemia. Nearly 60–70% of cerebral ischemia clinically reported is due to embolism (formation of a clot) in the heart or in a large artery [78]. Figure 3 portrays focal and global cerebral ischemic conditions that arise due to oxygen deficiency in the brain.

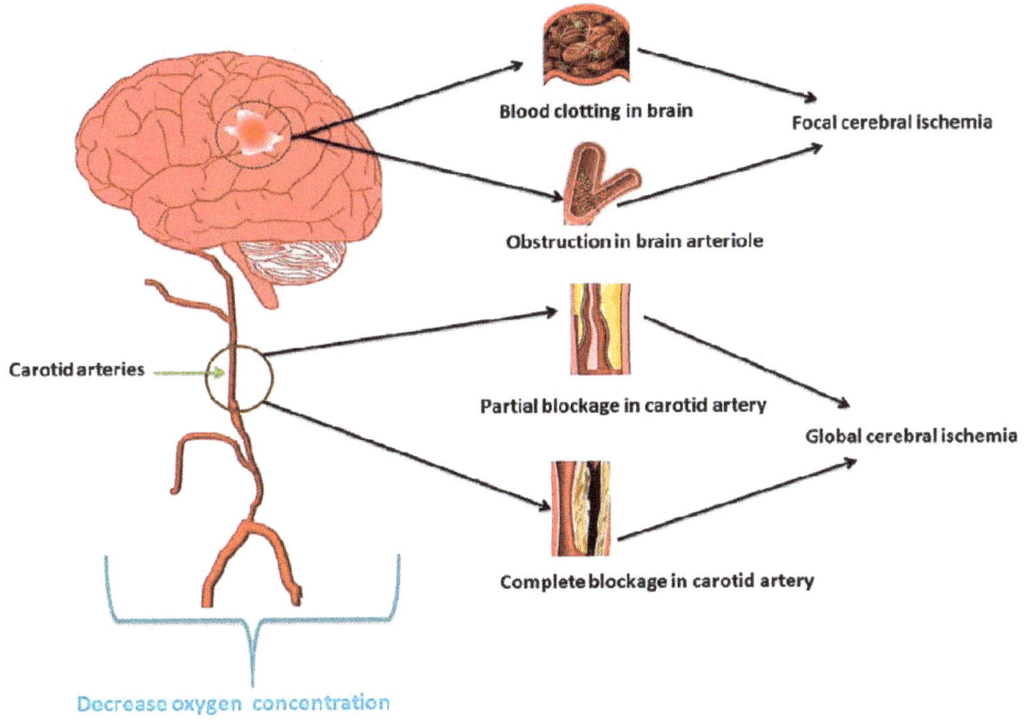

Figure 3. Cerebral ischemia developed due to decreased oxygen concentration in carotid arteries.

Vitamin D3 entrapped nanoemulsion was formulated using a blend of Tween 20, PEG400 and oleic acid for the improvement in cerebral ischemia. The targeting potential of formulated nanoemulsion was analysed through gamma scintigraphy in a rat model. The developed nanoemulsion possessed an average globular size of ~49.29 nm with a positive 13.7 mV zeta potential. The stable thermodynamic preparation was found to have a permeation coefficient of 7.8 cm/h after 3 h in sheep nasal mucosa. Further, analysed radiometry and gamma scintigraphy displayed an efficient percentage deposition of ^{99m}Tc-vitamin D3 nanoemulsion across nasal mucosa compared to IV-administered solution (0.8%). The magnetic resonance imaging on the ischemic rat model confirmed the promising antioxidant action of developed nanoemulsion through the intranasal pathway [79].

The neuroprotective property of antioxidant safranal was explored in a focal ischemic model that contained major issues, including neurobehavioral loss, hippocampal cell loss and release of oxidative stress markers. Sadeghnia et al. (2017) developed a safranal-loaded nanoemulsion for intranasal delivery to overcome the above-said issues related to the cerebral ischemic rat model. The study significantly demonstrated a reduction in neurological, hippocampal cell loss and thiobarbituric acid reactive substances (TBARS). Moreover, marked increases in antioxidant capacity and SH content were also observed, suggesting a potential role of antioxidant herbal safranal in neuroprotection, free radical suppression and the treatment of cerebral ischemic reperfusion [80]. Antioxidant thymoquinone has poor aqueous solubility and bioavailability. Its neuroprotective action and potential to ameliorate cerebral ischemia have attracted researchers to explore the design of nanoemulsion. Mucoadhesive thymoquinone-loaded nanoemulsion was prepared via the ionic–gelation method. The formulated system comprised small globules (average size ~94.8 nm) with negative zeta potential (−13.5 mV). Viscosity and percentage drug content were reported as~110 cp and 99.86%, respectively.

The developed bioanalytical method displayed comparatively enhanced biodistribution and brain bioavailability after nasal route delivery than the intravenous pathway. Thymoquinone brain targeting potential was observed up to 89.97% after post-intranasal delivery of nanoemulsion. A performed neurobehavioral activity on the middle cerebral artery occlusion-persuaded ischemic rat model revealed the potential for antioxidant thymoquinone-embedded nanoemulsion to treat cerebral ischemia [81]. Table 4 lists some intranasal nanoemulsions applied for the control of cerebral ischemia.

Table 4. Recently reported nanoemulsions for the management of cerebral ischemia.

Therapeutic Agent	Purpose	Evaluated Parameters	Outcomes	Application	Ref.
Safranal	Nanoemulsion containing antioxidant safranal was prepared to reduce oxidative stress in brain injury	Optimized mean size 89.64 nm Zetapotential −11.39 mV Drug content 98.47% Viscosity 124 cp	Improved locomotor activity, grip strength and antioxidant activity observed. Significantly decrease in glutathione reductase, superoxide dismutase and lipid peroxidation in model rat brain.	Treatment for cerebral ischemia-reperfusion injury	[82]
Chaxiong volatile oil	Thermosensitive intranasal in situ nanoemulsion gel was developed for brain targeting	Mean particle size ~21.02 nm, PDI 0.14 Negative zeta potential −20.4 mV pH 4.52 Viscosity 32.5 mV, gelling strength 42–47 s Mucoadhesive strength 5.2×10^2 dy/cm^2 Release kinetics Ritger–Peppas model.	Both nanoemulsion and in situ gel had potential to lessen neurological deficit score in ischemic rat model. Cerebral infarction size was reduced, which improved ischemic stroke.	Cerebral ischemic stroke	[83]

Table 4. *Cont.*

Therapeutic Agent	Purpose	Evaluated Parameters	Outcomes	Application	Ref.
6-Gingerol	Mucoadhesive intranasal nanoemulsion was designed to improve brain bioavailability and Neuro-protectiveness effect of 6-gingerol	Average particle size ~94.89 nm, Narrow PDI 0.129 Zetapotential +1.892 mV Retention time 1.27 min	Nanoemulsion preparedwith antioxidant 6-gingerol and lauroglycol 90 was then converted in to mucoadhesive with chitosan. Improved Cmax and AUC after intranasal administration exhibited and histopathological assay displayed reduction in infarction volume in induced Ischemic model.	Treatment of cerebral ischemia	[84]
Tenofovir disoproxil fumarate	Stable nucleotide reductase inhibitor tenofovir disproxil fumerate-loaded nanoemulsion was developed and optimized by phase diagram to target neuro-AIDS in brain.	Phase diagram for optimization of surfactant and co-surfactant (45%). Zeta potential (−18.7 mV) Average globular size (156.2 nm) PDI (0.463) Drug diffusion (egg membrane) 74.98% and sheep nasal mucosa (75.98%) after 3 h.	Developed nanoemulsion increased surface area and provided lipophilicity to the formulation. This alternate strategy provided quick onset of action against viral infection for all age groups sufferers.	Neuro-AIDS treatment	[85]

3.4. Nanoemulsions in Brain Infections

Brain infections may occur in the regions of the cerebrum, cerebellum, spinal cord and associated nerves. In this context, encephalitis, meningitis and abscess are declared as medical emergencies and their long-term sequelae may result in substantial mortality. Bloodborne pathogens, head trauma and skull fractures can mediate the opening of tight junctions between the CNS and other nerves. Further, neurosurgical procedures and medical device implantation (exterior drainage tube, shunt) may cause infection due to microbial colonization and thus behaveas the foci of infection [86]. Rinaldi et al. (2020) suggested a successful nose-to-brain intranasal delivery route for the management of fatal meningitis and encephalitis. Essential oils composed of nanoemulsions (mean average diameter 100 nm with PDI 0.2) were chosen to transport at the infected site of the central nervous system. In this context, antibacterial oils extracted from *Thymus vulgaris* and *Syzygium aromaticum* were individually incorporated in the chitosan-coated nanoemulsions (C-TV-NEs and C-SA-NEs) and investigated against multidrug-resistant methicillin-susceptible *S. aureus* and carbapenem-resistant *A. baumannii* and *K.pneumonia*. Nanoemulsions, i.e., TV-NEs and SA-Nes, exhibited negative zeta potentials as −40 mV and −30 mV, respectively, that were converted into positive charges after chitosan coating by virtue of electrostatic interaction. However, the mean droplet size and PDI were amplified after the coating of the mucoadhesive chitosan polymer.

Intranasally administered nanoemulsions appeared more suitable for the potential curing of brain infections caused by Gram-negative bacteria compared to the intravenously delivered high dose of the formulation, presenting an efficient alternative therapy for the cure of serious meningitis and encephalitis [87]. A blend of essential oils is widely used to manage different virus infections caused by human rhinovirus, bovine rotavirus, herpes virus, H5N1 and HIV. Essential oils, such as thyme oil, eucalyptol, borneol, alpha terpineol and sage oil, are reported to be included in the formulation of nanoemulsion. Their mixture initiates nucleoprotein trafficking abnormalities in the surface protein of the virus, hence interfering or masking the virion envelope and blocking virus internalization [88]. A myriad of viruses, including poliovirus, rabies, herpes simplex and HIV, can reach the CNS through an intraneural path, causing encephalitis and brain abscesses, whereas meningitis is spread through bacterial pathogen at the subarachnoid space. Any breaches or damages (necrosis, microhemorrhage) of the BBB mechanical obstruction (infected RBCs, WBCs or platelets) and excessive production of cytokines disturb the structural architecture (tight junction) of BBB that led to the brain infection. Several reported pathogens, including a wide range of bacteria, fungi, viruses, cerebral malaria and spirochetes, are extensive causes of CNS or brain infections [89].

The immunologically privileged central nervous system (CNS) remains the residing site for human immunodeficiency virus 1. Poor permeability across the tight junctions of BBB leads to inadequate and limited delivery of most of the anti-HIV therapeutics. Therefore, nanotechnology-based formulations have attracted medical scientists and researchers to design novel strategies for the management of neuro-AIDS. Saquinavir mesylate, a protease inhibitor, has been explored as an antiretroviral drug, but due to poor solubility and bioavailability (4%), its use is limited. O/W intranasal nanoemulsion containing saquinavir mesylate was formulated via a spontaneous emulsification method to enhance CNS bioavailability. The developed nanoemulsion was thermodynamically stable on analysing through freeze–thaw and heating–cooling cycles. The obtained low PDI (0.078) and smaller globular diameter (176.3 nm) indicated the development of monodispersed nanoemulsion that would be suitable for brain targeting via intranasal delivery. Estimated optimum zeta potential (−10.3 mV), pH (5.8) and refractive index (1.412) depicted good stability, non-irritancy and the compatibility of nanoemulsion. Higher-percentage drug permeation and permeability coefficients after 4 h on intranasal administration were observed as 76.96% and 0.51 cm/h, respectively, compared to the pure drug (26.73%, 0.17 cm/h).

The in vivo study in sheep nasal mucosa displayed higher drug permeation and biodistribution rate of nanoemulsion on comparing with its suspension. Further, the cilia-toxicity study depicted no prominent adverse action on the sheep nasal mucosa. Gamma scintigraphy images demonstrated higher drug transportation region in the rat brain that concluded its efficiency for treating neuro-AIDS by reducing the devastating viral load from reservoir sites [90].

3.5. Nanoemulsions in Migraine and Cerebral Vasospasm

Another neuron-disabling disorder, 'migraine', is characterized by intense throbbing headaches in one of the halves of the brain. One of the surveys demonstrated that migraine is the sixth most prevailing brain disorder and affects approximately 15% of sufferers worldwide [91].

Recurrent episodes of pain in unilateral headaches are allied with other visual, auditory and autonomic nervous disorders. Mostly, females are more affected due to their poor lifestyle. The pain is so intense and unbearable that the patient needs quick relief from it [92]. Poor solubility, erratic absorption, inadequate permeation and fewerpenetration properties of most of the conventional therapies across BBB led to poor efficacy in the brain region. Delayed gastric emptying, first-pass metabolism, slow onset of action, nausea and vomiting and the intake of high doses are undesirable issues associated with anti-migraine drugs if given orally.

The intranasal route is more effective for the delivery of anti-migraine therapeutics owing to the involvement of olfactory and trigeminal nerves. These anatomical features are potentially involved in the greater distribution of drugs without facing issues of first-pass metabolism [93].

Rizatriptan, a serotonin 5HT 1B/1D receptor agonist, exhibits only a 40% bioavailability upon oral delivery. An intranasal preparation of rizatriptan benzoate nanoemulsion improved brain tissue deposition and offered a non-invasive alternative approach for brain targeting. Biodistribution through the olfactory pathway directly delivers the drug to the central nervous system [94]. Another well-tolerated anti-migraine drug, zolmitriptan, also possesses low bioavailability after oral administration. Ahigh-dose prescription in conventional preparation causes serious side effects, including irregular heart rhythm, stroke and Raynaud's disease. Ebtsam et al., 2017 formulated a mucoadhesive nanoemulsion for the direct nose-to-brain transportation of zolmitriptan. The improved and quick onset of action is desired to alleviate acute migraine fulfilled with this formulation. An added 0.3% of chitosan acted as a mucoadhesive agent in the preparation and increased the residing time and drug permeation across associated nasal mucosa, hence assisting the transport of the drug to the brain tissues. Performed in vivo studies displayed shorter Tmax and higher $AUC_{0\text{–}8}$ of intranasally administered nanoemulsion compared to intravenous or nasal zolmitriptan solution [95]. Sumatriptan has been widely employed for the management of severe and painful migraine for decades. High hydrophilicity and less mucoadhesive features limit its application through the nasal route. Ribeiro et al. 2020 have developed a novel nanoemulsion containing sumatriptan, copaiba oil and organic biopolymers. i.e., alginate, pullulan, xanthan and pectin. Mean particle size, PDI and zeta potential of nanoemulsion were observed as approximately 120 nm, 0.2 and −25 mV, respectively.

The developed alginate-based nanoemulsion exhibited long term storage stability (1 year) due to optimised zeta potential. In vitro study depicted extended release from sumatriptan-loaded alginate nanoemulsion upto 24 h. The in vivo toxicity in the zebrafish model revealed no evidence of mortality and other cardiac toxicity and did not disturb spontaneous changes in zebrafish larvae, hence showing a promising alternative approach for the alleviation of brain diseases, including migraine [96].

Cerebral vasospasm is defined as the temporary narrowing or thinning of cerebral arteries. More often, subarachnoid haemorrhage and traumatic brain injuries result in cerebral vasospasm.

The appearance of new focal neurological signs, inflammation, microcirculatory failure, bipolar disorders and loss of consciousness re-associated aspects are addressed with subarachnoid haemorrhage [97]. Cerebral vasospasm is the leading cause of mortality if not identified and treated immediately. It typically affects concerned blood vessels at the skull base that influences arterial contraction, blood pressure and blood viscosity. Figure 4 defines different causes of cerebral vasospasm and its different signs and symptoms that have appeared in the sufferer.

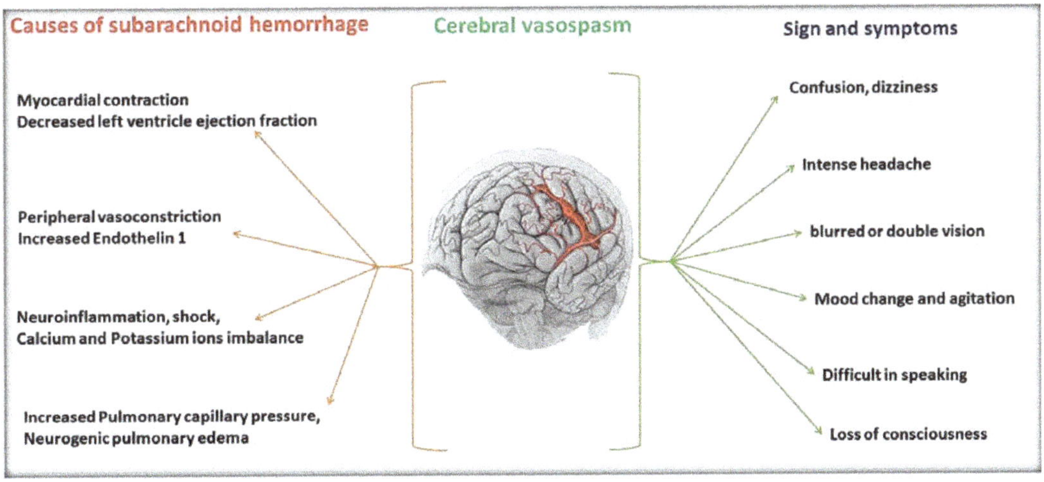

Figure 4. Various causes and symptoms of cerebral vasospasm.

Endovascular therapies are recommended for the management of severe vasospasm cases. However, the radiological and computed tomography evidence suggested that approximately 20% of cases of vasospasm can be treated via performing angiograms within a week of aneurysmal rupture. Some clinical reports also described that the level of nitric oxide was found to be decreased due to vasoconstrictor endothelin and extra vascular oxyhaemoglobin [98].

Cerebral vasospasm is often evident after aneurysmal subarachnoid haemorrhage. However, in some instances, it is observed after skull surgery associated with intractable epilepsy and amygdala-hippocampectomy. Carbamazepine 400 mg is recommended to overcome developed issues in this case. Dhobale et al., 2018 formulated carbamazepine nanoemulsion for efficient brain targeting via the intranasal route. A spontaneous emulsification method was employed for the formulation of nanoemulsion utilizing Capmul MCM (oil) and TWEEN 80/PEG-600 (3:1) as surfactant and co-surfactant, respectively. The developed nanoemulsion contained a smaller globular size of ~71.7 nm with a 0.256 $\pm$ 0.002 polydispersity index. The in vivo study depicted efficient drug distribution in 5 h [99]. Survival from cerebral vasospasm or aneurysmal subrachanoid haemorrhage was improved. FDA-approved nimodine is a calcium channel blocker and is widely prescribed clinically to treat subarachnoid haemorrhage. In this series, fasudil (Rho-kinase inhibitor), nicardipine (calcium channel blocker), statins (hypolipidemic), clazosentan (endothelin receptor antagonist), cilostazol (platelet aggregation factor and heparin (anticoagulant) has also been added as effective therapeutics in clinical practice [100].

Ahmad et al. have discussed a novel bioimaging tool. i.e., targeting probe that is based on an on–off signal and significantly traces the translocation of the therapeutic agent throughout the nose-to-brain pathway. A cargo was developed containing environmentally responsive dye P2 and P4, caumarin 6 and DiR conventional probes. Translocation of nanoemulsion (prepared with Labrafac®CC/ WL1349 and solutol®HS15) after intranasal administration was evident either via bioimaging or through histopathological examination in rats. Outcomes concluded that nanoemulsions (coated with chitosan and naked) with a mean particle size of 100 nm exhibited higher retention duration in the nasal cavity with slower clearance compared to larger particles. Weak P2 and P4 signals were also observed in the region of the olfactory bulb for coated nanoemulsions of particle size 100 nm. Moreover, nanoemulsion particle sizes of more than 900 nm were unable to reach the olfactory bulb. Signals obtained from caumarin 6 and DiR represented significant transportation of nanoemulsion in the brain region.

The importance of particle size and their cytoprotective efficiency under the circumstances of oxygen–glucose deprivation and reoxygenation were also signified by Varlamova and coworkers in 2022. It was reported that the average particle size of selenium (50 nm) could induce calcium ion responses and inhibit apoptosis in the brain cortex cells. At the same time, upon increasing particle size, i.e., 100 nm and 400 nm, exhibited induction of calcium ion oscillation and mixed pattern of calcium signals, respectively. Hence, cryoprotective action under oxygen–glucose deprivation and reoxygenation conditions can be provided at the size range 50–400 nm that is required to protect brain tissues from ischemic conditions through modulation of the calcium ions signal system of astrocytes [101,102].

4. Future Prospects and Conclusions

Although intranasal nanoemulsions successfully deliver therapeutics with minimum side effects at the site, several formulation aspects (type of oil, polymer and surfactant/cosurfactant), process stabilizing parameters (methodology, temperature, rheology, particle size) and pharmacokinetic issues (solubility, absorption, bioavailability) always create challenges for the competent treatment of brain disorders. High molecular weight containing hydrophilic therapeutics poses problems for nanoemulsion formulation. Enormous work on the preparation and evaluation of intranasal nanoemulsions has been reported within 5 years for the management of myriad brain-associated ailments. Nanoemulsions prepared with mucoadhesive polymers and in situ nanoemulgels confine fast nasal clearance and improve residing time and permeation of bioactive across the nose-to-brain pathway.

Mucoadhesive starch-based polymers (potato starch, wheat starch, rice starch), gaur gum, xanthan gum and tragacanth are not only safe, cost-effective and eco-friendly but also are proven natural components that make intimate contact between nasomucosal membrane and nanoemulsions system. Similarly, in situ nanoemulgels embedded with nanocarriers comprise stimuli-sensitive (pH, temperature, ionic concentration) ingredients that respond to the environment of the nasal cavity and augment the viscosity of the nanoemulsion system. Therefore, surged biodistribution and improved contact time across nasomucosal membranes overcome shortcomings of nanoemulsions and offer a safe and desirable benchmark for targeting brain diseases. Although mucoadhesive and in situ nanoemulgels are fascinating strategies, formulation aspects (osmolarity, pH, buffer capacity, viscosity, drug concentration) and physiological parameters (absorption, nasal blood flow, mucociliary clearance, enzymatic action, pathophysiological issue) are still to be monitored for the effective delivery of therapeutics via the nose-to-brain route.

Currently, nose-to-brain targeted rationally designed nanoemulsions have been proven as a successful alternate approach for the precise alleviation of countless brain-associated disorders. Adequate endothelial cell internalization by nanoemulsion via transcytosis and endocytosis mechanisms mediate desirable therapeutic effects with minimum adverse effects. Associated olfactory and trigeminal pathways facilitate biodistribution and bypass issues with the BBB barrier and drug first-pass metabolism. Exploring these virtues, intra-nasal nanoemulsions have proven brain-targeting cargos with their broad array of applications. Recently, targeting probes that contain environment-responsive dyes have been employed for the tracking of bio translocation of the nanoemulsion via fluorescence bioimaging technology. This strategy validates the efficacy of the nanoemulsion with evidence.

Author Contributions: Conceptualization, review and editing: K.P., original draft preparation: S.K.M. All authors have read and agreed to the published version of the manuscript.

Funding: This research received no external funding.

Data Availability Statement: No new data is generated.

Acknowledgments: The authors thankfully acknowledge the amenities provided by the School of Pharmaceutical Sciences, Chhatrapati Shahu Ji Maharaj University, Kanpur, for consenting to all working facilities for drafting this manuscript. Our sincere acknowledgments also go to the Faculty of Pharmacy, Uttar Pradesh University of Medical Sciences, Saifai, Etawah, for their generous assistance in assembling the data presented in this review article.

Conflicts of Interest: The authors declare no conflict of interest.

References

1. Liu, P.; Jiang, C. Brain-targeting drug delivery systems. *Wiley Interdiscip. Rev. Nanomed. Nanobiotechnol.* **2022**, *14*, e1818. [CrossRef]
2. Pajouhesh, H.; Lenz, G.R. Medicinal chemical properties of successful central nervous system drugs. *NeuroRx* **2005**, *2*, 541–553. [CrossRef]
3. Daneman, R.; Prat, A. The blood-brain barrier. *Cold Spring Harb. Perspect. Biol.* **2015**, *7*, a020412. [CrossRef] [PubMed]
4. Banks, W.A. From blood-brain barrier to blood-brain interface: New opportunities for CNS drug delivery. *Nat. Rev. Drug. Discov.* **2016**, *15*, 275–292. [CrossRef]
5. Kaushik, A.; Jayant, R.D.; Bhardwaj, V.; Nair, M. Personalized nano medicine for CNS diseases. *Drug. Discov. Today* **2018**, *23*, 1007–1015. [CrossRef]
6. Jaiswal, M.; Dudhe, R.; Sharma, P.K. Nanoemulsion: An advanced mode of drug delivery system. *3 Biotech.* **2015**, *5*, 123–127. [CrossRef]
7. Dhuria, S.V.; Hanson, L.R.; Frey, W.H. Intranasal delivery to the central nervous system: Mechanisms and experimental considerations. *J. Pharm. Sci.* **2010**, *99*, 1654–1673. [CrossRef] [PubMed]
8. Miyake, M.M.; Bleier, B.S. The blood-brain barrier and nasal drug delivery to the central nervous system. *Am. J. Rhinol. Allergy* **2015**, *29*, 124–127. [CrossRef] [PubMed]
9. Jones, N. The nose and paranasal sinuses physiology and anatomy. *Adv. Drug. Deliv. Rev.* **2001**, *51*, 5–19. [CrossRef]

10. Pardeshi, C.V.; Belgamwar, V.S. Direct nose to brain drug delivery via integrated nerve pathways by passing the blood-brainbarrier: An excellent platform for brain targeting. *Expert. Opin. Drug. Deliv.* **2013**, *10*, 957–972. [CrossRef]
11. Rassu, G.; Soddu, E.; Cossu, M.; Brundu, A.; Cerri, G.; Marchetti, N.; Ferraro, L.; Regan, R.F.; Giunchedi, P.; Gavini, E.; et al. Solid microparticles based on chitosan or methyl-β-cyclodextrin: A first formulative approach to increase the nose-to-brain transport of deferoxamine mesylate. *J. Control. Release* **2015**, *201*, 68–77. [CrossRef]
12. Erdő, F.; Bors, L.A.; Farkas, D.; Bajza, Á.; Gizurarson, S. Evaluation of intranasal delivery route of drug administration for brain targeting. *Brain Res. Bull.* **2018**, *143*, 155–170. [CrossRef]
13. Mistry, A.; Stolnik, S.; Illum, L. Nose-to-brain delivery: Investigation of the transport of nanoparticles with different surface characteristics and sizes in excised porcine olfactory epithelium. *Mol. Pharm.* **2015**, *12*, 2755–2766. [CrossRef] [PubMed]
14. Quintana, D.S.; Guastella, A.J.; Westlye, L.T.; Andreassen, O.A. The promise and pitfalls of intranasally administering psychopharmacological agents for the treatment of psychiatric disorders. *Mol. Psychiatry* **2016**, *21*, 29–38. [CrossRef] [PubMed]
15. Tan, C.; McClements, D.J. Application of advanced emulsion technology in the food industry: A review and critical evaluation. *Foods* **2021**, *10*, 812. [CrossRef] [PubMed]
16. Siasios, I.; Kapsalaki, E.Z.; Fountas, K.N. Cerebral vasospasm pharmacological treatment: An update. *Neurol. Res. Int.* **2013**, *2013*, 571328. [CrossRef]
17. Shen, R.; Yang, X.; Lin, D. pH sensitive double-layered emulsions stabilized by bacterial cellulose nanofibers/soy protein isolate/chitosan complex enhanced the bio accessibility of curcumin: In vitro study. *Food Chem.* **2023**, *402*, 134262. [CrossRef]
18. Piacentini, E.; Figoli, A.; Giorno, L.; Drioli, E. Membrane Emulsification. In *Comprehensive Membrane Science and Engineering*; Drioli, E., Ed.; Elsevier: Amsterdam, The Netherlands, 2010; pp. 47–78.
19. Guzmán, E.; Ortega, F.; Rubio, R.G. Pickering Emulsions: A Novel Tool for Cosmetic Formulators. *Cosmetics* **2022**, *9*, 68. [CrossRef]
20. Gao, H.; Ma, L.; Cheng, C.; Liu, J.; Liang, R.; Zou, L.; Liu, W.; McClements, D.J. Review of recent advances in the preparation, properties, and applications of high internal phase emulsions. *Trends Food Sci. Technol.* **2021**, *112*, 36–49. [CrossRef]
21. Costa, C.P.; Moreira, J.N.; Sousa, L.J.M.; Silva, A.C. Intranasal delivery of nanostructured lipid carriers, solid lipid nanoparticles and nanoemulsions: A current overview of in vivo studies. *Acta Pharm. Sin. B* **2021**, *11*, 925–940. [CrossRef] [PubMed]
22. Choudhury, H.; Gorain, B.; Karmakar, S.; Biswas, E.; Dey, G.; Barik, R.; Mandal, M.; Pal, T.K. Improvement of cellular uptake, in vitro antitumor activity and sustained release profile with increased bioavailability from a nanoemulsion platform. *Int. J. Pharm.* **2014**, *460*, 131–143. [CrossRef]
23. Edmond, J. Essential polyunsaturated fatty acids and the barrier to the brain: The components of a model for transport. *J. Mol. Neurosci.* **2001**, *16*, 181–193. [CrossRef] [PubMed]
24. Singh, Y.; Meher, J.G.; Raval, K.; Khan, F.A.; Chaurasia, M.; Jain, N.K.; Chourasia, M.K. Nanoemulsion: Concepts, development and applications in drug delivery. *J. Control. Release* **2017**, *252*, 28–49. [CrossRef] [PubMed]
25. Kumar, G.; Virmani, T.; Pathak, K.; Kamaly, O.A.; Saleh, A. Central Composite Design Implemented Azilsartan Medoxomil Loaded Nanoemulsion to Improve Its Aqueous Solubility and Intestinal Permeability: In Vitro and Ex Vivo Evaluation. *Pharmaceuticals* **2022**, *15*, 1343. [CrossRef]
26. Morrison, E.E.; Costanzo, R.M. Morphology of olfactory epithelium in humans and other vertebrates. *Microsc. Res. Tech.* **1992**, *23*, 49–61. [CrossRef] [PubMed]
27. Phukan, K.; Nandy, M.; Sharma, R.B.; Sharma, H.K. Nanosized Drug Delivery Systems for Direct Nose to Brain Targeting: A Review. *Recent. Pat. Drug. Deliv. Formul.* **2016**, *10*, 156–164. [CrossRef]
28. Djupesland, P.G.; Messina, J.C.; Mahmoud, R.A. The nasal approach to delivering treatment for brain diseases: An anatomic, physiologic, and delivery technology overview. *Ther. Deliv.* **2014**, *5*, 709–733. [CrossRef]
29. Hosny, K.M.; Banjar, Z.M. The formulation of a nasal nanoemulsion zaleplon in situ gel for the treatment of insomnia. *Expert Opin. Drug. Deliv.* **2013**, *10*, 1033–1041. [CrossRef]
30. Chatterjee, B.; Gorain, B.; Mohananaidu, K.; Sengupta, P.; Mandal, U.K.; Choudhury, H. Targeted drug delivery to the brain via intranasal nanoemulsion: Available proof of concept and existing challenges. *Int. J. Pharm.* **2019**, *565*, 258–268. [CrossRef]
31. Bonferoni, M.C.; Rossi, S.; Sandri, G.; Ferrari, F.; Gavini, E.; Rassu, G.; Giunchedi, P. Nanoemulsions for "Nose-to-Brain" Drug Delivery. *Pharmaceutics* **2019**, *11*, 84. [CrossRef]
32. Izquierdo, P.; Esquena, J.; Tadros, T.F.; Dederen, C.; Garcia, M.J.; Azemar, N.; Solans, C. Formation and stability of nanoemulsions prepared using the phase inversion temperature method. *Langmuir* **2002**, *18*, 26–30. [CrossRef]
33. Sood, S.; Jain, K.; Gowthamarajan, K. Optimization of curcumin nanoemulsion for intranasal delivery using design of experiment and its toxicity assessment. *Colloids Surf. B Bio. Interfaces* **2014**, *113*, 330–337. [CrossRef]
34. Bahadur, S.; Pardhi, D.M.; Rautio, J.; Rosenholm, J.M.; Pathak, K. Intranasal nanoemulsions for direct nose-to-brain delivery of actives for CNS disorders. *Pharmaceutics* **2020**, *12*, 1230. [CrossRef]
35. Samaridou, E.; Alonso, M.J. Nose-to-brain peptide delivery—The potential of nano technology. *Bioorg. Med. Chem.* **2018**, *26*, 2888–2905. [CrossRef] [PubMed]
36. Dalpiaz, A.; Fogagnolo, M.; Ferraro, L.; Capuzzo, A.; Pavan, B.; Rassu, G.; Salis, A.; Giunchedi, P.; Gavini, E. Nasal chitosan microparticles target a zidovudine prodrug to brain HIV sanctuaries. *Antiviral. Res.* **2015**, *123*, 146–157. [CrossRef] [PubMed]
37. Diedrich, C.; Camargo, Z.I.; Schineider, M.C.; Taise, F.M.; Maissar, K.N.; Badea, I.; Mara, M.R. Mucoadhesive nanoemulsion enhances brain bioavailability of luteolin after intranasal administration and induces apoptosis to SH-SY5Y neuroblastoma cells. *Int. J. Pharm.* **2022**, *626*, 122142. [CrossRef]

38. Rassu, G.; Porcu, E.P.; Fancello, S.; Obinu, A.; Senes, N.; Galleri, G.; Migheli, R.; Gavini, E.; Giunchedi, P. Intranasal delivery of genistein-loaded nanoparticles as a potential preventive system against neurodegenerative disorders. *Pharmaceutics* **2018**, *11*, 8. [CrossRef]
39. Gupta, A.; Eral, H.B.; Hatton, T.A.; Doyle, P.S. Nanoemulsions: Formation, properties and applications. *Soft Matter* **2016**, *12*, 2826–2841. [CrossRef]
40. Prakash, R.T.; Thiagarajan, P. Nanoemulsions for drug delivery through different routes. *Res. Biotechnol.* **2011**, *2*, 1–13.
41. Mehta, M.; Adem, A.; Sabbagh, M.N. New acetylcholinesterase inhibitors for Alzheimer's disease. *Int. J. Alzheimer's Dis.* **2012**, *2012*, 728983. [CrossRef]
42. Jiang, Y.; Jiang, Y.; Ding, Z.; Yu, Q. Investigation of the "Nose-to-Brain" pathways in intranasal Hup Anan emulsions and evaluation of their in vivo pharmacokinetics and brain-targeting ability. *Int. J. Nanomed.* **2022**, *17*, 3443–3456. [CrossRef] [PubMed]
43. Parikh, H.R.; Patel, J.R. Nanoemulsions for intranasal delivery of riluzole to improve brain bioavailability: Formulation development and pharmacokinetic studies. *Curr. Drug. Deliv.* **2016**, *13*, 1130–1143. [CrossRef]
44. Usama, A.M.; Vyas, P.; Vohora, D.; Kumar, S.P.; Nigam, K.; Dang, S.; Ali, J.; Baboota, S. Amelioration of oxidative stress utilizing nanoemulsion loaded with bromocriptine and glutathione for the management of Parkinson's disease. *Int. J. Pharm.* **2022**, *618*, 121683. [CrossRef]
45. Patel, R.J.; Parikh, R.H. Intranasal delivery of topiramate nanoemulsion: Pharmacodynamic, pharmacokinetic and brain uptake studies. *Int. J. Pharm.* **2020**, *585*, 119486. [CrossRef] [PubMed]
46. Kaur, A.; Nigam, K.; Srivastava, S.; Tyagi, A.; Dang, S. Memantine nanoemulsion: A new approach to treat alzheimer's disease. *J. Microencapsul.* **2020**, *37*, 355–365. [CrossRef]
47. Kaur, A.; Nigam, K.; Bhatnagar, I.; Sukhpal, H.; Awasthy, S.; Shanka r, S.; Tyagi, A.; Dang, S. Treatment of Alzheimer's disease using donepezil nanoemulsion: An intranasal approach. *Drug. Deliv. Transl. Res.* **2020**, *10*, 1862–1875. [CrossRef] [PubMed]
48. Jiang, Y.; Liu, C.; Zhai, W.; Zhuang, N.; Han, T.; Ding, Z. The optimization design of lactoferrin loaded hup A nanoemulsion for targeted drug transport via intranasal route. *Int. J. Nanomed.* **2019**, *14*, 9217–9234. [CrossRef] [PubMed]
49. Kotta, S.; Mubarak, A.H.; Badr-Eldin, S.M.; Alhakamy, N.A.; Md, S. Coconut oil-based resveratrol nanoemulsion: Optimization using response surface methodology, stability assessment and pharmacokinetic evaluation. *Food Chem.* **2021**, *357*, 129721. [CrossRef]
50. Das, S.S.; Sarkar, A.; Chabattula, S.C.; Verma, P.R.P.; Nazir, A.; Gupta, P.K.; Ruokolainen, J.; Kesari, K.K.; Singh, S.K. Food-grade quercetin-loaded nanoemulsion ameliorates effects associated with Parkinson's disease and cancer: Studies employing atransgenic *C. elegans* model and human cancer cell lines. *Antioxidants* **2022**, *11*, 1378. [CrossRef]
51. Gaba, B.; Khan, T.; Haider, M.F.; Alam, T.; Baboota, S.; Parvez, S.; Ali, J. Vitamin Eloaded naringenin nanoemulsion via intranasal delivery for the management of oxidative stress in a 6-OHDA Parkinson's disease model. *Biomed Res. Int.* **2019**, *2019*, 2382563. [CrossRef]
52. Kumar, S.; Ali, J.; Baboota, S. Design Expert(®) Supported optimization and predictive analysis of selegiline nanoemulsion via the olfactory region with enhanced behavioural performance in Parkinson's disease. *Nanotechnology* **2016**, *27*, 435101. [CrossRef] [PubMed]
53. Nasr, M. Development of An optimized hyaluronic acid-based lipidic nanoemulsion co-encapsulating two polyphenols for nose to brain delivery. *Drug. Deliv.* **2016**, *23*, 1444–1452. [CrossRef] [PubMed]
54. Ahmad, N.; Ahmad, R.; Alam, M.A.; Ahmad, F.J.; Amir, M. Impact of ultrasonication techniques on the preparation of novel amiloride-nanoemulsion used for intranasal delivery in the treatment of epilepsy. *Artif. Cells Nanomed. Biotechnol.* **2018**, *46*, S192–S207. [CrossRef]
55. Iqbal, R.; Ahmed, S.; Jain, G.K.; Vohora, D. Design and development of letrozolen an emulsion: A comparative evaluation of brain targeted nanoemulsion with free letrozole against status epilepticus and neurodegeneration in mice. *Int. J. Pharm.* **2019**, *565*, 20–32. [CrossRef] [PubMed]
56. El-Zaafarany, G.M.; Soliman, M.E.; Mansour, S.; Cespi, M.; Palmieri, G.F.; Illum, L.; Casettari, L.; Awad, G.A.S. A Tailored Thermosensitive PLGA-PEG-PLGA/emulsomes composite for enhanced oxcarbazepine brain delivery via the nasal route. *Pharmaceutics* **2018**, *10*, 217. [CrossRef] [PubMed]
57. Md, S.; Gan, S.Y.; Haw, Y.H.; Ho, C.L.; Wong, S.; Choudhury, H. In vitro neuroprotective effects of naringenin nanoemulsion against β-amyloid toxicity through the regulation of amyloid genesis and tau phosphorylation. *Int. J. Biol. Macromol.* **2018**, *118*, 1211–1219. [CrossRef]
58. Ramires Júnior, O.V.; Alves, B.D.S.; Barros, P.A.B.; Rodrigues, J.L.; Ferreira, S.P.; Monteiro, L.K.S.; Araújo, G.M.S.; Fernandes, S.S.; Vaz, G.R.; Dora, C.L.; et al. Nanoemulsion improves the neuroprotective effects of curcumin in an experimental model of Parkinson's disease. *Neurotox. Res.* **2021**, *39*, 787–799. [CrossRef]
59. Geetha, K.M.; Shankar, J.; Wilson, B. Neuroprotective effect of chia seed oil nanoemulsion against rotenone induced motor impairment and oxidative stress in mice model of Parkinson's disease. *Adv. Tradit. Med.* **2022**. [CrossRef]
60. Pandey, Y.R.; Kumar, S.; Gupta, B.K.; Ali, J.; Baboota, S. Intranasal delivery of paroxetine nanoemulsion via the olfactory region for the management of depression: Formulation, behavioural and biochemical estimation. *Nanotechnology* **2016**, *27*, 025102. [CrossRef]

61. Samiun, W.S.; Ashari, S.E.; Salim, N.; Ahmad, S. Optimization of processing parameters of nanoemulsion containing aripiprazole using response surface methodology. *Int. J. Nanomed.* **2020**, *15*, 1585–1594. [CrossRef]
62. Kumbhar, S.A.; Kokare, C.R.; Shrivastava, B.; Gorain, B.; Choudhury, H. Preparation, characterization, and optimization of asenapine maleate mucoadhesive nanoemulsion using Box-Behnken design: In vitro and in vivo studies for brain targeting. *Int. J. Pharm.* **2020**, *586*, 119499. [CrossRef] [PubMed]
63. Khan, S.; Patil, K.; Yeole, P.; Gaikwad, R. Brain targeting studies on buspirone hydrochloride after intranasal administration of mucoadhesive formulation in rats. *J. Pharm. Pharmacol.* **2009**, *61*, 669–675. [CrossRef] [PubMed]
64. Ahmad, N.; Khalid, M.S.; Al Ramadhan, A.M.; Alaradi, M.Z.; Al Hammad, M.R.; Ansari, K.; Alqurashi, Y.D.; Khan, M.F.; Albassam, A.A.; Ansari, M.J.; et al. Preparation of melatonin novel-mucoadhesive nanoemulsion used in the treatment of depression. *Polym. Bull.* **2022**, 1–40. [CrossRef]
65. Saenzdel Burgo, L.; Hernández, R.M.; Orive, G.; Pedraz, J.L. Nanotherapeutic approaches for brain cancer management. *Nanomedicine* **2014**, *10*, 905–919. [CrossRef] [PubMed]
66. Chu, L.; Wang, A.; Ni, L.; Yan, X.; Song, Y.; Zhao, M.; Sun, K.; Mu, H.; Liu, S.; Wu, Z.; et al. Nose-to-brain delivery of temozolomide-loaded PLGA nano particles functionalized with anti-EPHA3 for glioblastoma targeting. *Drug. Deliv.* **2018**, *25*, 1634–1641. [CrossRef]
67. Yan, A.; Joachims, M.L.; Thompson, L.F.; Miller, A.D.; Canoll, P.D.; Bynoe, M.S. CD73 promotes glioblastoma pathogenesis and enhances its chemoresistance via A2B adenosine receptor signaling. *J. Neurosci.* **2019**, *39*, 4387–4402. [CrossRef]
68. Chaskis, E.; Luce, S.; Goldman, S.; Sadeghi, N.; Melot, C.; De Witte, O.; Devriendt, D.; Lefranc, F. Early postsurgical temozolomide treatment in newly diagnosed bad prognosis glioblastoma patients: Feasibility study. *Bull. Cancer* **2018**, *105*, 664–670. [CrossRef]
69. Ban, M.M.; Chakote, V.R.; Dhembre, G.N.; Rajguru, J.R.; Joshi, D.A. In-situ gel for nasal drug delivery. *Int. J. Dev. Res.* **2013**, *8*, 18763–18769.
70. Bayanati, M.; Khosroshahi, A.G.; Alvandi, M.; Mahboobian, M.M. Fabrication of a Thermosensitive In Situ Gel Nanoemulsion for Nose to Brain Delivery of Temozolomide. *J. Nanomater.* **2021**, *2021*, 1–11. [CrossRef]
71. Colombo, M.; Figueiró, F.; de Fraga, D.A.; Teixeira, H.F.; Battastini, A.M.O.; Koester, L.S. Kaempferol-loaded mucoadhesive nanoemulsion for intranasal administration reduces glioma growth in vitro. *Int. J. Pharm.* **2018**, *543*, 214–223. [CrossRef]
72. Mahajan, H.S.; Patil, N.D. Nanoemulsion containing a synergistic combination of curcumin and quercetin for nose-to-brain delivery: In vitro and in vivo studies. *Asian Pac. J. Trop. Biomed.* **2021**, *11*, 510–518. [CrossRef]
73. Azambuja, J.H.; Schuh, R.S.; Michels, L.R.; Gelsleichter, N.E.; Beckenkamp, L.R.; Iser, I.C.; Lenz, G.S.; de Oliveira, F.H.; Venturin, G.; Greggio, S.; et al. Nasal Administration of cationic nanoemulsions as *CD73-siRNA* delivery system for glioblastoma to treatment: A new therapeutical approach. *Mol. Neurobiol.* **2020**, *57*, 635–649. [CrossRef]
74. Qu, Y.; Li, A.; Ma, L.; Iqbal, S.; Sun, X.; Ma, W.; Li, C.; Zheng, D.; Xu, Z.; Zhao, Z.; et al. Nose-to-brain delivery of disulfiram nanoemulsion in situ gel formulation for glioblastoma targeting therapy. *Int. J. Pharm.* **2021**, *597*, 120250. [CrossRef]
75. Savale, S.K. Formulation and evaluation of quercetin nanoemulsions for treatment of brain tumor via intranasal pathway. *Asian J. Biomater. Res.* **2017**, *3*, 28–32.
76. Desai, A.; Vyas, T.; Amiji, M. Cytotoxicity and apoptosis enhancement in brain tumor cells upon coadministration of paclitaxel and ceramide in nanoemulsion formulations. *J. Pharm. Sci.* **2008**, *97*, 2745–2756. [CrossRef]
77. Qu, Y.; Sun, X.; Ma, L.; Li, C.; Xu, Z.; Ma, W.; Zhou, Y.; Zhao, Z.; Ma, D. Therapeutic effect of disulfiram inclusion complex embedded in hydroxypropyl-β-cyclodextrin on intracranial glioma-bearing male rats via intranasal route. *Eur. J. Pharm. Sci.* **2021**, *156*, 105590. [CrossRef]
78. Martini, S.R.; Kent, T.A. Ischemic Stroke. In *Cardiology Secrets, Ved.*; Glenn, N.L., Ed.; Elsevier: Philadelphia, PA, USA, 2018; pp. 493–504.
79. Kumar, M.; Nishad, D.K.; Kumar, A.; Bhatnagar, A.; Karwasra, R.; Khanna, K.S.K.; Sharma, D.; Dua, K.; Mudaliyar, V.; Sharma, N. Enhancement in brain uptake of vitamin D3 nanoemulsion for treatment of cerebral ischemia: Formulation, gamma scintigraphy andefficacy study in transient middle cerebral artery occlusion rat models. *J. Microencapsul.* **2020**, *37*, 492–501. [CrossRef] [PubMed]
80. Sadeghnia, H.R.; Shaterzadeh, H.; Forouzanfar, F.; Hosseinzadeh, H. Neuroprotective effect of safranal, an active ingredient of Crocus sativus, in a rat model of transient cerebral ischemia. *Folia Neuropathol.* **2017**, *55*, 206–213. [CrossRef] [PubMed]
81. Ahmad, N.; Ahmad, R.; Alam, M.A.; Samim, M.; Iqbal, Z.; Ahmad, F.J. Quantification and evaluation of thymoquinone loadedmucoadhesivenanoemulsionfortreatmentofcerebralischemia. *Int. J. Biol. Macromol.* **2016**, *88*, 320–332. [CrossRef] [PubMed]
82. Ahmad, N.; Ahmad, R.; Abbas, N.A.; Ashafaq, M.; Alam, M.A.; Ahmad, F.J.; Al-Ghamdi, M.S. The effect of safranal loaded mucoadhesive nanoemulsion on oxidative stress markers in cerebral ischemia. *Artif. Cells Nanomed. Biotechnol.* **2017**, *45*, 775–787. [CrossRef]
83. Huang, C.; Wang, C.; Zhang, W.; Yang, T.; Xia, M.; Lei, X.; Peng, Y.; Wu, Y.; Feng, J.; Li, D.; et al. Preparation, In vitro and in vivo evaluation of nanoemulsion in situ gel for transnasal delivery of traditional Chinese medicine volatile oil from *Ligusticumsinense* Oliv.cv. Chaxiong. *Molecules* **2022**, *27*, 7644. [CrossRef]
84. Niyaz, A.; Rizwan, A.; Mohd, A.; Md, A.; Alam, M.Z.A.; Abuzer, A.; Ahmad, A.; Ashraf, K. Ischemic brain treated with 6-gingerol loaded mucoadhesive nanoemulsion via intranasal delivery and their comparative pharmacokinetic effect in brain. *J. Drug. Deliv. Sci. Technol.* **2021**, *61*, 102130.

85. Nemade, S.M.; Kakad, S.P.; Kshirsagar, S.J.; Padole, T.R. Development of nanoemulsion of antiviral drug for brain targeting in the treatment of neuro-AIDS. *Beni-Suef Univ. J. Basic. Appl. Sci.* **2022**, *11*, 138. [CrossRef]
86. Solomon, I.H. Molecular and Histologic Diagnosis of Central Nervous System Infections. *Surg. Pathol. Clin.* **2020**, *13*, 277–289. [CrossRef] [PubMed]
87. Rinaldi, F.; Oliva, A.; Sabatino, M.; Imbriano, A.; Hanieh, P.N.; Garzoli, S.; Mastroianni, C.M.; DeAngelis, M.; Miele, M.C.; Arnaut, M.; et al. Antimicrobial essential oil formulation: Chitosan coated nanoemulsions for nose to brain delivery. *Pharmaceutics* **2020**, *12*, 678. [CrossRef] [PubMed]
88. Franklyne, J.S.; Gopinath, P.M.; Mukherjee, A.; Chandrasekaran, N. Nanoemulsions: The rising star of antiviral therapeutics and nano delivery system-current status and prospects. *Curr. Opin. Colloid. Interface Sci.* **2021**, *54*, 101458. [CrossRef]
89. Giovane, R.A.; Lavender, P.D. Central Nervous System Infections. *Prim. Care* **2018**, *45*, 505–518. [CrossRef]
90. Hitendra, S.; Mahajan, M.S.; Mahajan, P.P.; Nerkar Agrawal, A. Nanoemulsion-based intranasal drug delivery system of saquinavir mesylate for brain targeting. *Drug. Deliv.* **2014**, *21*, 148–154.
91. Vos, T.; Abajobir, A.A.; Abate, K.A. Global, regional, and national incidence, prevalence, and years lived with disability for 328 diseases and injuries for 195 countries,1990–2016: A systematic analysis for the Global Burden of Disease Study 2016. *Lancet* **2017**, *390*, 1211–1259. [CrossRef]
92. Chen, P.K.; Wang, S.J. Non-headache symptoms in migraine patients. *F1000Research* **2018**, *7*, 188. [CrossRef]
93. Tanna, V.; Sawarkar, S.P.; Ravikumar, P. Exploring nose to brain nano delivery for effective management of migraine. *Curr. Drug. Deliv.* **2023**, *20*, 144–157. [CrossRef] [PubMed]
94. Bhanushali, R.S.; Gatne, M.M.; Gaikwad, R.V.; Bajaj, A.N.; Morde, M.A. Nanoemulsion based intranasal delivery of antimigraine drugs for nose to brain targeting. *Indian J. Pharm. Sci.* **2009**, *71*, 707–709.
95. Abdou, E.M.; Kandil, S.M.; Miniawy, H.M.F.E. Brain targeting efficiency of antimigraine drug loaded mucoadhesive intranasal nanoemulsion. *Int. J. Pharm.* **2017**, *529*, 667–677. [CrossRef] [PubMed]
96. Ribeiro, L.N.M.; Rodrigues da Silva, G.H.; Couto, V.M.; Castro, S.R.; Breitkreitz, M.C.; Martinez, C.S.; Igartúa, D.E.; Prieto, M.J.; de Paula, E. Functional hybrid nanoemulsions for sumatriptan intranasal delivery. *Front. Chem.* **2020**, *8*, 589503. [CrossRef]
97. Chan, A.; Choi, E.; Yuki, I.; Suzuki, S.; Golshani, K.; Chen, J.; Hsu, F. Cerebral vasospasm after subarachnoid hemorrhage: Developing treatments. *Brain Hemorrhages* **2021**, *2*, 15–23. [CrossRef]
98. Huang, S.; Huang, Z.; Fu, Z.; Shi, Y.; Dai, Q.; Tang, S.; Gu, Y.; Xu, Y.; Chen, J.; Wu, X.; et al. A novel drug delivery carrier comprised of nimodipine drug solution and a nanoemulsion: Preparation, characterization, invitro, and in vivo studies. *Int. J. Nanomed.* **2020**, *15*, 1161–1172. [CrossRef]
99. Dhobale, A. Formulation and evaluation of carbamazepine nanoemulsion for brain targeted drug delivery via intranasal route. *Indo Am. J. Pharm. Sci.* **2018**, *8*, 1437–1452.
100. Maruhashi, T.; Higashi, Y. An overview of pharmacotherapy for cerebral vasospasm and delayed cerebral ischemia after subarachnoid hemorrhage. *Expert Opin. Pharmacother.* **2021**, *22*, 1601–1614. [CrossRef]
101. Ahmad, E.; Feng, Y.; Qi, J.; Fan, W.; Ma, Y.; He, H.; Xia, F.; Dong, X.; Zhao, W.; Lu, Y.; et al. Evidence of nose-to-brain delivery of nanoemulsions: Cargoes but not vehicles. *Nanoscale* **2017**, *9*, 1174–1183. [CrossRef]
102. Varlamova, E.G.; Gudkov, S.V.; Plotnikov, E.Y.; Turovsky, E.A. Size-Dependent Cytoprotective Effects of Selenium Nanoparticles during Oxygen-Glucose Deprivation in Brain Cortical Cells. *Int. J. Mol. Sci.* **2022**, *23*, 7464. [CrossRef]

colloids and interfaces

MDPI

Review

Recent Advances in Improving the Bioavailability of Hydrophobic/Lipophilic Drugs and Their Delivery via Self-Emulsifying Formulations

Rakesh Kumar Ameta [1], Kunjal Soni [1] and Ajaya Bhattarai [2,3,*]

[1] Sri M. M. Patel Institute of Sciences & Research, Kadi Sarva Vishwavidhyalaya, Gandhinagar 382023, India
[2] Department of Chemistry, Mahendra Morang Adarsh Multiple Campus, Tribhuvan University, Biratnagar 56613, Nepal
[3] Department of Chemistry, Indian Institute of Technology, Madras 600036, India
* Correspondence: ajaya.bhattarai@mmamc.tu.edu.np

Abstract: Formulations based on emulsions for enhancing hydrophobic and lipophilic drug delivery and its bioavailability have attracted a lot of interest. As potential therapeutic agents, they are integrated with inert oils, emulsions, surfactant solubility, liposomes, etc.; drug delivering systems that use emulsion formations have emerged as a unique and commercially achievable accession to override the issue of less oral bioavailability in connection with hydrophobic and lipophilic drugs. As an ideal isotropic oil mixture of surfactants and co-solvents, it self-emulsifies and forms fine oil in water emulsions when acquainted with aqueous material. As droplets rapidly pass through the stomach, fine oil promotes the vast spread of the drug all over the GI (gastrointestinal tract) and conquers the slow disintegration commonly seen in solid drug forms. The current status of advancement in technologies for drug carrying has promulgated the expansion of innovative drug carriers for the controlled release of self-emulsifying pellets, tablets, capsules, microspheres, etc., which got a boost for drug delivery usage with self-emulsification. The present review article includes various kinds of formulations based on the size of particles and excipients utilized in emulsion formation for drug delivery mechanisms and the increase in the bioavailability of lipophilic/hydrophobic drugs in the present time.

Keywords: lipophilic drugs; hydrophobic drugs; emulsion

Citation: Ameta, R.K.; Soni, K.; Bhattarai, A. Recent Advances in Improving the Bioavailability of Hydrophobic/Lipophilic Drugs and Their Delivery via Self-Emulsifying Formulations. *Colloids Interfaces* **2023**, *7*, 16. https://doi.org/10.3390/colloids7010016

Academic Editors: César Burgos-Díaz, Mauricio Opazo-Navarrete, Eduardo Morales and Ramón G. Rubio

Received: 29 December 2022
Revised: 11 February 2023
Accepted: 13 February 2023
Published: 17 February 2023

1. Introduction

The improvements in combinatorial chemistry show an enormous rise in a variety of less water-soluble drugs. At least forty novel pharmacologically active lipophilic/hydrophobic moieties exhibit very low aqueous solubility. Nevertheless, a unique challenge regarding drugs is presented to pharmaceutical scientists: orally administrated drugs have innate low aqueous solubility, leading to inadequate oral bioavailability with higher inter- and intra-subject changeability and scarcity of dose proportionality [1]. Numerous formulation perspectives are currently being applied to handle challenges related to formulations of biopharmaceutical class system (BCS) drugs; this includes compound pre-dissolution in suitable solvents followed by capsule filing with this formulation [2], or as solid solution formulations that utilize water-soluble polymers [3]. Although these perspectives will help resolve the matter related to the primary dissolution of drug matter in a liquid phase inside the GI tract up to specific proportions, momentous restrictions such as the precipitation issue of drug molecules in the dispersal of formulations during the crystallization of drugs in the polymer-based matrix are still unsolved. For the same reasons, assessment of physical stability is critical and has been evaluated using techniques such as X-ray crystallography or differential scanning calorimetry. Several formulation modes, such as carrier technology, provide an innovative methodology for enhancing solubility in drug molecules with low solubilities. Advancements in oral drug molecule bioavailability use lipid-based formulations that have now become attraction points. Perhaps the most adaptable excipient class members presently available are lipids, providing a strong eventuality as a formulator in improving and controlling the lipophilic drug's absorption where ordinary formulation methods failed or when the drug molecule itself is an oil molecule (i.e., Dronabinol, ethyl icosapentate). In addition, with a low affinity for precipitation of lipophilic drugs in the GI tract during dilution, such formulations will favor partitioning kinetics in the lipid droplets to be retained [4]. The literature review indicates that the application of carrier technology is a perspective of scientific interest in lipid-based oral formulations and strengthens the ambidexterity in addressing the issues complementary to oral drug delivery of poorly soluble molecules [2–6]. Novel methods such as self-emulsification modes have also intensified the solubility of inadequately soluble drugs and have some advantages. The introduction of this self-emulsification concept and present-day advances in polymer science have led to application advancements with lipid-based self-emulsifying formulations in various drug delivery views comprising drug targeting. This article endeavors to review the far-reaching awareness of emulsion-forming drug delivery systems (DDS) for the bioavailability enhancement of hydrophobic/lipophilic drugs by cherishing numerous formulations. The present-day architectural innovations and the advancement of self-nano-emulsifying and self-micro-emulsifying formulations have also been considered. Thus, this review's main focus is on improving the bioavailability or solubility of hydrophobic/lipophilic drugs via self-emulsifying formulations (SEF).

2. Emulsion Concept and Types of Emulsions

An isotropic and transparent solution from an oil mixture and co-solvent, with surfactant and co-surfactant, is emulsified with gentle agitation equivalent to that experienced in the GI tract; this is known as SEF. Spontaneous emulsification is recognized for this solution in aqueous GI fluids in the presence of oral administration. Bile secretion is stimulated by this triglyceride (emulsified oil), which further emulsifies oil droplets containing the drug. Lipases and co-lipases, secreted from various portions such as the salivary gland, pancreas, and gastric mucosa, then metabolize these lipid droplets and hydrolyze the triglycerides by forming free fatty acids and di- and mono-glycerides. Additionally, these molecules get solubilized when they pass through the GI tract. In due course, emulsion droplets are formed in various sizes, along with mixed micelles and vesicular structures containing phospholipids, bile salts, and cholesterol [5]. The synthesis of chylomicron occurs in lymphatics, ensuring enhanced drug absorption. The bioavailability intensifies formulations' self-emulsification characteristics primarily in connection with confident in vivo features, such as inhibiting cellular efflux mechanisms and retaining the drugs from circulation; this is because of the attachment of several lipidic excipients with a particular drug, a decrease in drug metabolism by the liver in the first pass, as well as its uptake in the lymphatic transport system. In addition, the formation of micellar suspensions and fine dispersions that restrict recrystallization and/or precipitation of drug molecules, where changes in the GI fluid begin because of the properties of several lipid components that favor upgraded drug absorption [6]. Usually, emulsion-forming drug delivery systems (EFDDS) are prepared as simple emulsions, whereas surfactants with a hydrophilic–lipophilic balance (HLB) < 12 SEFs are formulated. Self-nano-emulsifying formulations (SNEFs) and self-micro-emulsifying formulations (SMEs) are acquired using surfactants of HLB > 12. Because of surface enhancement for dispersion, the formulations contain improved dissolution (drugs with poor solubility) and high stability. Due to this, independent drug absorption from bile secretion ensures a speedy shift over that of less soluble drugs in blood. Additionally, their formulations have specific, definite characteristics associated with upgraded drug delivery systems. Thus, the emulsion contains hydrophobic and hydrophilic parts.

3. Excipients for Self-Emulsifying Formulations

Based on studies, the process of self-emulsification is precise to the kind of surfactant/oil pair utilized, the concentration of surfactant, the ratio of oil/surfactant, and the temperature at which the self-emulsification materialized. These salient findings have been supported by the fact that only selective combinations of pharmaceutical excipients led to effective systems of self-emulsifying therapeutics. Numerous remarkable components are used in EFDDS, such as the following.

3.1. Oils

The most natural support for lipid vehicles is available from consistent edible oils. However, they have poor dissolving properties for enormous amounts of hydrophobic drugs, and limitations for superior self-micro emulsification considerably decrease their use in SEFs. Vegetable oils that are altered or hydrolyzed have a widespread role in successive SEFs attributed to their biocompatibility. Naturally occurring di- and triglycerides have been used exponentially as excipients with susceptibility for degradation (a crucial pathway for releasing drug molecules from EFDDS-based formulations). At present, triglycerides with a medium chain are being replaced by new semi-synthetic medium-chain-containing triglycerides, with molecules such as Gelucire. Robust emulsification systems are formed based on their ability for notable fluidity and the remarkable solubilizing prospect capabilities of self-emulsification. Oils and fats, such as corn, olive, palm, soya bean oils, and animal fats, either digestible or non-digestible, may be used as oil phases in EFDDS [7].

3.2. Surfactants or Emulsifiers

Based on the critical packing parameter and hydrophilic–lipophilic balance, the screening of surfactants can be carried out. For the formation of EFDDS, non-ionic surfactants are often chosen because of their limited toxicity; furthermore, they have reduced critical micelle concentrations compared with their ionic complements [8]. High HLB-value-containing surfactants are commonly used in forming EFDDS, including polysorbate 80, poloxamers, Gelucire (HLB 10), sorbitan monooleate (Span 80), cremophor EL, hexadecyltrimethylammonium bromide, sodium lauryl sulphate, and bis2-Ethylhexyl sulfosuccinate. Additionally, fatty alcohols and famous surfactants, such as cetyl and stearyl, lauryl, glyceryl, and esters of fatty acids, are also incorporated [9]. Surfactants occurring naturally are also endorsed for SEFs; the most commonly used surfactant is lecithin, with the due reason of the most significant biocompatibility. It has phosphatidylcholine as a fundamental component, having an amphiphilic structure and water-solubilizing properties. To get stable SEFs, the commonly used surfactant concentration is 30 to 60% *w*/*w*.

A higher surfactant concentration (~60%) may cause selective, reversible alterations in intestinal wall permeability or GI tract irritability. The subsequent hydrophilicity and higher HLB of surfactants are essential for the prompt formulation of *o*/*w* droplets and/or for the sudden spread of the formulate in an aqueous condition, yielding exceptional self-emulsifying/dispersing achievement. By nature, surface-active agents are amphiphilic and generally dissolve more remarkably than hydrophobic drugs. This property is of great importance as it prevents precipitation through the GI lumen, and drug molecules have prolonged presence in a solubilized form, an essential phase for effective absorption [10]. Recent reports show that the surfactant digestion mechanism affects the performance of SEFs, as the dissolving environment alterations can cause precipitation of a little less water-soluble drugs.

Additionally, more information is needed for degradation molecule formation from surfactants and their interaction with fatty acids and phospholipids, bile salts, and dietary lipids such as endogenous lipids. This may play a vital role in maintaining a solution with poorly water-soluble drugs, and an essential part in forming mixed micelles may be compromised [11–13]. Considering all these data reported, information about the impacts of non-ionic surfactants that may show inhibition of triglyceride digestion is vital for lipid-based formulation development. In addition, the susceptibility of surfactants towards digestion through pancreatic enzymes is an essential factor that should be considered in formulation development.

A simple mechanism/preparation of a self-emulsifying system is depicted in Figure 1, which could help the reader understand this.

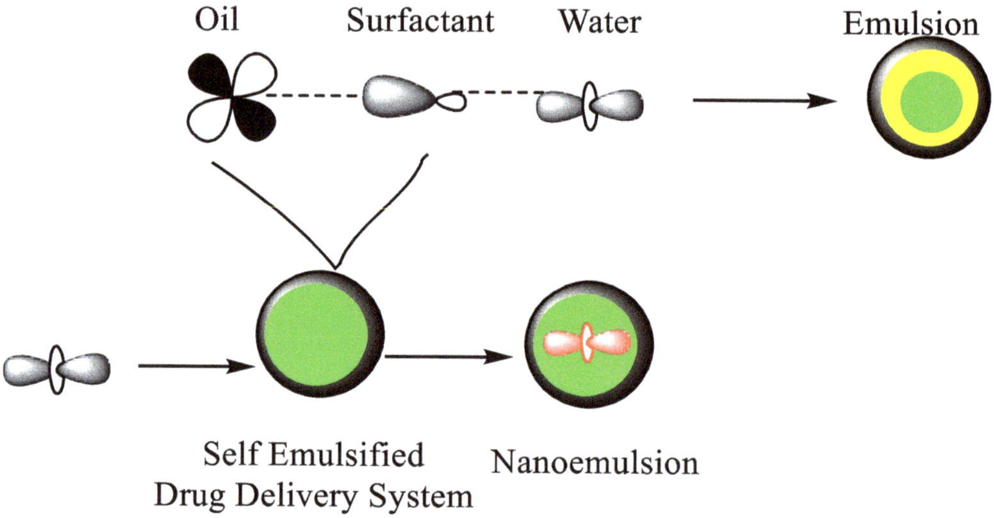

Figure 1. Pictorial representation of a self-emulsified drug delivery system and nanoemulsion.

3.3. Co-Surfactants/Co-Solvents

Frequently, a cosurfactant is added in the self-emulsifying formulations to increase interfacial area and dispersion entropy and decrease free energy at its minimum and interfacial tension [8]; on the cause of amphiphilic nature, this is substantially accumulated by a cosurfactant at an interfacial layer that increases interfacial film fluidity through surfactant monolayer penetration. Pentanol, hexanol, and octanol, like short and medium-chain alcohols, are preferred cosurfactants known to form self-emulsifying formulations spontaneously. Apart from cosurfactants, transcutol (diethylene glycol mono ethylene ether), triacetin (an acetylated derivative of glycerol), polyethene glycol, propylene glycol, propylene carbonate, glycofurol (tetrahydro furfuryl alcohol polyethene glycol ether), etc., like several co-solvents, are convenient for hydrophobic drug dissolution in lipid bases. Recently, with varying fatty acids, polyglycolyzed glycerides (PGG) and polyethene glycol (PEG) in aggregation with vegetable oils have also been reported for use in hydrophobic drug solubilization and emulsification [7,14,15].

Thus, the structure and stability of *w*/*o* and *o*/*w* types of emulsions are mainly based on the constituents of the emulsifier and how much hydrophobicity or hydrophilicity it has.

4. Recent Studies

4.1. Self-Stabilized Pickering Emulsion

A novel high-pressure homogenization technique for silybin oral bioavailability enhancement has been developed for silybin nanocrystal self-stabilized Pickering emulsion (SN-SSPE). The impacts of drug content and homogenization pressure on SN-SSPE formation were also evaluated. Using SEM (scanning electron micrograph), atomic force microscopy (AFM), and confocal laser scanning microscopy, the structure, size, and morphology of PE droplets were identified. Investigation of in vivo oral bioavailability and SN-SSPE release in vitro was also carried out. The results revealed that when homogenization pressure is scaled up to 100 MPa, the silybin nanocrystals' (SN-NC) particle size decreases. A stable silybin Pickering emulsion might be formed when silybin content reaches 300 mg or above; thus, surfaces of oil droplets are entirely covered by sufficient SN-NC, and a self-stabilized Pickering emulsion is formed. A core-shell arrangement composed of a core of SN-NC shell and oil was seen when the SN-SSPE emulsion droplet was 27.3 ± 3.1 μm. A stability of about 40 days or more has been recorded for SN-SSPE. SN-SSPE showed a faster in vitro release rate compared with silybin coarse powder. However, it is similar to the suspension of SN-NCS. Intragastric administration showed a 2.5-fold and 3.6-fold incremental peak concentration of SN-SSPE of silybin compared with SN-NCS and coarse powder of silybin in rats. Furthermore, there were 1.6-fold and 4.0-fold increases in the AUC of SN-SSPE concerning SN-NCS and silybin coarse powder. From the results, it has been confirmed that silybin nanocrystals could stabilize the Pickering emulsion of silybin and increase oral bioavailability. The self-stabilized Pickering emulsion of drug nanocrystals has encouraged a system for poorly soluble drugs for oral drug delivery [16]. For a controlled drug delivery system, (PLGA) (Poly lactide-co-glycolide) microparticles are often utilized. To formulate PLGA microparticles, standard emulsion methods have been used. However, they show less loading capacity, more precisely for poorly soluble drugs in organic solvents. A template of water-soluble polymers and nanocrystal technology was used to manufacture nanocrystal-loaded microparticles with enhanced encapsulation and drug-loading efficacy for extended breviscapine delivery, as reported by Hong Wang et al. [17]. A precipitation-ultrasonication method is used to prepare breviscapine nanocrystals, which are cast using water-soluble polymer mould load further into PLGA microparticles. These disc-like particles were characterized and compared with spherical particles by emulsion-solvent evaporation. Through confocal laser scanning microscopy (CLSM) and X-ray powder diffraction (XRPD), analysis of the highly dispersed breviscapine state in microparticle confirmation is carried out. The drug significantly affects the efficiency of breviscapine and its loading capacity in PLGA microparticles and liberation mechanism through loading percentage and fabrication methodology. When both template and nanocrystal methods were enforced, drug loading and encapsulation potential was enhanced by 2.4% to 15.3%, and 48.5% to 91.9%, respectively.

4.2. Drug Loaded Nanoemulsions

On the other hand, as drug loading is increased, loading efficiency is reduced. An initial release by bursting in all microparticles has been seen that later slows down to 28 days and is further followed by erosion-acceleration phase release, which supports sustained delivery over a month for breviscapine. Stable serum drug level after intramuscular microparticle injection in rats was observed even after 30 days. Therefore, nanocrystals of less-soluble drug-loaded PLGA microparticles provide a supportive way for long-term therapeutic outcome characterization desirable in vivo and in vitro accomplishment. An approved safe herbal drug for several hepatic disorders is Silymarin. However, poor oral bioavailability is its major limitation. Silymarin-loaded nanoemulsions could be prepared using the high-pressure homogenization (HPH) method [18]. Capryol 90 as the oil phase, Solutol HS 15 as a surfactant, and Transcutol HP as a co-surfactant were selected accordingly. Design-based quality has been employed for optimized nanoemulsions in conditions of several cycles, processing pressure, and (Smix) surfactant/co-surfactant mixture amount. Globule size, polydispersity index (PDI), zeta potential, transmittance, and percentage in vitro drug release of the optimized formulation were found as 50.02 ± 4.5 nm, 0.45 ± 0.02, −31.49 mV, 100.00 ± 2.21% and 90.00 ± 1.83%, respectively. The apparent permeability coefficient (Papp) has been enhanced by nanoemulsion, as shown in everted gut sac studies. Silymarin Papp in nanoemulsion and the oral suspension was 1.00×10^{-5} cm/h with a flux of 0.422 μg/cm^2/h, and 6.30×10^{-6} cm/h with a flux of 0.254 μg/cm^2/h at 2 h, respectively. The enhancement in Silymarin bioavailability in nanoemulsion was compared with its oral suspension. Silymarin nanoemulsion could be an excellent oral delivery system with improved oral bioavailability that was found to be significant ($p < 0.05$) in a pharmacokinetic study.

As a unique lipid excipient class, phosphorylated tocopherols have exhibited potential in pharmaceutical utilization. They can make poorly water-soluble drugs soluble, which demonstrates a potential advantage in enhancing drug bioavailability where solubility is a limiting factor. A formulation of CoQ10, combined from medium-chain triglyceride (MCT) and phosphorylated tocopherols, TPM, and their in vivo and in vitro functionality was compared with tocopherol-based additional alternative excipients studied. CoQ10 was less soluble in digesting MCT during in vitro digestion experimentation as an anticipated molecule. TPM addition facilitated improved CoQ10 solubilization as vitamin E TPGS did. Several other derivatives of tocopherol, such as tocopherol and tocopherol acetate, were found to be less impactful at active solubilizing during digestion. In vitro solubility trends were preserved during in vivo CoQ10 bioavailability after oral administration in rats, where TPGS and TPM formulations provide nearly double exposure of MCT apart. Simultaneously, the overall exposure is reduced by the addition of other tocopherol derivatives. The resultant condition reveals TPM as a potent new solubilizing excipient for drugs with low water-solubility in oral drug delivery [19]. Δ9-tetrahydrocannabinol (THC) and cannabidiol (CBD), a lipophilic phytocannabinoid, represent therapeutic potential in numerous medical situations. Both compounds have low water solubility and are subjected to comprehensive first-pass metabolism in the GI, leading to limited oral bioavailability of hardly up to 9%. An advanced drug delivery system with lipid-based self-emulsification has been developed by Irina Cherniakov et al. and termed an advanced pre-concentrate of Pro-NanoLiposphere (PNL). Lipids and emulsifying excipients of GRAS compose PNL and are known for solubility enhancement with decreased phase I metabolism of lipophilically active compounds. When panels are incorporated with a natural absorption improver, they become advanced PNLs. They are natural phenolic compounds and alkaloids reported as inhibiting phase I and II metabolism processes. Hence the use of these advanced-PNLs on THC and CBD oral bioavailability has been explored. A 6-fold increase is reported in AUC compared with CBD solution when CBD-piperine-PNL has been orally administered, indicated as the most potent screened formulation. Similar data was found during THC-piperine-PNL-based pharmacokinetic experiments, showing a 9.3-fold increment in AUC compared with the THC solution. The synthesized Piperine-PNL can synchronize piperine with THC or CBD delivery to the enterocyte site. The bioavailability of CBD and THC has increased due to this co-localization and the effect on pre-enterocyte and enterocyte levels during the absorption process. The further amplification in THC and CBD absorption is incorporated by piperine into PNL. It plays a role in phase I and phase II metabolism inhibition by piperine with an addition to P-gp and phase I metabolism by PNL. These offbeat results put forward the way for piperine-PNL to deliver less soluble, highly metabolized drugs that cannot be orally administered at present [20].

4.3. *Nano Lipospheres Formulation*

In pomegranate, ellagic acid is a predominantly bioactive compound with low bioavailability. A food-grade system from self-nano emulsification has developed that enhanced the dissolving and absorption of ellagic acid. Pseudo-turning phase images and solubility assays have revealed that the components are suitable for formulation. The optimal formulation has been achieved with polyethylene glycol, polysorbate, and capric triacylglycerol/caprylic at 45/45/10 wt.%. A fine nanoemulsion was yielded from controlled stirring and optimized formulation, with an average droplet size of 120 nm. With the formulation, the ellagic acid dissolution was remarkably increased. Based on the pharmacokinetics study carried out in rats, ellagic acid's bioavailability was 3.2 and 6.6-fold higher compared with its aqueous suspensions and pomegranate extract. A novel strategy has been developed to deliver ellagic acid with a self-nano-emulsifying method for developing dietary supplement products and ellagic acid-containing functional foods [21]. The most rapidly growing therapeutic segment is biologics, but it has limitations of low stability. It has an alternative delivery system for personal administration; however, it has particular physiological challenges that prompt protein susceptibility and function loss. Protein formulation in biomaterials, such as electrospun fiber, can resize these barriers. Still, optimization of such platforms is required for protein stability and maintenance of bioactivity during the formulation process. An emulsion electrospinning method has been developed for protein loading into Eudragit L100 fibers for perioral delivery. Alkaline phosphatase and horse reddish peroxidase encapsulation lead to higher efficiency into fibers and pH-specific release. Protein bioavailability recovery has been enhanced by aqueous emulsion phase reduction and hydrophilic polymer excipient inclusion. Hannah Frizzell et al. demonstrated that protein formulation in lyophilized electrospun fibers increases therapeutic compounds' shelf life compared with aquatic storage. Thus, a novel promising dosage form of biotherapeutics for perioral delivery has been available from the platform [22]. In another work, designing a new octa-arginine (R8) altered with (LE) a lipid emulsion system of the lipophilic drug disulfiram (DSF) for ocular delivery was the target purpose. On corneal permeation, R8 presence and lipid emulsion particle sizing (DSF-LE1, DSF-LE2, DSF-LE3) with DSF loading and altered with R8 (DSF-LE1-R8 and DSF-LE2-R8) was formulated. There was a change in zeta potential from negative to positive values for lipid emulsions after the modification of R8.

DSF-LE1-R8 fabricates the strongest mucoadhesion from different compositions of mucoadhesion studied. R8 altered lipid emulsion (DSF-LE1-R8) with nano-sized particles showed ocular distribution in vivo and corneal penetration in vitro, with high permeability, and the most significant DDC amount distribution in visual tissues. More homogeneous fluorescence was displayed by LE1-R8 when LE1-R8 was labelled with Coumarin-6 and deep penetration in the cornea compared with other formulations at different time frames. Furthermore, LE-R8 could be transported across the corneal epithelium apart from its paracellular routes by using transcellular ways because of an induced update on a cause of R8 modification and confirmation received by using confocal laser scanning microscopy. It was also reported that DSF-LE1-R8 exhibits a marked anti-cataract effect from evaluation. Hence, nano-sized particles with R8 alterations in lipid emulsions were proposed as a significant ocular delivery method to enhance penetration in the cornea and DSF visual delivery [23]. Capsule designing with insulin-like hydrophilic drugs for the preservation of its biological activity and stability through double emulsion methodology and its entrapment into biodegradable microcapsules in the following manner with xanthan and chitosan gum complexes containing shells was devotedly investigated by Mutaliyeva et al. Several formation factors such as biopolymer and oil type, stabilizer, and its concentration, aqueous phase internal content, volume fraction, regime mixing, and time were also evaluated. The complex's effects on the emulsion formation process, characteristics, and stability of resultant emulsions were interrogated using interfacial charge (zeta-potential) and the size distribution (DLS). The prepared capsules were analyzed using size distribution, zeta potential, and microscopic characterizations. Insulin release kinetics was monitored through UV–vis spectroscopy, and reports suggested that sustainability enhancement was progressive [24].

4.4. W/O/W Double Emulsion Formulation

Sesamol, the phenolic compound and degradative product of sesamolin, has poor bioavailability but is recognized for its anti-inflammatory properties. An attempt was made to increase its bioavailability through mixed phosphatidylcholine micelles encapsulation. Sesamol solubilization and entrapment can be seen in PCS (phosphatidylcholine mixed micelles), having a 3.0 nm particle size with 96% efficiency. PCS showed lower comparative fluorescence intensity when it was compared with free sesamol. PCS cellular uptake, bioaccessibility, and transport across cell monolayer was 1.2-fold, 8.58%, and 1.5 times improved compared with FS. By using lipoxygenase inhibition and an LPS-treated RAW 264.7 cell line, the FS and impact of PCS regarding its anti-inflammatory action were studied. The iNOS protein expression downregulation (27%), ROS (32%), NO (20%), and inhibition of lipoxygenase with 31.24 μM (IC_{50}) were affected by PCS in comparison to FS [25]. Omid Shamsara et al. loaded piroxicam (PX) into multi-layered oil-water emulsions and stabilized using the complexes of β-lactoglobulin (β-L) and pectin, in which homogenized sunflower oil was used as a primary emulsion with a β-L solution containing PX. These droplets get stabilized by a subordinate layer of pectin. The low-methoxyl sunflower pectin (LMSP), low-methoxyl citrus pectin (LMCP), high-methoxyl apple pectin (HMAP), and high-methoxyl citrus pectin (HMCP), were respectively utilized to produce emulsion droplets with a secondary or double layer. Creaming stability, PX entrapping efficiency (%), and droplet mean size of emulsion (D43) have been determined. Trend or PX release was examined and used for the zero-order kinetic experiment. The output of such experimental work has suggested that citrus pectins with NaCl and β-L/high-methoxyl-apple-stabilized emulsions were found with maximum stability, with good PX loading capacity compared with stabilized emulsions by a β-L complex of pectin in the absence of NaCl. The mean droplet size of double-layered emulsions increased at a high pectin fraction and reduced by a low β-L fraction [26].

A potent bioactive molecule such as betulinic acid (BA) is recognized for therapeutic action. Yet, it has limited efficacy because of poor solubility and low bioavailability. Harwansh and coworkers developed BA-loaded nanoemulsions with increased hepatoprotective activity and bioavailability. Using the BA-NE1 procedure, the nanoemulsion was formulated containing surfactants such as labrasol, olive oil, aqueous phase, and co-surfactant, such as plural isostearate, in the convenient ratio optimized. Its characterization was done through several parameters, such as the size of the droplet, refractive index, zeta potential, FTIR, UV-spectrophotometry, TEM, and stability studies. A droplet size of about 150.3 nm with negative zeta potential such as −10.2 mV of this emulsion was evaluated. Pharmacokinetic limits such as Cmax (96.29 ngmL^{-1}), AUC0-t ∞ (2540.35 nghmL^{-1}), the elimination half-life (11.35 h), Tmax (12.32 h), and relative bioavailability (440.48%F) were also investigated and compared with BA. The hepatic serum marker levels and antioxidant enzymes concerning CCl4-intoxicated groups (** $p < 0.05$ and *** $p < 0.01$) were significantly restored by BA-NE1. Accordingly, the study also reveals that the BA-loaded nanoemulsion could improve hepatoprotective activity because of increased solubilization and enhanced oral bioavailability [27]. There is an advanced formulation with better biocompatibility, stability, and higher loading of hydrophobic drugs in submicron emulsions (SEs) from sterilization with an autoclave. To increase the targeting and uptake of tumor cells, SEs get altered by target moieties and a positive charge. Cationic DocSEs (DocCSEs), docetaxel-loaded SEs (DocSEs), and peptide-RLT-modified DocCSEs targeted by low-density lipoprotein receptor (LDLR) were formulated. A particle size of 182.2 ± 10 nm and loading efficiency of docetaxel (Doc) 98% with a zeta potential of 39.62 ± 2.41 mV was reported for optimized RLT-DocCSEs. They showed 96 h of sustained release and were found stable for 2 months at 4 degrees Celsius. Significantly more cell apoptosis and RLT-DocCSEs have caused inhibition of cells compared with DocCSEs and DocSEs. RLT-DocCSEs showed greater cellular uptake with slow elimination from DocCSEs and DocSEs [28].

4.5. Polymeric Emulsifier Containing Formulation

Polyethylene glycol (PEG), lactide (LA), and ε-caprolactone (CL) were derived as amphiphilic bioresorbable copolymers studied for their emulsification and degradation properties. With monomethoxy PEG, lipophilic 20 wt.% PCL, PLACL block, PLA, and 80 wt.% PEG (hydrophilic) block comprising polymers were formulated with the LA and/or CL ring-opening polymerization process. These emulsifiers have analogous capabilities for stabilizing squalane/water interfaces in emulsification as they possess equivalent hydrophilic–lipophilic balance (HLB) values. Polymer degradation within the emulsion and in the aqueous phase at 37 degrees Celsius to mimic conditions of the human body was carried out. According to the result, the polymer degradability was found to cause instability in the emulsion. In addition, emulsion polymer matrices exhibit lower degradative rates than in an aqueous phase from corresponding polymers. The characteristics in pharmaceutical applications are of keen interest, particularly for sustained delivery mechanism designing [29]. A study aimed to develop a re-dispersible dry emulsion that contains simvastatin, a model drug with lipophilic, low water solubility properties; they used a fluid bed coating methodology. This represented manufacturing of dry-emulsion-mode-formulated pellets, in which a dry emulsion layer was applied to a neutral core. As an oily phase, 1-oleoyl-rac-glycerol, a preliminary formulated material, was selected because of its higher drug solubility and potent bioavailability possibilities. Tween 20, mannitol, and HPMC were used as solid surfactants and carriers. The experimental design was used more specifically for mixture design to get the optimal formulation composition. The initial responses used as formulation optimization parameters were the stability and ability to reconstitute the emulsion. On optimization, the formulation represented slender-sized droplet distribution at reconstitution, high strength, satisfactory drug encapsulation, and an increase in dissolution possessions as compared with a pure drug and a non-lipid-based tablet. Uniform morphological data for the functional layer and separated droplets with simvastatin and uniform distribution of size and coated pellets with a circular shape were derived from image analysis using Raman spectroscopy and scanning electron microscopy. The work revealed the evidential design concept of re-dispersible dry emulsions using the fluid bed layer technique [30].

4.6. Nano-Precipitation/Dry Formulations

From another work, nanoemulsion from surfactant-free Pickering formulations can liberate a drug with enhanced oral bioavailability at specific pH. By using the nano-precipitation method, magnesium hydroxide-based stabilizing nanoparticles were obtained. The $Mg(OH)_2$ nanoparticles stabilized oil-in-water Pickering nanoemulsions were prepared using a high-energy procedure and sonication probe. The effect of all formulating properties, composition, and the $Mg(OH)_2$ nanoparticles' size on the physicochemical parameters of Pickering nanoemulsions was explored with experimental processes. By using transmission electron microscopy and DLS, the formation was characterized. Moreover, the $Mg(OH)_2$ was solubilized in an acid medium as an advantage that leads to nanoemulsion destabilization and oral release of active components. It is revealed from the acid-releasing work (pH = 1.2) that an increase in release is due to the loading of nanodroplets with the saturation of concentration. At pH = 6.8 (an alkaline media), ibuprofen is significantly released from saturated nanoemulsions in an acid medium. These nanoemulsions not only prompt drug bioavailability but also protect patients from acid medicine side effects through the basic features of hydroxides. Additionally, hydroxides increase pH when present in the stomach; enhancing the release of ibuprofen is greatly affected by pH for solubility [31].

The drug atorvastatin calcium (ATV) is less bioavailable. A dry emulsion method was orally utilized with lyophilized disintegrating tablet development to improve its dissolution in vitro and performance in vivo. Under proper homogenization, the emulsions were formulated using a collapse protectant (glycine) as an aqueous phase, 4% alginate/gelatin-containing mannitol, and synperonic PE/P 84 (surfactant) as an oil phase. The impact of the emulsion formulation parameters was investigated for the prepared tablets' friability, in vitro dissolution, and disintegration time for tablets to the drug. The outcomes revealed the important impact of matrix emulsifier types and the former on disintegration time. From a study of in vitro dissolution, the ATV rate of dissolution was enhanced from lyophilized-dry-emulsion-tablets (LDET) in comparison with a plain drug. Optimized ATV-loaded LDET was studied for DSC and XRD, and the results proved drug presence in the amorphous form. From the SEM images, the intact, non-collapsible, porous-structured LDET was seen to have a complete ATV crystallinity loss. When high-fatty rats were administrated with ATV-loaded LDET, the serum and tissue levels were found to be significantly decreased [32]. The polymers that have grown extensively in the last decades are known to be smart polymers because of their extensive uses for drug targets with controlled drug delivery methods. Based on this concept, Chekuri Ashok et al. used *Albizia lebbeck* L. seed polysaccharide (ALPS) to design and make the preparation of smart releasing emulsion (o/w). Similarly, the physicochemical properties, such as the capacity of emulsion (EC), viscosity, stability of emulsion (ES), polydispersity index (PDI) zeta potential, and related parameters were examined. The EC/ES was found to increase with the increase in ALPS concentration. Using factorial design possibilities, emulsion formulations were statistically oriented. The shear-thinning behavior was seen in all the emulsions. The polydispersity index and zeta potential were recorded at 0.232–1.000 and −35.83 mV to −19.00 mV, respectively.

4.7. Solid-Self-Nano/Micro-Emulsifying-Drug-Delivery-Systems

Moreover, the cumulative percent drug release at 8 h from the emulsions was between 30.19–82.65%. The zero-order release kinetics was observed for the drug release profile. So, as a conclusion, the ALPS could be used as a smart polymer and a natural emulsifier to prepare pH-sensitive emulsions for drug delivery systems [33]. Parth Sharma and coworkers demonstrated that various solid-self-nano-emulsifying-drug-delivery-systems (S-SNEDDS) could be prepared using porous hydrophobic and hydrophilic carriers to enhance the simvastatin (SIM) bioavailability and dissolution rate. When 0.1% SIM is used to prepare SNEDDS containing Labrafil M 1944 CS, Tween-80, and ethanol, the resultant droplet is 40.69 nm. Aerosil-200, Syloid XDP 3150, Micro Crystalline Cellulose PH102, Syloid 244FP, and lactose were applied as hydrophobic carriers, whereas CMC, sodium carboxy methyl cellulose, hydroxyl propyl-β-cyclodextrin, and polyvinyl alcohol were used as hydrophilic carriers. Through biopharmaceutical, micrometric, and stability studies, S-SNEDDS was characterized. The S-SNEDDS and liquid-SNEDDS of Aerosil 200 have shown significant superiority on unprocessed and marketed SIM from in vitro dissolution determinations. The crystalline SIM was revealed through a DSC, XRD, and scanning electron microscope when present in altered amorphous SNEDDS preparations, where Aerosil 200 was used as a carrier. In addition, pharmacokinetic studies carried out on rats demonstrated an increase in time of 0.5 h for (Tmax) maximal concentration, maximal concentration (Cmax) by 3.75 folds, mean residence time with 1.22 h, (AUC0-t) area under the curve by 1.54 times, AUC0-∞ with 2.10 folds, and bioavailability by 3.28 folds. All these findings support the superiority of developed S-SNEDDS over the market formulation. It can be concluded that such S-SNEDDS enhances the SIM's bioavailability and dissolution rate [34].

Another piece of work sought to investigate excipient influences on liquid self-micro-emulsifying-drug-delivery-systems (SMEDDS) and l-tetrahydropalmatine (l-THP) containing SMEDDS laden in the pellet properties. In addition, the study was extended with a rabbit model to compare l-THP suspension and such SMEDDS bioavailability. Capryol 90 and surfactant mis were interrogated in the SMEDDS formulation at their optimum ratio. In pellet-SMEDDS, l-THP showed an amorphous state proved by powder X-ray diffractometry. When a pharmacokinetic study using LCMS spectrometry was carried out in a rabbit model, the results revealed that SMEDDS enhances l-THP oral bioavailability by 198.63% compared with the l-THP suspension. The study also demonstrated that there was no noteworthy difference from the original liquid SMEDDS, the ultimate mean concentration (Cmax), and the relative mean bioavailability of pellet-SMEDDS [35]. Another potent anti-inflammatory agent, *Boswellia serrate* gum resin, has been vastly used in ancient medicines. However, its efficacy must be evaluated, as it has low oral bioavailability. Hence, a self-nano emulsifying system (SNES) has been used to improve the systemic concentration of boswellic acids. The (KBA) 11-keto-β-boswellic acid and (AKBA) acetyl-11-keto-β-boswellic acid are the most biologically active constituents of boswellic acids used for indication and evaluation of the efficiency of the self-emulsifying system. In a lipolysis study, KBA and AKBA bio-accessibility and aqueous solubility were reported to increase by 2.3 and 2.7-fold, respectively. A noteworthy increase in the oral bioavailability of these two acids was recorded by more than two, as compared with bulk oil suspension from an in vivo pharmacokinetic study. Thus, SNES is an effective oral formulation with more storage stability to improve the bioavailability of boswellic acids [36].

4.8. Self-Nanoemulsifying System to Improve the Oral Bioavailability

Curcumin and coumarin are famous for their broad spectra pharmacological and biological activities, such as antioxidant, anti-inflammatory, anticancer, and antimicrobial, with impaired therapeutic administrations due to poor solubility and less stability in water. Based on the encapsulation efficacy (EE; the loaded amount of drug into formulations) of such drugs, their bioavailability is assessed. Hence, work was carried out to overcome these limitations by enhancing bioavailability with nano-encapsulated emulsions. Aminated nano cellulose (ANC) particles can stabilize the PE through an oil/water-based process for complete factorial optimization design from different components of the oil phase with Tween 80 and medium-chain triglyceride (MCT) for nanoemulsions. The PEs and nanoemulsion formulations were obtained with a particle size of $\leq$150 nm. Along with zeta potentials, the storage time, pH, and ANC concentration on emulsion stability as influencing factors were examined. Curcumin and coumarin's EE were found to be > 90%. The kinetic profiles of the encapsulated PEs' release demonstrated a sustained release with possible increased bioavailability. Curcumin-encapsulated PE showed a high release percentage compared with coumarin. Curcumin and coumarin-loaded PEs were also examined for antimicrobial as well as anticancer potential using Gram (+)/(−) bacteria and fungi and L929 and MCF-7 (human cell lines), respectively, in in vitro cytotoxicity determination. The results proved that PE curcumin and coumarin are significant inhibitors of microbial growth and prevention from cancer [37]. Ibuprofen is a potent analgesic and a non-steroidal anti-inflammatory drug, where the administration can pose side effects or reactions in the body, such as ulcers or bleeding, and can lead to increased stomach or intestinal perforation risk. In a report, Yiping Deng et al. reported IBU nanoparticles (IBU-NPs) formulation by emulsion-solvent-freeze-drying/evaporation to improve its solubility. Under optimum conditions, IBUNPs were produced with a 216.9 $\pm$ 10.7 nm particle size, which was characterized by DSC, X-ray, SEM techniques, equilibrium solubility, in vitro transdermal rate, and transdermal bioavailability. Morphological features of IBU-NPs revealed porous clusters. From the analysis of prepared IBU-NPs, a low crystallinity was observed. Chloroform and ethanol residual amounts were 9.6 and 170 ppm, respectively, lesser than the class II ICH limit. The IBU-NPs chemical structure was retained, but IBU-NPs, after preparation, underwent amorphous states, reported from measurement analysis. Compared with transdermal and oral raw IBU, IBU transdermal bioavailability was significantly enhanced by the IBU-NP group. Moreover, IBU-NP transdermal gel exhibited a stable cooling rate and longer cooling duration in febrile rats. Even at low and mid doses, better efficacy was given by IBU-NP transdermal gel than oral IBU. The results concluded that for transdermal delivery formulations, IBU-NPs could be appliable and have potent value for non-oral administration [38]. In different kinds of cancer treatment, colloidal particles (CPs) are developing materials in drug transport as they have rapid effectiveness and biosafety. Rose PLGA/Bengal CPs were formulated by W/O/W emulsion and layer-by-layer electrostatic adsorption. Furthermore, they were evaluated and characterized as having potential for breast cancer treatment. They were also examined for the efficacy of zeta potential, drug release kinetics, size, HCC70 (negative breast cancer cell line), and cell viability inhibition. The outcomes revealed that all kinds of CPs may be an alternative to routine cancer treatment as having enhanced retention (EPR)/permeation impacts solid tumors. However, CPs with W/O/W double emulsion showed delivery times of up to 60% in 2 days, which is more suitable, whereas layer-by-layer CPs displayed a t release of 50% in just 90 min. Cell viability could be decreased with both types of CPs, which was encouraging for in vivo testing models that can help prove their feasibility and efficacy against triple-negative breast cancer treatment [39].

With high-pressure homogenization, corn oil and polysorbate 80 tea polyphenols (TP) were emulsified. The 99.42 $\pm$ 1.25 nm sized droplet for O/W TP nanoemulsion was reported on the formulation. In storage at 4/25/40 °C, TP nanoemulsion was found to be stable. A simulated digestion assay in an in vitro study found that (−)-epigallocatechin gallate (EGCG) bio-accessibility was enhanced in a nanoemulsion than in aqueous solution. However, (−)-epigallocatechin (EGC), (−)-gallocatechin gallate (GCG), and (−)-epicatechin (EC) bio-accessibilities were significantly decreased. A rat-fed study with an aqueous solution and TP nanoemulsion showed considerably low plasma concentrations of EGCG and

EGC. The data confirmed that a nanoemulsion for the tea polyphenols delivery might improve the absorption of EGCG through controlled release [40]. Heba Elmotasem and group aimed to innovate an effective oral sustained release of caffeine, a water-soluble drug. Due to its rapid absorption and elimination, caffeine is frequently used in administration to get an excellent therapeutic outcome. So, a *w/o* Pickering emulsion with caffeine incorporation was constituted from stabilized wheat germ oil using synthetic MgO nanoparticles (MgO NPS). Based on antioxidant hepatoprotective abilities, and anticarcinogenic, the components of the emulsion were selected. Such NPs of MgO were formed via the sol-gel process, and further characterization was done by TEM, cytotoxicity, contact angle, and X-ray diffractometry analysis. This Pickering emulsion stabilized by MgO NPs and conventional MgO particles was compared. Both methods evaluated caffeine release, stability, and droplet size. Earlier, a droplet size of 665.9 ± 90 nm was found stable in phase separation for 2 months. F1 could afford caffeine's sustained release following zero order kinetics that reached 70% within 48 h. About 36% hepatocellular carcinoma (HEPG2) growth inhibition has been shown by 100 ppm of F1. CCl4-intoxicated rats were used for in vivo and histopathological examinations. Liver enzymes (ALT and AST), inflammation marker (protein kinase C), and oxidative stress biomarkers targeted biochemical analysis suggested that the selected formula induces satisfactory hepatoprotection. The procedure has an economical approach in multiple therapies. It is safe, effective, and sustained levels of caffeine [41].

A study was carried out to examine the bacterial cellulose (BC) potential of melatonin (MLT) oral administration. It is a natural hormone and has issues such as poor solubility with low oral bioavailability. Bacterial cellulose got oxidized after sulfuric acid hydrolyzation to produce bacterial cellulose nanofiber suspension (BCNs). The emulsion solvent evaporation technique prepared melatonin-loaded BCNs (MLT-BCNs). These were characterized by XRD, FTIR, DSC, thermal analysis, SEM, fluorescence microscopy (FM), and instrument tools. The resulting data indicated that in BCNs, fibers became shorter and thinner than with BC. Both MLT-BCNs and BCNs have impressive thermodynamic stability; MLT was homogeneously distributed in MLT-BCNs. MLT-BCNs showed more speedy dissolution MLT rates compared with MLT in SGF and SIF, which are commercially available. The cumulative release rate of dissolution was found to be approximately 2.1 times that from MLT (commercially available), whereas in rats, it was 2.4 times more for the same. Hence, a promising delivery could be provided by MLT-BCNs with improved bioavailability and dissolution in the oral administration of MLT [42]. In vivo, cyclosporine ophthalmic emulsion (COE) with a spherical size distribution was studied; applied shear performance was affected by several physicochemical parameters such as zeta potential, pH, surface tension, and osmolality by the function of viscosity profile. A study was carried out using a modeling approach to predict drug bioavailability from COE to the conjunctiva, tear film breakup time, and cornea in human subjects as a function of the vehicle physicochemical properties such as surface tension, osmolality, and viscosity. From the bioavailability predictions, it was found that geometric mean ratios for test-to-reference in comparison to qualitatively and quantitatively formulations showed minor sensitivity. In contrast, the individual predictions were found to be sensitive to conjunctival permeability variations and corneal. The tear film breakup time from baseline values was found to be too sensitive to viscosity, showed slight susceptibility for surface tension, and was insensitive towards osmolality, as concluded from the parameter sensitivity analysis results. In addition, further enhancements in the modeling framework will develop the study to be more helpful in future prospects of COE bioequivalence in strong generic drug compounds [43].

Yamasaki et al. aimed to improve the oral bioavailability of praziquantel in conjugation with human serum albumin (HSA). These were prepared using praziquantel in an oil solution and HSA aqueous solution by spray drying. Amorphous praziquantel containing multiple smooth corrugated particles was aggregated and almost equivalent to the theoretical dosing. In an aqueous medium, the Praziquantel solubility was enhanced in the physical mixture and prepared particles. Moreover, the dissolution rate was also increased in the event of particles and not in physical combination. Hence, HAS addition increases the dissolution rate when spraying through emulsification. The produced particles (HSA/praziquantel = 1/1 w/w) have higher concentration-time curve (AUC) values, of about two times, and maximum plasma concentration (Cmax) than raw praziquantel values, reported from a pharmacokinetic study. The particles' oral bioavailability was enhanced and was considered because of the increased dissolution rate. The process of praziquantel-HSA particle production could advance the oral bioavailability of other hydrophobic drugs [44]. The lipidic biocompatible and safe molecules are in high demand for self-micro-emulsifying drug delivery systems (SMEDDS). A work targeted to study oral mucosal irritation was reported to see the application of erucic acid-based bicephalous hydrolipid (BHL) in SMEDDS as an oil phase using Efavirenz (EFA), with poor bioavailability and water solubility of the drug. It showed higher drug loading efficiency, about 80.35 $\pm$ 3.1%, with 0.23 $\pm$ 0.031 polydispersity index (PDI). EFA SMEDDS was also examined for standard stability tests and revealed that it was highly stable. EFA SMEDDS in vitro dissolution profile manifested > 95% release of drug in an hour and substantially enhanced bioavailability in vivo, at almost six times more than a drug in plain suspension. From the data outcomes, it was concluded that BHL could effectively be used as an oil phase in SMEDDS to improve the BCS Class II drug's solubility and bioavailability. In addition, it holds possibilities as a novel excipient for enhancements of solubility and bioavailability [45]. Candesartan cilexetil drug delivery faces a significant limitation of poor oral bioavailability, mainly due to its low solubility in aqueous solution and intestinal P-glycoprotein (P-GP) transporters effluxions. Yet, the P-gp extent role in decreased candesartan cilexetil oral bioavailability is cryptic. A study was carried out where previously developed candesartan cilexetil-loaded self-nano-emulsifying drug delivery system (SNEDDS) was examined for its ability to enhance oral bioavailability through intestinal P-GP transporters inhibition. P-GP–mediated efflux has some role in decreased candesartan cilexetil oral bioavailability despite whether or not the developed SNEDDS showed P-GP inhibition activity. Alternately, SNEDDS formulation with high surfactant concentration demonstrated a remarkable challenge in broad applications, specifically to chronically administered drugs. As the period of treatment increases, a reduction in intestinal mucosal damage was recorded from toxicity studies having acute and subacute designing. The surfactant-induced mucosal damage was found reversible from the observations. Hence, a sound delivery system with improved oral bioavailability could result from developed SNEDDS against chronically administered drugs [46].

4.9. Water Picric Emulsion Formulations

PE has received extensive attention for the encapsulation of lipophilic guests in the field of food and biomedicine. Although PE stabilities and demulsification control allow the release of species that have been encapsulated, they retain a challenge for the gastrointestinal tract. In a word, natural kaolinite with phosphatidylcholine was altered to prepare phosphatidylcholine-kaolinite to act as an emulsifier in stabilizing medium-chain triglyceride (MCT)/water PE that encapsulates curcumin. This was done to study curcumin-loaded-MCT-water PE for curcumin bioavailability and emulsification in a cell uptake assay and simultaneous implementation of intestinal digestion. The results suggested that phosphatidylcholine-kaolinite wettability would be modified by regulatory alteration temperature so the emulsion stability might be prevented. In the gastric acid presence condition, the prepared phosphatidylcholine-kaolinite has a contact angle of 123° of three-phase, which has optimal values for improved stabilization of the MCT/water PE. A condensed shell composition formed on the emulsion droplet surface from phosphatidylcholine-kaolinite, when it was dispersed in the W/O interface, controls demulsification efficiency from the release of encapsulated curcumin. After a period of 120 min of imitation gastric digestion, only 18.9% of the curcumin was released because of MCT/water PE demulsification. On the contrary, it was entirely removed after 150 min of simulated intestinal digestion, as expected. The PE stabilization by phosphatidylcholine-kaolinite is an encouraging transport carrier for drugs or lipophilic foods for improved bioavailability [47].

Na Man et al. aimed at the preparation of a myricitrin-loaded self-micro-emulsifying drug delivery system (MSMEDDS) for the enhancement of myricitrin's low oral bioavailability. MSMEDDS consisting of oil phase (ethyl oleate), (surfactant) Cremophor EL35, and, as a co-surfactant, dimethyl carbinol was prepared. The particle size, encapsulation efficiency, and zeta potential were used to characterize MSMEDDS. Prepared MSMEDDS exhibited a 21.68 ± 0.15 nm droplet with -ve zeta potential and high encapsulation efficiency of about −23.17 ± 1.03 mV and 92.73%, respectively. M-SMEDDS are able to release myricitrin more significantly than free myricitrin, which was revealed from an in vitro release study. M-SMEDDS has a 2.47-fold increased relative oral bioavailability compared with free drugs. Both in vivo and in vitro studies demonstrated that M-SMEDDS could enhance the solubility of myricitrin and its oral bioavailability, providing preliminary confirmation for further M-SMEDDS [48] applications through clinical research. Darunavir-loaded lipid nanoemulsion was formulated in a research work to enhance its oral bioavailability and improve brain uptake. Darunavir was prepared from several lipid nanoemulsion batches by high-pressure homogenization using egg lecithin, Tween 80, and soya bean oil. DNE-3 was an optimized batch with a 109.5 nm globule size, −41.1 mV zeta potential, 93% entrapment efficiency, and 98% creaming volume. It was stable for 1 month at 4 °C with inconsiderable changes in spherical size and zeta potential ($p > 0.05$). Pharmacokinetics studies of in vivo male Wistar rats showed 223% Darunavir bioavailability compared with the suspension of the drug. DNE-3 has a two-fold higher Cmax and brain uptake than in suspensions found in an organ biodistribution study. Darunavir's increased bioavailability in nanoemulsions could lower the side effects related to the dose. Furthermore, because of high organ distribution, HIV reservoir organs have Darunavir passive uptake [49]. Amphiphilic bacterial cellulose nanocrystals (ABCNs) interfacial assembly by the PE method was suggested to enhance the compatibility with hydrophobic drugs and alginate. Biosynthesized bacterial cellulose was hydrolyzed by sulfuric acid to prepare BCNs used in particulate emulsifiers, considering alfacalcidol dissolved in CH_2Cl_2 as a model drug in the oil phase. The ultrasonic dispersion PE of O/W was formulated by and later well-dispersed in a solution containing alginate. By this, beads of drug-loaded alginate composite were prepared auspiciously by external gelation. BCNs have good colloidal properties, and a flocculated fibril network could be formed, beneficial for stabilizing Pickering emulsions revealed from results. BCNs have irreversible adsorption at the O/W interface, which could preserve PE droplets against Ostwald coalescence and ripening upon dispersing in an alginate-rich solution. Amphiphilic BCNs interfacial assembly and alginate composite beads hydrogel shells formed due to external gelation attain loading and sustained release of alfacalcidol. The release curves of alfacalcidol and the release mechanism from composite beads were significantly fitted in the Korsmeyer–Peppas model when associated with non-Fickian transport. Additionally, subsequent alginate composite beads can exhibit less cytotoxicity and advanced competencies for osteoblast differentiation [50].

Recent studies have revealed that to treat type 2 diabetes, metformin hydrochloride (Met) is a primitive drug and has the potential for Alzheimer's disease reduction. Met (B-Met-W/O/W SE) containing borneol W/O/W composite submicron emulsion was prepared with the expectation of better bioavailability, prolonged circulation time in vivo, and Met drug brain targeting. The optimized formulation has a mean droplet size of 386.5 nm, 0.219 polydispersity index, and 87.26% composite encapsulation potency. Met collaborated with carriers in B-Met-W/O/W SE, confirmed from FTIR analysis. Met in the B-Met-W/O/W SE in vitro release delivery system was slower than the Met-free drug. The AUC 1.27, MRT 2.49, and t1/2 of the B-Met-W/O/W SE system are 4.02-fold higher compared with the Met-free drugs, revealed from rat pharmacokinetic studies. B-Met-W/O/W SE system drug-targeting index to brain tissue was also more than that of the Met-W/O/W SE system and Met-free drug. The results concluded that the B-Met-W/O/W SE drug delivery system had encouraged candidature for clinical Alzheimer's disease treatment [51].

4.10. Nano System-Containing Formulations (NSCF)

NSCFs have recently shown significant advancement in nanotechnology research for active agents and drug delivery to human skin. A protein drug was extracted from tissues of medicinal leech and examined for kinetic stability, isotopic nanoemulsion formulation for topical delivery with negligible surfactant and co-surfactant amounts, and optimal stability and solubility were investigated in a study. Nanoemulsion formation and its stability were affected by oil phase physical properties. Several factors, such as oil content and type (sesame oil and olive oil), were evaluated for their impact on protein nano emulsion particle size and stability. In addition, protein nanoemulsion with optimized formulation was characterized for zeta potential, pH, viscosity, refractive index, droplet size, and transmission electron microscopy (TEM). For selecting the best formulation, stability studies were also carried out. The results concluded that an increase in sesame oil and olive oil concentration yielded nanoemulsions with some properties, such as higher stability and small-sized droplets. However, an olive oil-based nanoemulsion was observed to have slight alterations in droplet size. Several experiments selected a 25% olive oil-containing nanoemulsion as an optimized formulation as it has a tiny droplet size (143.1 nm), high zeta potential (−33.3 mV), and low polydispersity index. No significant changes were observed at 4 °C for a 30-day storage duration in pH, viscosity, and droplet size. The technique also showed that the nanoemulsion selected was physically stable. Additionally, the particles have a spherical shape, morphologically confirmed from TEM studies. So, in conclusion, nanoemulsions of protein drugs have been proven as promising novel formulations and can improve protein drug stability significantly [52]. Sadaf Chaudhary et al. demonstrated that a self-nano-emulsifying drug delivery system (SNEDDS) improves Nabumetone (NBT) oral bioavailability and anti-inflammatory impact. NBT has poor solubility in aqueous drugs exploring less bioavailability when orally administrated. The preparation of NBT-SNEDDS was done using a pseudo-ternary phase diagram and polyethylene glycol-400 (PEG-400), Capryol-90, and Tween-80. SNEDDS components were curtained, and for deionized water, PEG-400, Tween-80, and Capryol-90 in an optimal ratio of 58:4:16:22 was found to be optimal. The prepared SNEDDS underwent characterization for anti-inflammatory and pharmacokinetics properties and compared with optimized NBT-SNEDDS and the marketed tablet suspension in rats. The SNEDDS drug had a 3.02-times higher oral bioavailability than the marketed tablet suspension. A significant increase in anti-inflammatory activity was presented by NBT-SNEDDS than commercial NBT products that were orally administered. NBT-SNEDDS could be a potent carrier for NBT oral dosing with improved bioavailability and therapeutic impacts suggested from the results [53]. Resveratrol (RVT) re-dispersible dry emulsion (DE) was prepared using caprylic/capric glyceride (CCG) and using low-methoxy pectin (LMP) as the lipid phase comprising component and emulsifier. A Box–Behnken design was used to optimize and examine redispersed emulsion size, spraying efficiency, and angle of repose from spray dryer pump speed and formulation effects. For the estimation of RVT-DE dissolution properties, redispersibility was used. LMP and CCG concentrations affect redispersed emulsion size explored from the results. Any change in concentration of LMP and CCG, i.e., increase or reduction, results in a small droplet size in the emulsion and rapid drug dissolution from RVT-DE and influences the repose angle. High CCG and low LMP concertation using RVT-DE generation explored less repose angle, suggesting suitable flow property. The prepared formulation was optimized within the design space with 7% *w*/*w* of CCG and 2.75% *w*/*w* of LMP when sprayed at a pump speed of 10.1 mL/min, significantly fulfilling all criteria, i.e., high spraying efficiency, good flow, and small redispersed size. From intact RVT, RVT in RVT-DE has extraordinarily high photostability [54].

Organogels used in pharmaceutical and food sciences have some technical issues, such as confined drug diffusion and insufficient proper gelating molecules. They are necessary features to design new products. The use of emulsions is an alternative for enhancing the technological properties of organogels. There is a need for more information about the permeability and bio-accessibility behavior of bioactive-loaded, organogel-based emulsions. To study the physical properties of betulin, curcumin, and quercetin, three different bioactive-loaded vegetable oil-containing organogel-based emulsions, experimental work was carried out for bio-accessibility and influence. Coconut oil, canola, and myverol were used as a gelator (10% w/w) to prepare organogels by mixing water (80 °C) and melting proper organogels at high shear conditions (20,000 rpm) Water-in-oil emulsions (at 5, 10 and 12.5 wt.% of water content) were formulated. Rheological tests (frequency, creep-compliance measurements, temperature sweeps, and amplitude), micrographs, particle size, and DSC analysis were performed. Lipolysis, bio-accessibility, in vitro digestion, and permeability assays on the Caco-2 cell culture were also investigated. Coconut oil-based organogels have poor emulsification properties [55]. A multiple W/O/W emulsion (ME) was formulated from clotrimazole (CLT) and examined for anticandidal agent efficacy against the marketed products. Physicochemical characterization was carried out from the estimated CLT-ME selected previously. Franz diffusion cells were used with human skin, sublingual and vaginal mucosae, and porcine buccal biological membranes to assess in vitro liberation and ex vivo permeation behavior. Antifungal activity was also tested against *Candida* strains. CLT-MEs with two different sizes, 29.206 and 47.678 μm, exhibited high zeta potential of −55.13 and −55.59 mV, with skin-compatible pH values of 6.47 and 6.42 and dependency on pH variation. CLT-MEs showed physicochemical stability at room temperature and were kept up to 180 days. CLT-MEs have exhibited pseudoplastic behavior with viscosities and hysteresis areas of 331 mPa·s and 286, with high spreadability properties to commercial products. An enhanced CLT release pattern was contributed with the ME system with a hyperbolic model following. Compared with commercial products, CLT with high skin permeation flux was from the ME system. Compared with commercial reference, more CLT amounts were retained in mucosae and skin. CLT-MEs have high antimycotic efficacy, so they could become an excellent tool for topical candidiasis treatments and clinical investigations [56].

José Soriano-Ruiz et al. conducted a study to concur bisdemethoxycurcumin (BDMC) low bioavailability and solubility by preparing a self-micro-emulsifying system loaded with BDMC (BDMC-SMEDDS). Pseudo-ternary phase diagrams (PTPDs), compatibility and solubility tests, and d-optimal concepts were used for formulation designing. In vitro assessment of fabricated BDMC-SMEDDS was done to determine entrapment efficiency (EE), droplet size (DS), morphology, drug stability, and release. Moreover, in vivo behavior examination was also done in rats after BDMC-SMEDDS oral administration. The resultant formulation contained BDMC (50 mg), ethyl oleate (EO, oil, 207.5 mg), Kolliphor EL (K-EL, as an emulsifier, 645.3 mg), and PEG 400 (co-emulsifier, 147.2 mg). Good stability of BDMC-SMEDDS was found, with a mean size of 21.25 ± 3.23 nm and 98.31 ± 0.32% EE. The optimal formulation was found to compose of Kolliphor EL (K-EL, emulsifier, 645.3 mg), PEG 400 (co-emulsifier, 147.2 mg), ethyl oleate (EO, oil, 207.5 mg), and BDMC (50 mg). The BDMC-SMEDDS with good stability had a mean size of 21.25 ± 3.23 nm and EE of 98.31 ± 0.32%. From BDMC-SMEDDS, around 70% of BDMC was released within 84 h than free BDMC, at <20%. In particular, the in vivo behavior of BDMC-SMEDDS showed that BDMC plasma concentration and AUC (0–12 h) were increased when compared with free BDMC. In all respects, BDMC-SMEDDS is potent in improving BDMC bioavailability and solubility and could be applicable in clinics [57]. Another study targeted luteolin's oral bioavailability and solubility improvement through supersaturated self-nano-emulsifying drug delivery system (S-SNEDDS) employment. SNEDDS formulation is composed of polyethylene glycol 400, caprylic/capric triglyceride, and castor oil hydrogenated with polyoxyl 35 with a ratio of 31.7:20.1:48.2 by weight. It was optimized and determined from pseudo-ternary phase diagrams, solubility studies, and central composite design. For luteolin-loaded SNEDDS at a 2% mass ratio, hydroxypropyl methylcellulose (HPMC) K4M was found to be an optimal precipitation inhibitor based on in vitro precipitation evaluations. A nanoemulsion having a 25.60 nm particle size was formed from luteolin S-SNEDDS and has a −10.2 mV zeta potential after dilution. Luteolin and HPMC K4M interactions were examined by differential scanning calorimetry (DSC), (FTIR) Fourier-transform infrared spectroscopy, powder X-ray diffraction (XRD), and 1H NMR spectroscopy.

Additionally, an outstanding 99% in vitro dissolution has been achieved by S-SNEDDS at pH 6.8 in phosphate buffer with 0.5% Tween 80. A S-SNEDDS pharmacokinetics study in vivo revealed a more noteworthy luteolin oral bioavailability enhancement (2.2-fold) in rats than in conventional SNEDDS. In conclusion, from the data illustrated, S-SNEDDS technology could be applicable at most minuscule for luteolin in promoting oral bioavailability and solubility for poorly water-soluble drugs [58]. For the treatment of osteoporosis, a novel cathepsin K inhibitor, HL235, has been designed and synthesized. To improve HL235 oral bioavailability, SMEDDS was designed to overcome HL235's low aqueous solubility. For the selection of a suitable oil, surfactant, and cosurfactant, a HL235 solubility study was conducted. Pseudo-ternary phase diagrams were prepared to identify the components' range in the isotropic environment and microemulsion region. To optimize the formulation of SMEDDS, desirability function and D-optimal mixture design were interrogated to get requisite physicochemical features, such as high solubilization capability and higher drug concentration after 15 min of dilution with simulated gastric fluid (SGF); this led to the formulation and optimization of HL235-loaded SMEDDS composed of surfactant (75.0% Tween 20), cosurfactant (20.0% Carbitol), and oil 5.0% Capmul MCM EP. The microemulsion formulated has been optimized, and the droplet size was 10.7 ± 1.6 nm with a spherical shape. Furthermore, there was a 3.22-fold elevated SMEDDS formulation relative oral bioavailability in rat pharmacokinetic studies than in its DMSO: PEG400 (8:92, *v*/*v*) solution. A promising approach could be made available by SMEDDS formulations on optimization by D-optimal mixture design to improve HL235 oral bioavailability [59].

Table 1 summarizes the different types of emulsion formulations and their impacts.

Table 1. Different types of emulsion formulation and their uses.

	Emulsion Type	Uses	Ref
1	Silybin nanocrystal self-stabilized Pickering emulsion, core-shell arrangement with 27.3 µm droplet, 40 days stability.	Increase oral bioavailability of silybin	[16]
2	(Poly lactide-co-glycolide) microparticles standard Emulsion	For a controlled drug delivery system	[17]
3	Silymarin-loaded nanoemulsions with Capryol 90	Improve oral bioavailability	[18]
4	Phosphorylated tocopherols-based emulsion	make poorly water-soluble drugs soluble	[19]
5	Polyethylene glycol, polysorbate, and capric triacylglycerol/caprylic-based emulsion	For delivering ellagic acid	[21]
6	Protein formulation in lyophilized electrospun fibers	Increases therapeutic compounds' shelf life	[22]
7	Capsule designing with insulin-like hydrophilic drugs	Preservation of biological activity and stability	[23]
8	Mixed phosphatidylcholine micelles encapsulation	Increase the bioavailability of sesamolin	[24]
9	Betulinic acid-loaded nanoemulsions with labrasol	Increase hepatoprotective activity and bioavailability of betulinic acid	[27]
10	Emulsion polymer matrices	Polymer degradability	[29]
11	Nanoemulsion from surfactant-free Pickering formulations	Enhancing oral bioavailability	[31]
12	Emulsion with collapse protectant (glycine) as an aqueous phase, 4% alginate/gelatin containing mannitol and synperonic PE/P 84 (surfactant) as an oil phase	Lyophilized disintegrating tablet	[32]
13	Smart-releasing emulsion (*o*/*w*)	Controlled drug delivery	[33]
14	Solid-self-nano-emulsifying-drug-delivery-systems	Enhancing the simvastatin bioavailability and dissolution rate	[34]
15	Self-nano emulsifying system	Improve the bioavailability of boswellic acids	[36]
16	Emulsion-solvent-freeze-drying/evaporation	Stable Ibuprofen nanoparticles formulation	[38]
17	W/O/W emulsion and layer-by-layer electrostatic adsorption	Formation of PLGA/Bengal colloidal particles	[39]
18	Produce bacterial cellulose nanofiber suspension	Enhancing oral bioavailability of bacterial cellulose	[42]
19	Emulsion based on praziquantel in an oil solution and HSA aqueous solution by spray drying	Improve the oral bioavailability of praziquantel	[44]

Table 1. *Cont.*

	Emulsion Type	Uses	Ref
20	Self-micro-emulsifying drug delivery systems	Improving the bioavailability and water solubility of the drug	[45]
21	Phosphatidylcholine-kaolinite acts as an emulsifier in stabilizing medium chain triglyceride (MCT)/water	Improved bioavailability of curcumin	[47]
22	Myricitrin-loaded self-micro-emulsifying drug delivery system	Myricitrin low oral bioavailability enhancement	[48]
23	Emulsion based on Amphiphilic bacterial cellulose nanocrystals	Enhance the compatibility with hydrophobic drugs and alginate	[50]
24	Self-nano-emulsifying drug delivery system	Improves Nabumetone (NBT) oral bioavailability and anti-inflammatory	[53]
25	W/O/W emulsion from clotrimazole	Anticandidal agent efficacy against marketed products	[56]
26	Self-micro-emulsifying system	Improving bioavailability and solubility of bisdemethoxycurcumin	[57]
27	Supersaturate self-nano-emulsifying drug delivery system	Luteolin's oral bioavailability and solubility improvement	[58]

5. Conclusions

The bioavailability and solubility of hydrophobic and lipophilic drugs is an issue for their oral administration, which is resolved using self-emulsifying formulations to a certain extent. Thus, this review article deals with emulsion-formulation-based drug delivery and improving the bioavailability of hydrophobic and lipophilic drugs. The dispersion or loading efficacy of the drug into emulsions depends upon the emulsion's constituents, such as oil, surfactants, and co-surfactants. In this regard, the nanoemulsions have been found effective for target drug delivery and improving the bioavailability of poorly water-soluble drugs. The present scenario of progression in technologies for drug carrying has broadcast the expansion of innovative drug carriers for the target release of self-emulsifying pellets, tablets, capsules, and microspheres, which has improved drug delivery with self-emulsification. Thus, the present review article will help readers working in the field of emulsion-based drug delivery with the increased bioavailability of lipophilic/hydrophobic drugs at the current time.

Author Contributions: R.K.A. wrote manuscript and prepared concept, K.S. checked grammar and plagiarism, A.B. proposed concept and executed the idea of this review article. All authors have read and agreed to the published version of the manuscript.

Funding: This research received no external funding.

Acknowledgments: Authors are greatly thankful to Kadi Sarva Vishwavidhyalaya, Gandhinagar, India, for support and infrastructure facilities. A.B. acknowledges TWAS-UNESCO Associateship-Ref. 3240321550 for providing opportunities to visit the Department of Chemistry, Indian Institute of Technology Madras, Chennai, India.

Conflicts of Interest: The authors declare no conflict of interest.

References

1. Stegemanna, S.; Leveillerb, F. When poor solubility becomes an issue: From early stage to proof of concept. *Eur. J. Pharm. Sci.* **2007**, *31*, 249. [CrossRef] [PubMed]
2. Ewart Cole, T.; Dominique, C.; Hassan, B. Challenges and opportunities in encapsulating liquid and semisolid formulations into capsules for oral administration. *Adv. Drug Deliv. Rev.* **2008**, *60*, 747. [CrossRef] [PubMed]
3. Fatouros, D.G.; Mullertz, A. In Vitro Lipid Digestion Models in Design of Drug Delivery Systems for Enhancing Oral Bioavailability. *Expert Opin. Drug Metab. Toxicol.* **2008**, *4*, 65. [CrossRef] [PubMed]
4. Jie, S.; Minjie, S.; Qineng, P.; Zhi, Y.; Wen, I. Incorporating liquid lipid in lipid nanoparticles for ocular drug delivery enhancement. *Nanotechnology* **2009**, *21*, 100.
5. Porter, C.J.; Charman, W.N. In vitro assessment of oral lipid based formulations. *Adv. Drug Deliv. Rev.* **2001**, *50*, 127. [CrossRef] [PubMed]

6. Christopher Porter, J.H.; Natalie Trevaskis, L.; William Charman, N. Lipids and lipid-based formulations: Optimizing the oral delivery of lipophilic drugs. *Nat. Rev. Drug Discov.* **2007**, *6*, 231. [CrossRef]
7. Bose, S.; Kulkarni, P.K. Preparation and in vitro evaluation of ibuprofen spherical agglomerates. *Indian J. Pharm. Educ.* **2002**, *36*, 184.
8. Shafiq, S.; Shakeel, F.; Talegaonkar, S.; Ahmad, F.J.; Khar, R.K.; Ali, M. Nanoemulsions as vehicles for transdermal delivery of aceclofenac. *Eur. J. Pharm. Biopharm.* **2007**, *66*, 227. [CrossRef]
9. Larsen, A.; Jannin, V.; Mullertz, A. Pharmaceutical surfactants in biorelevant media: Impact on lipolysis and solubility of a poorly soluble model compound: Danazol. In Proceedings of the 5th World Meeting in Pharmaceutics, Biopharmaceutics and Pharmaceutical Technology, Geneva, Switzerland, 2 March 2006.
10. Panayiotis Constantinides, P. Lipid Microemulsions for Improving Drug Dissolution and Oral Absorption: Physical and Biopharmaceutical Aspects. *Pharm. Res.* **1995**, *12*, 1561–1572. [CrossRef]
11. Sachin, A.; Roshan, G.; Yuvraj, R.; Beny, B. Lipid Based Drug Delivery System: An Approach to Enhance Bioavailability of Poorly Water Soluble Drugs. *Int. J. Phar. Pharm. Res.* **2020**, *19*, 93–714.
12. Devalapally, H.; Silchenko, S.; Zhou, F.; McDade, J.; Goloverda, G.; Owen, A.; Hidalgo, I.J. Evaluation of a nanoemulsion formulation strategy for oral bioavailability enhancement of danazol in rats and dogs. *J. Pharm. Sci.* **2013**, *102*, 3808–3815. [CrossRef] [PubMed]
13. Fernandez, S.; Chevrier, S.; Ritter, N.; Mahler, B.; Demarne, F.; Carrire, F.; Jannin, V. In vitro gastrointestinal lipolysis of four formulations of piroxicam and cinnarizine with the self-emulsifying excipients Labrasol® and Gelucire® 44/14. *Pharm. Res.* **2009**, *26*, 1901. [CrossRef] [PubMed]
14. Nazzal, S.; Smalyukh, I.I.; Lavrentovich, O.D.; Khan, M.A. Preparation and in vitro characterization of a eutectic based semisolid self-nano emulsified drug delivery system. *Int. J. Pharm.* **2002**, *235*, 247–265. [CrossRef] [PubMed]
15. Halbaut, L.; Berbe, C.; del Pozo, A. An investigation into physical and chemical properties of semi-solid selfemulsifying systems for hard gelatin capsules. *Int. J. Pharm.* **1996**, *130*, 203–212. [CrossRef]
16. Newton, J.M.; Pinto, M.R.; Podczeck, F. The preparation of pellets containing a surfactant or a mixture of mono-and di-gylcerides by extrusion/spheronization. *Eur. J. Pharm. Sci.* **2007**, *30*, 333. [CrossRef] [PubMed]
17. Chun Gwon, P.; Beom Kang, H.; Se-Na, K.; Seung Ho, L.; Hye Rim, H.; Young Bin, C. Nanostructured mucoadhesive microparticles to enhance oral drug bioavailability. *J. Ind. Eng. Chem.* **2017**, *54*, 262–269.
18. Hong, W.; Guangxing, Z.; Xueqin, M.; Yanhua, L.; Jun, F.; Kinam, P.; Wenping, W. Enhanced encapsulation and bioavailability of breviscapine in PLGA microparticles by nanocrystal and water-soluble polymer template techniques. *Eur. J. Pharm. Biopharm.* **2017**, *115*, 177–185.
19. Amrita, N.; Babar, I.; Shobhit, K.; Shrestha, S.; Javed, A.; Sanjula, B. Quality by design based silymarin nanoemulsion for enhancement of oral bioavailability. *J. Drug Deliv. Sci. Technol.* **2017**, *40*, 35–44.
20. Anna Pham, C.; Paul, G.; Roksan, L.; Gisela, R.; Ben Boyd, J. A new lipid excipient, phosphorylated tocopherol mixture, TPM enhances the solubilization and oral bioavailability of poorly water soluble CoQ10 in a lipid formulation. *J. Control. Release* **2017**, *268*, 400–406. [CrossRef]
21. Irina, C.; Dvora, I.; Abraham Domb, J.; Amnon, H. The effect of Pro NanoLipospheres (PNL) formulation containing natural absorption enhancers on the oral bioavailability of delta-9-tetrahydrocannabinol (THC) and cannabidiol (CBD) in a rat model. *Eur. J. Pharm. Sci.* **2017**, *109*, 21–30.
22. Shang-Ta, W.; Chien-Te, C.; Nan-Wei, S. A food-grade self-nanoemulsifying delivery system for enhancing oral bioavailability of ellagic acid. *J. Funct. Foods* **2017**, *34*, 207–215.
23. Hannah, F.; Tiffany Ohlsen, J.; Kim Woodrow, A. Protein-loaded emulsion electrospun fibers optimized for bioactivity retention and pH-controlled release for peroral delivery of biologic therapeutics. *Int. J. Pharm.* **2017**, *533*, 99–110.
24. Chang, L.; Qi, L.; Wei, H.; Changlu, N.; Chunjuan, Z.; Tonghua, X.; Tongying, J.; Siling, W. Octa-arginine modified lipid emulsions as a potential ocular delivery system for disulfiram: A study of the corneal permeation, transcorneal mechanism and anti-cataract effect. *Colloids Surf. B Biointerfaces* **2017**, *160*, 305–314.
25. Mutaliyeva, B.; Grigoriev, D.; Madybekova, G.; Sharipova, A.; Aidarova, S.; Saparbekova, A.; Miller, R. Microencapsulation of insulin and its release using w/o/w double emulsion method. *Colloids Surf. A Physicochem. Eng. Asp.* **2017**, *521*, 147–152. [CrossRef]
26. Yashaswini, P.S.; Nawneet Kurrey, K.; Sridevi Singh, A. Encapsulation of sesamol in phosphatidyl choline micelles: Enhanced bioavailability and anti-inflammatory activity. *Food Chem.* **2017**, *228*, 330–337. [CrossRef] [PubMed]
27. Omid, S.; Seid Mahdi, J.; Zayniddin Kamarovich, M. Development of double layered emulsion droplets with pectin/β-lactoglobulin complex for bioactive delivery purposes. *J. Mol. Liq.* **2017**, *243*, 144–150.
28. Ranjit Harwansh, K.; Pulok Mukherjee, K.; Sayan, B. Nanoemulsion as a novel carrier system for improvement of betulinic acid oral bioavailability and hepatoprotective activity. *J. Mol. Liq.* **2017**, *237*, 361–371. [CrossRef]
29. Pengchao, S.; Nan, Z.; Haiying, H.; Qian, L.; Xuexiao, Z.; Qian, S.; Yongxing, Z. Low density lipoprotein peptide conjugated submicron emulsions for combating prostate cancer. *Biomed. Pharmacother.* **2017**, *86*, 612–619.
30. Chiung-Yi, H.; Ming-His, H. Emulsifying properties and degradation characteristics of bioresorbable polymeric emulsifiers in aqueous solution and oil-in-water emulsion. *Polym. Degrad. Stab.* **2017**, *139*, 138–142.

31. Mitja, P.; Luka, P.; Matevz, L.; Rok, D. A redispersible dry emulsion system with simvastatin prepared via fluid bed layering as a means of dissolution enhancement of a lipophilic drug. *Int. J. Pharm.* **2018**, *549*, 325–334.
32. Papa Mady, S.; Nicolas, A.; Ysia Idoux, G.; Sidy Dienga, M.; Nadia, M.; Said, E.; Mounibe, D.; Thierry Vandamme, F. Pickering nanoemulsion as a nanocarrier for pH-triggered drug release. *Int. J. Pharm.* **2018**, *549*, 299–305.
33. Alaa Salama, H.; Mona, B.; Sally El, A. Experimentally designed lyophilized dry emulsion tablets for enhancing the antihyperlipidemic activity of atorvastatin calcium: Preparation, in-vitro evaluation and in-vivo assessment. *Eur. J. Pharm. Sci.* **2018**, *112*, 52–62. [CrossRef] [PubMed]
34. Chekuri, A.; Kumar, V.; Jayaram Kumar, K. Formulation and optimization of pH sensitive drug releasing O/W emulsions using Albizia lebbeck L. seed polysaccharide. *Int. J. Biol. Macromol.* **2018**, *116*, 239–246.
35. Parth, S.; Sachin Kumar, S.; Narendra Kumar, P.; Sarvi Yadav, R.; Palak, B.; Bimlesh, K.; Monica, G.; Saurabh, S.; Surajpal, V.; Ankit Kumar, Y.; et al. Impact of solid carriers and spray drying on pre/post-compression properties, dissolution rate and bioavailability of solid self-nanoemulsifying drug delivery system loaded with simvastatin. *Powder Technol.* **2018**, *338*, 836–846.
36. Nguyen-Thach, T.; Cao-Son, T.; Thi-Minh-Hue, P.; Hoang-Anh, N.; Tran-Linh, N.; Sang-Cheol, C.; Dinh-Duc, N.; Thi-Bich-Huong, B. Development of solidified self-microemulsifying drug delivery systems containing l-tetrahydropalmatine: Design of experiment approach and bioavailability comparison. *Int. J. Pharm.* **2018**, *537*, 9–21.
37. Yuwen, T.; Yike, J.; Shuqi, Z.; Colin, L.C.; Qingrong, H. Self-nanoemulsifying system (SNES) enhanced oral bioavailability of boswellic acids. *J. Funct. Foods* **2018**, *40*, 520–526.
38. Fahanwi Asabuwa, N.; Sevinc Ilkar, E.; Ufuk, Y. Pickering emulsions stabilized nanocellulosic-based nanoparticles for coumarin and curcumin nanoencapsulations: In vitro release, anticancer and antimicrobial activities. *Carbohydr. Polym.* **2018**, *201*, 317–328. [CrossRef]
39. Yiping, D.; Fengjian, Y.; Xiuhua, Z.; Lu, W.; Weiwei, W.; Chang, Z.; Mingfang, W. Improving the skin penetration and antifebrile activity of ibuprofen by preparing nanoparticles using emulsion solvent evaporation method. *Eur. J. Pharm. Sci.* **2018**, *114*, 293–302.
40. María Loya-Castro, F.; Mariana, S.; Dante Sánchez-Ramírez, R.; Rossina, D.; Noe, E.; Paola Oceguera-Basurto, E.; Edgar Figueroa-Ochoa, B.; Antonio, Q.; Aliciadel Toro, A.; Antonio, T.; et al. Preparation of PLGA/Rose Bengal colloidal particles by double emulsion and layer-by-layer for breast cancer treatment. *J. Colloid Interface Sci.* **2018**, *518*, 22–129. [CrossRef]
41. Yunru, P.; Qilu, M.; Jie, Z.; Bo, C.; Junjun, X.; Piaopiao, L.; Liang, Z.; Ruyan, H. Nanoemulsion delivery system of tea polyphenols enhanced the bioavailability of catechins in rats. *Food Chem.* **2018**, *242*, 527–532.
42. Heba, E.; Hala Farag, K.; Abeer Salama, A.A. In vitro and in vivo evaluation of an oral sustained release hepatoprotective Caffeine loaded w/o Pickering emulsion formula—Containing wheat germ oil and stabilized by magnesium oxide nanoparticles. *Int. J. Pharm.* **2018**, *547*, 83–96.
43. Yuanyuan, L.; Xiuhua, Z.; Yanjie, L.; Jianhang, Y.; Qian, Z.; Lingling, W.; Weiwei, W.; Qilei, Y.; Bingxue, L. Melatonin loaded with bacterial cellulose nanofiber by Pickering-emulsion solvent evaporation for enhanced dissolution and bioavailability. *Int. J. Pharm.* **2019**, *559*, 393–401.
44. Ross Walenga, L.; Andrew Babiskin, H.; Xinyuan, Z.; Mohammad, A.; Liang, Z.; Robert Lionberger, A. Impact of Vehicle Physicochemical Properties on Modeling-Based Predictions of Cyclosporine Ophthalmic Emulsion Bioavailability and Tear Film Breakup Time. *J. Pharm. Sci.* **2019**, *108*, 620–629. [CrossRef] [PubMed]
45. Keishi, Y.; Kazuaki, T.; Koji, N.; Masaki, O.; Hakaru, S. Enhanced dissolution and oral bioavailability of praziquantel by emulsification with human serum albumin followed by spray drying. *Eur. J. Pharm. Sci.* **2019**, *139*, 105064.
46. Kapil Chaudhari, S.; Krishnacharya Akamanchi, G. Novel bicephalous heterolipid based self-microemulsifying drug delivery system for solubility and bioavailability enhancement of efavirenz. *Int. J. Pharm.* **2019**, *560*, 205–218. [CrossRef]
47. Khaled Aboul, F.; Ayat Allam, A.; El-Badry, M.; Ahmed El-Sayed, M. A Self-Nanoemulsifying Drug Delivery System for Enhancing the Oral Bioavailability of Candesartan Cilexetil: Ex Vivo and In Vivo Evaluation. *J. Pharm. Sci.* **2019**, *108*, 3599–3608.
48. Qi, T.; Xiangli, X.; Cunjun, L.; Bowen, Z.; Xiaolong, C.; Guozhen, Z.; Chunhui, Z.; Linjiang, W. Medium-chain triglyceride/water Pickering emulsion stabilized by phosphatidylcholine-kaolinite for encapsulation and controlled release of curcumin. *Colloids Surf. B Biointerfaces* **2019**, *183*, 110414.
49. Na, M.; Qilong, W.; Huihua, L.; Michael Adu, F.; Congyong, S.; Kangyi, Z.; Qiuxuan, Y.; Qiuyu, W.; Hao, J.; Elmurat, T.; et al. Improved oral bioavailability of myricitrin by liquid self-microemulsifying drug delivery systems. *J. Drug Deliv. Sci. Technol.* **2019**, *52*, 597–606.
50. Jagruti, D.; Hetal, T. Enhanced oral bioavailability and brain uptake of Darunavir using lipid nanoemulsion formulation. *Colloids Surf. B Biointerfaces* **2019**, *175*, 143–149.
51. Huiqiong, Y.; Xiuqiong, C.; Zaifeng, S.; Wei, Z.; Yue, W.; Chaoran, K.; Qiang, L. Entrapment of bacterial cellulose nanocrystals stabilized Pickering emulsions droplets in alginate beads for hydrophobic drug delivery. *Colloids Surf. B Biointerfaces* **2019**, *177*, 112–120.
52. Lufeng, H.; Xin, L.; Youmei, B.; Craig Duvall, L.; Caiyun, Z.; Weidong, C.; Can, P. Preparation, preliminary pharmacokinetic and brain targeting study of metformin encapsulated W/O/W composite submicron emulsions promoted by borneol. *Eur. J. Pharm. Sci.* **2019**, *133*, 160–166.
53. Mohamadi Saani, S.; Abdolalizadeh, J.; Zeinali Heris, S. Ultrasonic/sonochemical synthesis and evaluation of nanostructured oil in water emulsions for topical delivery of protein drugs. *Ultrason. Sonochem.* **2019**, *55*, 86–95. [CrossRef] [PubMed]

54. Sadaf, C.; Mohd, A.; Yasmin, S.; Mohd Abul, K. Self-nanoemulsifying drug delivery system of nabumetone improved its oral bioavailability and anti-inflammatory effects in rat model. *J. Drug Deliv. Sci. Technol.* **2019**, *51*, 736–745.
55. Pontip, B.; Suchada, P.; Pornsak, S. Improving dissolution and photostability of resveratrol using redispersible dry emulsion: Application of design space for optimizing formulation and spray-drying process. *J. Drug Deliv. Sci. Technol.* **2019**, *51*, 411–418.
56. Ojeda-SernaI, E.; Rocha-Guzmán, N.E.; Gallegos-Infante, J.A.; Cháirez-Ramírez, M.H.; Rosas-Flores, W.; Pérez-Martínez, J.D.; Moreno-Jiménez, M.R.; González-Laredo, R.F. Water-in-oil organogel based emulsions as a tool for increasing bioaccessibility and cell permeability of poorly water-soluble nutraceuticals. *Food Res. Int.* **2019**, *120*, 415–424. [CrossRef] [PubMed]
57. José Soriano-Ruiz, L.; Joaquim Suner, C.; Ana Calpena-Campmany, C.; NuriaBozal-de, F.; LydaHalbaut, B.; Boix-Montanes, A.; Eliana Souto, B.; Clares-Naveros, B. Clotrimazole multiple W/O/W emulsion as anticandidal agent: Characterization and evaluation on skin and mucosae. *Colloids Surf. B Biointerfaces* **2019**, *175*, 166–174. [CrossRef] [PubMed]
58. Jian, L.; Qilong, W.; Omari-Siaw, E.; Adu-Frimpong, M.; Jing, L.; Ximing, X.; Jiangnan, Y. Enhanced oral bioavailability of Bisdemethoxycurcumin-loaded self-microemulsifying drug delivery system: Formulation design, in vitro and in vivo evaluation. *Int. J. Pharm.* **2020**, *590*, 119887.
59. Nan, Z.; Fei, Z.; Shuo, X.; Kaiqing, Y.; Wenjing, W.; Weisan, P. Formulation and evaluation of luteolin supersaturatable self-nanoemulsifying drug delivery system (S-SNEDDS) for enhanced oral bioavailability. *J. Drug Deliv. Sci. Technol.* **2020**, *58*, 101783.

colloids and interfaces

MDPI

Article

A Microfluidic Approach to Investigate the Contact Force Needed for Successful Contact-Mediated Nucleation

Gina Kaysan, Theresa Hirsch, Konrad Dubil and Matthias Kind *

Institute for Thermal Process Engineering, Karlsruhe Institute of Technology (KIT), 76131 Karlsruhe, Germany
* Correspondence: matthias.kind@kit.edu; Tel.: +49-721-6084-2390

Abstract: Emulsions with crystalline dispersed phase fractions are becoming increasingly important in the pharmaceutical, chemical, and life science industries. They can be produced by using two-stage melt emulsification processes. The completeness of the crystallization step is of particular importance as it influences the properties, quality, and shelf life of the products. Subcooled, liquid droplets in agitated vessels may contact an already crystallized particle, leading to so-called contact-mediated nucleation (CMN). Energetically, CMN is a more favorable mechanism than spontaneous nucleation. The CMN happens regularly because melt emulsions are stirred during production and storage. It is assumed that three main factors influence the efficiency of CNM, those being collision frequency, contact time, and contact force. Not all contacts lead to successful nucleation of the liquid droplet, therefore, we used microfluidic experiments with inline measurements of the differential pressure to investigate the minimum contact force needed for successful nucleation. Numerical simulations were performed to support the experimental data obtained. We were able to show that the minimum contact force needed for CMN increases with increasing surfactant concentration in the aqueous phase.

Keywords: contact-mediated nucleation; contact force; emulsion; crystallization; CFD

Citation: Kaysan, G.; Hirsch, T.; Dubil, K.; Kind, M. A Microfluidic Approach to Investigate the Contact Force Needed for Successful Contact-Mediated Nucleation. *Colloids Interfaces* **2023**, *7*, 12. https://doi.org/10.3390/colloids7010012

Academic Editors: César Burgos-Díaz, Mauricio Opazo-Navarrete and Eduardo Morales

Received: 27 December 2022
Revised: 26 January 2023
Accepted: 26 January 2023
Published: 31 January 2023

1. Introduction

The kinetics of crystallization in terms of the nucleation probability of a droplet and the influence of the surfactant on crystallization has been studied in recent years, especially in the field of solid-liquid nanoparticles [1–6].

Contacts between subcooled, liquid droplets and already crystallized droplets, i.e., particles, can occur during the crystallization of melt emulsions and their storage at rest or in stirred vessels. These contacts may lead to contact-mediated nucleation (CMN), resulting in the crystallization of the subcooled droplets. Direct contact with the blank interfaces of the colliding droplet and particle seems to be needed for CMN. Adsorbed surfactants may shield the droplets from direct contact with each other and hinder CMN. Moreover, a minimum contact time and a minimum relative velocity of the two collision partners must be overcome so that the liquid droplet is inoculated by the particle [7].

In this study, we aim to determine the minimum contact force needed for CMN as a function of the aqueous surfactant concentration.

McClements et al. [8] formulated the hypothesis of CMN for quiescent emulsions due to Brownian motion, to explain their observation that crystallization of the droplets of an n-hexadecane-in-water emulsion with Tween®20 (TW20) as the surfactant accelerated when already solidified droplets were present. Emulsions with only subcooled, liquid droplets did not crystallize or crystallized negligibly slowly [9].

The hypothesis of CMN was strengthened in [10], where the crystallization progress (time-resolved change of the solid fraction of the dispersed phase) of several emulsions with 50% liquid and 50% solid dispersed phase fractions was investigated over a period of 175 h. Spectroscopic nuclear magnetic resonance measurements were used to study the crystallization progress of n-hexadecane-in-water emulsions with TW20 contents between 0 and 14 wt-% added to the continuous phase. Emulsions with already solidified droplets continued to crystallize even at low subcooling, whereas emulsions with completely liquid or solidified droplets did not change their number of solid particles over time and the droplet size distribution of all the emulsions remained constant. Since no other external forces had been applied to the emulsion which could have induced nucleation, CMN was assumed.

In addition to the influence of solid and liquid dispersed phase fractions, different observations had been made regarding the impact of the surfactant concentration. Dickinson et al. [9] and McClements et al. [10] demonstrated that increasing the surfactant concentration was associated with faster crystallization due to an increased rate of CMN. Different approaches are known to explain the impact of surfactants on CMN. They assumed that, for example, micelles in the continuous phase may form a bridge between the approaching reaction partners (subcooled, liquid droplet and crystalline, solid particle), containing a small concentration of oil and surfactant molecules forming a transient connection [9]. Another possible explanation was depletion flocculation, which is enhanced by increasing aqueous surfactant concentrations [10]. Additionally, Povey et al. [11] described a significant impact of the type of surfactant on the CMN. In contrast to what Dickinson et al. and McClements et al., Kaysan et al. [12] found, by a targeted contact between a liquid and a solidified droplet in a microfluidic setup, a reduction in the nucleation efficiency (percentage of collisions that led to the crystallization of the subcooled, liquid droplet compared to the total number of experiments) of the CMN when micelles were present in the continuous phase.

In this work, a differential pressure sensor is connected to a microfluidic chip. By measuring the pressure drop Δp during the CMN, we aim to determine the force needed for CMN due to the fluid field F_c. At a constant relative velocity of the droplet and the particle, the minimum force needed for nucleation $F_{CMN,min}$ increases only due to F_c, as the force caused by the impulse F_i should be the same for all experiments (compare Equation (8)). It is therefore crucial to understand the flow and pressure pattern within the microfluidic channel.

The pressure drop within a straight pipe section can be determined by

$$\Delta p = f \cdot \frac{L}{d_h} \cdot \frac{\rho u^2}{2}, \quad (1)$$

where f represents the so-called friction factor, L the channel length, d_h the hydraulic diameter of the pipe, ρ the fluid density, and u the average velocity. f is dependent on the Reynolds number [13], the flow properties, the channel geometry, and the roughness of the walls [14]. Here, the Reynolds number is defined as

$$Re = \frac{u\,\rho\,d_h}{\eta}, \quad (2)$$

where η represents the dynamic viscosity of the fluid. The hydraulic diameter is calculated as

$$d_h = \frac{4\,A_c}{P_W}, \quad (3)$$

where A_c represents the cross-sectional area of the flow channel and P_W is the wetted perimeter [14]. Regarding noncircular channel cross-sections and laminar flows, the pipe friction coefficient is calculated using φ, which depends on the geometry of the channel:

$$f = \varphi \frac{64}{Re}. \quad (4)$$

According to [15], φ = 0.92 for the channel used in this work with a width W = 300 μm and depth D = 200 μm. The pressure loss is directly proportional to the fluid velocity in laminar pipe flows. This results in a calculation of the pressure drop as

$$\Delta p = \varphi \frac{64}{Re} \cdot \frac{L\rho u^2}{2d_h} = \varphi \frac{32\eta u L}{d_h^2}. \tag{5}$$

The wall roughness in conventional pipes has a negligible impact on the pressure drop during laminar flows. However, the roughness in microfluidic systems may obstruct a significant part of the flow channel, which is discussed in the next section [14].

If mini- or microchannels are utilized instead of conventional channels, not all assumptions and equations mentioned above may be used. Kandlikar and Grande [16] proposed a classification of the different kinds of channels depending on their hydraulic diameter d_h. The channel is classified as a microchannel for 10 μm < $d_h \leq$ 200 μm and as a minichannel for 200 μm < $d_h \leq$ 3 mm.

Length-related effects become more important on a smaller scale, such as the entrance length of a flow [17]. Chan et al. [18] investigated fluid films between molecularly smooth plates at distances in the nanometer range and found that the conventional Navier-Stokes equations are still valid up to a distance of 50 nm. The liquid can no longer be regarded as a continuum when the film thickness is less than ten molecular layers (5 nm) [18]. Therefore, continuum mechanical behavior can still be assumed for fluid flows in the micrometer range.

Qu et al. [19] and Xu et al. [20] carried out numerical and experimental tests to investigate flow development and pressure loss in micro- and minichannels. They found that the conventional Navier-Stokes equations also predict the flow in microfluidics. Re in mini- and microchannels are significantly lower than in commercial pipe systems due to smaller d_h and u. This means that the frictional forces dominate and, simultaneously, the inertial forces are weak. However, the critical Reynolds number Re_{crit}, describing the transition from laminar to turbulent flow, must be adjusted in microfluidics. The transition from laminar to turbulent flow may start below Re_{crit} = 2300, due to the impact of surface roughness, as determined for conventional pipe flows [14]. This transition already takes place for $Re_{crit} = 300 - 900$ for hydraulic channel diameters $d_h = 30 - 344$ μm [21]. In this work, d_h = 240 μm. Kandlikar [14] also considered the impact of the wall roughness on Re_{crit}. With the relative roughness ϵ of our channel (compare Figure 3, ϵ ~0.02), Re_{crit} ~1800. Compared to both values, our experiments are strictly laminar (compare Appendix A Tables A1 and A2).

According to Mirmanto et al. [22], the friction factors of microfluidics in fully developed flows (either laminar or turbulent) are consistent with those in conventional theory. Ghajar et al. [23] reviewed the existing literature and came to the same conclusion. In addition, they found that the relative roughness of the channels in microfluidics has a major influence. Steinke and Kandlikar [24] compared the friction factors presented in the literature with those from theory and noted that the values are very similar. They explained deviations by inaccuracies related to the irregularity of the channel dimensions.

In addition, droplets moving along the channel can impact Δp. Monodisperse droplets can be generated in microfluidic setups using a T-junction where two fluids meet (e.g., [7,25,26]). Due to the chemical change of the hydrophobicity of the channel walls, droplets can be formed which are not in contact with the channel walls and, therefore, do not fill the channel but almost reach its width and height and form rounded ends, i.e., caps [27]. The volume of a droplet in a rectangular channel can be approximated according to Musterd et al. [27].

If the droplets flow along the channel together with the continuous phase, an increased pressure loss is likely compared to the simple single-phase flow. Fuerstmann et al. [28] list the following parameters for this pressure loss: the number of droplets, their total length (body and cap), the aspect ratio of the channel, the emulsifier concentration, and the viscosity of the continuous phase. When the Reynolds number is low, transition zones between these areas can be neglected [28]. Considering the flow around the droplet bodies, this can be further subdivided into fluid moving through a thin gap between the droplet and the channel walls, as well as a part that flows through the gap formed by the round droplets and the rectangular channel corners (bypass flow). According to Ransohoff and Radke [29], the contribution to the pressure drop by the thin films between the flowing drop and the wall can be neglected because the pressure drop in the corners is predominant.

Fuerstmann et al. [28] considered different concentrations of the emulsifier to calculate the pressure drop in a channel with flowing liquid droplets. Regardless of whether no emulsifier, a very low ($c < 0.01$ critical micelle concentration [cmc]) or a very high concentration c of emulsifier ($c > 1000\ cmc$) were present, the pressure drops to a higher degree in the area of the caps rather than over the bodies of the droplets. Therefore, the number of droplets in the channel is decisive, as more droplets also mean a higher number of caps.

The pressure drop across the body of the flowing droplets was the greatest for intermediate emulsifier concentrations (1–2 orders of magnitude of cmc), which is why the total length of the droplets was most important in this case.

One possible reason for the different components of the pressure drop is the dependence of the flow velocity within the corners on the concentration of the emulsifier. If no emulsifier is present, the flow in the corners is negligible and the droplets moved at the same speed as the continuous fluid. At intermediate concentrations of surfactant, the bypass flow is fast, causing the continuous phase to move 1.2 times faster than the droplets. At $c > 1000\ cmc$, the flow velocity within the corners decreased again [28].

To the best of our knowledge, no investigations have been made that deal with a particle partially blocking the cross-sectional area of a rectangular channel with laminar flow.

In this work, the minimum contact force $F_{CMN,min}$ needed for successful CMN will be investigated. $F_{CMN,min}$ is a combination of the force due to the impulse of the decelerating droplet F_i and the force due to the fluid field F_c (compare Equation (8)). F_c is determined in microfluidic experiments by analyzing the pressure drop between the inlet and the outlet of a microfluidic chip during the crystallization of a liquid droplet due to contact with an already crystallized and immobile particle. A detailed understanding of the different pressure drop contributions that can influence the total pressure drop during the process is necessary for the interpretation of the data and testing of the setup. The influence of particle geometry on the pressure loss, which has barely been considered so far, will be especially investigated in more detail with both, experiments and simulations.

2. Materials and Methods

2.1. Microfluidic Setup

The production and set up of the microfluidic chip used are described elsewhere [7,25]. Rectangular channels with a width of $W = 300\ \mu m\ {}^{+10}_{-52}\ \mu m$ and depth of $D = 200\ \mu m \pm 20\ \mu m$ were milled into a polycarbonate chip. A differential pressure sensor (Deltabar S PMD70, Endress + Hauser, Reinach, Switzerland) was connected to the inlet and outlet of the microfluidic channel (Figure 1).

The differential pressure sensor gives an analogous output current signal to the signal transducer (General Industrial Controls Private Limited, Pune, Maharashtra, India), which converts the current into a voltage signal. This signal is then transferred to the computer using an analogous digital converter (Measurement Computing Corporation, Norton, MA, USA) and the measurement data are recorded using the software DAQamiTM v4.2.1 (Measurement Computing, Norton, MA, USA). The calibration of the sensor led to a linear relationship between the voltage output signal and the actual differential pressure. The signal data were smoothed with OriginPro 2021 (OriginLab, Northampton, MA, USA) using the 'floating average' method, averaging ten measurement points. Compensation measurements without the polycarbonate chip were carried out and the influences of valves, connectors, and hose connections were determined to consider only the pressure drop of the main channel. These corrections are always considered for all Δp data presented.

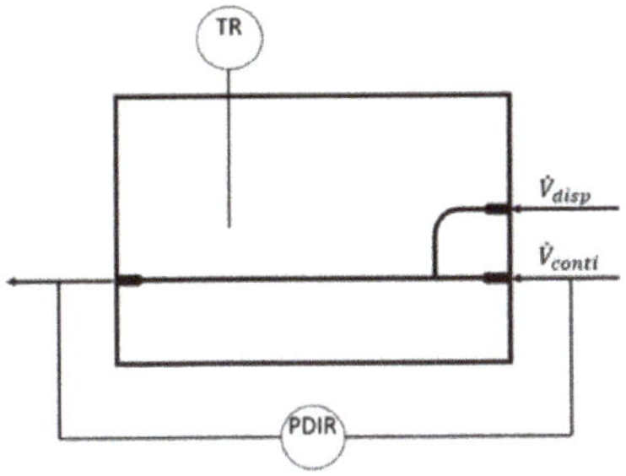

(a)

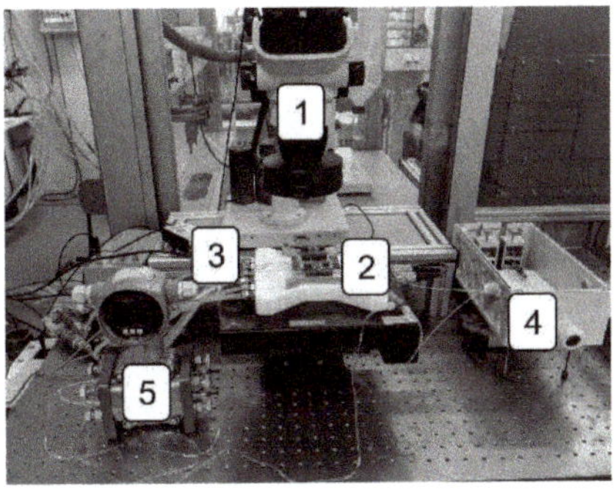

(b)

Figure 1. (**a**) Schematic drawing of the connection of the pressure sensor to the microfluidic chip. $\dot{V}_{disp}$ and $\dot{V}_{conti}$ show where the dispersed and continuous phases were connected to the microfluidic chip. The channel where the dispersed phase ($\dot{V}_{disp}$) entered the system could be closed to increase the accuracy of the measurements of Δp. The dispersed phase is forming droplets at the T-junction, where the two streams meet. (**b**) Experimental setup: (1) stereo microscope with polarization filter and high-speed camera, (2) microfluidic chip on top of the tempering unit, (3) in- and outlet of coolant, (4) syringe pump, and (5) differential pressure sensor.

Ultrapure water was used (electrical conductivity 0.057 μS cm^{-1}, OmniTap, Stakpure GmbH, Niederahr, Germany) as a continuous phase. The dispersed phase was the organic substance n-hexadecane (Hexadecane ReagentPlus®, 99%, Sigma-Aldrich, St. Louis, MO, USA) and for some experiments, the droplets were additionally stabilized with the surfactant Tween®20 (TW20, Merck KGaA, Darmstadt, Germany) at a concentration of either $\tilde{c}_{TW20}$ = 0.24 mol m^{-3} (~4 *cmc*) or 0.41 mol m^{-3} (~8 *cmc*). The surfactant was dissolved in the continuous phase. The production of TW20 includes esterification and further chemical reactions, therefore, its purity may differ from batch to batch. The same bottle of TW20 was used for all experiments to exclude any fluctuations in purity and composition and guarantee comparable experimental conditions. The *cmc* of TW20 is given by Linke et al. [30] as 0.059 mol m^{-3} at 298 K. The melting point of n-hexadecane was determined previously as $\vartheta_{m,hex}$ = 18.6 °C [7]. Volume flows of the continuous phase between 50 μL h^{-1} and 500 μL h^{-1} were investigated. This resulted in Reynolds numbers between 0.05 and 0.52 (compare Appendix A Table A2) and, consequently, a strictly laminar flow.

Figure 2 summarizes the procedures to investigate Δp for the following different setups: The channel is filled with continuous phase only (1A), liquid droplets are moving along the minichannel (2A), the channel is partly blocked by a solid particle (2B), or the CMN itself (2C). A more detailed description of the experimental procedure to introduce droplets in the microfluidic channel and of the temperature profile used for CMN can be found in [7].

Whereas the liquid droplet can move along the channel, the solid particle no longer changes its position, which finally enables the direct contact of the particle and the droplet at a given droplet velocity and subcooling ΔT. The continuous phase flows 1.01 to 1.1 times faster than the droplets moving through the channel.

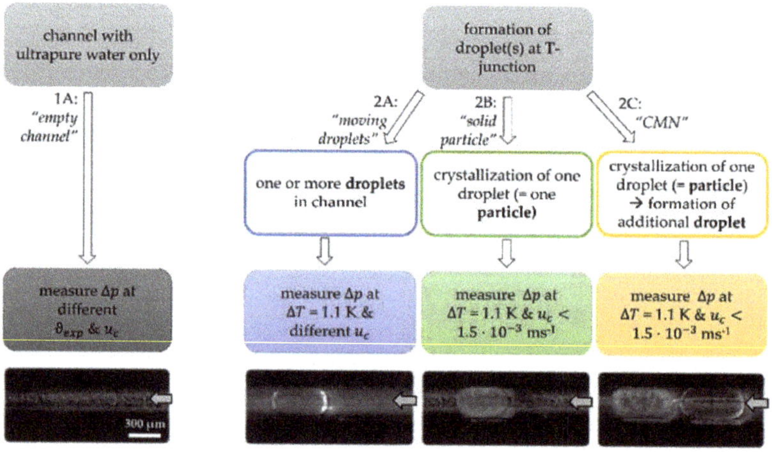

Figure 2. The experimental approach to measure the differential pressure Δp for the different setups: Channel filled with continuous phase only (1A, Section 3.1), liquid droplet(s) in the channel (2A, Section 3.3), solid particle partly blocking the channel (2B, Section 3.2), and during the CMN (2C, Section 3.4). u_c hereby represents the velocity of the continuous phase and ΔT the subcooling. The latter is calculated as $\Delta T = \vartheta_{m,hex} - \vartheta_{exp}$, with ϑ_{exp} representing the temperature of the experiment. The arrow in the pictures indicates the direction of the flow of the continuous phase.

The wall roughness (Figure 3) mainly impacts the flow in the mini- and microchannels. Therefore, the structure of the minichannel was investigated optically using a digital microscope (VHX-700, Keyence, Osaka, Japan).

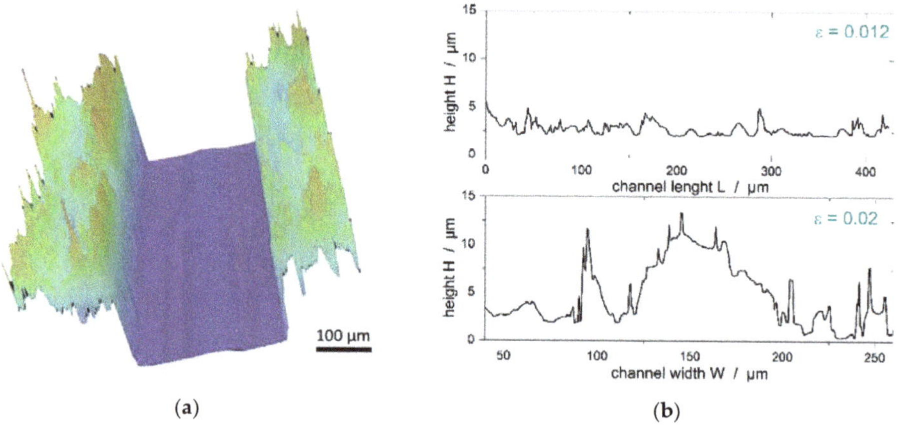

Figure 3. (**a**) Microscopic three-dimensional record of the microfluidic channel. (**b**) Roughness along the channel length (top) and width (bottom). ϵ indicates the relative roughness.

An evaluation of the arithmetical mean height of the channel resulted in an absolute roughness of R_a = 4.2 µm ± 1.5 µm (averaged over 11 datasets across the whole channel width and length). The relative roughness ϵ differed mainly between the determination across the channel width and along the channel length, which is a result of the milling process.

2.2. Numeric Flow Simulations

All simulations were carried out using the open-source simulation software *OpenFOAM* (Version 6) [31]. The geometry was generated using the computer-aided design and drafting tool *Salome* (Version 9.3.0) [32]. Three different parameters of the particle implemented were adjusted (Figure 4):

- length of the particle body l_{body},
- length of the caps of the particles l_{cap}, and
- distance between the particle boundary and the channel wall, i.e., film thickness h_f.

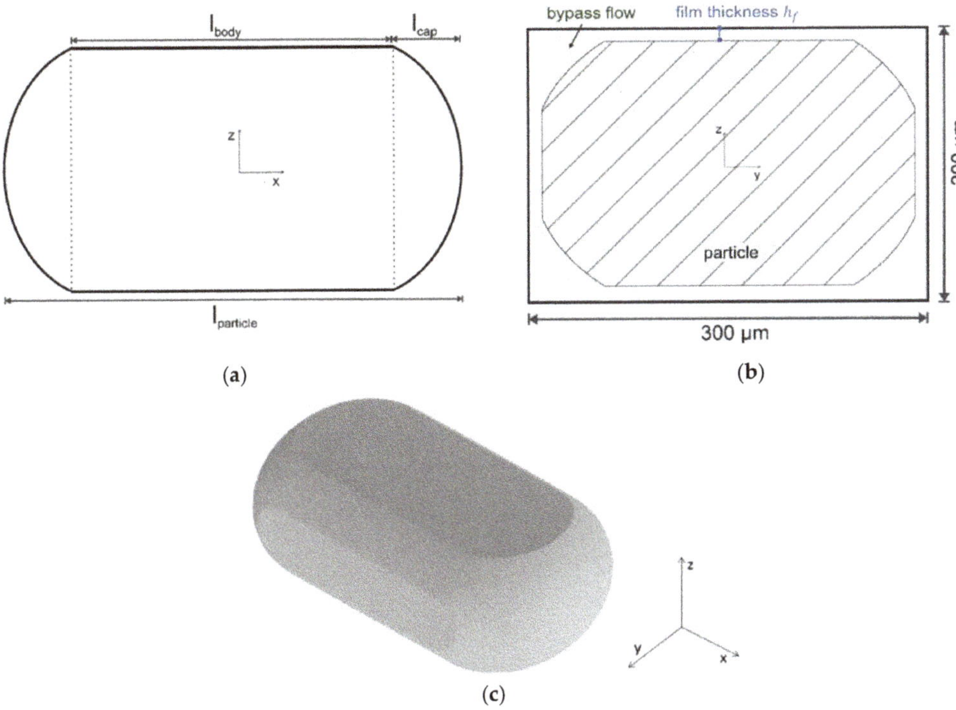

Figure 4. (**a**) Two-dimensional schematic drawing of a particle used for the simulations. (**b**) Cross-sectional drawing across the microfluidic channel. It can be seen that the particle does not stay in contact with the channel walls (due to the ionization of the walls), resulting in bypass and wall flow. (**c**) Three-dimensional presentation of the particle that was cut from the channel to do all the simulations.

The particle was integrated into the center of the channel resulting in equal film thicknesses on all sides of the object. We ensured in preliminary tests that the length of the channel (1.6 mm) was sufficient to establish a fully developed flow. This was also checked after the integration of the particle whose length corresponded to 14–34% of the channel length. Therefore, the pressure drop could be divided into an empty channel contribution, which was extrapolated to the experimental channel length of 5 cm, and the contribution of the particle. The dimensions of the channel were 300 × 200 µm ($W \times D$).

A grid convergence study was carried out to analyze the influence of the numerical mesh on the simulation results. The grid convergence index (GCI) proposed by Roache [33] was calculated to quantify the discretization errors. It is based on the Richardson extrapolation and dependent on the order of convergence. The base cell sizes tested were 15 µm, 11.6 µm, and 9 µm (Table 1).

Table 1. The results of the grid convergence study and corresponding grid convergence index (GCI) values. A base cell size of 11.6 µm was chosen as a compromise between the duration of the simulations and their accuracy.

	Δp/Pa			GCI	
	15 µm	**11.6 µm**	**9 µm**	$GCI_{15,11.6}$	$GCI_{11.6,9}$
empty channel (no particle)	44.21	44.63	44.94	0.0329	0.0242

The solver *simpleFoam* was used to solve the Navier-Stokes equations at an unknown pressure field, as an incompressible, steady-state flow of a Newtonian fluid was assumed. A tool available in the *OpenFOAM* software (*SnappyHexMesh*) was used to generate the mesh. The mesh was refined up to three times depending on the distance to the particle and the channel walls to adequately resolve gradients that form near the solid surfaces (Figure 5).

Figure 5. Example of a mesh generated by *SnappyHexMesh*, which was used for the simulative parameter studies of the influence of the particle shape on the overall pressure drop. To achieve reliable data, the refinement differs depending on the relative location of the particle and the channel walls. In this case, the film thickness between the particle and the channel was 0.

The mesh refinement is relative to the base grid. All cells whose centers had a distance of 40 µm from the channel wall and 50 µm away from the particle surface were refined by a factor of 2. Those cells whose centers were closer than 3 µm to the surface of the particle were refined 3 times.

A mapped boundary condition with varying average values was chosen as the boundary condition for the velocity at the inlet. This means that the velocity profile at the outlet was averaged across the outlet surface and then impressed at the inlet, where the new profile was calculated concerning the averaged value. The channel was long enough to form a fully developed flow field, therefore, the velocity gradient was set to 0 at the outlet. A fixed value of 0 was given at the wall of the channel and the wall of the particle. A gradient of zero was chosen for the inlet, the wall of the channel, and the particle for the pressure. The pressure at the outlet was set to a constant value.

3. Results and Discussion

With this study, we aimed to investigate the minimal contact force $F_{CMN,min}$ needed for a successful CMN. This is of crucial importance to further understand the mechanisms behind CMN and how it might happen during the industrial production of melt emulsions. In analogy to [34], CMN is influenced by the external flow field [12,35], as the collision frequency and the contact force F in stirred systems are proportional to the apparent shear rate $\dot{\gamma}_{app}$, whereas the contact or interaction time t_c is inversely proportional to $\dot{\gamma}_{app}$.

Microfluidic experiments are promising in gaining further insights into the induction times t_{ind} [7] and minimal forces for successful CMN $F_{CMN,min}$ (this study) needed to induce nucleation (Table 2).

Table 2. Comparison of the parameters influencing the efficiency of contact-mediated nucleation (CMN) in stirred systems and the parameters that are investigated in the microfluidic system.

Parameters Influencing CMN in Stirred Systems [1]	Parameters Investigated in the Microfluidic Approach	Condition for Successful CMN in Stirred Systems
Collision frequency	-	-
Contact force F	minimum contact force for CMN $F_{CMN,min}$	$F \geq F_{CMN,min}$
Contact time t_c	induction time t_{ind}	$t_c \geq t_{ind}$

[1] Due to the external flow, in analogy to [34].

This publication aims to verify or falsify the following hypothesis:

Theorem 1. *The minimum contact force $F_{CMN,min}$ needed to induce nucleation due to the contact of a solid particle with a subcooled droplet increases with the increasing surfactant concentration as the disjoining pressure increases.*

To investigate the stated hypothesis, firstly, the impact of the empty channel, moving liquid droplets and immobile solid particles in the channel geometry on the overall pressure drop was estimated.

3.1. Empty Channel

The flow of water through an empty rectangular channel was studied to validate the implementation of the differential pressure sensor by comparing the experimental measurements with the numeric simulations (Figure 6). The experimental data will also be compared to data presented in the literature to indicate their accuracy (Figure 7).

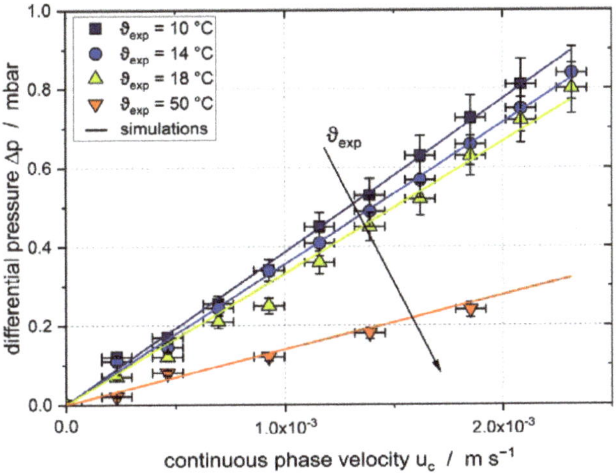

Figure 6. Experimental and simulated differential pressure values for the rectangular microfluidic channel with pure water at temperatures ϑ_{exp} between 10 °C and 50 °C flowing through it.

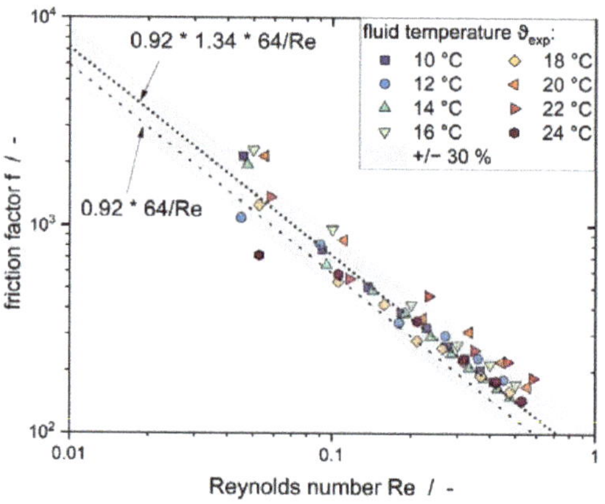

Figure 7. The so-called Moody diagram shows the friction factor f as a function of the Reynolds number Re. The friction factor was calculated according to Equation (1), Re according to Equation (2). The grey region represents the range of $\pm$ 30% of the fitted data points ($f = 0.92 \cdot 1.34 \cdot 64/Re$). 0.92 hereby represents the correction factor for using a rectangular channel instead of a round cross-section [15]. The additional correction factor of 1.34 is a result of the roughness of the channel walls.

When the temperature was increased, the pressure drop through the channel decreased at all velocities of the continuous phase. There is a good agreement between the results of the experimental and the simulative works. Moreover, the data points are in good agreement with the theoretically calculated values (Figure 7).

Moody [36] described the friction factor as being independent of the relative roughness ϵ in the laminar flow regime for $\epsilon < 0.05$. Therefore, the description of the laminar part should follow $f = 64/Re$. Considering the different channel geometry (rectangular instead of round cross-section), $f = 0.92 \cdot 64/Re$ [15] should describe the experimental data. Nonetheless, not only does the geometry of the channel play a major role for mini- and microchannels, but different authors described that increasing the relative roughness resulted in an increasing friction factor, even in laminar flow, due to a significant change of the free cross-sectional area of the channel (e.g., [23,36–39]). This could explain the shift of the experimental friction factor compared to the theoretical friction factor. Consequently, the experimental data points were fitted by introducing another correction parameter that indicates the deviation of the rough channel walls from smooth ones. The fit (Figure 7) led to $f = 0.92 \cdot 1.34 \cdot 64/Re$. As the latter formula is able to describe the data with a coefficient of determination $R^2 > 0.98$, this indicates that inertial effects and the redirection of the flow due to the wall roughness can be neglected.

The results presented show that the experimental setup can generally be used to determine the pressure drop in the microfluidic channel. In the following step, a solid particle will be introduced into the channel. The simulations aim to determine the influence of the film thickness between the particle and the wall as well as in the channel corners (area of bypass flow) on the overall pressure drop.

3.2. Solid Particle in Channel

Simulations were carried out for particles with different lengths to outline the impact of a solid particle in the microfluidic channel on Δp (Figure 8). The cap of the particles used for these simulations was constructed following the images taken from the experiments. This resulted in a cap length of 45 µm. In the first approach, a film thickness of 2 µm was assumed. It is shown later that this assumption leads to comparable results for the numeric simulations and the experiments.

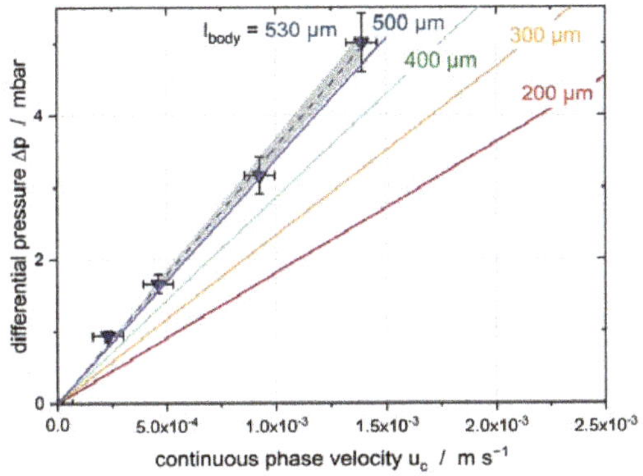

Figure 8. Experimental (data points) and simulations (lines) at ϑ_{exp} = 17.5 °C to outline the impact of the body length l_{body}. The experimental data points were fitted linearly and the grey region represent the 95% confidence interval of this fit.

It can be seen that Δp increases with an increasing flow velocity of the continuous phase u_c and an increase in the particle length. The experimental validation of the simulative results has been successful. A simulative parameter study was performed to further understand the impacts of the particle dimensions (h_f, l_{body}, l_{cap}) (Figure 9).

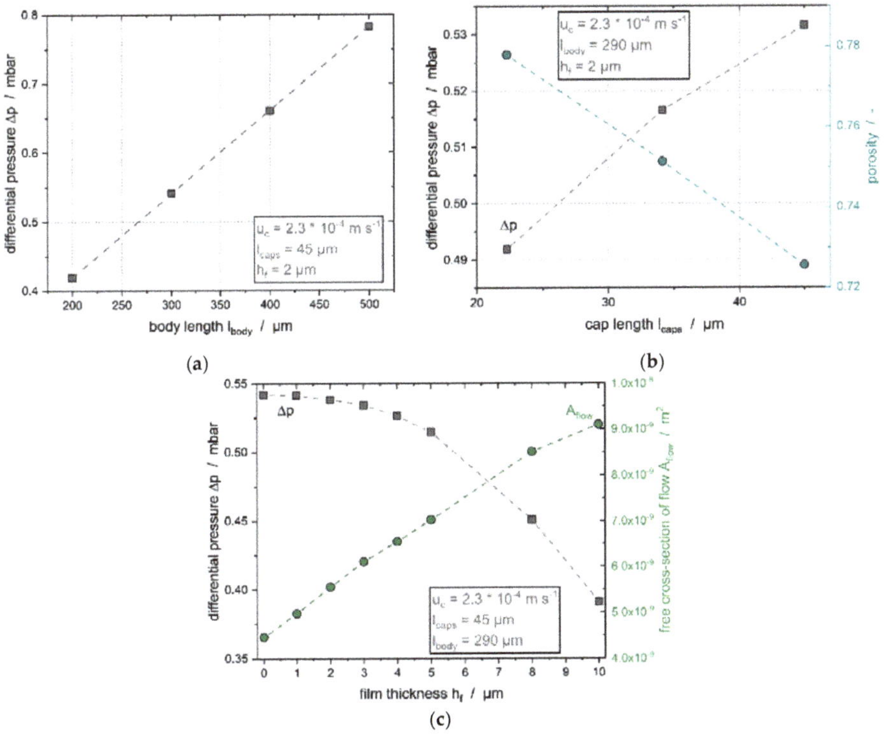

Figure 9. Influence of (**a**) body length l_{body}, (**b**) cap length l_{cap} and (**c**) film thickness h_f on the pressure drop in the minichannel. The lines presented are guides for the eye.

With increasing l_{body} and l_{cap}, Δp increased along the microfluidic channel. For l_{body} and l_{cap} this increase can be explained by the reduction in porosity. The linear increase of Δp with increasing l_{body} follows expectations. When the thickness of the liquid film between the particle and the channel wall h_f was increased, Δp along the channel decreased as A_{flow} increased. With h_f increasing from 0 µm to 5 µm, the relative deviation is only about 6%. As this is smaller than the accuracy of the experimental measurements and a good agreement between the experiments and the simulation was achieved with a thickness of 2 µm (compare Figure 8), the film thickness in the experiments should be between 0 µm and 5 µm.

The thickness of the film separating a moving droplet in a microchannel from the wall is mentioned to be about 1% to 5% of the half height of the channel for Capillary numbers of the droplet smaller than $Ca_d < 0.01$ [40–42] (compare Appendix A Table A2). In our work, the half-height is assumed to be half of d_h, leading to a film thickness between 1.2 µm and 6 µm. This is in good agreement with the simulation of the particle in the flow field and the corresponding experimental validation, although the assumption was made for liquid droplets and not for solid particles. As the particle decreases in size due to the solidification, a slightly larger film thickness would be reasonable.

The simulative Δp and u_c were tracked along three lines in the minichannel to investigate the individual impact of the caps and the body on Δp (Figure 10).

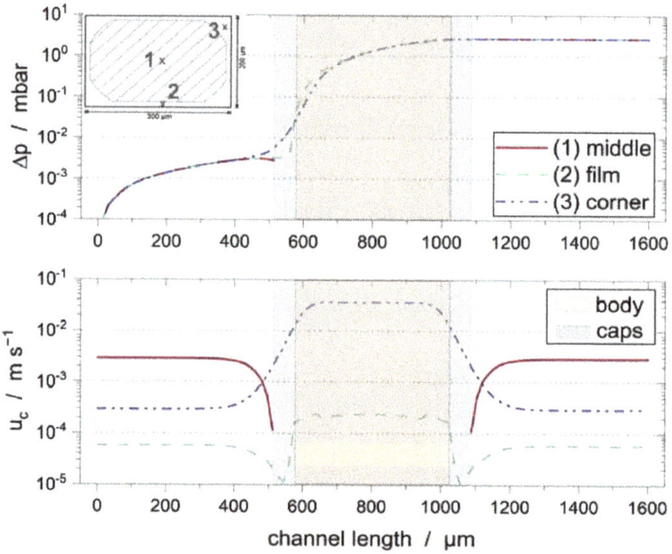

Figure 10. Differential pressure drop (**top**) and the velocity of the continuous phase u_c (**bottom**) along different lines (top left corner) in the microfluidic channel at $\vartheta_{exp} = 20$ °C and $u_c = 1.4 \cdot 10^{-3}$ m s^{-1}. The orange region represents the body of the particle ($l_{body} = 580$ µm) and the grey regions of the two caps ($l_{cap} = 45$ µm). Please notice the different scales (logarithmic and linear) of the axis of ordinates.

It becomes visible that the main fraction of Δp decreases along the body of the particle, and the impact of the particle on the overall Δp is larger than the Δp along the empty channel. In addition, the pressure drop along the corners dominates the overall Δp. This is in good agreement with the findings of [29]. The main pressure drop for liquid droplets without surfactant was found to be in the plugs, i.e., sections between the droplets, for inviscid droplets, and along the body for viscous droplets [43]. Fuerstman et al. [28] found that, by introducing an intermediate concentration of surfactant (1 to 2 orders of magnitude of *cmc*), the pressure drop along a bubble is dominated by the loss along the body length. The pressure was found to drop most rapidly along the caps for no or a high concentration of surfactant. The two main differences between the data found in literature and our results are that we investigated a stationary solid particle instead of a movable liquid or gaseous droplet/bubble.

3.3. Moving Droplets

The last step before evaluating the contact force needed for inoculation is to investigate how liquid droplets that move through the channel impact Δp (Figure 11). Various authors discuss the impact of moving droplets on Δp and describe an increasing Δp when introducing moving droplets into the microfluidic channel [43,44].

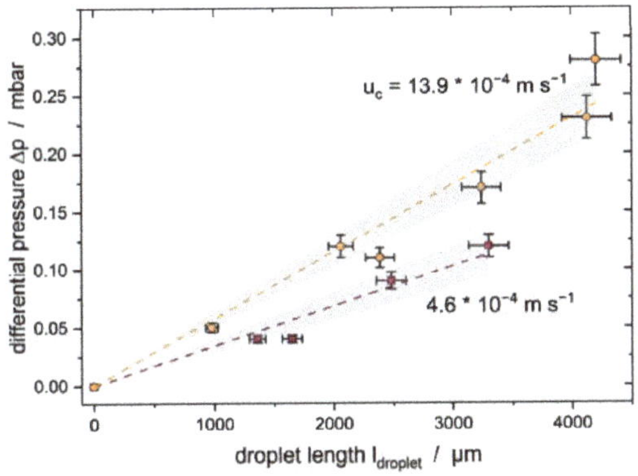

Figure 11. Experimental results of droplets moving along the microfluidic channel for two different velocities of the continuous phase u_c. The grey areas represent the 95% confidence intervals of the linear fits.

At a constant velocity of the continuous phase u_c, Δp increases linearly with the total length of the droplets inside the channel. This points out that the moving droplets also influence the overall pressure loss, which was also described in the literature previously [43,44]. This result can be explained by the observation that the liquid droplets moved slightly slower than the adjusted velocity of the continuous phase. According to [45], the difference between the velocity of the continuous phase and that of the droplet should be no larger than 6%. In our experiments, the droplets were 1–10% slower than the continuous phase, which is in good agreement with the literature [45]. Regarding the impact of a solid particle on Δp, there is a difference in an order of magnitude (compare Figure 8). Nonetheless, the impact of moving droplets on the overall Δp must be considered when finally evaluating the CMN and calculating the force needed to trigger crystallization.

3.4. Differential Pressure during Crystallization

Having gained knowledge about the impacts of droplets and particles on the overall pressure loss, the results for the CMN can now be evaluated. Figure 12 shows an exemplary measurement of Δp during an experimental performance of a CMN.

Different phases can be found for Δp as a function of time t during the experimental investigation of the CMN:

- $t < 0$: One solid particle can already be found in the channel. The corresponding Δp_{ec+1pc} is the sum of the pressure loss due to the fluid flow of the continuous phase around the particle and the empty channel.
- Yellow region: In addition to the solid particle, one liquid droplet is formed at the T-junction of the channel and is inserted into the main channel.
- Orange region: The formation of the liquid droplet is finished, and the droplet is moving towards the solid particle along the rectangular channel. As the previous results stated, the moving droplet resulted in an additional pressure loss. That is

why $\Delta p > \Delta p_{ec+1pc}$ in this region. The first contact between the liquid and the solid particles happened at the end of this region.

- The time between the first visible contact and the visible start of the crystallization is called the induction time and describes the time needed for the successful inoculation of the subcooled, liquid droplet. We were able to show in a previous work that the induction time is a function of the aqueous surfactant concentration and the relative velocity between the two collision partners [7]. Moreover, nucleation is possible because the solid particle seems to have a partial interfacial coverage with surfactant molecules compared to the fully covered interface of liquid droplets [46]. The surfactant molecules are moveable on the interface of the droplet, therefore, a molecular contact between these two partners can be given.
- Green region: After the crystallization starts, the crystal strands grow through the subcooled droplet until the droplet is completely crystallized. The speed of growth strongly depends on the subcooling (here: $\Delta T \sim 1.1$ K). The speed of growth increases as the temperature decreases. The increase of Δp can be explained by the change in the elasticity of the droplet as the latter becomes solid and due to the deformation of the liquid part of the droplet as it is pushed to the particle.
- As soon as the droplet is completely crystallized, Δp becomes constant at the level indicated as Δp_{ec+2pc} (empty channel + two solid particles). As has been shown previously, the pressure loss due to the particle depends strongly on the size of the particle.

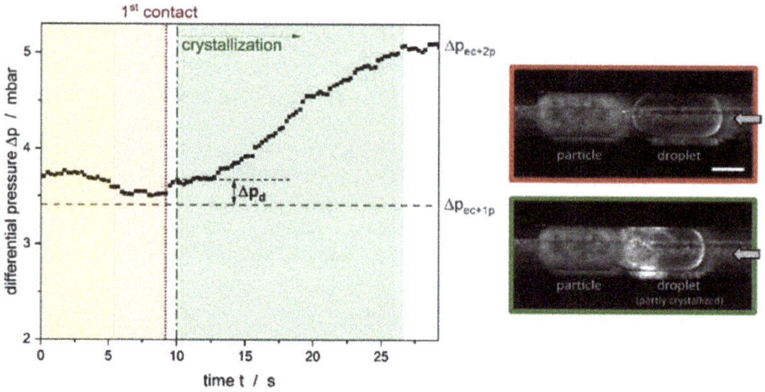

Figure 12. Exemplary measurement data for the pressure drop as a function of time during the formation of the droplet at the T-junction (yellow), the movement of the droplet along the channel (orange), the first contact of particle and droplet as well as the crystallization of the droplet (green). Δp_{ec+1p} and Δp_{ec+2p} represent the differential pressures measured for the system when one or two solid particles were circulated by the continuous phase. The two microscopic images show the droplets moving toward the solid particle (red frame) and the crystallization of the droplet after contact with the solid particle (green frame). The white bar represents 200 μm. The arrows in the pictures indicate the flow direction of the continuous phase and, consequently, also of the subcooled droplet.

The pressure difference Δp_d between Δp_{ec+1pc} and the beginning of the crystallization can be used to estimate the force needed for successful inoculation after the contact F_c:

$$F_c = \Delta p_d \cdot A_{channel}. \tag{6}$$

$A_{channel}$ hereby represents the cross-section of the channel, here $A_{channel} = 6 \cdot 10^{-8}$ m^2. Δp_d needed for nucleation was investigated as a function of the surfactant concentration at different u_c (Figure 13).

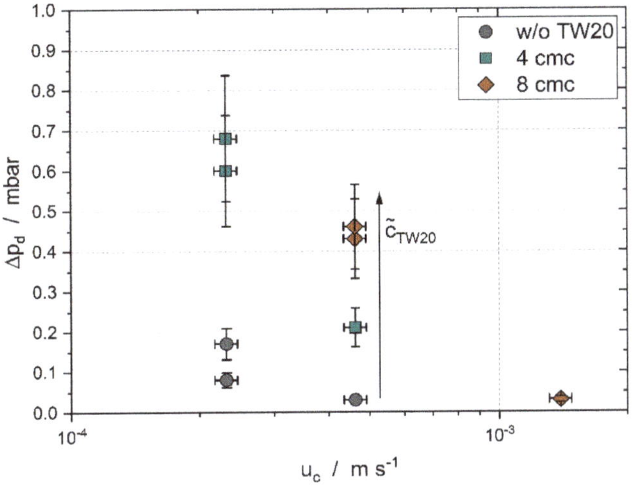

Figure 13. Differential pressure Δp_d needed to inoculate the liquid droplet after it came into contact with the solid particle with (4 or 8 *cmc*) and without (*w*/*o*) TW20 as the surfactant. Δp_d correlates linearly with the force needed for CMN due to the flow field (Equation (6)).

At a constant relative velocity of the droplet and the particle Δu, Figure 13 shows that Δp_d increases with increasing surfactant concentration. The velocity of the continuous phase u_c has a linear relationship to Δu as the latter equals the velocity of the liquid droplet in this setup. For the two lower surfactant concentrations (w/o TW20 and 4 *cmc*), the induction time t_{ind} was <<1 s for $u_c > 4.6 \cdot 10^{-4}$ m s^{-1} and, consequently, no Δp_d could be determined, but all collisions led to nucleation of the subcooled droplet.

In addition to F_c, the force due to the impulse of the decelerating droplet F_i must be considered to determine the overall contact force needed for successful CMN:

$$F_i = m_d \cdot \frac{du_d}{dt}. \tag{7}$$

The minimal and maximal droplet volumes were used to estimate the mass of the droplets $m_d = 1.2 \cdot 10^{-8} - 3.4 \cdot 10^{-8}$ kg. The time the droplet needed for deceleration was assumed to be 0.01 s as this was the highest possible resolution of the camera. Any possible deformation of the droplet was not considered. As the time the droplet needed for decelerating cannot be estimated properly, this points out the limitation of the estimation of F_i. Nonetheless, higher droplet velocities must lead to an increase of F_i (Figure 14).

Both forces presented must be added together to determine the overall minimum force needed for CMN $F_{CMN,min}$:

$$F_{CMN,min} = F_c + F_i. \tag{8}$$

The increase of $F_{CMN,min}$ with increasing surfactant concentration might be explained by an increase in the aqueous number of micelles. This is in good agreement with, for example, Dudek et al. [47] or Kaysan et al. [7], who found longer mean coalescence times or mean induction times for CMN when increasing $\tilde{c}_{TW20}$ in the aqueous phase. Both results lead to the conclusion that the aqueous number of micelles impacts the efficiency of CMN. For $\tilde{c}_{TW20} = 0.41$ mol m^{-3} (8 *cmc*), there was no crystallization detectable for $u_c < 4.6 \cdot 10^{-4}$ m s^{-1}, which also can be explained by the large number of micelles that hinder the contact between the two collision partners. Before the nucleation could happen, the droplet surrounded the particle and was finely emulsified into smaller droplets behind it.

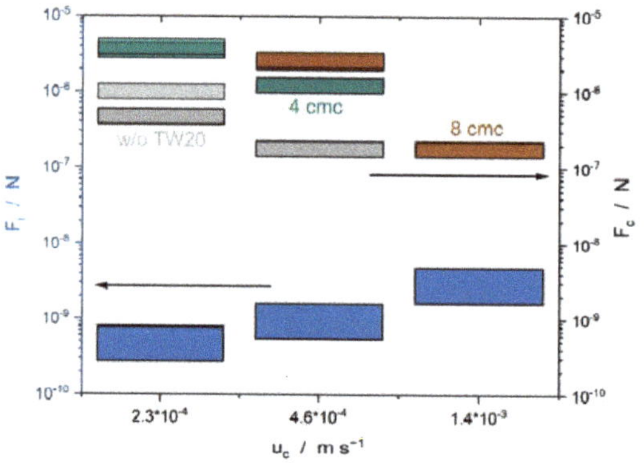

Figure 14. Force F_i acting on the droplet due to the impulse (Equation (7)), and the force due to the flow field after the contact F_c (Equation (6)) for three different velocities of the continuous phase and three aqueous surfactant concentrations $\tilde{c}_{TW20}$ (without (w/o) TW20, 4 cmc or 8 cmc). A cross-sectional area of $A_{channel} = 6 \cdot 10^{-8}$ m^2 was used for the calculations.

Proof of Theorem 1. Measuring the differential pressure during the CMN in a microfluidic channel in combination with numeric simulations led to the verification of Theorem 1: Increasing minimum contact forces $F_{CMN,min}$ were determined with increasing surfactant concentrations. □

The calculation of F_i is prone to errors due to experimental limitations, therefore, the comparison of different velocities with each other must be considered with caution. On the one hand, the decrease of F_c with increasing Δu (shown as an increase in u_c) might be a result of increasing F_i and, consequently, only fewer additional forces are needed induce CMN. $F_{CMN,min}$ could further be reduced by the destruction and displacement of micelles at increasing velocities of the approaching subcooled droplet. The latter hypothesis is strengthened by Christov et al. [48], who found that TW20 micelles can be dissolved by the application of shear stress. Kinoshita et al. [49], for example, succeeded in demonstrating that internal flows are formed in droplets moving through a rectangular channel. The flow on the surface of the droplet is opposite to the flow direction. This could mean for an interface loaded with surfactant molecules that with higher droplet velocities, the surfactant molecules on the surface are also displaced more strongly and accumulate at the rear part of the droplets due to their low solubility in the oil phase. This would result in a decrease of F_c and, consequently, $F_{CMN,min}$ because fewer surfactant molecules must be displaced.

4. Conclusions

This work investigated the pressure loss in a microfluidic channel. Experimental results have been compared with numerical values for a rectangular microfluidic channel that had water flowing through it. For the first time, the impact of a solid particle on the pressure loss was examined during a numerical parameter study and revealed that the film thickness between the particle and the wall should be up to 2% of the hydraulic diameter of the channel. The highest fluid velocities around the particle were found in the bypass or gusset flows in the corners of the rectangular channel. Moreover, the length of the bodies of the droplets had the largest impact on the pressure drop along the channel.

Regarding the CMN, the force needed for a successful inoculation during the CMN has been estimated for the first time. We were able to prove that the addition of surfactant increased the force needed or even hindered the CMN completely. Further investigations will now focus on the investigation of charged surfactants as they increase the disjoining pressure and, therefore, the contact force needed for CMN should increase at the same time.

Author Contributions: Conceptualization, G.K.; methodology, G.K. and K.D.; software, G.K. and K.D.; validation, G.K.; formal analysis, G.K.; investigation, G.K., T.H. and K.D.; resources, M.K.; data curation, G.K., T.H. and K.D.; writing—original draft preparation, G.K.; writing—review and editing, K.D., T.H. and M.K.; visualization, G.K.; supervision, M.K. All authors have read and agreed to the published version of the manuscript.

Funding: This research received no external funding.

Institutional Review Board Statement: Not applicable.

Informed Consent Statement: Not applicable.

Data Availability Statement: All important data are included in the paper. The raw data are available upon request.

Acknowledgments: The authors want to thank Max Renaud from the Institute of Thermal Process Engineering for the fabrication of the raw polycarbonate chip and for helping to integrate the differential pressure sensor. Furthermore, the help of Martin Kansy with the setup of the numeric parameter study is acknowledged. We acknowledge support by the KIT-Publication Fund of the Karlsruhe Institute of Technology.

Conflicts of Interest: The authors declare no conflict of interest.

Appendix A. Calculation of Dimensionless Numbers

- Reynolds number *Re*

Re is meant to be the prominent example of dimensionless key figures in the field of fluid dynamics [13,50]. The inertia force for *Re* << 1 can be neglected, and the Navier-Stokes equation simplifies to a Stokes form.

- Capillary number *Ca*

The ratio between viscous friction forces and surface forces can be described using *Ca*:

$$Ca = \frac{\eta \cdot u}{\gamma}. \tag{A1}$$

Especially regarding the microfluidic system, *Ca* is a relevant criterium to, for example, describe the droplet formation at the T-junction or the thickness of the liquid film [51].

Table A1. Fluid properties at ϑ_{exp} = 17.5 °C used for the calculations of the dimensionless numbers.

Fluid Properties		Symbol	Value
Density/kg m^{-3}	water	ρ	998.7
	n-hexadecane		775.1
Dynamic viscosity/mPa s	water	η	1.07
	n-hexadecane		3.70 [1]
Interfacial tension/ N m^{-1}	with TW20	γ	0.004 [2]
	without TW20		0.047 [3]

[1] Own measurements, for $\vartheta > \vartheta_{m,hex}$ and linear approximation to ϑ_{exp} = 17.5 °C ($<\vartheta_{m,hex}$); [2] own measurements, in analogy to [7]; [3] [52].

Table A2. Important dimensionless numbers for the fluid velocities in the microfluidic channel. The fluid properties listed in Table A1 were used for the calculations.

Fluid Velocity u_c/m s^{-1}	Droplet Capillary Number Ca_d		Reynolds Number *Re*
	without TW20	with TW20	
$2.3 \cdot 10^{-4}$	$1.8 \cdot 10^{-5}$	$2.1 \cdot 10^{-4}$	$5.2 \cdot 10^{-2}$
$4.6 \cdot 10^{-4}$	$3.6 \cdot 10^{-5}$	$4.3 \cdot 10^{-4}$	$1.0 \cdot 10^{-1}$
$9.3 \cdot 10^{-4}$	$7.3 \cdot 10^{-5}$	$8.6 \cdot 10^{-4}$	$2.1 \cdot 10^{-1}$
$1.4 \cdot 10^{-3}$	$1.1 \cdot 10^{-4}$	$1.3 \cdot 10^{-3}$	$3.1 \cdot 10^{-1}$
$1.9 \cdot 10^{-3}$	$1.5 \cdot 10^{-4}$	$1.7 \cdot 10^{-3}$	$4.2 \cdot 10^{-1}$
$2.3 \cdot 10^{-3}$	$1.8 \cdot 10^{-4}$	$2.1 \cdot 10^{-3}$	$5.2 \cdot 10^{-1}$

References

1. Bunjes, H.; Koch, M.H.J.; Westesen, K. Influence of emulsifiers on the crystallization of solid lipid nanoparticles. *J. Pharm. Sci.* **2003**, *92*, 1509–1520. [CrossRef] [PubMed]
2. Kalnin, D.; Schafer, O.; Amenitsch, H.; Ollivon, M. Fat Crystallization in Emulsion: Influence of Emulsifier Concentration on Triacylglycerol Crystal Growth and Polymorphism. *Cryst. Growth Des.* **2004**, *4*, 1283–1293. [CrossRef]
3. Maruyama, J.M.; Soares, F.A.S.D.M.; D'Agostinho, N.R.; Gonçalves, M.I.A.; Gioielli, L.A.; Da Silva, R.C. Effects of emulsifier addition on the crystallization and melting behavior of palm olein and coconut oil. *J. Agric. Food Chem.* **2014**, *62*, 2253–2263. [CrossRef] [PubMed]
4. Miskandar, M.S.; Man, Y.C.; Rahman, R.A.; Aini, I.N.; Yusoff, M. Effects of Emulsifiers on Crystal Behavior of Palm Oil Blends On Slow Crystallization. *J. Food Lipids* **2007**, *14*, 1–18. [CrossRef]
5. Kashchiev, D.; Clausse, D.; Jolivet-Dalmazzone, C. Crystallization and Critical Supercooling of Disperse Liquids. *J. Colloid Interface Sci.* **1994**, *165*, 148–153. [CrossRef]
6. Mastai, Y. *Advanced Topics in Crystallization*; IntechOpen: London, UK, 2015; ISBN 9789535121251.
7. Kaysan, G.; Rica, A.; Guthausen, G.; Kind, M. Contact-Mediated Nucleation of Subcooled Droplets in Melt Emulsions: A Microfluidic Approach. *Crystals* **2021**, *11*, 1471. [CrossRef]
8. McClements, J.D.; Dickinson, E.; Povey, M. Crystallization in hydrocarbon-in-water emulsions containing a mixture of solid and liquid droplets. *Chem. Phys. Lett.* **1990**, *172*, 449–452. [CrossRef]
9. Dickinson, E.; Kruizenga, F.-J.; Povey, M.J.; van der Molen, M. Crystallization in oil-in-water emulsions containing liquid and solid droplets. *Colloids Surf. A Physicochem. Eng. Asp.* **1993**, *81*, 273–279. [CrossRef]
10. McClements, D.J.; Dungan, S.R. Effect of Colloidal Interactions on the Rate of Interdroplet Heterogeneous Nucleation in Oil-in-Water Emulsions. *J. Colloid Interface Sci.* **1997**, *186*, 17–28. [CrossRef]
11. Povey, M.J.W.; Awad, T.S.; Huo, R.; Ding, Y. Quasi-isothermal crystallisation kinetics, non-classical nucleation and surfactant-dependent crystallisation of emulsions. *Eur. J. Lipid Sci. Technol.* **2009**, *111*, 236–242. [CrossRef]
12. Kaysan, G.; Schork, N.; Herberger, S.; Guthausen, G.; Kind, M. Contact-mediated nucleation in melt emulsions investigated by rheo-nuclear magnetic resonance. *Magn. Reason. Chem.* **2022**, *60*, 615–627. [CrossRef] [PubMed]
13. Reynolds, O. An experimental investigation of the circumstances which determine whether the motion of water shall be direct or sinuous, and of the law of resistance in parallel channels. *Phil. Trans. R. Soc.* **1883**, *174*, 935–982. [CrossRef]
14. Kandlikar, S.G. Single-phase liquid flow in minichannels and microchannels. In *Heat Transfer and Fluid Flow in Minichannels and Microchannels*; Kandlikar, S.G., Garimella, S., Li, D., Colin, S., King, M., Eds.; Elsevier: London, UK, 2006; pp. 87–136. ISBN 9780080445274.
15. Kast, W.; Gaddis, E.S.; Wirth, K.-E.; Stichlmair, J. L1 Pressure Drop in Single Phase Flow. In *VDI Heat Atlas*; VDI e.V., Ed.; Springer: Berlin/Heidelberg, Germany, 2010; pp. 1053–1116. ISBN 978-3-540-77876-9.
16. Kandlikar, S.G.; Grande, W.J. Evolution of Microchannel Flow Passages–Thermohydraulic Performance and Fabrication Technology. *Heat Transf. Eng.* **2003**, *24*, 3–17. [CrossRef]
17. Kockmann, N. *Transport Phenomena in Micro Process Engineering*; Springer: Berlin/Heidelberg, Germany, 2008; ISBN 978-3-540-74616-4.
18. Chan, D.Y.C.; Horn, R.G. The drainage of thin liquid films between solid surfaces. *J. Chem. Phys.* **1985**, *83*, 5311–5324. [CrossRef]
19. Qu, W.; Mudawar, I.; Lee, S.-Y.; Wereley, S.T. Experimental and Computational Investigation of Flow Development and Pressure Drop in a Rectangular Micro-channel. *J. Electron. Packag.* **2006**, *128*, 1–9. [CrossRef]
20. Xu, B.; Ooti, K.T.; Wong, N.T.; Choi, W.K. Experimental investigation of flow friction for liquid flow in microchannels. Int. Commun. *Heat Mass Transf.* **2000**, *27*, 1165–1176. [CrossRef]
21. Mohiuddin Mala, G.; Li, D. Flow characteristics of water in microtubes. *Int. J. Heat Fluid Flow* **1999**, *20*, 142–148. [CrossRef]
22. Mirmanto, M. Prediction and Measurement of Pressure Drop of Water Flowing in a Rectangular Microchannel. *DTM* **2017**, *3*, 75–82. [CrossRef]
23. Ghajar, A.J.; Tang, C.C.; Cook, W.L. Experimental Investigation of Friction Factor in the Transition Region for Water Flow in Minitubes and Microtubes. *Heat Transf. Eng.* **2010**, *31*, 646–657. [CrossRef]
24. Steinke, M.E.; Kandlikar, S.G. Single-phase liquid friction factors in microchannels. *Int. J. Therm. Sci.* **2006**, *45*, 1073–1083. [CrossRef]
25. Selzer, D.; Spiegel, B.; Kind, M. A Generic Polycarbonate Based Microfluidic Tool to Study Crystal Nucleation in Microdroplets. *J. Cryst. Process Technol.* **2018**, *8*, 1–17. [CrossRef]
26. Xu, J.H.; Li, S.W.; Tan, J.; Wang, Y.J.; Luo, G.S. Preparation of highly monodisperse droplet in a T-junction microfluidic device. *AIChE J.* **2006**, *52*, 3005–3010. [CrossRef]
27. Musterd, M.; van Steijn, V.; Kleijn, C.R.; Kreutzer, M.T. Calculating the volume of elongated bubbles and droplets in microchannels from a top view image. *RSC Adv.* **2015**, *5*, 16042–16049. [CrossRef]
28. Fuerstman, M.J.; Lai, A.; Thurlow, M.E.; Shevkoplyas, S.S.; Stone, H.A.; Whitesides, G.M. The pressure drop along rectangular microchannels containing bubbles. *Lab Chip* **2007**, *7*, 1479–1489. [CrossRef] [PubMed]
29. Ransohoff, T.; Radke, C. Laminar flow of a wetting liquid along the corners of a predominantly gas-occupied noncircular pore. *J. Colloid Interface Sci.* **1988**, *121*, 392–401. [CrossRef]

30. Linke, D. Detergents. In *Guide to Protein Purification*, 2nd ed.; Burgess, R., Deutscher, M., Eds.; Elsevier: London, UK, 2009; pp. 603–617. ISBN 9780123745361.
31. Weller, H.G.; Tabor, G.; Jasak, H.; Fureby, C. A tensorial approach to computational continuum mechanics using object-oriented techniques. *Comput. Phys.* **1998**, *12*, 620–631. [CrossRef]
32. *Salome*, Version 9.3.0.; Open Cascade SA: Guyancourt, France, 2019.
33. Roache, P.J. Perspective: A Method for Uniform Reporting of Grid Refinement Studies. *J. Fluids Eng.* **1994**, *116*, 405–413. [CrossRef]
34. Chesters, A.A. The modelling of coalescence processes in fluid-liquid dispersions: A review of current understanding. *Chem. Eng. Res. Des.* **1991**, *69*, 259–270.
35. Kaysan, G.; Spiegel, B.; Guthausen, G.; Kind, M. Influence of Shear Flow on the Crystallization of Organic Melt Emulsions—A Rheo-Nuclear Magnetic Resonance Investigation. *Chem. Eng. Technol.* **2020**, *43*, 1699–1705. [CrossRef]
36. Moody, L.F. Friction Factors for Pipe Flow. *Trans. Am. Soc. Mech. Eng.* **1944**, *66*, 671–678. [CrossRef]
37. Krishnamoorthy, C.; Ghajar, A.J. Single-Phase Friction Factor in Micro-Tubes: A Critical Review of Measurement, Instrumentation and Data Reduction Techniques From 1991–2006. In Proceedings of the ASME 2007 5th International Conference on Nanochannels, Microchannels, and Minichannels, Puebla, Mexico, 18–20 June 2007; pp. 813–825, ISBN 0-7918-4272-X.
38. Tam, L.M.; Tam, H.K.; Ghajar, A.J.; Ng, W.S.; Wong, I.W.; Leong, K.F.; Wu, C.K. The Effect of Inner Surface Roughness and Heating on Friction Factor in Horizontal Micro-Tubes. In Proceedings of the ASME-JSME-KSME 2011 Joint Fluids Engineering Conference: Symposia—Parts A, B, C, and D, Hamamatsu, Japan, 24–29 July 2011; Volume 1, pp. 2971–2978, ISBN 978-0-7918-4440-3.
39. Gloss, D.; Herwig, H. Microchannel Roughness Effects: A Close-Up View. *Heat Transf. Eng.* **2009**, *30*, 62–69. [CrossRef]
40. Reinelt, D.A.; Saffman, P.G. The Penetration of a Finger into a Viscous Fluid in a Channel and Tube. *SIAM J. Sci. Stat. Comput.* **1985**, *6*, 542–561. [CrossRef]
41. Hazel, A.L.; Heil, M. The steady propagation of a semi-infinite bubble into a tube of elliptical or rectangular cross-section. *J. Fluid Mech.* **2002**, *470*, 91–114. [CrossRef]
42. Schwartz, L.W.; Princen, H.M.; Kiss, A.D. On the motion of bubbles in capillary tubes. *J. Fluid Mech.* **1986**, *172*, 259–275. [CrossRef]
43. Baroud, C.N.; Gallaire, F.; Dangla, R. Dynamics of microfluidic droplets. *Lab Chip* **2010**, *10*, 2032–2045. [CrossRef] [PubMed]
44. Adzima, B.J.; Velankar, S.S. Pressure drops for droplet flows in microfluidic channels. *J. Micromech. Microeng.* **2006**, *16*, 1504–1510. [CrossRef]
45. Jousse, F.; Lian, G.; Janes, R.; Melrose, J. Compact model for multi-phase liquid-liquid flows in micro-fluidic devices. *Lab Chip* **2005**, *5*, 646–656. [CrossRef]
46. Kaysan, G.; Kräling, R.; Meier, M.; Nirschl, H.; Guthausen, G.; Kind, M. Investigation of the surfactant distribution in oil-in-water emulsions during the crystallization of the dispersed phase via nuclear magnetic resonance relaxometry and diffusometry. *Magn. Reson. Chem.* **2022**, *60*, 1131–1147. [CrossRef]
47. Dudek, M.; Fernandes, D.; Helno Herø, E.; Øye, G. Microfluidic method for determining drop-drop coalescence and contact times in flow. Colloids Surf. A Physicochem. *Eng. Asp.* **2020**, *586*, 124265. [CrossRef]
48. Christov, N.C.; Danov, K.D.; Zeng, Y.; Kralchevsky, P.A.; von Klitzing, R. Oscillatory structural forces due to nonionic surfactant micelles: Data by colloidal-probe AFM vs theory. *Langmuir* **2010**, *26*, 915–923. [CrossRef]
49. Kinoshita, H.; Kaneda, S.; Fujii, T.; Oshima, M. Three-dimensional measurement and visualization of internal flow of a moving droplet using confocal micro-PIV. *Lab Chip* **2007**, *7*, 338–346. [CrossRef] [PubMed]
50. Rott, N. Note on the History of the Reynolds Number. *Annu. Rev. Fluid Mech.* **1990**, *22*, 1–12. [CrossRef]
51. Bretherton, F.P. The motion of long bubbles in tubes. *J. Fluid Mech.* **1961**, *10*, 166–188. [CrossRef]
52. van der Graaf, S.; Schroën, C.G.P.H.; van der Sman, R.G.M.; Boom, R.M. Influence of dynamic interfacial tension on droplet formation during membrane emulsification. *J. Colloid Interface Sci.* **2004**, *277*, 456–463. [CrossRef] [PubMed]

colloids and interfaces

MDPI

Article

Preparation of Hydrophobic Monolithic Supermacroporous Cryogel Particles for the Separation of Stabilized Oil-in-Water Emulsion

Hayato Takase [1], Nozomi Watanabe [1], Koichiro Shiomori [2], Yukihiro Okamoto [1], Endang Ciptawati [1,3], Hideki Matsune [2] and Hiroshi Umakoshi [1,*]

1 Division of Chemical Engineering, Graduate School of Engineering Science, Osaka University, 1-3 Machikaneyama-cho, Toyonaka 560-8531, Japan
2 Department of Applied Chemistry, University of Miyazaki, 1-1 Gakuenkibanadai-nishi, Miyazaki 899-2192, Japan
3 Department of Chemistry, Faculty of Mathematics and Natural Science, Universitas Negeri Malang, Jl. Semarang 5, Malang 65145, Indonesia
* Correspondence: b-ice@cheng.es.osaka-u.ac.jp; Tel./Fax: +81-6-6850-6286

Abstract: Here, we prepared hydrophobic cryogel particles with monolithic supermacropores based on poly-trimethylolpropane trimethacrylate (pTrim) by combining the inverse Leidenfrost effect and cryo-polymerization technique. The hydrophobic cryogel particles prepared by adopting this method demonstrated the separation of the stabilized O/W emulsion with surfactant. The prepared cryogel particles were characterized in terms of macroscopic shape and porous structure. It was found that the cryogel particles had a narrow size distribution and a monolithic supermacroporous structure. The hydrophobicity of the cryogel particles was confirmed by placing aqueous and organic droplets on the particles. Where the organic droplet was immediately adsorbed into the particles, the aqueous droplet remained on the surface of the particle due to repelling force. In addition, after it adsorbed the organic droplet the particle was observed, and the organic solvent was diffused into the entire particle. It was indicated that monolithic pores were distributed from the surface to the interior. Regarding the application of the hydrophobic cryogel particles, we demonstrated the separation of a stabilized oil-in-water emulsion, resulting in the successful removal of the organic solvent from the emulsion.

Keywords: hydrophobic cryogel particles; supermacropores; inverse Leidenfrost effect; droplet method; separation of emulsion

Citation: Takase, H.; Watanabe, N.; Shiomori, K.; Okamoto, Y.; Ciptawati, E.; Matsune, H.; Umakoshi, H. Preparation of Hydrophobic Monolithic Supermacroporous Cryogel Particles for the Separation of Stabilized Oil-in-Water Emulsion. *Colloids Interfaces* **2023**, *7*, 9. https://doi.org/10.3390/colloids7010009

Academic Editors: César Burgos-Díaz, Mauricio Opazo-Navarrete and Eduardo Morales

Received: 12 December 2022
Revised: 10 January 2023
Accepted: 12 January 2023
Published: 17 January 2023

1. Introduction

Water pollution caused by oil spill accidents and the leakage of industrial wastewater causes serious environmental problems [1]. Wasted oil disperses as an emulsion in a heterogenous system; therefore, the removal of oil is difficult compared to that of immiscible oil, which spontaneously separates from aqueous solution [2]. Especially, oil-in-water emulsion including surfactants can be regarded as a semi-equilibrium state with high stability in the dispersed state, and it is difficult to remove its oil fraction from the system [3]. Thus, the oil–water separation process is important for treating oily wastewater containing surfactants and other substances. As the major oil–water separation process, the gravity-driven separation, centrifugal precipitation, coagulation, and chemical demulsification are well known [4–7]. However, these separation techniques can be restricted because of low-cost performance with complex treatment and high energy consumption [8]. The membrane separation method uses hydrophobic materials to adsorb oil and are relatively inexpensive, making them a useful method for oil-removing strategies compared to the conventional treatment [9].

Despite the fact that they were discovered more than 30 years ago, cryogels have only recently gained widespread attention due to their extraordinary properties. Cryogels are extremely tough gels that can withstand significant deformation, such as elongation and torsion, and can be squeezed almost entirely without crack propagation [10–12]. In recent years, many cryogenically structured polymeric materials have been of significant scientific and applied interest in various fields, such as separation, waste-water treatment, biotechnology, and tissue engineering [13–17]. Ihlenburg et al. demonstrated that a sulfobetaine-based cryogel adsorbs methyl orange preferentially from mixed dye solutions. Methyl orange (MO)/methyl blue (MB) mixtures can be separated by the selective adsorption of the MO to the cryogels, while the MB remains in solution [18]. Cryogel is a porous material obtained by the polymerization of the polymer precursor, a solution containing monomers, a cross-linker, and polymerization initiators, under a frozen temperature (cryo-polymerization) [19–21]. In the process of cryo-polymerization, the phase separation occurs between the unfrozen region (concentrated monomers) and the frozen one (ice crystals), both inhabited by the polymer precursor [22]. Polymerization can occur only in the unfrozen regions of the system, where frozen solvent crystals act as porogens [10]. After the polymerization, the porous structure in the range of the μ-meter scale can be obtained by melting and drying porogen (ice crystals), which is referred to as a "supermacroporous structure" [23]. The nature of the solvent, the type (monomer or polymer precursors), the temperature of the cryogenic process, the rate of freezing/thawing dynamics, and some other factors influence the properties of polymeric cryogels [24]. Therefore, it is necessary to consider the polymer composition, such as hydrophobic monomers and hydrophilic monomers, and the preparation method according to the purpose of application. Consequently, the hydrophobic cryogel, which prefers oil to water, is useful for the adsorption and separation of organic material from aqueous solutions [25].

The monolithic porous structure has a large specific area and high porosity; therefore, it is applicable in chromatographic separations [26]. Since the monolithic porous structure forms a hierarchical porous structure with an interconnected network of the polymer wall, it supplies efficient diffusive mass transfer. The properties of monolithic porous structure are classified into several parameters. For the application and optimization of monolithic porous materials as separation substances, the characterization of their properties is required. The parameters include (i) the size distribution of the particle diameter, (ii) the standard deviation of the particle diameter, (iii) the coefficient of variation of particle, (iv) the bulk density, and (v) the size distribution of the pore size. Furthermore, the macroscopic shape of the cryogel's (a) cylinder, (b) sheet, (c) disk, and (d) particle is an essential factor in the design of separation devices such as the particle-packed column or monolithic column chromatography. The various shapes of the cryogel can be conventionally prepared by performing cryo-polymerization in a mold vessel. Cryogel particles can be prepared by the polymerization method using emulsion as a mold; cryo-polymerization occurs at a phase containing polymer precursor [27,28]. Due to the acceleration of cryogel research, various preparation methodologies and polymer components have been developed [29,30]. The preparation of the feed system, freezing, incubating the gelation system in a frozen state, and thawing the frozen sample are all required stages in fabricating the cryogel [11]. As a different preparation method, we recently developed a combination technique, i.e., the inverse Leidenfrost effect and the cryo-polymerization technique (the iLF cryo-method) [31,32].

In this study, we first adopted the iLF cryo-method to create hydrophobic monolithic supermacroporous cryogel particles. The cryogel particles revealed the narrow size distribution of the diameter, and numerous pores with monolithic supermacroporous structures. Furthermore, the properties such as hydrophobicity and adsorption preference were characterized by using dyed solvents. Unlike the aqueous droplet, the organic droplet was immediately adsorbed into the particles. The cross section of the cryogel particle after the adsorption of the organic solvent showed that the entire particle was stained with dye, indicating that the oil diffused thoroughly into the interior of the particle. For the separation of the oil-in-water emulsions stabilized with surfactants, the cryogel particles successfully removed emulsions from the system.

2. Materials and Methods

2.1. Materials

Trimethylolpropane trimethacrylate (Trim), benzoyl peroxide (BPO), etlyl-4-dimethyla minobenzoate (EDMAB), and nile red were purchased from Tokyo Chemical Industry Ltd. Acetic acid, toluene, n-hexane, *N,N′*-methylenebisacrylamide (MBA), liquid paraffine, sudan (IV), Coomassie brilliant blue G-250 (CBB), and polyoxyethylene sorbitan monooleate (Tween 80) were purchased from Wako Pure Chemical Industry Ltd., (Osaka, Japan). Ultrapure water (conductive 18.2 MΩ cm) was purchased from a Merck Millipore. All materials were used as received without further treatment.

2.2. Preparation of Hydrophobic Cryogel Particles

The hydrophobic cryogel particles were prepared as described in previous works [31,32]. Briefly, the cryogel particles were prepared via a 2-step process: (1) the preparation of frozen droplets by utilizing the iLF effect and (2) cryo-polymerization under a frozen temperature. Trim used as monomer (Figure 1) and MBA used as cross-linker were dissolved in acetic acid. Then, BPO and EDMAB were added as initiators and the polymer precursor was obtained. Subsequently, the obtained polymer precursor was dropped onto liquid nitrogen and frozen droplets were prepared. These droplets were transferred to liquid paraffine at −15 °C and polymerized overnight. After cryo-polymerization, hydrophobic cryogel particles were washed with n-hexane and lyophilized.

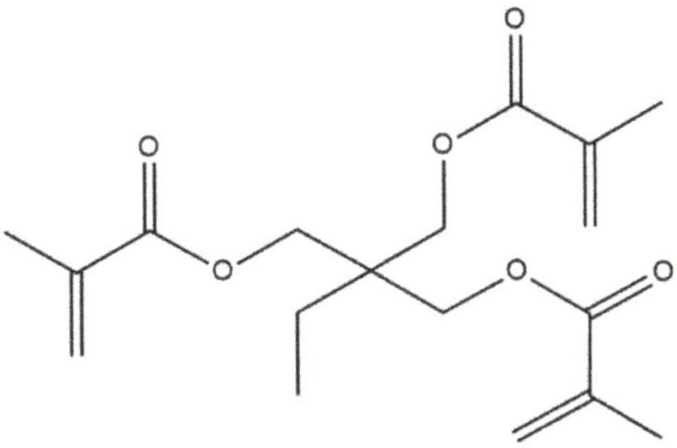

Figure 1. Chemical structure of trimethylolpropane trimethacrylate (Trim).

2.3. Observation of Macroscopic Shape

The overall cryogel particles in the bottle of glass were observed as photographic images by a charge-coupled device (CCD, iPhone 12 MJNJ3J/A) camera. In order to evaluate the size distribution of particles, the diameter of particles was measured from optical microscope images. The mean diameter of the particles (D_v) and standard deviation (σ) were calculated from the optical microscope images by using image J software. Furthermore, the coefficient of variation (C_v) was calculated according to Equation (1).

$$C_v = \frac{\sigma}{D_v} \times 100\ [\%] \quad (1)$$

Here, the tapped bulk density (ρ_{bulk}) was calculated according to Equation (2).

$$\rho_{bulk} = \frac{W_{particles}}{V_{cylinder}}\ [\text{g/mL}] \quad (2)$$

The $W_{particles}$ and $V_{cylinder}$ were represented as the weight of the particles and the volume of the particles in the graduated cylinder, respectively. The scale of the graduated cylinder with particles was read after tapping 100 times [33].

2.4. Characterization of Porous Properties

The surface of the porous structure of the cryogen particle was observed by scanning electronic microscope (Hitachi, SU3500, Tokyo, Japan, SEM), operated at 20 keV. As pre-treatment, gold was coated on cryogel for 30 s by spattering. The pore diameters were measured according to the size distribution that was analyzed using image J software.

2.5. Testing Hydrophobicity of Cryogel Particles

The hydrophobicity of cryogel particles was tested by using aqueous and organic droplets. The water was dyed with CBB as a blue-colored aqueous solution, and the toluene was dyed with Nile red as a red-colored organic solution. Then, the droplets placed on the cryogel particles were observed. Subsequently, the adsorption behavior of an organic solvent by cryogel particles was investigated. The red-dyed organic solution was separated from the water collecting at the top, and the cryogel particles were added to the water. A CCD camera was used to record the adsorption behavior over time three and ten seconds after the particle was added. Furthermore, the cryogel particle was cut in half after being adsorbed with the dyed organic solution, and the cross-section was observed.

2.6. Separation of Stabilized Oil-in-Water Emulsion

The oil-in-water emulsion stabilized with a surfactant was prepared by mixing 990 µL of n-hexane, 100 mg of Tween 80, and 99 mL of ultra-pure water under ultrasonication at room temperature (around 25–30 °C) and atmospheric pressure. After preparation, the sample was stored for over 24 h at room temperature and atmospheric pressure. Then, the stabilization of O/W emulsion (cloudy) was visually observed, and the light scattering was confirmed via laser at dark place (Tyndall phenomenon). The characterization of the emulsion after separation was recorded by photograph and using the optical microscope image. To investigate the separation of stabilized oil-in-water emulsion by cryogel particles, the organic phase of the emulsion was dyed with Sudan IV during the preparation of oil-in-water emulsion [34]. Then, the emulsion was separated by mixing the emulsion and cryogel particles in the glass bottle and kept for 3 h at room temperature. After that, the particles were removed from the solution by filter and measured the UV-Vis spectra by spectrophotometer (Shimadzu UV-1800, Kyoto, Japan).

3. Results and Discussion

3.1. Macroscopic Shape of pTrim Cryogel Particles

The pTrim cryogel particles were prepared by combining the inverse Leidenfrost effect and cryo-polymerization technique (iLF cryo-method) [31,32]. By adding the water droplets containing polymer precursor into an extremely low temperature bath, the inverse Leidenfrost phenomenon was induced, and frozen droplets with a spherical shape were formed on the liquid nitrogen. The frozen droplet was then polymerized under a frozen temperature (−15 °C). After cryo-polymerization, the pTrim cryogel particles were obtained by thawing and lyophilization. Figure 2 shows the photo image of pTrim cryogel particles (Figure 2a) and the size distribution of the particles through the observation of an optical microscope (Figure 2b). As shown in Figure 2a, the pTrim cryogel particles had a spherical shape with a white color. Subsequently, after the diameter of more than 200 individual particles was observed from the optical microscopic images, the size distribution was then determined. Figure 2b shows the size distribution of pTrim cryogel particles. As a result, the pTrim cryogel particle, prepared by our iLF-cryo method, was distributed in the range of 700–2300 µm and has a high frequency of around 1600 µm. Herein, the basic properties of pTrim cryogel particles were summarized in Table 1; the value of the mean diameter, D_v, and its standard deviation, σ, were 1654 µm and 220, respectively. Furthermore, the variation coefficient, C_v, was low, at a value of 13 %. Hence, pTrim cryogel particles have a narrow size distribution, and the monodispersiblitiy of the particle was high in comparison with that of the conventional droplet [35]. The tapped bulk density, ρ_{bulk}, was calculated from the bulk-particle volume in a graduated cylinder and from the weight of the particle; the ρ_{bulk} value was 0.3 g/mL. Comprehensively, the macroscopic properties of pTrim cryogel particles adopted by the iLF cryo-method were characterized on the basis of the factors D_v, σ, C_v, and ρ_{bulk}.

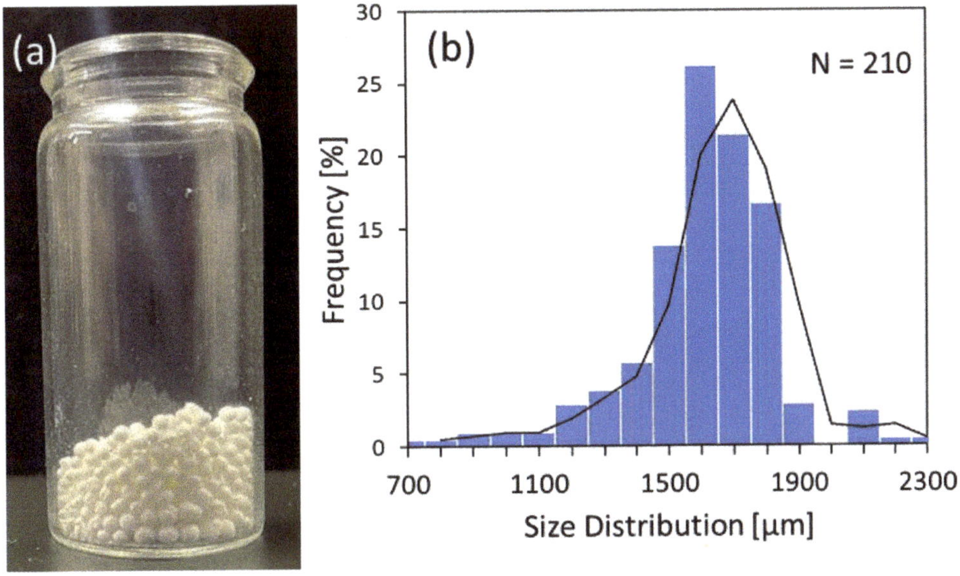

Figure 2. Macroscopic shape of pTrim cryogel particles' (**a**) appearance; and of (**b**) size distribution of particle diameter.

Table 1. Particle properties of pTrim cryogel particles.

D_V [a] [μm]	σ [b] [-]	C_v [c] [%]	ρ_{bulk} [d] [g/mL]
1654	220	13	0.3

[a] Mean diameter of the particles, [b] stand devitation, [c] coefficient of variation, and [d] bulk density.

3.2. Characterization of Porous Properties

In general, cryogels have a monolithic porous structure with supermacropores in the range of the μ-meter scale. During the cryo-polymerization process, the crystals supplied from the solvent (i.e., acetic acid) were employed as porogens; the porous structure was therefore formed. As shown in Figure 3, the porous structure of the surface pTrim cryogel particle was observed via SEM analysis. The observations were performed for the surface section of the pTrim cryogel particle, as described in Figure 3a. From the overall image, the particles were found to have a rough surface (Figure 3b). Subsequently, the surface section of the particle was observed from the high-magnification SEM image to characterize its porous structure (Figure 3c). It was found that the pTrim cryogel particle had a monolithic supermacroporous structure on its surface. Interestingly, the polymer wall of the porous structure appeared to have a smooth surface. The fine morphology of the porous structure (i.e., surface roughness, undulation) is changed by employing different porogens [36]. Herein, the pore diameter was measured from 47 pore diameters by a high-magnification SEM image to characterize the porous property. As shown in Figure 3d, the size distribution of the porous diameter was evaluated. The porous diameters were distributed in the range of 1–12 μm, and the mean porous diameter d_p was 4.7 μm. The results suggest that pTrim cryogel particles have a unique porous structure on their surface that is monolithic with a supermacroporous structure.

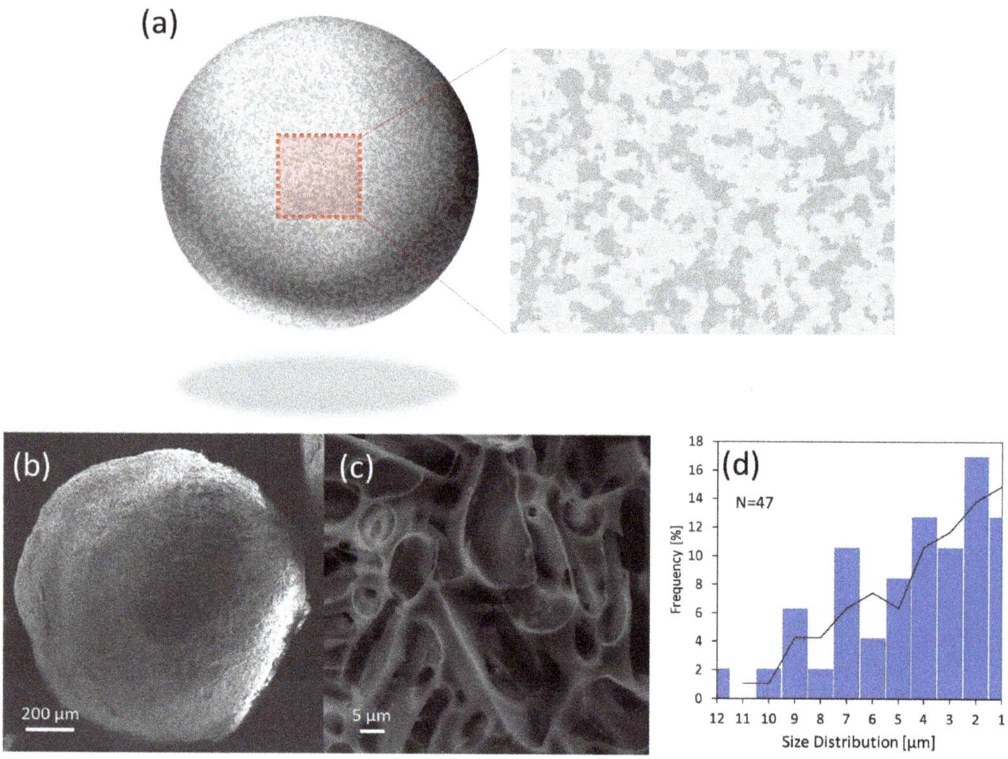

Figure 3. Characterization porous structure of pTrim cryogel particle. (**a**) Schematic illustration of the morphological observation overall particle and a surface section of high magnification. (**b**) Overall particle SEM image and (**c**) surface section at high magnification. (**d**) Size distribution of pores diameter.

3.3. *Hydrophobicity of pTrim Cryogel Particles*

The hydrophobicity of polymer particles can be investigated by confirming the behavior of aqueous or organic droplets when they are contacting the polymer particles. When the aqueous droplet is placed on the hydrophilic polymer, the droplet penetrates the hydrophilic polymer, while the organic droplet is repelled. In the case of the hydrophobic polymer, where the aqueous droplet is repelled, and the organic droplet penetrates the hydrophobic polymer. As shown in Figure 4, the hydrophobicity of pTrim cryogel particles was investigated. Herein, the behavior of droplets, aqueous (dyed as blue), and organic (dyed as red) was placed on the pTrim cryogel particles and observed (Figure 4a). As the result, the aqueous droplet was repelled from the pTrim cryogel particles, whereas the organic droplet penetrated the particles. As stated previously, the cryogel particles presented here comprise pTrim and thus have a hydrophobic surface. Subsequently, the adsorption of organic solvent in the oil–water system by pTrim cryogel particles was examined in Figure 4b,c. Cyrogel particles were added to water containing the red-dyed organic solvent. As shown in Figure 4b, three seconds after the addition of the particles, the particles floated to the liquid surface, resulting in aggregate formation. Furthermore, the particles rapidly started to adsorb the dyed organic solvent. Eventually, the particles were found to adsorb all organic solvents in the beaker after only 10 s (Figure 4c). After the adsorption of organic solvent, pTrim cryogel particles were removed from the water and cut in half; the appearance of cross section was then then observed (Figure 4d). It was confirmed that the pTrim cryogel particle was colored red, including the interior side from the image of the cross section. Hence, the adsorbed organic solvent was diffused through the interior of the particle. It is expected that the porous structure of the pTrim cryogel particle has pores running from the exterior to the interior of the particle and monolithic structure. These results are supported by the evidence that the monolithic porous structure was formed in the pTrim cryogel particle (Figure 3). The pTrim cryogel particles were thus shown to have the potential to be applied as a separation material because of their permeability function with the expansive monolithic porous structure.

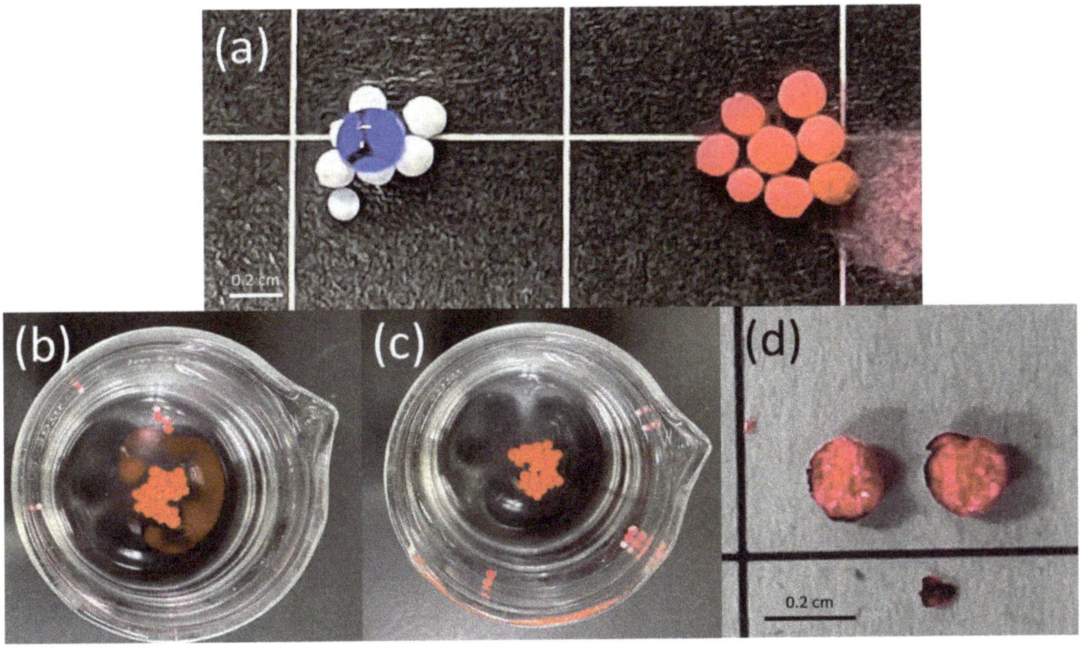

Figure 4. Hydrophobicity of pTrim cryogel particles (**a**) left; aqueous droplet on the particles, right; organic droplet on the particles. The adsorption of organic solvent from water by pTrim cryogel particles (**b**) 3 s and (**c**) 10 s. (**d**) Appearance cross section of the particle after adsorption organic solvent.

3.4. Separation of Stabilized Oil-in-Water Emulsion

Based on the characteristics of pTrim cryogel and its fundamental properties, it is possible to recover the oil-in-water emulsion from the aqueous solution. Typically, the oil-in-water emulsion can be formed by dispersing and stirring a small amount of oil in water. The formed oil-in-water emulsion is unstable in long-term and becomes a two-phase system which is thermodynamically stable. However, via the adsorption of a surfactant on the interface, the emulsion is stabilized and exists in a long-term heterogeneous system. It is difficult to remove the organic solvent from the stabilized emulsion because the emulsion does not spontaneously cause phase separation. Herein, the separation of stabilized oil-in-water emulsion with surfactant was performed by using pTrim cryogel particles in Figure 5. The emulsions were observed in a bulk-scale image and optical microscope images before the separation, and in a bulk-scale image and optical one after separation (Figure 5a–d). From bulk-scale images (Figure 5a,c), the solution was deeply cloudy in white before the addition of the cryogel particles, while the solution became transparent after that. The color change of these solutions indicates that the organic solvent was dispersed and became a heterogeneous system as a colloidal state (before the addition of cryogel particles), and then the colloid was removed (after the particle addition). As shown in Figure 5b,d, the states of solution were observed by using an optical microscope, the emulsions with several μ-meter scales were dispersed in the solution (before the particle addition), and then there was no colloidal emulsion (after the particle addition). It is possible that the slight turbidity of the solution after the particle addition could be caused by dispersed microemulsions that could not be observed with optical microscope. Subsequently, the separation property of pTrim cryogel particles was evaluated from UV-Vis spectra of the emulsion in which the organic phase was dyed with a probe Figure 5e. In the spectrum before the particle addition, the characteristic absorbance peak derived from the dye dissolved in the organic phase was detected. On the contrary, there was no absorbance peak in the spectra in the supernatant after the particle addition. The separation was induced simply by adding the cryogel particles with macroporous and hydrophobic nature through the interaction between the hydrophobicity of pTrim and the organic solvent as colloidal state dispersed in the solution. As a result, the separation of stabilized oil-in-water emulsion with surfactant has been demonstrated.

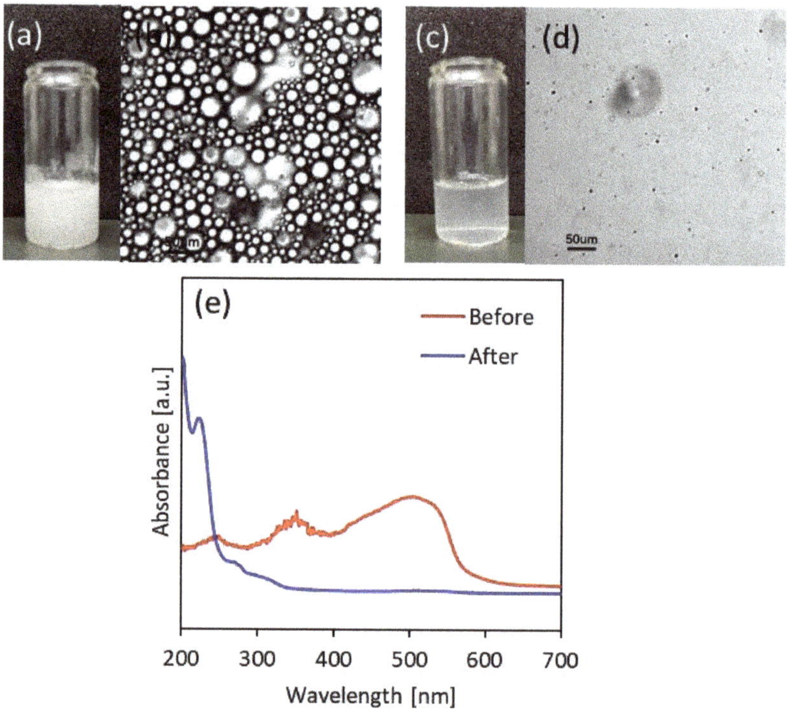

Figure 5. Separation of stabilized emulsion by using pTrim cryogel particles. Before separation of (**a**) photograph of the emulsion in bulk scale and (**b**) optical microscope image, and after separation of (**c**) photograph of the emulsion in bulk scale and (**d**) optical microscope image. (**e**) UV-vis spectra of emulsion.

4. Conclusions

In summary, the pTrim-based cryogel particles were prepared by adopting our developed method, the iLF cryo-method. The optical microscope and SEM were used to characterize the particle properties (i) D_v, (ii) ρ, (iii) C_v, and (iv) σ_{bulk}, and (v) the size distribution of pore size. In addition, the cryogel particle was found to have a monolithic supermacroporous structure. Subsequently, the hydrophobicity of cryogel particles was tested by placing aqueous and organic droplets on the particles, where the aqueous droplet remained on the surface of the particle by a hydrophilic repulsive force and the organic droplet was immediately adsorbed into the particle due to the hydrophobic surface of the particles. Furthermore, it was found that the adsorbed organic solvent was diffused into the entire particle by observation cross-section of the cryogel particle. The separation of stabilized oil-in-water emulsion by using cryogel particles was performed. As a result, in the case study, the efficient removal of stabilized oil-in-water emulsion has been demonstrated. This cryogel particle material can be applied for the separation of oil phase in the emulsion state.

Author Contributions: The manuscript was written through the contributions of all of the authors. All authors have read and agreed to the published version of the manuscript.

Funding: This work was primarily supported by the Japan Society for the Promotion of Science (JSPS) KAKENHI Grant-in-Aids for Scientific Research (A) (26249116, 21H04628), Grant-in-Aids for Scientific Research (B) (18H02005), Grant-in-Aids for Scientific Research (C) (20K05198), and JST SPRING (JPMJSP2138).

Institutional Review Board Statement: Not applicable.

Informed Consent Statement: Not applicable.

Data Availability Statement: Not applicable.

Conflicts of Interest: The authors declare no conflict of interest.

References

1. Doshi, B.; Sillanpää, M.; Kalliola, S. A Review of Bio-Based Materials for Oil Spill Treatment. *Water Res.* **2018**, *135*, 262–277. [CrossRef] [PubMed]
2. Tian, Y.; Zhou, J.; He, C.; He, L.; Li, X.; Sui, H. The Formation, Stabilization and Separation of Oil–Water Emulsions: A Review. *Processes* **2022**, *10*, 738. [CrossRef]
3. Wu, Z.; Liu, H.; Wang, X.; Zhang, Z. Emulsification and Improved Oil Recovery with Viscosity Reducer during Steam Injection Process for Heavy Oil. *J. Ind. Eng. Chem.* **2018**, *61*, 348–355. [CrossRef]
4. Yu, Y.; Chen, H.; Liu, Y.; Craig, V.S.J.; Lai, Z. Selective Separation of Oil and Water with Mesh Membranes by Capillarity. *Adv. Colloid Interface Sci.* **2016**, *235*, 46–55. [CrossRef]
5. Cambiella, A.; Benito, J.M.; Pazos, C.; Coca, J. Centrifugal Separation Efficiency in the Treatment of Waste Emulsified Oils. *Chem. Eng. Res. Des.* **2006**, *84*, 69–76. [CrossRef]
6. Kulkarni, P.S.; Patel, S.U.; Chase, G.G. Layered Hydrophilic/Hydrophobic Fiber Media for Water-in-Oil Coalescence. *Sep. Purif. Technol.* **2012**, *85*, 157–164. [CrossRef]
7. Abdulrazzaq Hadi, A.; Abdulkhabeer Ali, A. Chemical Demulsification Techniques in Oil Refineries: A Review. *Mater. Today Proc.* **2022**, *53*, 58–64. [CrossRef]
8. Ammann, S.; Ammann, A.; Ravotti, R.; Fischer, L.J.; Stamatiou, A.; Worlitschek, J. Effective Separation of Awater in Oil Emulsion from a Direct Contact Latent Heat Storage System. *Energies* **2018**, *11*, 2264. [CrossRef]
9. Goodarzi, F.; Zendehboudi, S. A Comprehensive Review on Emulsions and Emulsion Stability in Chemical and Energy Industries. *Can. J. Chem. Eng.* **2019**, *97*, 281–309. [CrossRef]
10. Lozinsky, V.I. A Brief History of Polymeric Cryogels. *Adv. Polym. Sci.* **2014**, *263*, 1–48. [CrossRef]
11. Lozinsky, V.I.; Okay, O. Basic Principles of Cryotropic Gelation, Polymeric Cryogels: Macroporous Gels with Remarkable Properties Book Series. *Adv. Polym. Sci.* **2014**, *264*, 49–101. [CrossRef]
12. Okay, O.; Lozinsky, V.I. Synthesis and Structure-Property Relationships of Cryogels. *Adv. Polym. Sci.* **2014**, *263*, 103–157. [CrossRef]
13. Lozinsky, V.I. Cryostructuring of Polymeric Systems. 50.† Cryogels and Cryotropic Gel-Formation: Terms and Definitions. *Gels* **2018**, *4*, 77. [CrossRef] [PubMed]
14. Baimenov, A.; Montagnaro, F.; Inglezakis, V.J.; Balsamo, M. Experimental and Modeling Studies of Sr^{2+} and Cs^{+} Adsorption on Cryogels and Comparison to Commercial Adsorbents. *Ind. Eng. Chem. Res.* **2022**, *61*, 8204–8219. [CrossRef]
15. Guo, F.; Wang, Y.; Chen, M.; Wang, C.; Kuang, S.; Pan, Q.; Ren, D.; Chen, Z. Lotus-Root-like Supermacroporous Cryogels with Superphilicity for Rapid Separation of Oil-in-Water Emulsions. *ACS Appl. Polym. Mater.* **2019**, *1*, 2273–2281. [CrossRef]
16. Juan, L.T.; Lin, S.H.; Wong, C.W.; Jeng, U.S.; Huang, C.F.; Hsu, S.H. Functionalized Cellulose Nanofibers as Crosslinkers to Produce Chitosan Self-Healing Hydrogel and Shape Memory Cryogel. *ACS Appl. Mater. Interfaces* **2022**, *14*, 36353–36365. [CrossRef]
17. Bencherif, S.A.; Sands, R.W.; Ali, O.A.; Li, W.A.; Lewin, S.A.; Braschler, T.M.; Shih, T.Y.; Verbeke, C.S.; Bhatta, D.; Dranoff, G.; et al. Injectable Cryogel-Based Whole-Cell Cancer Vaccines. *Nat. Commun.* **2015**, *6*, 7556. [CrossRef]
18. Ihlenburg, R.B.J.; Lehnen, A.C.; Koetz, J.; Taubert, A. Sulfobetaine Cryogels for Preferential Adsorption of Methyl Orange from Mixed Dye Solutions. *Polymers* **2021**, *13*, 208. [CrossRef]

19. Kirsebom, H.; Rata, G.; Topgaard, D.; Mattiasson, B.; Galaev, I.Y. Mechanism of Cryopolymerization: Diffusion-Controlled Polymerization in a Nonfrozen Microphase. An NMR Study. *Macromolecules* **2009**, *42*, 5208–5214. [CrossRef]
20. Memic, A.; Colombani, T.; Eggermont, L.J.; Rezaeeyazdi, M.; Steingold, J.; Rogers, Z.J.; Navare, K.J.; Mohammed, H.S.; Bencherif, S.A. Latest Advances in Cryogel Technology for Biomedical Applications. *Adv. Ther.* **2019**, *2*, 1800114. [CrossRef]
21. Plieva, F.M.; Galaev, I.Y.; Noppe, W.; Mattiasson, B. Cryogel Applications in Microbiology. *Trends Microbiol.* **2008**, *16*, 543–551. [CrossRef] [PubMed]
22. Kumar, A.; Mishra, R.; Reinwald, Y.; Bhat, S. Cryogels: Freezing Unveiled by Thawing. *Mater. Today* **2010**, *13*, 42–44. [CrossRef]
23. Kumar, A.; Srivastava, A. Cell Separation Using Cryogel-Based Affinity Chromatography. *Nat. Protoc.* **2010**, *5*, 1737–1747. [CrossRef] [PubMed]
24. Lozinsky, V.I. Cryogels on the Basis of Natural and Synthetic Polymers: Preparation, Properties and Application. *Usp. Khim.* **2002**, *71*, 579–584. [CrossRef]
25. Chen, X.; Sui, W.; Ren, D.; Ding, Y.; Zhu, X.; Chen, Z. Synthesis of Hydrophobic Polymeric Cryogels with Supermacroporous Structure. *Macromol. Mater. Eng.* **2016**, *301*, 659–664. [CrossRef]
26. Pfaunmiller, E.L.; Paulemond, M.L.; Dupper, C.M.; Hage, D.S. Affinity Monolith Chromatography: A Review of Principles and Recent Analytical Applications. *Anal. Bioanal. Chem.* **2013**, *405*, 2133–2145. [CrossRef]
27. Zhan, X.Y.; Lu, D.P.; Lin, D.Q.; Yao, S.J. Preparation and Characterization of Supermacroporous Polyacrylamide Cryogel Beads for Biotechnological Application. *J. Appl. Polym. Sci.* **2013**, *130*, 3082–3089. [CrossRef]
28. Yun, J.; Dafoe, J.T.; Peterson, E.; Xu, L.; Yao, S.J.; Daugulis, A.J. Rapid Freezing Cryo-Polymerization and Microchannel Liquid-Flow Focusing for Cryogel Beads: Adsorbent Preparation and Characterization of Supermacroporous Bead-Packed Bed. *J. Chromatogr. A* **2013**, *1284*, 148–154. [CrossRef]
29. Hiramure, Y.; Suga, K.; Umakoshi, H.; Matsumoto, J.; Shiomori, K. Preparation and Characterization of Poly-*N*-Isopropylacrylamide Cryogels Containing Liposomes and Their Adsorption Properties of Tryptophan. *Solvent Extr. Res. Dev.* **2018**, *25*, 37–46. [CrossRef]
30. Plieva, F.; Xiao, H.; Galaev, I.Y.; Bergenståhl, B.; Mattiasson, B. Macroporous Elastic Polyacrylamide Gels Prepared at Subzero Temperatures: Control of Porous Structure. *J. Mater. Chem.* **2006**, *16*, 4065–4073. [CrossRef]
31. Takase, H.; Shiomori, K.; Okamoto, Y.; Watanabe, N.; Matsune, H.; Umakoshi, H. Micro Sponge Balls: Preparation and Characterization of Sponge-like Cryogel Particles of Poly (2-Hydroxyethyl Methacrylate) via the Inverse Leidenfrost Effect. *ACS Appl. Polym. Mater.* **2022**, *4*, 7081–7089. [CrossRef]
32. Takase, H.; Watanabe, N.; Shiomori, K.; Okamoto, Y.; Matsune, H.; Umakoshi, H. Versatility of the Preparation Method for Macroporous Cryogel Particles Utilizing the Inverse Leidenfrost Effect. *ACS Omega* **2023**, *8*, 829–834. [CrossRef] [PubMed]
33. Nayak, A.; Jain, S.K.; Pandey, R.S. Controlling Release of Metformin HCl through Incorporation into Stomach Specific Floating Alginate Beads. *Mol. Pharm.* **2011**, *8*, 2273–2281. [CrossRef] [PubMed]
34. Wang, C.F.; Huang, H.C.; Chen, L.T. Protonated Melamine Sponge for Effective Oil/Water Separation. *Sci. Rep.* **2015**, *5*, 14294. [CrossRef]
35. Merakchi, A.; Bettayeb, S.; Drouiche, N.; Adour, L.; Lounici, H. Cross-Linking and Modification of Sodium Alginate Biopolymer for Dye Removal in Aqueous Solution. *Polym. Bull.* **2019**, *76*, 3535–3554. [CrossRef]
36. Kirsebom, H.; Mattiasson, B. Cryostructuration as a Tool for Preparing Highly Porous Polymer Materials. *Polym. Chem.* **2011**, *2*, 1059–1062. [CrossRef]

colloids and interfaces

MDPI

Article

Casein-Hydrolysate-Loaded W/O Emulsion Preparation as the Primary Emulsion of Double Emulsions: Effects of Varied Phase Fractions, Emulsifier Types, and Concentrations

Pelin Salum [1], Çağla Ulubaş [1], Onur Güven [2], Levent Yurdaer Aydemir [1] and Zafer Erbay [1,*]

1 Department of Food Engineering, Faculty of Engineering, Adana Alparslan Turkes Science and Technology University, Adana 01250, Turkey
2 Department of Mining Engineering, Faculty of Engineering, Adana Alparslan Turkes Science and Technology University, Adana 01250, Turkey
* Correspondence: zafererbay@yahoo.com; Tel.: +90-322-4550000 (ext. 2080)

Abstract: Stable primary emulsion formation in which different parameters such as viscosity and droplet size come into prominence for their characterization is a key factor in W/O/W emulsions. In this study, different emulsifiers (Crill™ 1, Crill™ 4, AMP, and PGPR) were studied to produce a casein-hydrolysate-loaded stable primary emulsion with lower viscosity and droplet size. Viscosity, electrical conductivity, particle size distribution, and emulsion stability were determined for three different dispersed phase ratios and three emulsifier concentrations. In 31 of the 36 examined emulsion systems, no electrical conductivity could be measured, indicating that appropriate emulsions were formed. While AMP-based emulsions showed non-Newtonian flow behaviors with high consistency coefficients, all PGPR-based emulsions and most of the Crill™-1- and -4-based ones were Newtonian fluids with relatively low viscosities (65.7–274.7 cP). The PGPR-based emulsions were stable for at least 5 days and had D(90) values lower than 2 μm, whereas Crill™-1- and -4-based emulsions had phase separation after 24 h and had minimum D(90) values of 6.8 μm. PGPR-based emulsions were found suitable and within PGPR-based emulsions, and the best formulation was determined by TOPSIS. Using 5% PGPR with a 25% dispersed phase ratio resulted in the highest relative closeness value. The results of this study showed that PGPR is a very effective emulsifier for stable casein-hydrolysate-loaded emulsion formations with low droplet size and viscosity.

Keywords: double emulsion; W/O emulsion; emulsifier; encapsulation

Citation: Salum, P.; Ulubaş, Ç.; Güven, O.; Aydemir, L.Y.; Erbay, Z. Casein-Hydrolysate-Loaded W/O Emulsion Preparation as the Primary Emulsion of Double Emulsions: Effects of Varied Phase Fractions, Emulsifier Types, and Concentrations. *Colloids Interfaces* **2023**, *7*, 1. https://doi.org/10.3390/colloids7010001

Academic Editors: César Burgos-Díaz, Mauricio Opazo-Navarrete and Eduardo Morales

Received: 23 November 2022
Revised: 21 December 2022
Accepted: 23 December 2022
Published: 28 December 2022

1. Introduction

In recent years, significant progress has been made in developing food-derived bioactive ingredients, and one important ingredient group of compounds in this context is bioactive peptides [1,2]. While bioactive peptides are inactive in protein structures, they can show a wide variety of physiological effects with their hormone-like properties when they are released from the protein structure by hydrolysis [3–5]. In general, these compounds contain 2–20 amino acid residues and show a variety of biological properties, depending on their structural properties, amino acid composition, sequence, and charge [6,7]. Milk proteins, especially caseins, are known to be a good source of bioactive peptides with anti-hypertensive, antidiabetic, antiobesity, antioxidant, immunomodulatory, mineral-binding, opioid, and antimicrobial properties [7–10].

However, there are several difficulties in the development of food products fortified with bioactive peptides [11]. The difficulties in the use of peptides arise from their low solubility, chemical and physical instability, undesirable flavor properties (especially bitter taste), and low bioavailability [12–14]. To overcome these problems, encapsulation technology presents promising solutions; however, encapsulation of bioactive peptides is a challenging area with its specific features [15]. Among encapsulation methods, a technique with high potential for the encapsulation of bioactive peptides is the use of double emulsions [15–18].

Double emulsions are liquid dispersion systems in which the droplets of one emulsion's dispersed phase contain smaller dispersed droplets [19–21]. While the internal emulsion, generally termed in the production as "primary emulsion", can be described as the dispersed phase of the double emulsion, the whole emulsion system including the outer continuous liquid phase is generally termed as "secondary emulsion". They can be in two main morphologies such as oil-in-water-in-oil (O/W/O) or water-in-oil-in-water (W/O/W), and the latter one is used more commonly in the literature [21]. A two-stage emulsification technique is commonly used in the preparation of W/O/W emulsions. In the first stage, a stable primary emulsion (W/O) is obtained using lipophilic emulsifiers. It is important to obtain small and monodispersed water droplets in the oil phase. In the second step, the primary emulsion is dispersed in an external water-based continuous phase containing hydrophilic emulsifiers and stabilizers [22]. W/O/W systems have been used for the encapsulation of several types of protein hydrolysates and bioactive peptides with high efficiency [1,15–18,23–27]. However, the main problem in food-related applications of W/O/W emulsions is their long-term instability due to the limited variety of food-grade substances that can be used as emulsifiers or stabilizers [28]. On the other hand, the characteristics of primary emulsion play a significant role in the production of stable W/O/W [29]. Therefore, one of the approaches for improving the stability of double emulsions is developing a primary W/O emulsion with enhanced stability [19].

In this context, one reason for the low stability of W/O emulsions is the high mobility of water droplets. In W/O emulsions, only steric forces stabilize the emulsion due to the low electrical conductivity of the continuous phase [30]. The ratio of the dispersed phase, the type and concentration of the emulsifier, the properties of the oil, the presence of osmolyte, and the mechanism and processing conditions of homogenization affect the emulsion droplet size, viscosity, and thus stability [25,31,32]. Another important point for the stability of W/O/W emulsions is the properties of the dispersed water phase of the primary emulsion. The composition of this water phase and the presence of other compounds influence the final stability of the double emulsion [33]. For instance, it is well known that peptides exhibit interfacial activity and this interfacial activity can lead to unstable emulsions due to the interaction of the peptides with the lipophilic emulsifier at the water/oil interface [24]. Additionally, the W/O needs to be dispersed with a high internal droplet yield and smaller droplets in the external water phase [28]. To achieve this, a reduction in the viscosity of the primary emulsion is an option. The primary emulsion with low viscosity supports the formation of small W/O droplets in a double emulsion with a low-energy homogenization process, and the primary emulsion droplets can be easily dispersed in the continuous water phase [24,34].

The present work aimed to produce a stable peptide-loaded W/O emulsion with low droplet size and viscosity that can be used as a primary emulsion in W/O/W double emulsions for the encapsulation of casein hydrolysates. For this purpose, different emulsifiers were used in the preparation of W/O primary emulsions at various emulsifier concentrations and dispersed phase ratios.

2. Materials and Methods

2.1. Materials

The skimmed raw cow's milk was obtained from Sarıçam Ali Baba'nın Çiftliği Milk and Dairy Products Company in Adana, Turkey and used in the preparation of acid casein. Alcalase® 2.4 L was obtained as a gift sample kindly provided by Novozymes (Bagsvaerd, Denmark). Sunflower oil was purchased from a local store. Sorbitan monolaurate (Crill™ 1), and sorbitan monooleate (Crill™ 4) were kindly supplied by Croda Chemicals (Snaith, UK). Ammonium phosphatide (AMP 4455) and polyglycerol polyricinoleate (PGPR 4150) were obtained as gift samples kindly provided by Palsgaard® (Juelsminde, Denmark). The chemical structure of the emulsifiers used in the present study was presented in Figure 1 [35–37] and the critical micelle concentrations of Crill™ 1, Crill™ 4, and PGPR were reported in the literature as 2.1×10^{-5}, 1.8×10^{-5}, and 9.0×10^{-3} mol/L, respectively [38,39]. Acetic acid, trisodium citrate, trisodium phosphate, and potassium sorbate were purchased from Sigma-Aldrich (St Louis, MO, USA).

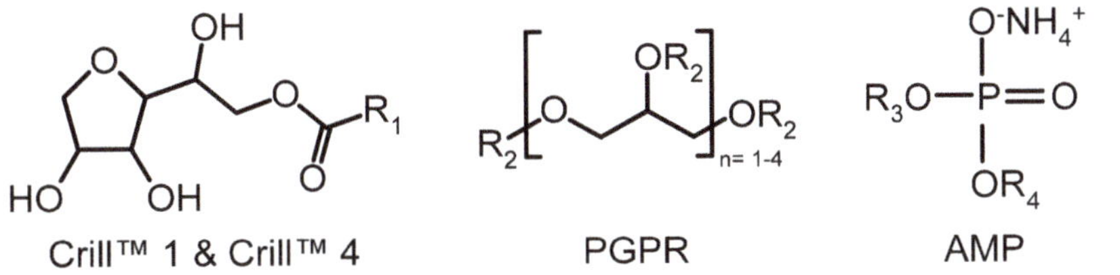

R$_1$: $C_{11}H_{23}$ (for Crill 1), $C_{17}H_{33}$ (for Crill 4)
R$_2$: H, or a fatty acyl group derived from poly condensed ricinoleic acid
n: the degree of polymerization of glycerol
R$_3$: Mono- or di-glyceride
R$_4$: H, or mono- or di-glyceride

Figure 1. The chemical structure of the emulsifiers used in the study (Crill™ 1 (sorbitan monolaurate), Crill™ 4 (sorbitan monooleate), PGPR 4150 (polyglycerol polyricinoleate), and AMP 4455 (ammonium phosphatide).

2.2. Casein Hydrolysate Production

First of all, casein was precipitated from skimmed milk by acidifying based on the method described in Sarode et al., (2016) with some modifications [40]. Skimmed milk was pasteurized by heating at 75 °C for 30 s in a heated, circulating water bath (IKA ICC Basic Eco 8). After pasteurization, milk was quickly cooled to 40 °C and kept at this temperature for 40 min. Then, milk was cooled to the precipitation temperature (35 °C). At this condition, 0.1 M acetic acid solution was slowly added with very gentle mixing until milk pH reached 4.6 and casein was precipitated by acidifying. Afterward, the mixture was heated to 50 °C and kept at this temperature for 15 min and whey was removed. Then, distilled water was added to the curd at 50 °C, and the curd was washed for 15 min with moderate stirring in a magnetic stirrer (IKA C-MAG HS 10). The washing water was removed using cheesecloth and the same washing process was repeated at 40 °C and 35 °C with distilled water. After each washing process, the excess water was removed from the casein curd using cheesecloth. The cheesecloth was hung up and left for 20 min. Finally, the cheesecloth was squeezed by hand for draining the whey, and acid casein was obtained.

The acid casein should be converted into a stable, homogenous fluid to provide effective hydrolysis. For this purpose, it was considered to bring the pH of acid casein to neutral pH with a phosphate buffer (0.1 M salt-free buffer containing NaH_2PO_4 and Na_2HPO_4 adjusted to pH 8.0). The incubations were carried out at a suitable temperature for enzyme activity (45 °C). Moreover, it was predicted that the hydrolysis process might continue for up to 72 h, and the highest acid casein concentration that would not lose its homogeneous dispersion structure, collapse, and/or adhere to the container walls in a shaking incubator for 72 h at 45 °C was determined during preliminary experiments. Accordingly, acid casein was crumbled by mixing at 7000 rpm for 60 s in a Thermomix TM31 (Vorwerk, Wuppertal, Germany) and phosphate buffer (containing 2% trisodium citrate, 1% trisodium phosphate salts, and 0.2% potassium sorbate on casein basis) was added to obtain a mixture with 1:3 acid casein: phosphate buffer ratio. Then, the mixture was heated to 80 °C with 4000 rpm mixing and the mixing process was continued for 15 min at 80 °C. Later on, the casein dispersion was rapidly cooled and 1.25% Alcalase was added. Samples were incubated at 45 °C for 8 h. At the end of the incubation, enzyme activity was terminated by a heat treatment at 90 °C for 15 min. Afterward, the hydrolysates were centrifuged at 8000 rpm for 10 min at room conditions and the supernatants were collected. Finally, the collected supernatant was diluted 1:1 with phosphate buffer and used as dispersed water phase in the emulsions. The pH of the diluted casein hydrolysate was 7.2.

2.3. Preparation of W/O Emulsions

An appropriate amount of sunflower oil and emulsifier were mixed with a magnetic stirrer at room temperature. Crill™ 1, Crill™ 4, polyglycerol polyricinoleate, and ammonium phosphatide were used as emulsifiers in the present study and their abbreviations were written as C1, C4, PGPR, and AMP, respectively. After dissolving the emulsifier, the hydrolysate was added to the mixture. Then the mixture was homogenized with a rotor-stator (Ultra-Turrax, T18, IKA, Königswinter, Germany) and the shear force was gradually increased as follows: 5000 rpm for 30 s; 10,000 rpm for 30 s; and 15,000 rpm for 5 min.

Preliminary experiments were carried out to determine the highest dispersed phase ratio and the lowest emulsifier concentration to be considered as the restrictions for the study. After emulsion preparation, the pictures of the emulsions were taken (Figure 2). First, an emulsion containing 50% *w*/*w* distilled water as a dispersed phase and 50% *w*/*w* oil phase (2% *w*/*w* PGPR, and 48% *w*/*w* sunflower oil) was prepared. A proper emulsion was obtained (Figure 2a). After that, the emulsion was produced by using casein hydrolysate instead of distilled water as the dispersed phase with the same production technique. In this case, W/O emulsion could not form (Figure 2b). The reason for this can be related to the surface activity of the peptides and/or variation in the osmotic pressure due to the buffer solution in the hydrolysate solution. It was reported in the literature that the interfacial activity of peptides might interfere with PGPR at the W/O interface, leading again to destabilized emulsions [27]. In order to understand the effect of the buffer, the emulsion was prepared using the buffer solution as the dispersed phase (Figure 2c). Although emulsion formation seemed to be improved, proper W/O emulsion could also not be created. After that, the internal phase ratio was reduced to 40% (Figure 2d) and a better emulsion formation seemed to occur, whereas a stable W/O emulsion could not be obtained similar to the emulsion in Figure 2c. Finally, the emulsifier ratio was increased to 3% and a stable emulsion was formed (Figure 2e). Therefore, the lowest emulsifier ratio and the dispersed phase ratio to be used in experiments were decided as 3% and 40%, respectively.

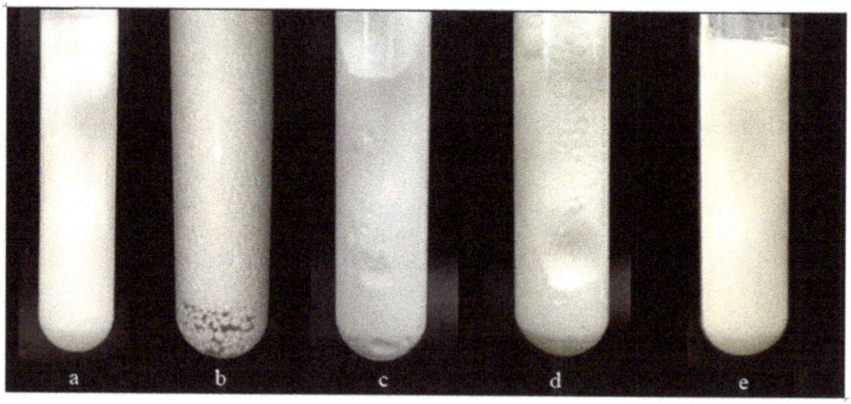

Figure 2. The emulsions containing (**a**) 50% *w*/*w* dispersed phase (distilled water) and 2% *w*/*w* PGPR, (**b**) 50% *w*/*w* dispersed phase (casein hydrolysate) and 2% *w*/*w* PGPR, (**c**) 50% *w*/*w* dispersed phase (buffer solution) and 2% *w*/*w* PGPR, (**d**) 40% *w*/*w* dispersed phase (casein hydrolysate) and 2% *w*/*w* PGPR, and (**e**) 40% *w*/*w* dispersed phase (casein hydrolysate) and 3% *w*/*w* PGPR.

Emulsions were also produced with other emulsifiers (Crill™ 1, Crill™ 4, and AMP) at the same conditions with the emulsion showed in Figure 2e. It has been observed that the emulsions produced with AMP present relatively high viscosity values. Therefore, the emulsifier concentrations were reduced for emulsion prepared with AMP. After preliminary studies, the emulsion formulations used in the present study, and their codes are given in Table 1. A total of 36 emulsions were produced with 3 different dispersed phase ratios (ΦW), and 3 emulsifier concentrations (ΦE) for 4 emulsifiers (Table 1). Each production was duplicated and samples were analyzed immediately after the emulsion preparations.

Table 1. Compositions of W/O emulsions used in the present study and their sample codes *.

	ΦE: 3% *w/w* (for AMP 1% *w/w*)	**ΦE: 5% *w/w* (for AMP 2% *w/w*)**	**ΦE: 7% *w/w* (for AMP 3% *w/w*)**
ΦW: 10% *w/w*	1	4	7
ΦW: 25% *w/w*	2	5	8
ΦW: 40% *w/w*	3	6	9

* ΦE: emulsifier concentration. ΦW: dispersed phase ratio.

2.4. Determination of Flow Behaviors

Flow behaviors of the emulsions was determined by viscometer (DV-II+ Pro Viscometer, Brookfield Engineering, Middleborough, MA, USA) coupled with a small sample adapter (SSA-13RD, Brookfield Engineering, Middleborough, MA, USA). The measurements were carried out at 35 °C and the temperatures of the samples were adjusted by a water circulation system (ICC Basic Eco 8, IKA, Staufen, Germany). Samples were analyzed with varying shear rates (in the range of 6.6–52.8 s^{-1}) and in addition to viscosity values, the flow behavior parameters (dimensionless flow behavior indices (n) and the consistency coefficients (K)) were calculated with the power-law model.

$$\tau = K(\dot{\gamma})^n \quad (1)$$

Here, τ is the shear stress, $\dot{\gamma}$ is the shear rate, K is the consistency coefficient, and n is the dimensionless flow behavior index.

2.5. Determination of Droplet Size Distribution by Microscopy-Assisted Digital Image Analysis

Droplet size distributions of emulsion samples were determined by a microscopy-assisted digital image analysis technique using Trainable Weka Segmentation as explained in detail by Salum et al., (2022) [41]. Firstly, micrographs were obtained by a compound microscope (M83EZ, OMAX Microscopes, Kent, WA, USA) combined with a 5-megapixel CMOS camera (A3550U, OMAX Microscopes, Kent, WA, USA). Briefly, 5 µL of the emulsion was diluted with 50 µL sunflower oil, dropped onto the microscope slide, and carefully covered with a coverslip. Immersion oil was dripped onto the coverslip and analysis was carried out with 100x/1.25 oil and a 160/0.17 objective lens. At least 120 photographs were taken from 5 separate microscope slides for each sample. Image analysis was performed using ImageJ/Fiji (ver.1.53c.) software. Trainable Weka Segmentation (TWS) (ver.3.2.35) plugin was used for the segmentation of emulsion droplets from the background. For this purpose, TWS was trained and a classifier was established. For the training, a total of 9 images were used which were selected from micrographs of emulsion prepared with PGPR, Crill™ 1, and Crill™ 4. After that, segmentations of the micrographs of the emulsions were performed using the trained TWS classifier. The images obtained as a result of TWS segmentation were turned into a single stack and this stack was converted to 8 bits and then transformed into the binary format. Afterward, the "Fill holes" and the "Open" commands were applied. Pixels corresponding to 10 µm were determined on the micrograph of the calibration slide and it was used as a scale for the particle size analyses. Eventually, the particle sizes of these images were calculated with a roundness value below 0.85. Over 8000 droplets were analyzed for each sample. By using the droplet size data, D(90), D [3, 2], and D [4, 3] values were calculated, which represent the equivalent volume diameters at 90%, and the area- and volume-weighted mean diameters, respectively.

2.6. Determination of Electrical Conductivity

Electrical conductivity values of the emulsions were determined with an electrical conductivity probe of pH/mV/EC/TDS/NaCl/Temp Bench Meter (MW 180Max, Milwaukee Instruments, Rocky Mount, NC, USA).

2.7. Monitoring the Emulsion Stability

The emulsion stability was observed visually by storing the emulsion samples at room temperature. For this purpose, emulsions were transferred to flat-bottom glass tubes and kept motionless. The emulsions were photographed for 5 days and the changes in their structure were evaluated. The emulsion stability was determined according to the occurrence of creaming or phase separation. Photographing was carried out in a box. The inner surface of the box was covered with a black cloth and the entrance of light from the outside into the box was prevented. A 7 W led light (daylight) was placed behind the test tubes to enhance the contrast during photographing.

2.8. Statistical Analysis

Experimental data were analyzed by performing ANOVA and Duncan post-hoc tests, and the statistical significance was $p < 0.05$. Moreover, principle component analysis (PCA) was performed to graphically show the relationship between the emulsions and their properties. SPSS statistical package program (SPSS ver. 22.0 for Windows, SPSS Inc., Chicago, IL, USA) and XLSTAT (Addinsoft, New York, NY, USA) were used in performing these statistical analyses.

The appropriate emulsion formulation according to the viscosity and droplet size values of the samples was determined using a multi-criteria decision analysis method called TOPSIS (the technique for order of preference by similarity to ideal solution). In this method, the chosen alternative should have the longest geometric distance from the negative ideal solution and the shortest geometric distance from the positive ideal solution [42]. While determining these positive and negative ideal solutions, first a matrix is constructed with the experimental results, and then the solution matrix is normalized and weighted. Briefly, the experimental results determine the boundary conditions for the ideal solutions [43]. The result of the TOPSIS was presented by the term called relative closeness (C) to the positive ideal solution. This value is between 0 and 1, and it is desirable to be close to 1.

3. Results and Discussion

3.1. Electrical Conductivity

Electrical conductivity measurement is a very useful method in evaluating the formation of W/O emulsions. While the electrical conductivity of the oil, which is the continuous/outer phase, is negligible, the dispersed/inner water phase (casein hydrolysate in the present study) shows a significant electrical conductivity. In this study, the electrical conductivity value measured for sunflower oil was 0.03 ± 0.00 µS/cm, and the same parameter for the casein hydrolysate prepared in buffer solution was detected as 13085 ± 135 µS/cm. The electrical conductivity of the emulsion is expected to be dominated by the continuous phase of the emulsion. Therefore, a negligible electrical conductivity value should be observed in an appropriate W/O emulsion and if a distinct electrical conductivity is observed in a sample, it means that the emulsion was not appropriately formed or lost its stability. In other words, a negligible electrical conductivity value indicates that the hydrophilic bioactive material is effectively encapsulated. In this context, the electrical conductivity values of the emulsions prepared in this study were checked immediately after the emulsion preparation. Among 36 samples, 31 samples did not have any electrical conductivity, only C1-3, C4-3, AMP-3, AMP-6, and AMP-9 had 2044 ± 151, 10400 ± 593, 1764 ± 151, 1675 ± 155, and 1228 ± 83 µS/cm, respectively. Moreover, rapid phase separation was also observed in these samples. Therefore, these samples were not analyzed in the continuation of the study.

3.2. Flow Behaviors

The flow behavior of emulsions is an important indicator of emulsion stability and properties. In this study, viscosity, flow behavior index, and consistency constant values were measured (Table 2), and higher dispersed phase ratios generally resulted in increased viscosity for W/O emulsions. In higher-concentration systems, the droplets begin to interact with each other through a combination of hydrodynamic and colloidal interactions [31].

Table 2. Flow behaviors of W/O emulsions produced using different emulsifiers with various emulsifier concentrations and dispersed phase ratios *.

Sample	n (-)	K (Pa.s ^{n})	Viscosity (cP)
Crill 1-1	098 ± 0.00 ^{b,y}	1.14 ± 0.00 ^{b,w}	80.8 ±0.73 ^{b,x}
Crill 1-2	0.90 ± 0.00 ^{a,x}	1.37 ± 0.04 ^{a,w}	69.5 ± 0.41 ^{a,w}
Crill 1-4	0.95 ± 0.00 ^{a,m,x}	1.18 ± 0.00^{c,k,w}	74.5 ± 0.03 ^{a,w}
Crill 1-5	0.85 ± 0.03 ^{a,l,x}	1.55 ± 0.17 ^{a,k,w}	96.6 ± 2.57 ^{c,x}
Crill 1-6	0.63 ± 0.00 ^{k,w}	2.66 ± 0.07 ^{l,w}	Non-Newtonian
Crill 1-7	0.95 ± 0.00 ^{a,l,y}	1.10 ± 0.01 ^{a,k,w}	92.7 ± 0.05 ^{c,x}
Crill 1-8	0.78 ± 0.07 ^{a,kl,x}	1.95 ± 0.44 ^{a,kl,w}	89.4 ± 0.49 ^{b,w}
Crill 1-9	0.70 ± 0.00 ^{k,w}	2.66 ± 0.07 ^{l,x}	Non-Newtonian
Crill 4-1	0.96 ± 0.00 ^{b,xy}	1.30 ± 0.02 ^{b,w}	65.7 ± 0.60 ^{a,w}
Crill 4-2	0.86 ± 0.00 ^{a,x}	1.37 ± 0.04 ^{a,w}	83.5 ± 2.82 ^{a,y}
Crill 4-4	0.97 ± 0.01 ^{b,m,xy}	1.15 ± 0.03 ^{a,k,w}	77.6 ± 0.23 ^{b,x}
Crill 4-5	0.86 ± 0.00 ^{a,l,x}	1.35 ± 0.01 ^{a,l,w}	80.8 ± 0.80 ^{a,w}
Crill 4-6	0.63 ± 0.01 ^{k,w}	2.62 ± 0.04 ^{m,w}	Non-Newtonian
Crill 4-7	0.88 ± 0.00 ^{a,l,x}	1.20 ± 0.03 ^{ab,k,w}	77.0 ± 0.62 ^{b,w}
Crill 4-8	0.81 ± 0.02 ^{a,kl,x}	1.67 ± 0.17 ^{a,k,w}	85.0 ± 1.09 ^{a,w}
Crill 4-9	0.71 ± 0.04 ^{k,w}	1.76 ± 0.26 ^{k,w}	Non-Newtonian
PGPR-1	0.93 ± 0.00 ^{a,k,x}	1.04 ± 0.03 ^{a,k,w}	83.9 ± 0.23 ^{a,k,y}
PGPR-2	1.04 ± 0.03 ^{a,l,y}	1.19 ± 0.11 ^{a,k,w}	132.5 ± 0.39 ^{a,l,y}
PGPR-3	0.98 ± 0.00 ^{a,kl}	2.79 ± 0.03 ^{a,l}	260.2 ± 0.95 ^{a,m}
PGPR-4	0.99 ± 0.01 ^{b,k,y}	1.04 ± 0.04 ^{a,k,w}	93.4 ± 0.33 ^{b,k,y}
PGPR-5	1.01 ± 0.01 ^{a,k,y}	1.36 ± 0.02 ^{a,l,w}	139.8 ± 0.44 ^{b,l,y}
PGPR-6	0.98 ± 0.00 ^{b,k,x}	2.83 ± 0.01 ^{a,m,w}	271.9 ± 0.58 ^{b,m}
PGPR-7	1.01 ± 0.00 ^{b,l,z}	1.02 ± 0.00 ^{a,k,w}	100.1 ± 1.26 ^{c,k,y}
PGPR-8	1.01 ± 0.01 ^{a,l,y}	1.40 ± 0.00 ^{a,l,w}	146.0 ± 1.48 ^{c,l,x}
PGPR-9	0.99 ± 0.00 ^{b,k,x}	2.84 ± 0.01 ^{a,m,x}	274.7 ± 1.56 ^{b,m}
AMP-1	0.43 ± 0.01 ^{a,w}	11.4 ± 0.38 ^{a,x}	Non-Newtonian
AMP-2	0.28 ± 0.01 ^{b,w}	95.3 ± 4.08 ^{a,x}	Non-Newtonian
AMP-4	0.45 ± 0.00 ^{ab,w}	10.1 ± 0.33 ^{a,x}	Non-Newtonian
AMP-5	0.21 ± 0.01 ^{a,w}	145.2 ± 9.25 ^{b,x}	Non-Newtonian
AMP-7	0.48 ± 0.01 ^{b,w}	9.5 ± 0.78 ^{a,x}	Non-Newtonian
AMP-8	0.31 ± 0.00 ^{b,w}	133.1 ± 4.62 ^{b,x}	Non-Newtonian

* Abbreviations are n, flow behavior index; K, consistency coefficient. Values are mean ± standard deviation of the analysis results and the same superscript letters ($^{a–c}$ for emulsifier concentration; $^{k–m}$ for dispersed phase ratio; $^{w–z}$ emulsifier type) indicate no significant difference between the samples produced at different conditions ($p > 0.05$).

In addition, the results show that the type of emulsifier had a significant effect on the viscosity and the flow behavior of the emulsions. Emulsions produced with PGPR behaved as Newtonian-type fluid and their viscosity increased with dispersed phase ratio and emulsifier concentration. Similarly, Jo et al., (2019) determined Newtonian flow properties in collagen peptide hydrolysate-loaded primary emulsions prepared using PGPR and also they noted that the viscosity increased with increasing internal phase [24]. In emulsions prepared with sorbitan esters (both Crill™ 1 and Crill™ 4), the dispersed phase ratio dominated the flow behavior of emulsions. It was found that samples 6 and 9, with the highest internal phase ratio (40%), had significantly higher K values and showed non-

Newtonian flow behaviors. It is reported that the Newtonian character of both W/O and O/W emulsions changes and the emulsions exhibit non-Newtonian behaviors with the increase in the dispersed phase ratio, especially before the phase inversion [44–46]. Samples prepared with AMP differed from the others in terms of flow behavior and consistency. All formulations produced with AMP had significantly higher K values compared to other emulsifiers even though they were used in lower ratios. Moreover, emulsions prepared with AMP showed non-Newtonian flow behavior regardless of the dispersed phase ratio. Shear thinning behavior of AMP-stabilized W/O were reported by Rivas et al., (2016), and also, higher viscosity values were obtained in AMP-stabilized W/O than in PGPR-stabilized ones [47]. According to the micrographs of emulsions prepared with AMP, droplets were not completely dispersed and seemed to be aggregated (Figure 3). Rivas et al., (2016) and Balcaen et al., (2017) observed the same trends with light microscopy [47,48]. Emulsions prepared using AMP were highly aggregated systems, while emulsions stabilized with PGPR had very small, individual droplets [47,48]. The emulsions prepared in this study were designed as the inner phase of double emulsions and were aimed to be of low viscosity. Since the viscosity of the emulsions produced with AMP was very high, the emulsions produced with this emulsifier were not included in the later parts of the study.

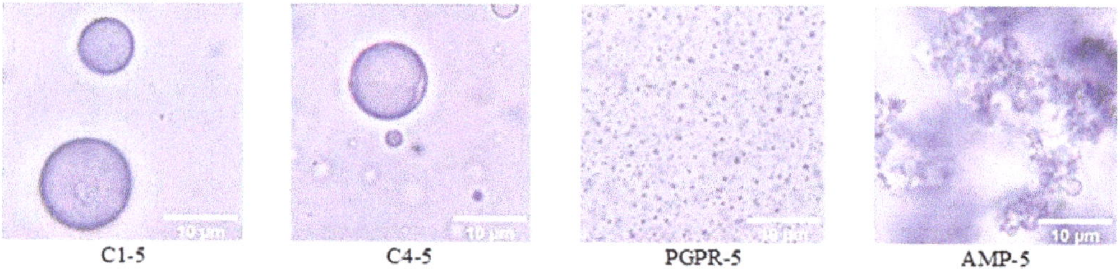

Figure 3. Micrographs of W/O emulsion produced with constant 25% dispersed phase ratio and medium-level emulsifier concentrations (5% for Crill™ 1, Crill™ 4, and PGPR and 2% for AMP).

3.3. *Emulsion Stability*

The gravimetric approach was used to determine the emulsion stability. W/O emulsions stabilized with Crill™ 1, Crill™ 4, and PGPR was examined through photographs taken during the 5-day storage period (Figure 4). It is observed that Crill™-1- and Crill™-4-stabilized emulsions had district phase separation after 24 h. However, no visual phase separation was detected in the PGPR-stabilized samples even after 5 days. PGPR produced significantly stable casein-hydrolysate-loaded W/O emulsions compared to others in all formulations. The highly effective applications of PGPR in stabilizing W/O emulsions have been reported by several authors in the literature [15,47,49]. In one of these studies, PGPR was used to produce peptide-loaded emulsions [15]. Ying et al., (2021) determined that the PGPR-stabilized soy-peptide-loaded W/O emulsion presented better stability than the Span-60- and lecithin-stabilized ones [15].

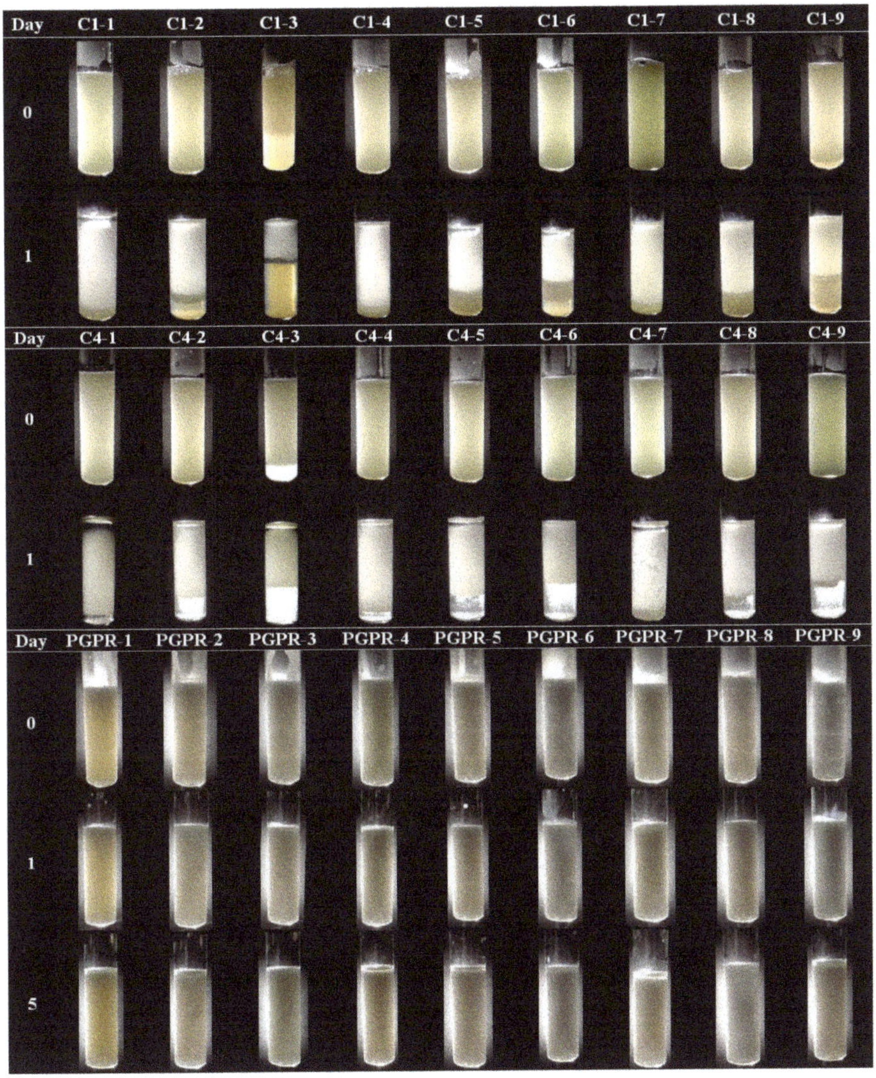

Figure 4. The gravimetric stability of W/O emulsions stabilized with Crill™ 1, Crill™ 4, and PGPR.

3.4. Droplet Size Distributions

The droplet size and size distribution in emulsions are important parameters on stability, as an increase in droplet size can lead to destabilization of the emulsions by flocculation, coalescence, or Oswalt ripening [31]. The droplet size distributions of emulsions can differ for various reasons, such as emulsification conditions, type and amount of emulsifier, and the ratio of dispersed and continuous phases [24]. As seen in Figure 5, the droplet size of the Crill™ 1 and Crill™ 4 samples were remarkably varied from the droplet sizes of the PGPR samples. This was also observed in micrographs (Figure 3). According to the comparison of the emulsifiers with the same formulations, D(90) values of Crill™-1- and Crill™-4-stabilized emulsions were found 9.5 to 20.8 and 6.7 to 21.4 times higher than PGPR stabilized emulsions, respectively. These results were in harmony with the emulsion stability results (Figure 4). While D(90) values in all PGPR-stabilized emulsions were below 2 μm, D [4, 3], and D [3, 2] values were less than 1 μm.

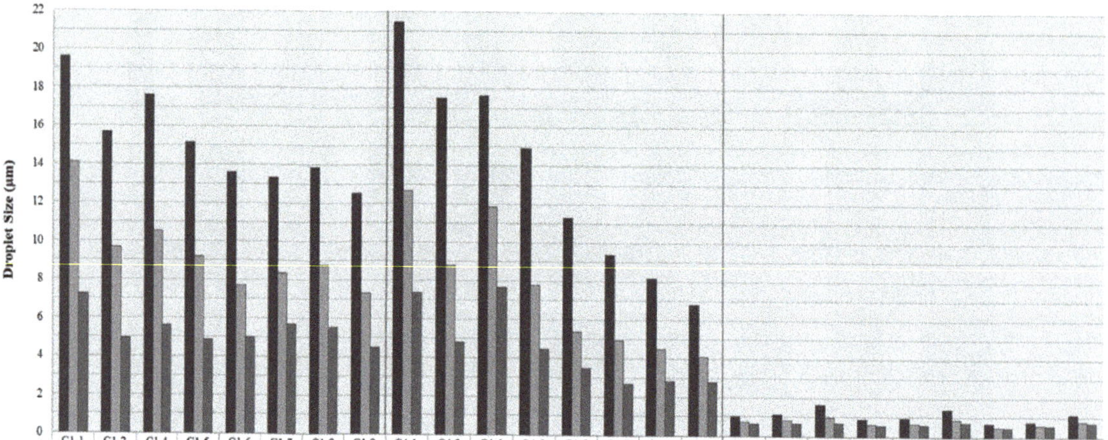

	C1-1	C1-2	C1-4	C1-5	C1-6	C1-7	C1-8	C1-9	C4-1	C4-2	C4-4	C4-5	C4-6	C4-7	C4-8	C4-9	PGPR-1	PGPR-2	PGPR-3	PGPR-4	PGPR-5	PGPR-6	PGPR-7	PGPR-8	PGPR-9
D (90)	19.61	15.66	17.62	15.12	13.57	13.32	13.87	12.53	21.45	17.52	17.66	14.89	11.30	9.40	8.15	6.80	1.00	1.11	1.63	0.88	0.95	1.40	0.67	0.75	1.13
D [4,3]	14.11	9.68	10.54	9.22	7.69	8.35	8.76	7.34	12.64	8.78	11.86	7.77	5.37	4.96	4.49	4.08	0.70	0.82	0.98	0.62	0.66	0.87	0.49	0.57	0.79
D [3,2]	7.24	4.95	5.62	4.87	5.00	5.70	5.54	4.49	7.37	4.81	7.64	4.46	3.45	2.70	2.84	2.81	0.62	0.65	0.68	0.56	0.59	0.71	0.45	0.53	0.70

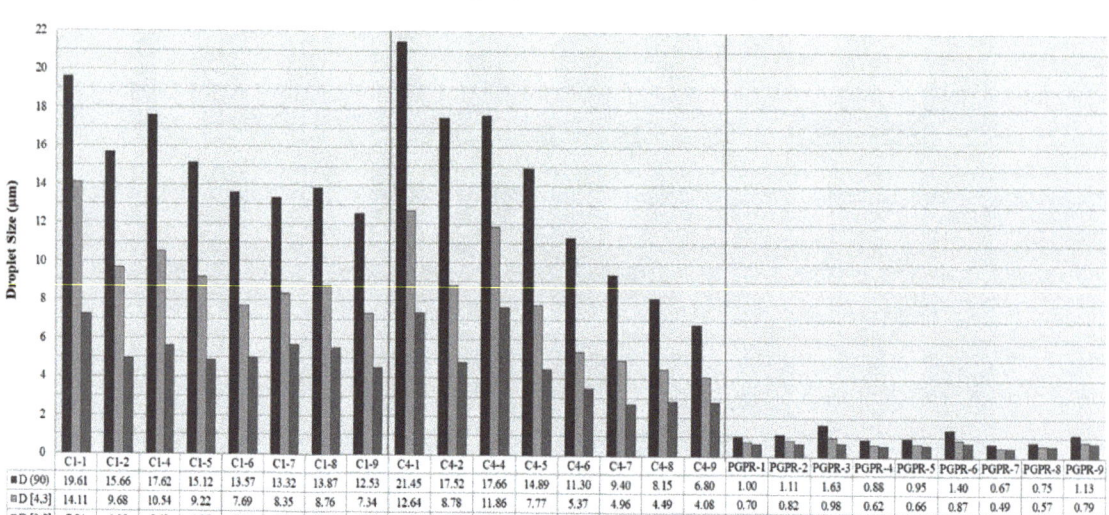

Figure 5. The droplet size parameters for W/O emulsions were prepared with different emulsifiers and formulations.

PCA was performed and a PCA bi-plot was drawn to show the closeness of the samples with droplet size parameters and consistency coefficient (Figure 6). The calculated principal components (F1 and F2) explained a very important part of the variations, e.g., 99.1% and 75.9% of the variations that could be represented by the component (F1) are shown on the x-axis alone. According to PCA analysis, all samples produced with PGPR were located away from the droplet size parameters along the x-axis. However, PGPR samples containing 40% internal phase were positioned at the positive F2; likewise, the consistency coefficient and other PGPR stabilized samples were positioned at the negative side of the y-axis. In addition, a negative correlation concerning F1 was found between the consistency coefficient and the droplet size parameters. This negative correlation may be due to several possible reasons. As the droplet size decreases, the average separation distance between the droplets also decreases, resulting in an increase in hydrodynamic interaction and viscosity. Furthermore, the increases in viscosity upon droplet size reduction may be due in part to an increase in the effective dispersed phase concentration. In other words, as the droplet size decreases, the thickness of the adsorbed emulsifier layer relative to the droplet size becomes more important. Finally, it is known that the polydispersity generally decreases with the decrease in droplet size and influences the flow behaviors [50].

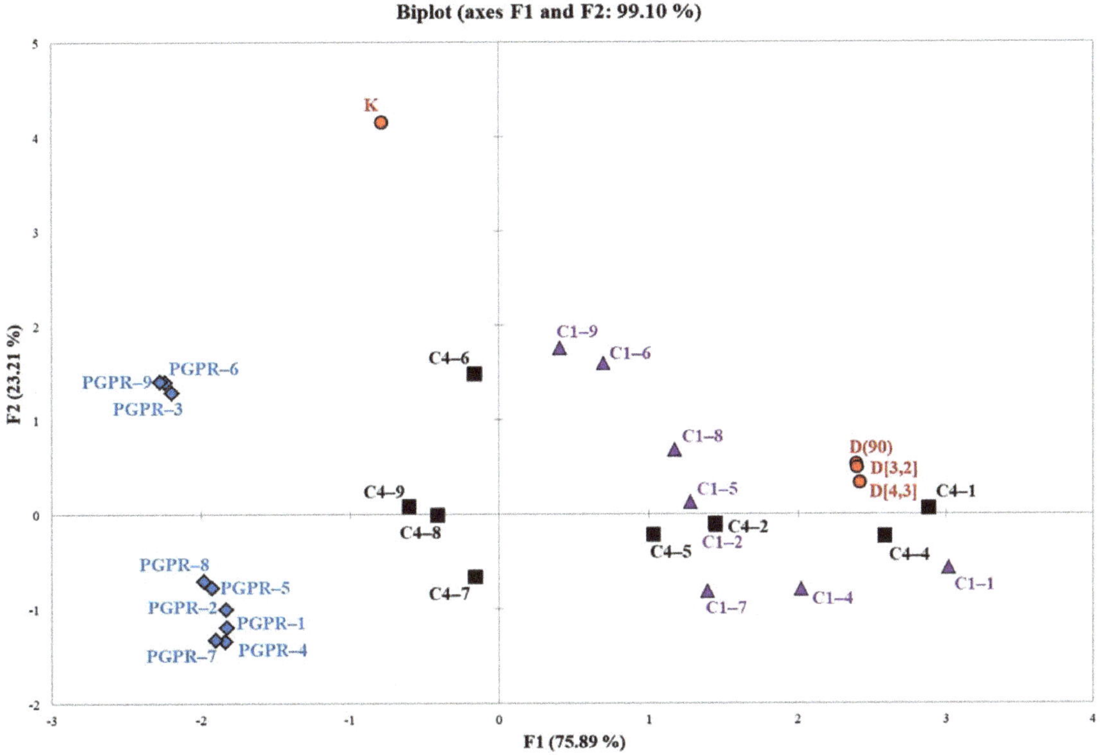

Figure 6. PCA bi-plot diagram for droplet size parameters and consistency coefficient for W/O emulsions prepared with different emulsifiers and formulations.

3.5. Determination of Appropriate Emulsion Formulation

In the present study, it was determined that all samples prepared with PGPR had low droplet sizes with high emulsion stability. Rivas et al., (2016) also highlighted that PGPR emulsions have mono-modal droplet size distribution with smaller droplets without networking tendency, and these emulsions showed Newtonian flow behavior with much higher stability to phase separation. As a result of these features, using PGPR in the formulation of the primary emulsion during the production of the double emulsion was suggested [47]. Considering that, PGPR was selected as the most suitable emulsifier for the preparation of primary emulsion.

Although the results in this study were evident and showed that PGPR would be appropriate for the use in the formation of a primary emulsion of the double emulsion, it is not clear in which formulation it would be preferred. For this purpose, TOPSIS analysis was performed to determine the most appropriate PGPR-stabilized emulsion prepared using low emulsifier concentrations, high dispersed phase ratio at low emulsion viscosity, and droplet size. In TOPSIS, dispersed phase ratio, emulsifier concentration, viscosity, and droplet size (D [4, 3]) parameters were used as responses, and their weights were decided as 20%, 20%, 30%, and 30%, respectively. The emulsion with the highest relative closeness value calculated by TOPSIS was selected. It was seen that three emulsion formulations (PGPR-1, PGPR-2, and PGPR-5) gave very close relative closeness values above 0.6, and PGPR-5 (25% dispersed phase ratio and 5% emulsifier concentration) had the highest value as 0.621 (see Table 3).

Table 3. Technique for order of preference by similarity to ideal solution (TOPSIS) similarity values for the emulsions prepared by PGPR 4150.

Code	Dispersed Phase (%)	PGPR (%)	Viscosity (cP)	D [4, 3] (μm)	Relative Closeness (-)
PGPR-1	10	3	83.9	0.704	0.611
PGPR-2	25	3	132.5	0.822	0.616
PGPR-3	40	3	260.2	0.976	0.430
PGPR-4	10	5	93.4	0.618	0.591
PGPR-5	25	5	139.8	0.662	0.621
PGPR-6	40	5	271.9	0.873	0.396
PGPR-7	10	7	100.1	0.488	0.568
PGPR-8	25	7	146.0	0.575	0.572
PGPR-9	40	7	274.7	0.795	0.381

4. Conclusions

The results of the present study showed that the selection of emulsifier type is an important cornerstone for obtaining more stable primary emulsions, which is critical for the stability of double emulsions. Thus, the results of experimental studies clearly showed that PGPR is the most suitable emulsifier type among other types due to its ability to form emulsions with relatively low viscosity and significantly high stability. The most appropriate formulation with PGPR-based primary emulsion was a 25% of dispersed phase ratio and 5% of emulsifier concentration in the dispersed (oil) phase. This emulsion formulation resulted in a viscosity of 139.8 cP and a D [4, 3] value of 0.662 μm. In summary, the detailed characterization of emulsions provides the true selection of emulsifier type in terms of more stable primary emulsion and accordingly high-quality double emulsions.

Author Contributions: Conceptualization, Z.E.; methodology, P.S. and Z.E.; validation, P.S. and Ç.U.; formal analysis, P.S., Ç.U. and O.G.; investigation, P.S. and Z.E.; resources, O.G., L.Y.A. and Z.E.; data curation, P.S. and Ç.U.; writing—original draft preparation, P.S. and Z.E.; writing—review and editing, O.G. and L.Y.A.; visualization, P.S.; supervision, L.Y.A.; project administration, Z.E.; funding acquisition, L.Y.A. and Z.E. All authors have read and agreed to the published version of the manuscript.

Funding: This work was supported by the Scientific and Technological Research Council of Turkey (TUBITAK) (Project No. 120O763).

Institutional Review Board Statement: Not applicable.

Informed Consent Statement: Not applicable.

Data Availability Statement: Not applicable.

Acknowledgments: The authors are also grateful to Novozymes for the enzymes provided and Palsgaard and Croda for the emulsifiers supplied.

Conflicts of Interest: The authors declare no conflict of interest.

References

1. Agrawal, H.; Joshi, R.; Gupta, M. Optimization of Pearl Millet-Derived Bioactive Peptide Microspheres with Double Emulsion Solvent Evaporation Technique and Its Release Characterization. *Food Struct.* **2021**, *29*, 100200. [CrossRef]
2. Peighambardoust, S.H.; Karami, Z.; Pateiro, M.; Lorenzo, J.M. A Review on Health-promoting, Biological, and Functional Aspects of Bioactive Peptides in Food Applications. *Biomolecules* **2021**, *11*, 631. [CrossRef] [PubMed]
3. Clare, D.A.; Swaisgood, H.E. Bioactive Milk Peptides: A Prospectus. *J. Dairy Sci.* **2000**, *83*, 1187–1195. [CrossRef] [PubMed]
4. Korhonen, H. Technology Options for New Nutritional Concepts. *Int. J. Dairy Technol.* **2002**, *55*, 79–88. [CrossRef]
5. Phelan, M.; Aherne, A.; Fitzgerald, R.J.; Brien, N.M.O. Casein-Derived Bioactive Peptides: Biological Effects, Industrial Uses, Safety Aspects and Regulatory Status. *Int. Dairy J.* **2009**, *19*, 643–654. [CrossRef]
6. Korhonen, H.; Pihlanto, A. Food-Derived Bioactive Peptides-Opportunities for Designing Future Foods. *Curr. Pharm. Des.* **2003**, *9*, 1297–1308. [CrossRef]

7. Xu, Q.; Yan, X.; Zhang, Y.; Wu, J. Current Understanding of Transport and Bioavailability of Bioactive Peptides Derived from Dairy Proteins: A Review. *Int. J. Food Sci. Technol.* **2019**, *54*, 1930–1941. [CrossRef]
8. Korhonen, H. Milk-Derived Bioactive Peptides: From Science to Applications. *J. Funct. Foods* **2009**, *1*, 177–187. [CrossRef]
9. Korhonen, H.; Pihlanto, A. Bioactive Peptides: Production and Functionality. *Int. Dairy J.* **2006**, *16*, 945–960. [CrossRef]
10. Nongonierma, A.B.; FitzGerald, R.J. The Scientific Evidence for the Role of Milk Protein-Derived Bioactive Peptides in Humans: A Review. *J. Funct. Foods* **2015**, *17*, 640–656. [CrossRef]
11. Gómez-Mascaraque, L.G.; Miralles, B.; Recio, I.; López-Rubio, A. Microencapsulation of a Whey Protein Hydrolysate within Micro-Hydrogels: Impact on Gastrointestinal Stability and Potential for Functional Yoghurt Development. *J. Funct. Foods* **2016**, *26*, 290–300. [CrossRef]
12. Giroldi, M.; Grambusch, I.M.; Lehn, D.N.; de Souza, C.F.V. Encapsulation of Dairy Protein Hydrolysates: Recent Trends and Future Prospects. *Dry. Technol.* **2021**, *39*, 1513–1528. [CrossRef]
13. McClements, D.J. Encapsulation, Protection, and Release of Hydrophilic Active Components: Potential and Limitations of Colloidal Delivery Systems. *Adv. Colloid Interface Sci.* **2015**, *219*, 27–53. [CrossRef]
14. Mohan, A.; Rajendran, S.R.C.K.; He, Q.S.; Bazinet, L.; Udenigwe, C.C. Encapsulation of Food Protein Hydrolysates and Peptides: A Review. *RSC Adv.* **2015**, *5*, 79270–79278. [CrossRef]
15. Ying, X.; Gao, J.; Lu, J.; Ma, C.; Lv, J.; Adhikari, B.; Wang, B. Preparation and Drying of Water-in-Oil-in-Water (W/O/W) Double Emulsion to Encapsulate Soy Peptides. *Food Res. Int.* **2021**, *141*, 110148. [CrossRef]
16. Yang, W.; Li, J.; Ren, D.; Cao, W.; Lin, H.; Qin, X.; Wu, L.; Zheng, H. Construction of a Water-in-Oil-in-Water (W/O/W) Double Emulsion System Based on Oyster Peptides and Characterisation of Freeze-Dried Products. *Int. J. Food Sci. Technol.* **2021**, *56*, 6635–6648. [CrossRef]
17. Giroux, H.J.; Robitaille, G.; Britten, M. Controlled Release of Casein-Derived Peptides in the Gastrointestinal Environment by Encapsulation in Water-in-Oil-in-Water Double Emulsions. *LWT Food Sci. Technol.* **2016**, *69*, 225–232. [CrossRef]
18. Giroux, H.J.; Shea, R.; Sabik, H.; Fustier, P.; Robitaille, G.; Britten, M. Effect of Oil Phase Properties on Peptide Release from Water-in-Oil-in-Water Emulsions in Gastrointestinal Conditions. *LWT Food Sci. Technol.* **2019**, *109*, 429–435. [CrossRef]
19. Garti, N.; Bisperink, C. Double Emulsions: Progress and Applications. *Curr. Opin. Colloid Interface Sci.* **1998**, *3*, 657–667. [CrossRef]
20. Leister, N.; Karbstein, H.P. Evaluating the Stability of Double Emulsions—A Review of the Measurement Techniques for the Systematic Investigation of Instability Mechanisms. *Colloids Interfaces* **2020**, *4*, 8. [CrossRef]
21. Muschiolik, G.; Dickinson, E. Double Emulsions Relevant to Food Systems: Preparation, Stability, and Applications. *Compr. Rev. Food Sci. Food Saf.* **2017**, *16*, 532–555. [CrossRef] [PubMed]
22. Garti, N. Double Emulsions-Scope, Limitations and New Achievements. *Colloids Surfaces A Physicochem. Eng. Asp.* **1997**, *123–124*, 233–246. [CrossRef]
23. Choi, M.J.; Choi, D.; Lee, J.; Jo, Y.J. Encapsulation of a Bioactive Peptide in a Formulation of W1/O/W2-Type Double Emulsions: Formation and Stability. *Food Struct.* **2020**, *25*, 100145. [CrossRef]
24. Jo, Y.J.; Karbstein, H.P.; Van Der Schaaf, U.S. Collagen Peptide-Loaded W1/O Single Emulsions and W1/O/W2 Double Emulsions: Influence of Collagen Peptide and Salt Concentration, Dispersed Phase Fraction and Type of Hydrophilic Emulsifier on Droplet Stability and Encapsulation Efficiency. *Food Funct.* **2019**, *10*, 3312–3323. [CrossRef] [PubMed]
25. Jo, Y.J.; van der Schaaf, U.S. Fabrication and Characterization of Double (W1/O/W2) Emulsions Loaded with Bioactive Peptide/Polysaccharide Complexes in the Internal Water (W1) Phase for Controllable Release of Bioactive Peptide. *Food Chem.* **2021**, *344*, 128619. [CrossRef]
26. Nuti, N.; Rottmann, P.; Stucki, A.; Koch, P.; Panke, S.; Dittrich, P.S. A Multiplexed Cell-Free Assay to Screen for Antimicrobial Peptides in Double Emulsion Droplets. *Angew. Chem. Int. Ed.* **2022**, *61*, e202114632. [CrossRef]
27. Patange, S.R.; Sabikhi, L.; Shelke, P.A.; Rathod, N.; Shaik, A.H.; Khetra, Y.; Kumar, M.H.S. Encapsulation of Dipeptidyl Peptidase-IV Inhibitory Peptides from Alpha-Lactalbumin Extracted from Milk of Gir Cows—A Bos Indicus Species. *Int. J. Dairy Technol.* **2022**, *75*, 575–587. [CrossRef]
28. Muschiolik, G. Multiple Emulsions for Food Use. *Curr. Opin. Colloid Interface Sci.* **2007**, *12*, 213–220. [CrossRef]
29. Lutz, R.; Aserin, A.; Wicker, L.; Garti, N. Release of Electrolytes from W/O/W Double Emulsions Stabilized by a Soluble Complex of Modified Pectin and Whey Protein Isolate. *Colloids Surfaces B Biointerfaces* **2009**, *74*, 178–185. [CrossRef]
30. Ushikubo, F.Y.; Cunha, R.L. Stability Mechanisms of Liquid Water-in-Oil Emulsions. *Food Hydrocoll.* **2014**, *34*, 145–153. [CrossRef]
31. McClements, D.J. *Food Emulsions: Principles, Practice and Techniques*, 3rd ed.; McClements, D.J., Ed.; CRC Press: Boca Raton, FL, USA, 2016.
32. Leister, N.; Yan, C.; Karbstein, H.P. Oil Droplet Coalescence in W/O/W Double Emulsions Examined in Models from Micrometer-to Millimeter-Sized Droplets. *Colloids Interfaces* **2022**, *6*, 12. [CrossRef]
33. Chevalier, R.C.; Gomes, A.; Cunha, R.L. Tailoring W/O Emulsions for Application as Inner Phase of W/O/W Emulsions: Modulation of the Aqueous Phase Composition. *J. Food Eng.* **2021**, *297*, 110482. [CrossRef]
34. Gaonkar, A.G. Surface and Interfacial Activities and Emulsion Characteristics of Some Food Hydrocolloids. *Food Hydrocoll.* **1991**, *5*, 329–337. [CrossRef]
35. Mortensen, A.; Aguilar, F.; Crebelli, R.; Domenico, A.D.; Dusemund, B.; Frutos, M.J.; Galtier, P.; Gott, D.; Gundert-Remy, U.; Leblanc, J.C.; et al. Scientific Opinion on the Re-Evaluation of Ammonium Phosphatides (E 442) as a Food Additive. *EFSA J.* **2016**, *14*, 4597. [CrossRef]

36. Mortensen, A.; Aguilar, F.; Crebelli, R.; Domenico, A.D.; Dusemund, B.; Frutos, M.J.; Galtier, P.; Gott, D.; Gundert-Remy, U.; Leblanc, J.C.; et al. Scientific Opinion on the Re-Evaluation of Sorbitan Monostearate (E 491), Sorbitan Tristearate (E 492), Sorbitan Monolaurate (E 493), Sorbitan Monooleate (E 494) and Sorbitan Monopalmitate (E 495) When Used as Food Additives. *EFSA J.* **2017**, *15*, 4788. [CrossRef]
37. Weyland, M.; Hartel, R. Emulsifiers in Confectionery. In *Food Emulsifiers and Their Applications*; Hasenhuettl, G.L., Hartel, R., Eds.; Springer: New York, NY, USA, 2008; pp. 285–306.
38. Peltonen, L.; Hirvonen, J.; Yliruusi, J. The Behavior of Sorbitan Surfactants at the Water-Oil Interface: Straight-Chained Hydrocarbons from Pentane to Dodecane as an Oil Phase. *J. Colloid Interface Sci.* **2001**, *240*, 272–276. [CrossRef]
39. Romero-Peña, M.; Ng, E.K.; Ghosh, S. Development of Thermally Stable Coarse Water-in-Oil Emulsions as Potential DNA Bioreactors. *J. Dispers. Sci. Technol.* **2021**, *42*, 2075–2084. [CrossRef]
40. Sarode, A.R.; Sawale, P.D.; Khedkar, C.D.; Kalyankar, S.D.; Pawshe, R.D. Casein and Caseinate: Methods of Manufacture. In *Encyclopedia of Food and Health*; Caballero, B., Finglas, P., Toldrá, F., Eds.; Elsevier: Amsterdam, The Netherlands, 2016; pp. 676–682. [CrossRef]
41. Salum, P.; Güven, O.; Aydemir, L.Y.; Erbay, Z. Microscopy-Assisted Digital Image Analysis with Trainable Weka Segmentation (TWS) for Emulsion Droplet Size Determination. *Coatings* **2022**, *12*, 364. [CrossRef]
42. Hwang, C.L.; Yoon, K. *Multiple Attribute Decision Making: Methods and Applications*; Springer: New York, NY, USA, 1981.
43. Himmetagaoglu, A.B.; Erbay, Z.; Cam, M. Production of Microencapsulated Cream: Impact of Wall Materials and Their Ratio. *Int. Dairy J.* **2018**, *83*, 20–27. [CrossRef]
44. Bains, U.; Pal, R. Rheology and Catastrophic Phase Inversion of Emulsions in the Presence of Starch Nanoparticles. *ChemEngineering* **2020**, *4*, 57. [CrossRef]
45. Bennett, K.E.; Davis, H.T.; Macosko, C.W.; Scriven, L.E. Microemulsion Rheology: Newtonian and Non-Newtonian Regimes. In Proceedings of the SPE Annual Technical Conference and Exhibition, San Antonio, TX, USA, 4–7 October 1981; p. SPE-10061-MS. [CrossRef]
46. Maffi, J.M.; Meira, G.R.; Estenoz, D.A. Mechanisms and Conditions That Affect Phase Inversion Processes: A Review. *Can. J. Chem. Eng.* **2021**, *99*, 178–208. [CrossRef]
47. Rivas, J.C.M.; Schneider, Y.; Rohm, H. Effect of Emulsifier Type on Physicochemical Properties of Water-in-Oil Emulsions for Confectionery Applications. *Int. J. Food Sci. Technol.* **2016**, *51*, 1026–1033. [CrossRef]
48. Balcaen, M.; Vermeir, L.; Van der Meeren, P. Influence of Protein Type on Polyglycerol Polyricinoleate Replacement in W/O/W (Water-in-Oil-in-Water) Double Emulsions for Food Applications. *Colloids Surfaces A Physicochem. Eng. Asp.* **2017**, *535*, 105–113. [CrossRef]
49. Zhang, J.; Reineccius, G.A. Preparation and Stability of W/O/W Emulsions Containing Sucrose as Weighting Agent. *Flavour Fragr. J.* **2016**, *31*, 51–56. [CrossRef]
50. Pal, R. Effect of Droplet Size on the Rheology of Emulsions. *AIChE J.* **1996**, *42*, 3181–3190. [CrossRef]

colloids and interfaces

Article

Effect of Enzymatic Hydrolysis on Solubility and Emulsifying Properties of Lupin Proteins (*Lupinus luteus*)

Mauricio Opazo-Navarrete [1,*], César Burgos-Díaz [1], Karla A. Garrido-Miranda [1] and Sergio Acuña-Nelson [2]

1 Agriaquaculture Nutritional Genomic Center (CGNA), Temuco 4780000, Chile
2 Departamento de Ingeniería en Alimentos, Universidad del Bío-Bío, Chillán 3780000, Chile
* Correspondence: mauricio.opazo@cgna.cl; Tel.: +56-452740408

Abstract: Solubility and emulsifying properties are important functional properties associated with proteins. However, many plant proteins have lower techno-functional properties, which limit their functional performance in many formulations. Therefore, the objective of this study was to investigate the effect of protein hydrolysis by commercial enzymes to improve their solubility and emulsifying properties. Lupin protein isolate (LPI) was hydrolyzed by 7 commercial proteases using different E/S ratios and hydrolysis times while the solubility and emulsifying properties were evaluated. The results showed that neutral and alkaline proteases are most efficient in hydrolyzing lupin proteins than acidic proteases. Among the proteases, Protamex® (alkaline protease) showed the highest DH values after 5 h of protein hydrolysis. Meanwhile, protein solubility of LPI hydrolysates was significantly higher ($p < 0.05$) than untreated LPI at all pH analyzed values. Moreover, the emulsifying capacity (EC) of undigested LPI was lower than most of the hydrolysates, except for acidic proteases, while emulsifying stability (ES) was significantly higher ($p < 0.05$) than most LPI hydrolysates by acidic proteases, except for LPI hydrolyzed with Acid Stable Protease with an E/S ratio of 0.04. In conclusion, the solubility, and emulsifying properties of lupin (*Lupinus luteus*) proteins can be improved by enzymatic hydrolysis using commercial enzymes.

Keywords: emulsifying properties; solubility; protein hydrolysis; lupin proteins

Citation: Opazo-Navarrete, M.; Burgos-Díaz, C.; Garrido-Miranda, K.A.; Acuña-Nelson, S. Effect of Enzymatic Hydrolysis on Solubility and Emulsifying Properties of Lupin Proteins (*Lupinus luteus*). *Colloids Interfaces* **2022**, *6*, 82. https://doi.org/10.3390/colloids6040082

Academic Editor: Reinhard Miller

Received: 27 October 2022
Accepted: 12 December 2022
Published: 15 December 2022

1. Introduction

In recent years, the food and pharmaceutical industries have turned their attention to plant-derived materials due to their considerable advantages over animal-derived ingredients, such as a low prevalence of infection and contamination, dietary beliefs and practices, vegetarian habits, versatility, and lower cost [1]. In addition, an important food trend is the formulation of foods directed to minimize the effects of animal-derived ingredients on the environment, human health, and animal welfare [2]. Thus, the consumption of plant protein ingredients, chiefly from legumes, has increased sharply in the last decade as an alternative to animal proteins [3]. Since most plant proteins are composed of a mixture of different protein fractions, which have different isoelectric points (pI) [4]. Therefore, the modification of their characteristics to enhance their functional properties is necessary [1].

Among functional properties, solubility and emulsifying properties are important protein-associated properties [5]. Protein solubility mainly determines the potential use of a protein-based ingredient in food formulations [6]. In addition, plant proteins are widely used as natural emulsifiers to stabilize oil-in-water (O/W) emulsions due to their amphiphilic properties, nutritional benefits, increased consumer acceptability, and low cost [7]. Thus, proteins are relevant due to the fact that they are the most important surface-active compounds in plant materials [8]. However, several plant proteins present low water-solubility and emulsifying properties in comparison to synthetic surfactants, which restrain their performance in many aqueous-based food formulations [9]. Therefore, some vegetable proteins must be modified to improve their functional properties and performance. In general, protein modifications can be reached by physical, chemical, and biological methods. To date, some studies have addressed extensive or mild protein hydrolysis as a way to tailor the functional properties of plant proteins [10–19]. Liu et al. [11] found that mild hydrolysis of rice protein isolates resulted in the enhancement of emulsifying properties. In addition, Eckert et al. [14] showed that the extensive hydrolysis of fava bean proteins enhanced their solubility, foaming, and emulsifying properties. However, the plant protein modifications depend on the processing type, pH, concentration, and original structure [20].

The enzymatic hydrolysis by proteases stands out to be the most auspicious method for protein modification to formulate tailor-made foods, particularly due to reactants and by-products being innocuous [1,21]. Also, this type of modification presents some advantages over other modification methods, such as it can be reached by moderate conditions with only a few by-products, the chemical composition of the protein being preserved, and the fast enzyme reaction time and specificity [1]. Proteases based on characteristic mechanistic features are classified into six groups: cysteine, aspartate, glutamate, serine, metallo, and threonine [22]. There is a wide variety of proteases in commercial use, which range from detergent additives to therapeutics [23]. However, proteases today are becoming increasingly used to improve the functional properties of proteins. Concerning improvements in the emulsifying capacity of enzymatic hydrolysis derivatives of proteins, it has been found to depend on the protein source and hydrolysis conditions, such as the degree of hydrolysis, the type of enzyme, and the pH of the medium [24]. Currently, some research has analyzed the emulsifying properties of different proteins hydrolyzed with enzymes. For instance, Xu et al. [25] showed that partially hydrolyzed rice protein enhanced its emulsion stability. Also, Padial-Domínguez et al. [24] found that partially hydrolyzed soy protein presented improved emulsifying properties than its counterpart. Otherwise, Avramenko et al. [7] found that partially hydrolyzed lentil protein reduced their solubility and emulsifying properties as a function of the degree of hydrolysis (%).

However, the effect of enzymatic hydrolysis on the functional properties of the legume family *Lupinus* (family of *Fabaceae*) has been scarcely investigated up to now. *Lupinus* is a promising raw material due to its high amount of proteins (33 to 40% of dry weight) and excellent amino acid profile [26–28]. Amongst the sweet lupin species (*Lupinus luteus*, *Lupinus angustifolius*, and *Lupinus albus*), yellow lupin (*Lupinus luteus*) possesses very high protein content and high fibre content [29]. Thus, the lupin sweet variety Alu*Prot*-CGNA® contains around 60% (dry weight) of proteins in dehulled seeds [30]. The main seed storage lupin proteins are globular proteins (globulins), which are divided into four protein families: α-conglutin (11S globulins), β-conglutin (7S globulins), γ-conglutin (7S basic globulins), and δ-conglutin (2S sulphur-rich albumins), of which only α-conglutin, β-conglutin, and δ-conglutin are found in the lupin protein isolate (LPI) [3].

Consequently, this study aims to research the effect of enzymatic modifications of lupin proteins on the solubility and emulsifying properties produced by some commercial enzymes. Thus, enzymatic hydrolysis of lupin proteins could enhance their emulsifying properties to be used as a natural emulsifier of O/W emulsions for food formulations. It is important to mention that the effect of protein modifications of the lupin protein-rich variety Alu*Prot*-CGNA® has not been previously investigated. Therefore, the knowledge acquired from this research could be utilized to select the appropriate enzyme and hydrolysis conditions for a specific food application.

2. Materials and Methods

2.1. Materials

Dehulled yellow lupin seeds (*Lupinus luteus*) variety Alu*Prot*-CGNA® were provided by CGNA (Agriaquaculture Nutritional Genomic Center, Temuco, Chile). Analytical grade chemicals such as hydrochloric acid (HCl), sodium hydroxide (NaOH), Tris, glycerol, glycine, sodium dodecyl sulphate (SDS), bromophenol blue (BPB), and 2-mercaptoethanol were acquired from Sigma (St. Louis, MO, USA). The information about commercial enzymes is listed in Table 1.

Table 1. Sources and properties of the commercial enzymes used.

Enzyme	Type	Biological Source	Supplier	Activity (Under Optimal Conditions)
Acid Stable Protease	Aspartic endopeptidase	*Aspergillus niger*	Bio-Cat (Troy, VA, USA)	4000 SAP/g
Fungal Protease A	Aspartic exo- and endopeptidase	*Aspergillus oryzae*	Bio-Cat (Troy, VA, USA)	1,000,000 HUT/g
Opti-Ziome™ p3 Hydrolyzer™	Aspartic exo- and endopeptidase	*Aspergillus oryzae, Aspergillus melleus*	Bio-Cat (Troy, VA, USA)	130,000 HUT/g
Neutral Protease	Metallo endopeptidase	*Bacillus subtilis*	Bio-Cat (Troy, VA, USA)	2,000,000 PC/g
Protamex®	Serine endopeptidase	*Bacillus licheniformis, Bacillus amyloliquefacies*	Novozymes A/S (Bagsværd, Denmark)	1.5 AU-N/g
Alcaline Protease L	Serine endopeptidase	*Bacillus licheniformis*	Bio-Cat (Troy, VA, USA)	625,000 DU/g
Alcalase® 2.4 L FG	Serine endopeptidase	*Bacillus licheniformis*	Novozymes A/S (Bagsværd, Denmark)	2.4 AU-A/g

2.2. Preparation of Lupin Protein Isolate (LPI)

Lupin protein isolate (LPI) from lupin seeds was obtained according to the procedure described by Burgos-Díaz et al. [31] with minor modifications. The lupin seeds were milled using a rotor mill (Fritsch Mill Pulverisette 14, Indar-Oberstein, Germany) at 10,000 rpm and sieved to 200 μm size to turn into flour. Subsequently, the flour was suspended in a purified water ratio of 1:10 (*w*/*v*), and the pH was adjusted to 9.0 ± 0.1 by the addition of 1 M NaOH and stirred for 1 h at room temperature. Later, the suspension was centrifuged using a laboratory-scale centrifuge (GYROZEN 1580R, Daejeon, Korea) at 3500 rpm for 15 min at 20 °C. The supernatant containing the extracted proteins was shifted to the isoelectric point (pI) at pH 4.6 ± 0.1 using 1 M HCl to precipitate proteins and stirred for 1 h at room temperature. Afterwards, the suspension was subjected to centrifugation (GYROZEN 1580R, Daejeon, Korea) to separate the proteins under the same conditions mentioned above. The precipitated protein was neutralized by re-suspending it in purified water (1:5, *w*/*v*), bringing the pH to a value of 7.0 ± 0.1 using 1 M NaOH and frozen to subsequently freeze-dried using a lyophilizer (Liobras, Liotop LP1280, São Carlos, Brazil). Finally, the LPI powder was kept in a plastic bag and stored at room temperature until later use.

2.3. Chemical Analysis of LPI

The chemical analysis of LPI samples was carried out through the use of a Dumas Nitrogen Analyser (Dumatherm® N Pro, Königswinter, Germany) to determine the nitrogen content of LPI and then multiplied by a conversion factor of 6.25. The moisture was measured using the gravimetric method according to NCh 841 of 78.

2.4. Enzymatic Hydrolysis of LPI

LPI samples were enzymatically hydrolyzed in a 0.5 L bioreactor with pH- and temperature-adjusted to the optimal conditions of each commercial protease (Table 2). LPI suspensions were prepared by diluting LPI in purified water (1.5:10, *w*/*v*) and stirring until complete hydration. Three enzyme-to-substrate ratios (E/S) were chosen (0.01, 0.02, and 0.04% mg enzyme/g protein). Later, the enzyme was added to protein suspension and stirred while temperature and pH were kept constant. The proteins were hydrolyzed from 1 to 5 h and, after hydrolysis, heated at 90 °C for 5 min for enzyme inactivation. Subsequently, the protein hydrolysates were cooled and brought to neutral pH (7.0 ± 0.1) for neutralization. Finally, the hydrolysates samples were lyophilizer (Liobras, Liotop LP1280, São Carlos, Brazil) and milled using a rotor mill (Fritsch Mill Pulverisette 14, Indar-Oberstein, Germany) at 10,000 rpm and sieved to 200 μm size to turn into flour. The control samples used in this study were treated under the same processing conditions without the addition of the enzymes. The hydrolysis and control samples were performed in triplicate.

Table 2. Characteristics of the protease preparations during the LPI hydrolysis.

Enzyme	E/S (%)	Temperature (°C)	pH Value	Time (h)
Acid Stable Protease	0.01/0.02/0.04	55	2.5	1/2/5
Fungal Protease A	0.01/0.02/0.04	60	3.0	1/2/5
Opti-Ziome™ P³ Hydrolyzer™	0.01/0.02/0.04	60	6.0	1/2/5
Neutral Protease	0.01/0.02/0.04	55	7.0	1/2/5
Protamex®	0.01/0.02/0.04	55	8.0	1/2/5
Alcaline Protease L	0.01/0.02/0.04	55	8.5	1/2/5
Alcalase® 2.4 L FG	0.01/0.02/0.04	70	9.0	1/2/5

2.5. SDS-PAGE Analysis

The molecular weight (MW) distribution of protein dispersions was done by reducing SDS-PAGE electrophoresis in a mini-vertical gel electrophoresis unit (Mini-PROTEAN® Tetra Cell, Bio-Rad Laboratories, Inc., Richmond, CA, USA) as described by Cepero-Betancourt et al. [32]. For analyses, the sample was diluted in sample buffer (1:1), which consisted of 0.5 M Tris–HCl; pH 6.8; 2% *v/v* SDS; 2.5% *v/v* glycerol; 0.2% *v/v* bromophenol blue; and 0.5% *v/v* 2-mercaptoethanol and incubated at 90 °C for 4 min using a digital dry bath (Accu Block™, Labnet International Inc., Edison, NJ, USA). The gel bands were visualized by Coomassie staining G-250, while gel images were visualized using a digital imaging system (NUGenius, Syngene, Cambridge, UK). An amount of 15 µL of each sample and 12 µL of pre-stained standard molecular weight marker (Precision Plus Protein Kaleidoscope, Bio-Rad Laboratories, Hercules, CA, USA) were loaded on a 12% Tris–HCl Mini-PROTEAN TGX Precast Gel (Bio-Rad Laboratories Inc., Hercules, CA, USA) and SDS-PAGE was carried out at a constant voltage of 200 V.

2.6. Degree of Hydrolysis (DH)

The degree of hydrolysis (DH) attained was determined by the OPA method, according to Opazo-Navarrete et al. [33]. The OPA reagent (100 mL) was prepared the same day and stored in a bottle protected from light. L-serine was used to prepare the standard curve (50–200 mg/L). For determination, 200 µL of the sample (or standard) was mixed with 1.5 mL of OPA reagent. Finally, the samples were measured after 3 min of reaction with the OPA reagent at 340 nm using an HT Multi-Detection Microplate reader (Biotek Instruments Inc., Winooski, VT, USA). To determine the DH values of hydrolyzed and unhydrolyzed samples, absorbance values were converted to free amino groups (mmol/L) utilizing the standard curve and subtracting the free amino groups that were already present in the protein samples. Finally, serine amino equivalents (N-Terminal Serine) were utilized to express the free amino groups. Thus, the DH values were estimated using the following equations:

$$\mathrm{DH}\ (\%) = \frac{\mathrm{h}}{\mathrm{h_{tot}}} \times 100 \quad (1)$$

$$\mathrm{h} = \frac{\mathrm{Serine\ NH_2} - \beta}{\alpha} \quad (2)$$

where the constant values α and β used were equal to 1 and 0.4, respectively [34]. The h_{tot} value was calculated based on the concentration of each amino acid present in the lupin proteins, which was 7.73 meq/g. All the measurements were done in triplicate.

2.7. Protein Solubility

The protein solubility (%) of the hydrolyzed and unhydrolyzed LPI samples was determined in triplicate according to the method described by Morr et al. [35] with some modifications. For each measurement, was prepared a 3% (*w/v*) protein suspension in purified water. The pH value of suspensions was adjusted to pH 5.0, 7.0, and 9.0 by adding 0.1 M HCl or 0.1 M NaOH as appropriate. Subsequently, the protein suspensions were stirred for 1 h at room temperature, centrifuged at 15,520× *g* for 15 min at 20 °C (GYROZEN

1580R, Daejeon, Korea), and the supernatant was transferred to another tube to separate from the non-dissolved fraction. The supernatant was frozen at −20 °C and freeze-dried using a lyophilizer (Liobras, Liotop LP1280, São Carlos, Brazil). Finally, the freeze-dried sample was weight, and the protein content was determined by Dumas according to the methodology described above (Section 2.3), while protein solubility was estimated as follows:

$$\text{Protein solubility } (\%) = \frac{\text{sample mass supernatant (mg)} \times \text{protein content supernantant } (\%)}{\text{sample mass (mg)} \times \text{protein content } (\%)} \times 100 \quad (3)$$

2.8. *Emulsifying Capacity (EC) and Emulsion Stability (ES)*

The emulsifying capacity (EC) and emulsion stability (ES) were determined according to Burgos-Díaz et al. [36] with minor modifications. Thus, 0.2 g of LPI powder was added to 20 mL of distilled water (1%, *w*/*v*) in a 50-mL Falcon tube and shaken for 1 h at room temperature through constant stirring using an orbital shaker (Multi Reax, Heidolph Instruments, Schwabach, Germany). Subsequently, the sample was kept at 4 °C overnight. Further, the sample was adjusted to pH 7, 20 mL of sunflower oil was added (1:1), and the sample was homogenized for 2.5 min at 10,000 rpm utilizing an Ultra-Turrax (IKA-Werke GmbH & Co. KG, Staufen, Germany). After, the sample was allowed to stand at room temperature for 1 h. The emulsions were transferred to the test tubes, while total and emulsion layer height was measured at 0 and 24 h. The EC_{24} was estimated as the relation between the height of the emulsion layer at 24 h (H_{EL}) and the total height of the sample (H_T). For its part, the ES was calculated by dividing the EC_{24} by the EC at the start time (EC_0).

$$\text{EC } (\%) = \frac{H_{EL}}{H_T} \times 100, \quad (4)$$

$$\text{ES}(\%) = \frac{EC_{24}}{EC_0}. \quad (5)$$

2.9. *Statistical Analysis*

Statistical analysis was performed using Statgraphics Centurion XVI version 16.1.11 software (Statistical Graphics Corp., Herndon, VA, USA). The data were subjected to a one-way analysis of variance (ANOVA) and Duncan's multiple range test (DMRT) to determine significance differences ($p < 0.05$). All results in this research were expressed as mean values ± standard deviations.

3. Results and Discussions

3.1. *Enzymatic Hydrolysis of Lupin Proteins*

The Dumas analysis of LPI and their hydrolysates presented a protein content of 94.3% ± 0.2, and the dry matter was 98.3%. The degree of hydrolysis (DH) analyzed by the OPA method showed that LPI samples without adding enzymes presented a DH value of 0.82% ± 0.01. DH values differed by every enzyme treatment according to their protease specificity. The DH values of all enzymes progressed with the advancement of the hydrolysis time (Figure 1). Meanwhile, the Neutral Protease showed a fast hydrolysis rate during the first 3 h, and the rate subsequently decreased until reaching a stationary. This is typical behaviour of enzymatic reactions where the rate of hydrolysis decreases, which is due to a decrease of peptide bonds, causing the proteases and their substrates to reach a saturation state [37]. However, acidic proteases and Alcaline Protease L showed a low hydrolysis rate with a linear behaviour. Therefore, different results within the same protease family might be due to substrate specificity.

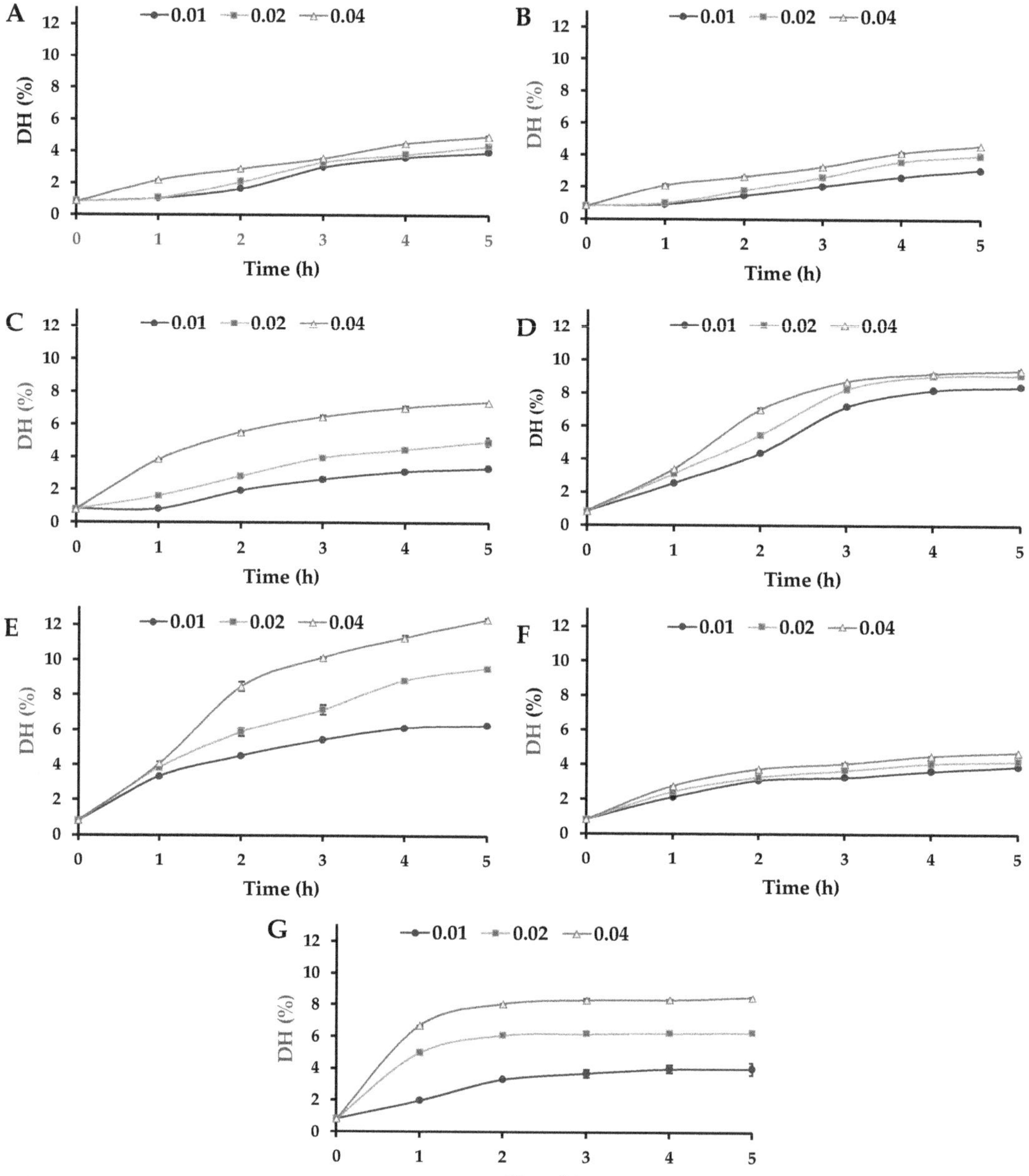

Figure 1. Degree of hydrolysis (DH) of LPI hydrolysates by different commercial enzymes. (**A**) Acid Stable Protease, (**B**) Fungal Protease A, (**C**) Opti-Ziome™ P³ Hydrolyzer™, (**D**) Neutral Protease, (**E**) Protamex®, (**F**) Alcaline Protease L, and (**G**) Alcalase® 2.4 L FG.

The DH was strongly dependent on the E/S ratio in most of the enzymes except for aspartic proteases (Acid Stable Protease and Fungal Protease A) and a serine protease (Alcaline Protease L), where the increase in the E/S ratio does not have a great influence in the DH values. Among the proteases, Protamex®, Neutral Protease, and Alcalase® 2.4 L

FG showed higher DH values after 5 h of protein hydrolysis. This result is different from those previously found in hydrolyzed pea protein isolate (PPI) using Protamex® at an E/S ratio of 0.5% [21]. In the mentioned study, after 2 h of hydrolysis reached a DH value of 4.15%, which is lower than those found in our study (8.51%) using an E/S ratio of 0.04%. In addition, García Arteaga et al. [21] showed a higher DH value for PPI hydrolyzed by Alcalase® 2.4 L FG (9.24%) for 2 h at an E/S ratio of 0.5%, which is a 12.5-times higher E/S ratio that used in this study. However, the protein hydrolysis of LPI by Alcalase® 2.4 L FG showed to be strongly dependent on the E/S ratio. Thus, higher DH values could be expected using the same E/S ratio.

On the other hand, acidic proteases showed lower DH values in comparison with alkaline and neutral proteases. Shu et al. [38] indicated that alkaline protease-assisted reactions exhibited higher DH values compared to neutral or acid enzymes from microbial, animal, or plant origin. However, the Alcaline Protease L exhibited lower DH values than the other alkaline enzymes. This could indicate a lower activity of Alcaline Protease L to hydrolyze lupin proteins. Meanwhile, the lower DH values of the acid proteases could be a consequence of the pHs utilized to hydrolyze lupin proteins with these enzymes (2.5 and 3.0) near the pI of the lupin proteins (4.5), which could result in a certain degree of aggregation and masking of cleavage sites. This phenomenon was previously described by Abdel-Hamid et al. [39] in a study on buffalo milk proteins. Nevertheless, Alcaline Protease L seems to be an exception due to the LPI hydrolyzed by this enzyme exhibiting lower DH values in comparison with other alkaline proteases, which could be a consequence of a lower activity at these E/S relations in comparison with the other alkaline proteases used in this study.

3.2. Molecular Weight Distributions (SDS-PAGE)

Along with the DH analysis, the molecular weight distribution of LPI hydrolysates under reducing conditions was utilized for the analysis of protein hydrolysis (Figure 2). The obtained SDS–PAGE analysis showed that enzymatic hydrolysis has an influence on molecular weight distribution, especially regarding high molecular weight fractions. Thus, all treatments hydrolyze the lupin proteins into smaller fragments with molecular sizes below 50 kD. However, acidic proteases (Acid Stable Protease and Fungal Protease A) hydrolyze the lupin proteins into polypeptides with molecular sizes below 20 kD (Figure 2A,B). Burgos-Díaz et al. [30] found that the major bands in the LPI variety Alu*Prot*-CGNA® are observed between 69 and 39 kD, while some low molecular weight bands are observed below 20 kD. In addition, they found that albumins have a high molecular weight of around 90 kD, α-conglutin has a molecular weight of around 50 kD, and β-conglutin shows a molecular weight of around 69 kD. Therefore, the results suggested that acid proteases can hydrolyze all fractions of LPI higher than 20 kD, while albumins and β-conglutin fractions are mainly hydrolyzed by neutral and alkaline proteases. This is an important issue due to polypeptides of β-conglutin with molecular weights >40 kD have been identified as the major allergens in lupin [19,40]. Thus, acid proteases showed to be the most effective enzymes for breakdown proteins into polypeptides. However, neutral proteases are most effective to hydrolyzed lupin proteins into amino acids (Figure 1).

According to the SDS-PAGE results, the E/S ratios seem to be an effect mainly in α- and β-conglutin fractions, while the time had an effect in the increase of the hydrolysis of the same fractions, which is evidenced by the faded of these bands. However, in this study, clear differences were observed among the different E/S ratios and hydrolysis times, which cannot be correlated with the DH results. These results are important because they could potentially be used to predict some functional properties of proteins.

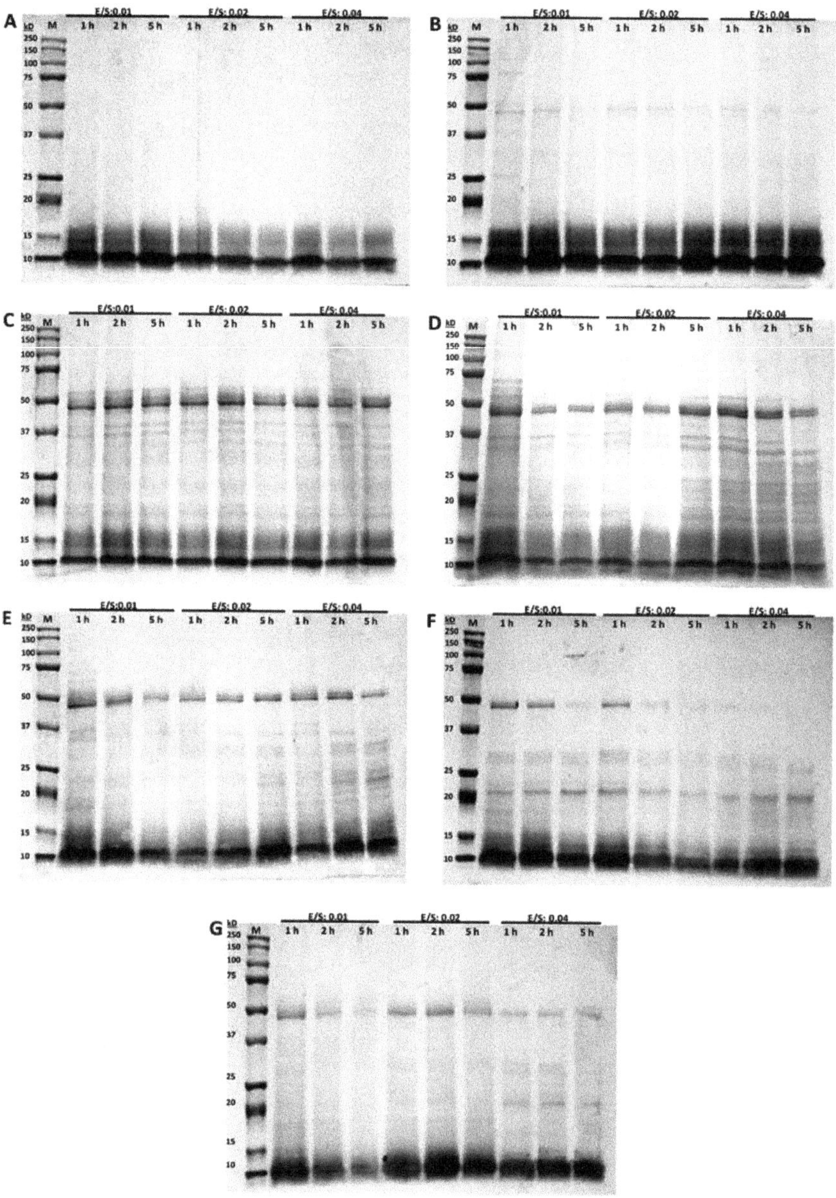

Figure 2. Molecular weight (kD) profiles of LPI hydrolysates by different commercial proteases at different E/S ratios and hydrolysis times as determined by SDS-PAGE under reducing conditions. (**A**) Acid Stable Protease, (**B**) Fungal Protease A, (**C**) Opti-Ziome™ P[3] Hydrolyzer™, (**D**) Neutral Protease, (**E**) Protamex®, (**F**) Alcaline Protease L, and (**G**) Alcalase® 2.4 L FG.

3.3. Functional Properties

3.3.1. Protein Solubility

The solubility of each LPI hydrolyzed at different E/S ratios was estimated as a function of pH at values of 5.0, 7.0, and 9.0 (Figure 3). Untreated LPI samples showed solubility of 24.4% ± 0.5 at pH 5.0, 50.1% ± 0.8 at pH 7.0, and 78.0% ± 1.5 at pH 9.0.

Meanwhile, protein solubility of all hydrolyzed LPI samples was significantly higher ($p < 0.05$) than untreated LPI at all pH analyzed values. Enzymatic hydrolysis provides better interaction of hydrophilic groups with the water molecules due to the decreased size of peptides, which increases protein solubility [18]. Thus, an increase in protein solubility could be a consequence of several facts, such as protein structural changes, the formation of small peptides and hydrophilic amino acids, and changes in the electrostatic forces [41].

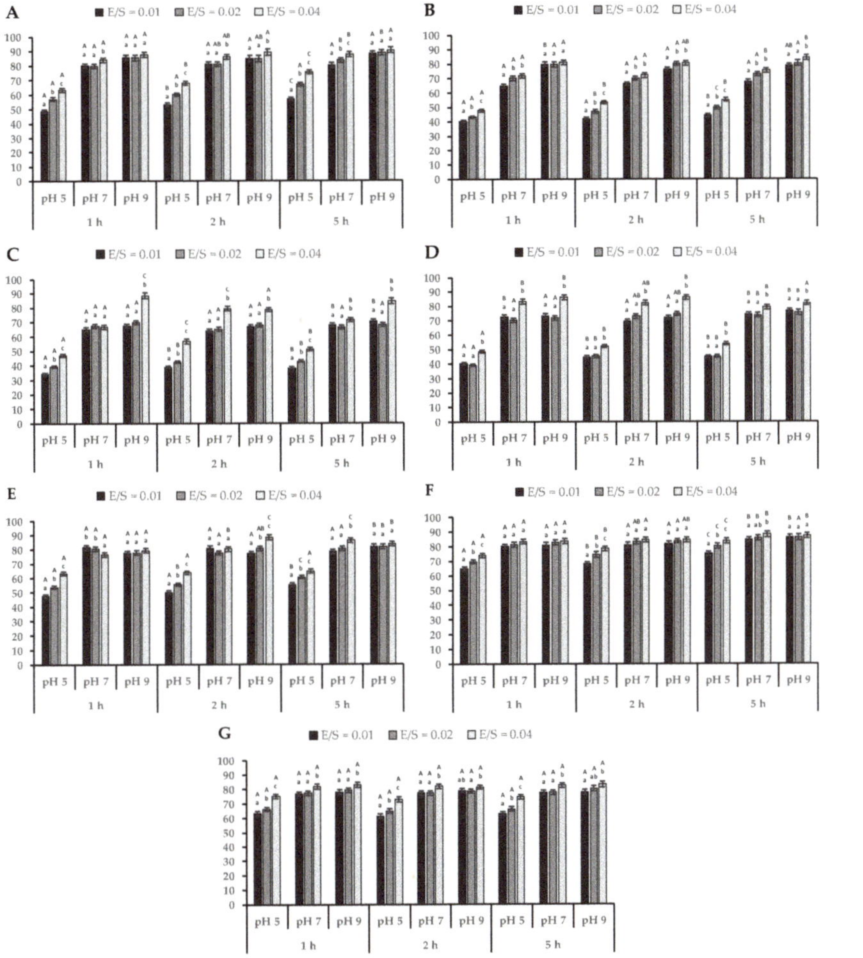

Figure 3. Protein solubility of the hydrolyzed lupin protein isolate (LPI) samples at different E/S (enzyme/substrate) ratios and hydrolysis times by different commercial proteases measured at different pHs. (**A**) Acid Stable Protease, (**B**) Fungal Protease A, (**C**) Opti-Ziome™ P³ Hydrolyzer™, (**D**) Neutral Protease, (**E**) Protamex®, (**F**) Alcaline Protease L, and (**G**) Alcalase® 2.4 L FG. Means with different capital letters in the same row indicate significant differences in each hydrolysis time ($p < 0.05$). Means with different lowercase letters in the same column indicate significant differences in each E/S ratio ($p < 0.05$).

The protein solubility of hydrolyzed LPI samples was strongly dependent on pH obtaining higher values at pH 9.0. Conversely, the lower values were obtained at pH 5.0, near the pI of lupin proteins. The decrease in solubility at low pH is a consequence of the lack of electric charge, which increases hydrophobic aggregation and precipitation [42].

The same was found by Schlegel et al. [19], where the LPI hydrolyzed by various proteases (Neutrase 0.8 L, Corolase 7089, Papain, and Alcalase® 2.4 L FG) exhibited higher protein solubility at pH 9.0 and lower at pH 5.0. The increase in protein solubility of hydrolyzed LPI in acidic conditions compared to unhydrolyzed LPI can be attributed to the protein hydrolysis generating low molecular weight soluble peptides [43]. Meanwhile, Schlegel et al. [19] indicated that the main influence that affects the protein solubility characteristics corresponds to hydrophobic interactions, which increase protein-protein interactions and decrease solubility. Protein solubility is an important property because an increase in the water-solubility of plant proteins is commonly correlated with some functional properties such as emulsifying, gelling, and foaming [11].

In addition, protein solubility at pH 5.0 was significantly dependent ($p < 0.05$) on the E/S ratio, proteolysis time, and commercial enzyme used to hydrolyze the LPI. Meanwhile, protein solubility at pH 7.0 for all hydrolyzed LPI samples varied between 64–67% by Fungal Protease A (E/S = 0.01) to 83–88% by Alcaline Protease L (E/S = 0.04). These results are higher than those previously found at neutral pH on faba bean protein (24.7%), chickpea (47%), and pea (63–75%) [14].

Finally, the protein solubility at pH 9.0 exhibited higher values reaching values near 90% for LPI hydrolyzed by Acid Stable Protease (E/S = 0.04). Hydrolysis by Acid Stable Protease produced polypeptides with molecular weight >20 kD (Figure 2). These results are higher than those found on rice proteins hydrolyzed by Alcalase and Papain, where values close to 50% were reached [11].

3.3.2. Emulsion Capacity (EC) and Emulsion Stability (ES) of Hydrolyzed LPI

Emulsification is one of the most important processes in the manufacturing of formulated foods. The oil-in-water (O/W) emulsions combining proteins and lipids with aqueous solutions are the most commonly used in the food industry [19]. The effect of protein hydrolysis on EC and ES of LPI is shown in Table 3. The EC of undigested LPI was 81.0% ± 1.7, which was lower than most of the hydrolysates, except for acidic proteases Acid Stable Protease and Fungal Protease A. Both acidic proteases showed lower DH values in comparison with the other proteases (Figure 1), which could be indicative that these enzymes could release more hydrophobic groups. Liu et al. [44] indicated that greater exposure by the hydrophobic groups of proteins could improve their emulsifying properties, mainly due to the enhancement of the electrostatic repulsive force between emulsion droplets. Shen et al. [45] reported that hydrophobic interactions enhanced interactions between proteins and then produced large aggregates, which increased in particle size. Furthermore, Zhang et al. [46] demonstrated that the formation of aggregates might decrease the flexibility of proteins to adsorb at the surface of lipid droplets, thus decreasing emulsifying properties. Pan et al. [47] linked the decrease in emulsifying properties to the particle size, where a larger particle size had a negative influence on EC and ES. In this study, there is an inverse correlation between the EC of proteins and protein solubility in hydrolyzed LPI samples, where less soluble samples showed a higher EC. These results are opposed to those found by Schlegel et al. [19], where a direct correlation was found between EC and protein solubility. They indicated that this behaviour was a consequence of a greater dissolution of proteins in the emulsion, which means that a higher amount of proteins is found in the oil-in-water interface during emulsification. The effect of the E/S ratio on EC did not show great changes after 2 h of hydrolysis. However, after 5 h of protein hydrolysis, an increase in E/S ratio presented a significant decrease ($p < 0.05$) in EC values, except for the hydrolyzed LPI samples with Opti-Ziome™ P^3 Hydrolyzer™ and Neutral Protease.

Table 3. Emulsifying properties of LPI hydrolyzed by different commercial enzymes.

Functional Property	Hydrolysis Time (h)	E/S (%)	Acid Stable Protease	Fungal Protease A	Opti-Ziome™ P³ Hydrolyzer™	Neutral Protease	Protamex®	Alcaline Protease L	Alcalase® 2.4 L FG
EC (%)	1	0.01	$85.7 \pm 0.6^{b,F}$	$83.3 \pm 1.5^{a,E}$	$73.3 \pm 1.2^{a,B}$	$80.0 \pm 0.0^{a,D}$	$78.0 \pm 1.0^{b,C}$	$70.3 \pm 0.6^{a,A}$	$78.0 \pm 1.0^{b,C}$
		0.02	$82.0 \pm 2.0^{a,C}$	$86.7 \pm 0.6^{a,D}$	$72.3 \pm 0.6^{a,A}$	$77.7 \pm 2.5^{a,B}$	$77.7 \pm 1.5^{b,B}$	$78.3 \pm 0.6^{b,B}$	$74.7 \pm 0.6^{a,A}$
		0.04	$83.3 \pm 1.5^{a,b,C}$	$82.7 \pm 3.2^{a,C}$	$71.7 \pm 1.5^{a,A}$	$82.7 \pm 4.0^{a,C}$	$75.0 \pm 0.1^{a,A,B}$	$77.7 \pm 2.1^{b,B}$	$75.3 \pm 0.6^{a,A,B}$
	2	0.01	$86.0 \pm 1.7^{b,E}$	$84.3 \pm 1.2^{a,D}$	$73.3 \pm 0.6^{a,A}$	$77.7 \pm 0.6^{a,B}$	$79.7 \pm 0.6^{b,C}$	$78.3 \pm 0.6^{a,B,C}$	$79.7 \pm 0.6^{b,C}$
		0.02	$86.3 \pm 1.5^{b,D}$	$82.0 \pm 1.0^{a,C}$	$74.3 \pm 0.6^{b,A}$	$79.0 \pm 1.0^{a,B}$	$79.0 \pm 1.0^{b,B}$	$77.0 \pm 2.6^{a,B}$	$79.0 \pm 1.0^{a,b,B}$
		0.04	$75.7 \pm 0.6^{a,A,B}$	$83.0 \pm 2.0^{a,C}$	$75.0 \pm 0.1^{b,A}$	$78.3 \pm 2.9^{a,B}$	$75.0 \pm 1.0^{a,A}$	$77.0 \pm 2.0^{a,A,B}$	$76.3 \pm 2.3^{a,A,B}$
	5	0.01	$86.0 \pm 2.0^{b,D}$	$84.7 \pm 1.5^{b,D}$	$71.7 \pm 1.2^{a,b,A}$	$75.3 \pm 1.2^{a,B}$	$84.0 \pm 1.7^{b,D}$	$80.3 \pm 0.6^{c,C}$	$84.0 \pm 1.7^{b,D}$
		0.02	$83.3 \pm 1.2^{b,D}$	$81.7 \pm 2.3^{a,b,D}$	$71.3 \pm 0.6^{a,B}$	$82.0 \pm 2.6^{b,B}$	$81.7 \pm 1.5^{b,D}$	$66.3 \pm 1.5^{b,A}$	$77.7 \pm 1.5^{a,C}$
		0.04	$80.3 \pm 0.6^{a,D}$	$80.3 \pm 1.2^{a,D}$	$73.0 \pm 0.0^{b,B}$	$85.3 \pm 1.2^{b,E}$	$77.7 \pm 1.5^{a,C}$	$61.3 \pm 2.3^{a,A}$	$79.3 \pm 1.2^{a,C,D}$
ES (%)	1	0.01	$81.3 \pm 1.1^{a,A}$	$88.0 \pm 2.2^{b,B}$	$95.9 \pm 1.4^{a,C}$	$97.1 \pm 4.0^{b,C,D}$	$98.7 \pm 1.3^{b,C,D}$	$90.6 \pm 1.9^{a,B}$	$99.5 \pm 0.8^{c,D}$
		0.02	$86.6 \pm 1.5^{b,B}$	$81.2 \pm 0.6^{a,A}$	$96.8 \pm 0.8^{a,E}$	$90.4 \pm 0.7^{a,b,C}$	$94.0 \pm 0.8^{a,D}$	$93.4 \pm 1.6^{a,D}$	$93.6 \pm 1.3^{b,D}$
		0.04	$84.8 \pm 2.2^{b,A}$	$84.7 \pm 2.3^{a,b,A}$	$97.7 \pm 1.6^{a,C}$	$84.8 \pm 4.4^{a,A}$	$91.0 \pm 3.3^{a,B}$	$96.6 \pm 0.8^{b,C}$	$88.9 \pm 0.8^{a,A,B}$
	2	0.01	$79.9 \pm 1.4^{a,A}$	$87.0 \pm 1.0^{a,B}$	$95.9 \pm 0.0^{a,E}$	$94.4 \pm 2.0^{b,D,E}$	$93.7 \pm 1.2^{a,D,E}$	$89.8 \pm 1.2^{a,C}$	$93.3 \pm 1.4^{a,D}$
		0.02	$82.3 \pm 2.4^{a,A}$	$88.2 \pm 1.5^{a,B}$	$95.5 \pm 3.3^{a,D}$	$94.9 \pm 0.1^{b,C,D}$	$92.4 \pm 2.1^{a,C,D}$	$91.8 \pm 1.7^{a,B,C}$	$92.4 \pm 2.1^{a,C,D}$
		0.04	$94.3 \pm 1.5^{b,B,C}$	$88.4 \pm 3.4^{a,A}$	$93.8 \pm 0.8^{a,B,C}$	$89.0 \pm 2.6^{a,A}$	$96.9 \pm 0.7^{b,C}$	$96.1 \pm 1.2^{b,C}$	$90.9 \pm 2.4^{a,A,B}$
	5	0.01	$81.4 \pm 1.0^{a,A}$	$85.5 \pm 2.5^{a,B}$	$97.7 \pm 0.8^{b,D}$	$93.1 \pm 0.9^{b,C}$	$97.9 \pm 1.9^{b,D}$	$91.3 \pm 1.3^{a,C}$	$92.1 \pm 1.7^{a,C}$
		0.02	$85.2 \pm 0.6^{b,A}$	$87.4 \pm 2.6^{a,A,B}$	$98.1 \pm 0.8^{b,E}$	$94.7 \pm 1.4^{b,D}$	$92.1 \pm 1.7^{a,C,D}$	$90.0 \pm 1.5^{a,B,C}$	$97.9 \pm 1.9^{b,E}$
		0.04	$87.6 \pm 0.1^{c,A}$	$88.0 \pm 1.6^{a,A,B}$	$93.2 \pm 2.4^{a,C}$	$87.5 \pm 1.3^{a,A}$	$89.8 \pm 1.7^{a,A,B,C}$	$91.3 \pm 4.0^{a,B,C}$	$93.3 \pm 0.7^{a,C}$

EC: emulsifying capacity; ES: emulsifying stability. Means with different capital letters in the same row indicate significant differences in each enzyme ($p < 0.05$). Means with different lowercase letters in the same column indicate significant differences in each hydrolysis time ($p < 0.05$).

The ES can be defined as the resistance of the system to changes in its physicochemical properties with time. The ES of LPI presented a value of 94.2% ± 0.8, which is significantly higher ($p < 0.05$) than most hydrolysate LPI samples by acidic proteases, except for LPI hydrolyzed with Acid Stable Protease with an E/S ratio of 0.04. The hydrolyzed samples of LPI with alkaline proteases showed higher ES values than acidic proteases in most of the studied samples, mainly in the samples hydrolyzed for 1 or 2 h at low E/S ratios. In general, hydrolysates with low DH showed lower ES values. This is contrary to what has been mentioned by other authors, who have found a negative correlation between DH and ES when hydrolyzed whey protein using Alcalase [48]. In this study, the high emulsion instability found could be a consequence of a larger number of low-molecular-weight peptides. However, SDS-PAGE analysis showed that acidic proteases form polypeptides of smaller molecular weight than alkaline proteases (Figure 2). Therefore, the current study is in agreement with the previous findings. Regarding the E/S ratio, there is no correlation with the ES values. This could be due to the fact that during protein hydrolysis of LPI, some peptides exhibited more hydrophobic than hydrophilic areas, while as the hydrolysis time progressed, this relation changed. In this regard, several reports have suggested that there is an optimum molecular size or chain length for peptides to improve functional properties and limited hydrolysis generally leads to improved functional properties [49]. Therefore, the degree of hydrolysis should be carefully studied to achieve an improvement in functional properties by enzymatic hydrolysis of LPI.

4. Conclusions

This study demonstrated that the solubility and emulsifying properties of lupin (*Lupinus luteus*) proteins could be enhanced by controlled enzymatic hydrolysis using some specific commercial enzymes. The results show that acidic proteases used in this study are the most efficient for hydrolyzing lupin proteins to obtain polypeptides of low molecular weight. However, the alkaline proteases studied showed to be more effective in achieving free amino groups from lupin proteins, except for Alcaline Protease L. The DH (%) values increased with the hydrolysis time, and the highest DH values were logged with Neutral Protease for the E/S ratio of 0.01 and Protamex® for E/S ratios of 0.02 and 0.04, which indicates that Protamex® has a greater ability to hydrolyze lupin proteins. The obtained results indicated that hydrolysis of LPI using the commercial acidic proteases studied can be used to improve the emulsifying capacity (EC) of lupin proteins. However, the protein hydrolysis by acidic proteases decreases the emulsifying stability (ES).

These findings demonstrate that enzymatic hydrolysis by commercial enzymes under controlled conditions is an effective way to improve the functional properties of lupin proteins and thus increase the potential use of lupin protein ingredients in food formulations.

Author Contributions: Conceptualization, M.O.-N. and C.B.-D.; formal analysis, M.O.-N., C.B.-D., K.A.G.-M. and S.A.-N.; funding acquisition, M.O.-N.; investigation, M.O.-N., C.B.-D., K.A.G.-M. and S.A.-N.; methodology, M.O.-N., C.B.-D. and S.A.-N.; project administration, M.O.-N.; resources, M.O.-N. and C.B.-D.; supervision, M.O.-N.; validation, M.O.-N.; visualization, K.A.G.-M.; writing—original draft, M.O.-N. and K.A.G.-M.; writing—review and editing, M.O.-N., C.B.-D., K.A.G.-M. and S.A.-N. All authors have read and agreed to the published version of the manuscript.

Funding: This research was funded by "Agencia Nacional de Investigación y Desarrollo" (ANID, Chile) through FONDECYT-REGULAR project N°1210136 and FONDECYT-POSTDOCTORAL 3220459.

Institutional Review Board Statement: Not applicable.

Informed Consent Statement: Not applicable.

Data Availability Statement: Not applicable.

Acknowledgments: We also acknowledge the Chilean Agency for Research and Development (ANID) and CGNA's chemical unit for their technical assistance.

Conflicts of Interest: The authors declare no conflict of interest.

References

1. Nikbakht Nasrabadi, M.; Sedaghat Doost, A.; Mezzenga, R. Modification Approaches of Plant-Based Proteins to Improve Their Techno-Functionality and Use in Food Products. *Food Hydrocoll.* **2021**, *118*, 106789. [CrossRef]
2. McClements, D.J.; Lu, J.; Grossmann, L. Proposed Methods for Testing and Comparing the Emulsifying Properties of Proteins from Animal, Plant, and Alternative Sources. *Colloids Interfaces* **2022**, *6*, 19. [CrossRef]
3. Burgos-Díaz, C.; Opazo-Navarrete, M.; Wandersleben, T.; Soto-Añual, M.; Barahona, T.; Bustamante, M. Chemical and Nutritional Evaluation of Protein-Rich Ingredients Obtained through a Technological Process from Yellow Lupin Seeds (*Lupinus luteus*). *Plant Foods Hum. Nutr.* **2019**, *74*, 508–517. [CrossRef] [PubMed]
4. Nikbakht Nasrabadi, M.; Goli, S.A.H.; Sedaghat Doost, A.; Dewettinck, K.; Van der Meeren, P. Bioparticles of Flaxseed Protein and Mucilage Enhance the Physical and Oxidative Stability of Flaxseed Oil Emulsions as a Potential Natural Alternative for Synthetic Surfactants. *Colloids Surf. B Biointerfaces* **2019**, *184*, 110489. [CrossRef]
5. Hu, G.J.; Zhao, Y.; Gao, Q.; Wang, X.W.; Zhang, J.W.; Peng, X.; Tanokura, M.; Xue, Y.L. Functional Properties of Chinese Yam (*Dioscorea opposita* Thunb. Cv. Baiyu) Soluble Protein. *J. Food Sci. Technol.* **2018**, *55*, 381–388. [CrossRef]
6. Rivera del Rio, A.; Opazo-Navarrete, M.; Cepero-Betancourt, Y.; Tabilo-Munizaga, G.; Boom, R.M.; Janssen, A.E.M. Heat-Induced Changes in Microstructure of Spray-Dried Plant Protein Isolates and Its Implications on in Vitro Gastric Digestion. *LWT Food Sci. Technol.* **2020**, *118*, 108795. [CrossRef]
7. Avramenko, N.A.; Low, N.H.; Nickerson, M.T. The Effects of Limited Enzymatic Hydrolysis on the Physicochemical and Emulsifying Properties of a Lentil Protein Isolate. *Food Res. Int.* **2013**, *51*, 162–169. [CrossRef]
8. Burgos-Díaz, C.; Mosi-Roa, Y.; Opazo-Navarrete, M.; Bustamante, M.; Garrido-Miranda, K. Comparative Study of Food-Grade Pickering Stabilizers Obtained from Agri-Food Byproducts: Chemical Characterization and Emulsifying Capacity. *Foods* **2022**, *11*, 2514. [CrossRef]
9. McClements, D.J.; Gumus, C.E. Natural Emulsifiers—Biosurfactants, Phospholipids, Biopolymers, and Colloidal Particles: Molecular and Physicochemical Basis of Functional Performance. *Adv. Colloid Interface Sci.* **2016**, *234*, 3–26. [CrossRef]
10. Chen, D.; Campanella, O.H. Limited Enzymatic Hydrolysis Induced Pea Protein Gelation at Low Protein Concentration with Less Heat Requirement. *Food Hydrocoll.* **2022**, *128*, 107547. [CrossRef]
11. Liu, N.; Lin, P.; Zhang, K.; Yao, X.; Li, D.; Yang, L.; Zhao, M. Combined Effects of Limited Enzymatic Hydrolysis and High Hydrostatic Pressure on the Structural and Emulsifying Properties of Rice Proteins. *Innov. Food Sci. Emerg. Technol.* **2022**, *77*, 102975. [CrossRef]

12. Gomes, M.H.G.; Kurozawa, L.E. Improvement of the Functional and Antioxidant Properties of Rice Protein by Enzymatic Hydrolysis for the Microencapsulation of Linseed Oil. *J. Food Eng.* **2020**, *267*, 109761. [CrossRef]
13. Al-Ruwaih, N.; Ahmed, J.; Mulla, M.F.; Arfat, Y.A. High-Pressure Assisted Enzymatic Proteolysis of Kidney Beans Protein Isolates and Characterization of Hydrolysates by Functional, Structural, Rheological and Antioxidant Properties. *LWT Food Sci. Technol.* **2019**, *100*, 231–236. [CrossRef]
14. Eckert, E.; Han, J.; Swallow, K.; Tian, Z.; Jarpa-Parra, M.; Chen, L. Effects of Enzymatic Hydrolysis and Ultrafiltration on Physicochemical and Functional Properties of Faba Bean Protein. *Cereal Chem.* **2019**, *96*, 725–741. [CrossRef]
15. Brückner-Gühmann, M.; Heiden-Hecht, T.; Sözer, N.; Drusch, S. Foaming Characteristics of Oat Protein and Modification by Partial Hydrolysis. *Eur. Food Res. Technol.* **2018**, *244*, 2095–2106. [CrossRef]
16. Chen, L.; Chen, J.; Yu, L.; Wu, K.; Zhao, M. Emulsification Performance and Interfacial Properties of Enzymically Hydrolyzed Peanut Protein Isolate Pretreated by Extrusion Cooking. *Food Hydrocoll.* **2018**, *77*, 607–616. [CrossRef]
17. Chen, L.; Chen, J.; Yu, L.; Wu, K. Improved Emulsifying Capabilities of Hydrolysates of Soy Protein Isolate Pretreated with High Pressure Microfluidization. *LWT Food Sci. Technol.* **2016**, *69*, 1–8. [CrossRef]
18. Wouters, A.G.B.; Rombouts, I.; Fierens, E.; Brijs, K.; Delcour, J.A. Relevance of the Functional Properties of Enzymatic Plant Protein Hydrolysates in Food Systems. *Compr. Rev. Food Sci. Food Saf.* **2016**, *15*, 786–800. [CrossRef]
19. Schlegel, K.; Sontheimer, K.; Hickisch, A.; Wani, A.A.; Eisner, P.; Schweiggert-Weisz, U. Enzymatic Hydrolysis of Lupin Protein Isolates—Changes in the Molecular Weight Distribution, Technofunctional Characteristics, and Sensory Attributes. *Food Sci. Nutr.* **2019**, *7*, 2747–2759. [CrossRef]
20. Muhoza, B.; Qi, B.; Harindintwali, J.D.; Farag Koko, M.Y.; Zhang, S.; Li, Y. Combined Plant Protein Modification and Complex Coacervation as a Sustainable Strategy to Produce Coacervates Encapsulating Bioactives. *Food Hydrocoll.* **2022**, *124*, 107239. [CrossRef]
21. García Arteaga, V.; Apéstegui Guardia, M.; Muranyi, I.; Eisner, P.; Schweiggert-Weisz, U. Effect of Enzymatic Hydrolysis on Molecular Weight Distribution, Techno-Functional Properties and Sensory Perception of Pea Protein Isolates. *Innov. Food Sci. Emerg. Technol.* **2020**, *65*, 102449. [CrossRef]
22. López-Otín, C.; Bond, J.S. Proteases: Multifunctional Enzymes in Life and Disease. *J. Biol. Chem.* **2008**, *283*, 30433–30437. [CrossRef] [PubMed]
23. Li, Q.; Yi, L.; Marek, P.; Iverson, B.L. Commercial Proteases: Present and Future. *FEBS Lett.* **2013**, *587*, 1155–1163. [CrossRef]
24. Padial-Domínguez, M.; Espejo-Carpio, F.J.; Pérez-Gálvez, R.; Guadix, A.; Guadix, E.M. Optimization of the Emulsifying Properties of Food Protein Hydrolysates for the Production of Fish Oil-in-Water Emulsions. *Foods* **2020**, *9*, 636. [CrossRef] [PubMed]
25. Xu, X.; Zhong, J.; Chen, J.; Liu, C.; Luo, L.; Luo, S.; Wu, L.; McClements, D.J. Effectiveness of Partially Hydrolyzed Rice Glutelin as a Food Emulsifier: Comparison to Whey Protein. *Food Chem.* **2016**, *213*, 700–707. [CrossRef] [PubMed]
26. Hickisch, A.; Bindl, K.; Vogel, R.F.; Toelstede, S. Thermal Treatment of Lupin-Based Milk Alternatives–Impact on Lupin Proteins and the Network of Respective Lupin-Based Yogurt Alternatives. *Food Res. Int.* **2016**, *89*, 850–859. [CrossRef]
27. Cabello-Hurtado, F.; Keller, J.; Ley, J.; Sanchez-Lucas, R.; Jorrín-Novo, J.V.; Aïnouche, A. Proteomics for Exploiting Diversity of Lupin Seed Storage Proteins and Their Use as Nutraceuticals for Health and Welfare. *J. Proteom.* **2016**, *143*, 57–68. [CrossRef]
28. Starkute, V.; Bartkiene, E.; Bartkevics, V.; Rusko, J.; Zadeike, D.; Juodeikiene, G. Amino Acids Profile and Antioxidant Activity of Different *Lupinus angustifolius* Seeds after Solid State and Submerged Fermentations. *J. Food Sci. Technol.* **2016**, *53*, 4141–4148. [CrossRef]
29. Thambiraj, S.R.; Phillips, M.; Koyyalamudi, S.R.; Reddy, N. Yellow Lupin (*Lupinus luteus* L.) Polysaccharides: Antioxidant, Immunomodulatory and Prebiotic Activities and Their Structural Characterisation. *Food Chem.* **2018**, *267*, 319–328. [CrossRef]
30. Burgos-Díaz, C.; Piornos, J.A.; Wandersleben, T.; Ogura, T.; Hernández, X.; Rubilar, M. Emulsifying and Foaming Properties of Different Protein Fractions Obtained from a Novel Lupin Variety Alu*Prot*-CGNA® (*Lupinus luteus*). *J. Food Sci.* **2016**, *81*, C1699–C1706. [CrossRef]
31. Burgos-Díaz, C.; Opazo-Navarrete, M.; Soto-Añual, M.; Leal-Calderón, F.; Bustamante, M. Food-Grade Pickering Emulsion as a Novel Astaxanthin Encapsulation System for Making Powder-Based Products: Evaluation of Astaxanthin Stability during Processing, Storage, and Its Bioaccessibility. *Food Res. Int.* **2020**, *134*, 109244. [CrossRef] [PubMed]
32. Cepero-Betancourt, Y.; Opazo-Navarrete, M.; Janssen, A.E.M.; Tabilo-Munizaga, G.; Pérez-Won, M. Effects of High Hydrostatic Pressure (HHP) on Protein Structure and Digestibility of Red Abalone (*Haliotis rufescens*) Muscle. *Innov. Food Sci. Emerg. Technol.* **2020**, *60*, 102282. [CrossRef]
33. Opazo-Navarrete, M.; Tagle Freire, D.; Boom, R.M.; Janssen, A.E.M. The Influence of Starch and Fibre on In Vitro Protein Digestibility of Dry Fractionated Quinoa Seed (*Riobamba Variety*). *Food Biophys.* **2019**, *14*, 49–59. [CrossRef]
34. Nielsen, P.M.; Petersen, D.; Dambmann, C. Improved Method for Determining Food Protein Degree of Hydrolysis. *J. Food Sci.* **2001**, *66*, 642–646. [CrossRef]
35. Morr, C.V.; German, B.; Kinsella, J.E.; Regenstein, J.M.; Buren, J.P.V.; Kilara, A.; Lewis, B.A.; Mangino, M.E. A Collaborative Study to Develop a Standardized Food Protein Solubility Procedure. *J. Food Sci.* **1985**, *50*, 1715–1718. [CrossRef]
36. Burgos-Díaz, C.; Rubilar, M.; Morales, E.; Medina, C.; Acevedo, F.; Marqués, A.M.; Shene, C. Naturally Occurring Protein–Polysaccharide Complexes from Linseed (*Linum usitatissimum*) as Bioemulsifiers. *Eur. J. Lipid Sci. Technol.* **2016**, *118*, 165–174. [CrossRef]

37. Yu, Y.; Fan, F.; Wu, D.; Yu, C.; Wang, Z.; Du, M. Antioxidant and ACE Inhibitory Activity of Enzymatic Hydrolysates from *Ruditapes philippinarum*. *Molecules* **2018**, *23*, 1189. [CrossRef]
38. Shu, G.; Huang, J.; Bao, C.; Meng, J.; Chen, H.; Cao, J. Effect of Different Proteases on the Degree of Hydrolysis and Angiotensin I-Converting Enzyme-Inhibitory Activity in Goat and Cow Milk. *Biomolecules* **2018**, *8*, 101. [CrossRef]
39. Abdel-Hamid, M.; Otte, J.; De Gobba, C.; Osman, A.; Hamad, E. Angiotensin I-Converting Enzyme Inhibitory Activity and Antioxidant Capacity of Bioactive Peptides Derived from Enzymatic Hydrolysis of Buffalo Milk Proteins. *Int. Dairy J.* **2017**, *66*, 91–98. [CrossRef]
40. Goggin, D.E.; Mir, G.; Smith, W.B.; Stuckey, M.; Smith, P.M.C. Proteomic Analysis of Lupin Seed Proteins to Identify Conglutin β as an Allergen, Lup an 1. *J. Agric. Food Chem.* **2008**, *56*, 6370–6377. [CrossRef]
41. Lam, A.C.Y.; Can Karaca, A.; Tyler, R.T.; Nickerson, M.T. Pea Protein Isolates: Structure, Extraction, and Functionality. *Food Rev. Int.* **2016**, *34*, 126–147. [CrossRef]
42. Kramer, R.M.; Shende, V.R.; Motl, N.; Pace, C.N.; Scholtz, J.M. Toward a Molecular Understanding of Protein Solubility: Increased Negative Surface Charge Correlates with Increased Solubility. *Biophys. J.* **2012**, *102*, 1907. [CrossRef] [PubMed]
43. Tsumura, K.; Saito, T.; Tsuge, K.; Ashida, H.; Kugimiya, W.; Inouye, K. Functional Properties of Soy Protein Hydrolysates Obtained by Selective Proteolysis. *LWT Food Sci. Technol.* **2005**, *38*, 255–261. [CrossRef]
44. Liu, Q.; Lu, Y.; Han, J.; Chen, Q.; Kong, B. Structure-Modification by Moderate Oxidation in Hydroxyl Radical-Generating Systems Promote the Emulsifying Properties of Soy Protein Isolate. *Food Struct.* **2015**, *6*, 21–28. [CrossRef]
45. Shen, H.; Stephen Elmore, J.; Zhao, M.; Sun, W. Effect of Oxidation on the Gel Properties of Porcine Myofibrillar Proteins and Their Binding Abilities with Selected Flavour Compounds. *Food Chem.* **2020**, *329*, 127032. [CrossRef]
46. Zhang, C.; Liu, H.; Xia, X.; Sun, F.; Kong, B. Effect of Ultrasound-Assisted Immersion Thawing on Emulsifying and Gelling Properties of Chicken Myofibrillar Protein. *LWT Food Sci. Technol.* **2021**, *142*, 111016. [CrossRef]
47. Pan, N.; Wan, W.; Du, X.; Kong, B.; Liu, Q.; Lv, H.; Xia, X.; Li, F. Mechanisms of Change in Emulsifying Capacity Induced by Protein Denaturation and Aggregation in Quick-Frozen Pork Patties with Different Fat Levels and Freeze–Thaw Cycles. *Foods* **2021**, *11*, 44. [CrossRef]
48. Tamm, F.; Gies, K.; Diekmann, S.; Serfert, Y.; Strunskus, T.; Brodkorb, A.; Drusch, S. Whey Protein Hydrolysates Reduce Autoxidation in Microencapsulated Long Chain Polyunsaturated Fatty Acids. *Eur. J. Lipid Sci. Technol.* **2015**, *117*, 1960–1970. [CrossRef]
49. Chen, C.; Chi, Y.-J.; Zhao, M.-Y.; Xu, Y. Influence of Degree of Hydrolysis on Functional Properties, Antioxidant and ACE Inhibitory Activities of Egg White Protein Hydrolysate. *Food Sci. Biotechnol.* **2012**, *21*, 27–34. [CrossRef]

colloids and interfaces

MDPI

Review

Review on the Antioxidant Activity of Phenolics in o/w Emulsions along with the Impact of a Few Important Factors on Their Interfacial Behaviour

Sotirios Kiokias [1,*] and Vassiliki Oreopoulou [2]

1 European Research Executive Agency (REA), Place Charles Rogier 16, 1210 Bruxelles, Belgium
2 Laboratory of Food Chemistry and Technology, School of Chemical Engineering, National Technical University of Athens, Iron Politechniou, 9, 15780 Athens, Greece
* Correspondence: sotirios.kiokias@ec.europa.eu

Abstract: This review paper focuses on the antioxidant properties of phenolic compounds in oil in water (o/w) emulsion systems. The authors first provide an overview of the most recent studies on the activity of common, naturally occurring phenolic compounds against the oxidative deterioration of o/w emulsions. A screening of the latest literature was subsequently performed with the aim to elucidate how specific parameters (polarity, pH, emulsifiers, and synergistic action) affect the phenolic interfacial distribution, which in turn determines their antioxidant potential in food emulsion systems. An understanding of the interfacial activity of phenolic antioxidants could be of interest to food scientists working on the development of novel food products enriched with functional ingredients. It would also provide further insight to health scientists exploring the potentially beneficial properties of phenolic antioxidants against the oxidative damage of amphiphilic biological membranes (which link to serious pathologic conditions).

Keywords: antioxidants; phenolics; emulsions; interfacial distribution

Citation: Kiokias, S.; Oreopoulou, V. Review on the Antioxidant Activity of Phenolics in o/w Emulsions along with the Impact of a Few Important Factors on Their Interfacial Behaviour. *Colloids Interfaces* **2022**, *6*, 79. https://doi.org/10.3390/colloids6040079

Academic Editors: César Burgos-Díaz, Mauricio Opazo-Navarrete and Eduardo Morales

Received: 13 October 2022
Accepted: 23 November 2022
Published: 12 December 2022

1. Introduction

Lipid oxidation is a serious problem for scientists since it adversely affects the product quality in the food, cosmetic, and pharmaceutical sectors [1]. Many common food products exist as oil-in-water (o/w) emulsions including beverages, dressings, sauces, soups, and desserts [2]. One of the main causes of the quality deterioration of these products is the oxidation of lipids [3]. The degradation of unsaturated lipids can lead to off-flavors, decreases the nutritional profile of foods, and may eventually generate toxic products [4]. In emulsified foods, lipid oxidation can occur rapidly due to their large surface area, with mechanisms that are more complex and not fully understood compared to bulk oils [5,6]. As a consequence, the food industry faces a serious problem due to the low oxidative stability of emulsified systems, which adversely impacts consumer safety and the economic viability of the products [7]. A better understanding of the endogenous and exogenous factors that monitor the microstructural and oxidative stability of food emulsions would help to maintain their desirable functional and sensory properties during the formulation, processing, and storage of relevant products [8]. In addition, emulsion systems generally mimic the amphiphilic nature of important biological membranes (such as lipoproteins) that are prone to oxidative degradation when attacked by singlet oxygen and free radicals [9]. This biochemical process eventually links to the development of serious human health conditions, such as aging, carcinogenesis, and cardiovascular diseases [10].

Over the last few years, there is a steady food market trend for the use of natural antioxidants, as a common strategy to slow down lipid oxidation in emulsified foods, increase their shelf life, and minimize bad odors [11,12]. A body of recent research [13,14] focuses on nano-emulsions that are increasingly used in various products in order to incorporate easily degradable bioactive compounds, protect them against oxidation, and enhance their bioavailability. In manufactured products (quite often heterogeneous systems containing lipids as emulsions or bulk phase) the efficiency of an antioxidant is determined not only by its chemical reactivity, but also by its physical properties and its interaction with other compounds present in the products [15].

With the term phenolics, we refer to a wide class of natural compounds (e.g., tocopherols, flavonoids, phenolic acids) with varying structures and antioxidant properties [16]. The application of phenolic compounds in various commercial products has attracted an increasing level of interest over the last few years from researchers in the food, pharmaceutical, and nutraceutical industries [17]. It is true, though, that certain structural characteristics (including the hydrophobic character of many phenolics) may affect their integration into real products as well as their general biological activities and their bio-absorption in human organisms [18]. Their potential to act as functional ingredients upon their addition in bulk or emulsified oils has been an area of emerging scientific interest in the last decade [19–21]. A few researchers in this field have focused on phenolic acids, their activity, health effects, and extraction from natural plant sources [21–23]. Phenolic acids are classified as hydroxybenzoic and hydroxycinnamic acids, and, depending on their structure, may present solubility in water or lipids, thus enabling their use in many products [7,14,24].

In this review paper, the authors focus on the antioxidant properties of phenolic antioxidants against the oxidative deterioration of food-relevant o/w model emulsions. The novelty of this paper mainly concerns the literature review of the most important interfacial factors that determine the antioxidant potential of phenolic compounds in food emulsion systems. The phenolic distribution is determined by solubility and partitioning behaviour in the different regions of the emulsion (oil core, aqueous phase, interface) and together with the mode of incorporation can significantly determine the antioxidant efficiency [25,26]. The interfacial synergistic activity of various phenolic antioxidants is also discussed as an additional important factor. Further to the technological importance of phenolics and the protection against oxidative deterioration as a result of their incorporation in real lipid-based products, there is also an additional aspect. The research findings from phenolic activity in interfacial systems could be of interest to clinical nutrition researchers who explore the potential effects of natural antioxidants against the oxidative damage of amphiphilic biological membranes.

In addition, the analysis of such research outcomes can feed the development of novel food emulsions and thereby trigger innovation and new market applications in the associated industrial sectors.

2. Overview of Studies Reporting on the Effect of Phenolics against the Oxidation of o/w Emulsions

Although most of the studies on phenolic antioxidants so far have focused on bulk oils, a body of recent research has examined the activity of various phenolic compounds in food-related o/w emulsions. In emulsified systems, the initial step of lipid oxidation takes place at the interface between the oil and water phases. Over the last few years, an increasing number of researchers have reported well-documented antioxidant activities of naturally occurring phenolics in o/w model systems [19,27].

Tocopherols are a class of natural compounds with a high nutritional value in the human diet because of their vitamin E activity. In addition, various authors have reported their well-established antioxidants activity in vitro (oil model systems) and in vivo (human clinical trials) [28,29]. Wang et al. [30], following oxidation experiments of flaxseed o/w emulsions, have noted a higher antioxidant efficiency of δ-tocopherol compared to α-tocopherol. The authors linked the superior activity of δ-tocopherol to its enhanced distribution in the interfacial area of the emulsion system. Another similar study—based on the autoxidation of shrimp o/w emulsions under storage at room temperature did not report any protective effect of α-tocopherol against oxidative destabilisation [31]. On the contrary, α-tocopherol when combined with chitooligosaccharides (at both 0.2 and

0.4 g/L concentrations) significantly delayed the oxidation process, in terms of both primary (conjugated dienes) and secondary (TBARs) oxidation products indicators [31]. Barouh et al. [15] further explored the activity of tocopherols in bulk oils and emulsion systems as a function of various physicochemical parameters (including their interaction with specific emulsifiers such as Tween 65, Tween 80, whey proteins, etc.).

Flavonoids comprise another class of widespread phenolic compounds with well-established antioxidant properties strongly related to their structure [32,33]. Common plant flavonoids such as quercetin, rutin, and hesperitin prolong the shelf life of o/w emulsions by retarding the oxidative deterioration and improving physical stability. However, only quercetin retained its antioxidant activity in the presence of ferric ions due to its structural configuration (3′4′ dihyrdoxy substitution of B ring and the presence o 3-OH) [34]. In fact, quercetin seems to be one of the most potent contributors to the antioxidant activity of plant extracts in emulsions [35]. However, quercetin has been reported to exert a pro-oxidant character in pure lipids and antioxidant activity in emulsions due to its hydroxyl group in position-3 [36]. Catechin acted as a better antioxidant than quercetin in the copper-induced oxidation of diluted, Tween-based, linoleic acid emulsion (0.02 M), while morin presented the best protective action [37]. Different metal ions may affect the activity of flavonoids in emulsions, where the pro-oxidant or antioxidant character may dependon the flavonoid structure and the metal type [38] Flavonoids have also been examined for the preparation of physically stable pickering emulsions, while also offering oxidative stability [39,40].

Plant extracts are rich in flavonoids and phenolic acids and exert a strong antioxidant activity in o/w emulsions that may be attributed to synergism among its components and also to partitioning in the aqueous phase (e.g., polar phenolic acids or flavonoid glycosides), oil-water interface (e.g., medium polarity phenolic acids and flavonoids), or the oil phase (e.g., flavonoid aglycons) of the emulsion [41,42]. Additionally, the presence of endogenous antioxidants such as α-tocopherol may affect the flavonoid activity in lipid systems; e.g., it showed a strong synergistic effect with quercetin in emulsions, and a clear antagonistic effect in bulk oil [43]. More information on the phenolic synergistic effects is given in Section 3.4.

The current paper has focused on various phenolic acids available in many natural sources (e.g., in olive oil, herbs, fruits) that have been widely explored in food systems [44]. Caffeic acid (CA) is perhaps the most well-known phenolic acid widely spread in the plant kingdom [45]. CA exerted a clear antioxidant effect when added in o/w emulsion systems stabilised by Tween and prepared with various vegetable oils [46,47]. Ferulic acid (FA) is another phenolic acid -commonly present in many plant seeds [48] that was reported to act as a strong antioxidant following its addition in a range of oil-based emulsions [49–51].

For gallic acid (GA), which is found in high amounts in tea and berries [52] there is some contradictory evidence in the literature about its protective action against oxidation of food emulsions. GA and its alkyl esters were found to exert a clear antioxidant effect against the oxidation of rapeseed o/w nano-emulsions stabilised by SDS [53] or even double emulsions by the use of encapsulation [54]. Additionally, Zhu et al. [55] concluded a strong inhibitory effect of GA and its alkyl esters against the formation of both primary (Peroxide Values) and secondary oxidation products (hexanal content). Propyl gallate exerted a superior activity compared to gallic acid and the other gallate esters. Other studies, however, did not report any protective effect of GA in o/w emulsions [56,57].

A large variety of plants and vegetables are abundant in p-coumaric acid (p-CA) [58]. Park et al. [59] reported that roasted rice hull extracts, with a high concentration of p-CA enhanced the oxidative stability of bulk oil and o/w emulsions at 60 °C. Rosmarinic acid (RA) is the main phenolic component in various edible and aromatic herbs of the Lamiaceae family (including *Rosmarinus officinalis*, *Origanum* spp., etc.) [60]. Choulitoudi et al. [19] observed that ethanol and ethyl acetate extracts of *Satureja thymbra*—articularly rich in RA—reduced by 75–80% the oxidation rate of sunflower o/w emulsions at chilling temperature (5 °C). Other researchers reported the antioxidant activity of RA in corn oil-

and soyabean oil-based emulsions [61,62] as well as in a model emulsion system based on linoleic acid [46].

Bakota et al. [63] incorporated either pure RA or RA-rich extract (from *Salvia officinalis* leaves), at a concentration of ~30 mg/g, into o/w emulsions and observed that both treatments were effective in suppressing lipid oxidation.

Vanillic acid (VA) is a phenolic acid found in several fruits, olives, and cereal grains. Keller et al. [64] observed a strong antioxidant character of VA during the autoxidation of Tween 40-based o/w systems, at pH 3.5. In addition, a few authors [30,65] reported that the addition of tannic acid (TA) enhanced the resistance of plant-based emulsions against both droplet aggregation and lipid oxidation as a result of its strong ferrous ion-binding properties.

Ascorbic acid is another phenolic compound widely present in nature and with a high nutritional value (also known as vitamin C). A number of scientists highlighted the important role of ascorbic acid in antioxidant synergies examined in both food and clinically relevant studies [66,67].

As will be discussed under Section 3.4, a shift from the prooxidant to an antioxidant potential of ascorbic acid in interfacial systems is vastly dependent on the presence of other phenolic compounds in the system (and in particular of tocopherols).

3. Interfacial Factors That Affect the Antioxidant Activity of Phenolic Compounds

Various authors examined a number of factors that can potentially influence the physical and oxidative stability of emulsions including type and concentration of oil phase and of emulsifiers, pH and ionic composition, emulsion droplet size, and interfacial properties [68,69]. Raikos et al. [70] noted that the physical location of antioxidants in o/w emulsions can have a significant influence on their free radical scavenging activity and ability to inhibit lipid oxidation. The partitioning of antioxidant molecules in emulsions has been reported to affect their activity [71]. According to Shahidi and Zhong [6], the efficacy of antioxidants in emulsified systems is determined by their polarity but also by their molecular size, mechanism of action, and the presence and type of emulsifiers. Kiokias and Oreopoulou [72] explored the antioxidant effect of carotenoids with varying polarities in sunflower oil-in-water emulsions. The authors attributed the superior antioxidant effect of polar carotenoids to their enhanced distribution in the interface of the emulsion system and claimed that less potent β-carotene is likely to be homogeneously dispersed in the oil droplets. On the contrary, the polar xanthophylls (lutein, and annatto carotenoids) may be located near the oil-water interface where their hydrophilic groups (-OH, COOH, etc.) are better orientated, and thereby more efficiently trap the AAPH-derived radicals attacking from the water phase of the emulsion.

The authors note that a similar mode of action is also expected to occur during the action of phenolic antioxidants in an emulsion system. Figure 1 provides a general picture of the expected distribution of an amphiphilic phenolic antioxidant in the emulsion phases. The alkyl chain (hydrophobic part) is mainly oriented towards the lipid core of the oil droplet, whereas the hydrophilic part (with OH/COOH groups) is preferentially located towards the interfacial region of the emulsion phase. The most important interfacial factors that determine the antioxidant potential of phenolic compounds in food emulsion systems are examined in the following sections.

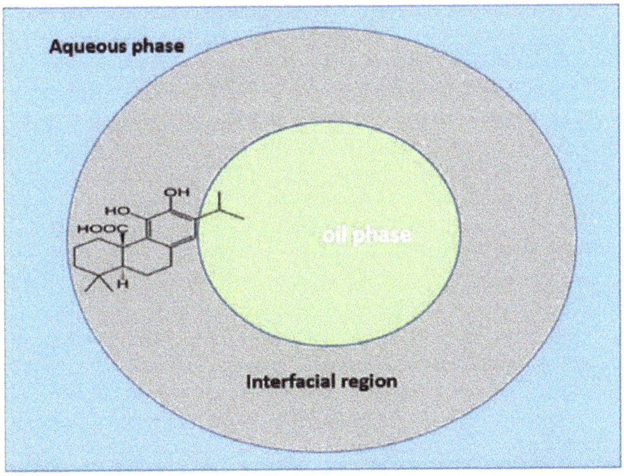

Figure 1. Expected partitioning of phenolic antioxidants between the different regions of an o/w emulsion system (alkyl chain towards the lipid core, hydrophilic part towards the interface).

3.1. *Effect of pH of the Aqueous Phase on the Interfacial Activity of Phenolic Compounds*

There is a general agreement about the important role of pH in monitoring the microstructural stability of food-relevant emulsion products [73]. However, how pH affects the oxidative destabilisation of emulsions is pending some further clarification due to contradictory evidence in the relevant literature [4].

Interestingly, Branco et al. [74] observed that a pH change from 3.0 to 7.0 increased the peroxide value (PV) and reduced the thiobaribituric acid reactive substance (TBARS) in o/w emulsions. The authors noted, though, that the presence of ascorbic acid (1 mmol/L) not only had no significant inhibitory effect on PV and TBARS ($p < 0.05$), but even triggered a prooxidant character. Costa et al. [75] noted that a modification of the pH in the aqueous phase could largely affect the interfacial distribution, and consequently the antioxidant activity, of phenolic acids.

Zhou and Elias [76] reported that the pH is capable of influencing the antioxidant activity of phenolics, along with transition metal properties, and is overall a factor to take into careful consideration in the production of food-relevant emulsion systems. The results of Sorensen et al. [77] indicated that the pH, especially in the presence of iron, had a greater impact on lipid oxidation than the size of the lipid droplets. CA promoted oxidation at pH 3 in the presence of iron, a fact attributed to its ability to reduce Fe^{3+} to Fe^{2+}, thereby propagating lipid oxidation and catalysing peroxide decomposition to the formation of secondary products. At pH 6, CA interacted and formed complexes with iron, as also observed for other phenolics, thus depressing PV formation, either with or without iron ions in the emulsion.

A similar study [78] concluded that in the presence of 25 μM Fe^{3+}, epigallocatechin gallate (EGCG), at a concentration of 400 μM, accelerated the oxidation of a Tween 80-stabilised flaxseed o/w emulsion at pH 3, but exhibited an antioxidant character at pH 7. Other researchers [79] observed that EGCG and epicatechin gallate (ECG)—even at lower concentrations (100 μM)—protected Tween-20 stabilised sunflower o/w emulsions from their oxidative degradation at pH 5.5, in the absence of Fe^{3+}. According to Kim and Choe [80], the oxidative degradation of soybean o/w emulsions proceeded faster at pH 4 in the absence of iron. Interestingly, the emulsions were more stable at the same pH compared to higher or lower acidity values when iron was present in the system. Upon addition of peppermint extract at pH 4, iron reduced and significantly delayed the degradation of total polyphenols (including RA and CA).

Tian et al. [81] reported that tea polyphenols, when added at a relatively low level, are located at the oil droplet surfaces in the whey protein-stabilised emulsions, because of their tendency to bind to the adsorbed protein molecules. Moreover, pH affects the potent synergistic actions of the phenolic acids with endogenous antioxidants as tocopherols. Kittipongpittaya et al. [82] observed that RA more effectively regenerated a-tocopherol at pH 7, due to its higher electron donation ability to the a-tocopheroxyl radicals at the oil-water interface, compared to pH 3. Losada-Barreiro et al. [83] investigated deeper the effects of pH on the partition constants and the distribution of GA and CA between the aqueous and interfacial regions of 10% corn oil/acidic water emulsions stabilised by Tween 20. The authors reported that a decrease in pH from 4 to 3 leads to an increase of GA distribution in the interface (from 20% to 60%), and of CA (from 50% to 75%) and concluded that acidity can exert a great impact on the partition of antioxidants between the emulsion phases. Similarly, a decrease in pH from 3.6 to 3 caused an increase in CA fraction in the interface from 90 to 95% [84]. These results may be attributed to the higher solubility of the phenolic ions (present at higher pH values) in the aqueous phase. A recent study [45] examined the antioxidant activity of GA and its alkyl esters in SDS-stabilised rapeseed o/w emulsions. GA, at pH 7, exerted a lower antioxidant effect than its ester derivatives. This was explained by the fact that GA at this pH is negatively charged and thereby repelled by the anionic SDS-coated interface, contrary to the electrostatically neutral ester derivatives. Furthermore, Zoric et al. [85] reported a lower stability of RA at pH 7.5 than at pH 2.5.

Overall, the pH of the aqueous phase plays an important role in determining the antioxidant effect of phenolic acids against the oxidative deterioration of o/w emulsion systems. In the absence of metals, a lower pH links to higher phenolic distribution in the interface leading to stronger antioxidant efficiencies. The situation, though, is more complex in the presence of transition metals in the emulsion system that seems to highly affect the role of pH on phenolic antioxidant activity. For instance, higher pH values overall induce metal-catalyzed oxidation by increasing the number of dissociated species. Another factor that can be crucial here is the formation of metal complexes with phenolic antioxidants and the change in oxidation potentials. Therefore, further investigation is needed to more precisely elucidate the mechanisms and actual effect of pH on the phenolic antioxidant activity in the presence of metals.

3.2. *Effect of the Emulsifier and of the Interfacial Concentration of Phenolics on Their Antioxidant Activity*

A few researchers have examined how certain features of the emulsifier may affect the lipid oxidation in food emulsions [86,87]. More specifically, the size and conformation of the emulsifier define the thickness of the emulsion droplet interface through the formation of biopolymer layers. In principle, the thicker the interfacial emulsifier layer, the more difficult for the free radicals to penetrate, thereby the better the antioxidant protection. This theory also explains why according to certain studies, low molecular weight emulsifiers (e.g., Tween, SDS) result in emulsions that are easier and faster oxidized compared to systems prepared under exactly the same experimental conditions but with larger molecular weight emulsifiers such as Brij 700 or certain proteins [88,89].

Among the possible environmental factors influencing the efficiency of phenolic antioxidants, the nature and concentration of surfactants was noted as the most important by McClements and Decker [90]. A few studies examined how emulsifiers may affect the partitioning of antioxidants in emulsions and thereby moderate their interfacial activities. Kiralan et al. [91] have suggested that surfactant micelles could increase the antioxidant activity of phenolics by changing their physical location. Other researchers in this field concluded that the location and concentration of phenolic antioxidants in the interface of an o/w model system are highly dependent on their molecular interaction with the present emulsifiers [26,62].

Da Silveira et al. [53] investigated the effect of the presence of surfactant micelles on the antioxidant efficiency of a homologous series of gallate phenolipids (with 0–16 carbon atoms alkyl chains: G0, G3, G8, G12, or G16) in o/w nano-emulsions. According to the results, the surfactant micelles hinder the antioxidant action of hydrophobic phenolipids, but not that of hydrophilic ones. Losada-Barreiro et al. [83] reported that a tenfold increase of the surfactant (Tween 20) volume enhanced the interfacial distribution of both GA (from 20% to 60%) and CA (from to 50% and 90%), thereby concluding that the emulsifier concentration largely determines the partition of phenolic acids in emulsified systems.

Keller et al. [64] examined the partitioning behaviour of VA following its addition in 30% o/w model systems stabilised by Tween 40 under low pH conditions (pH of aqueous phase: 3.5). The authors reported that when oil, water, and emulsifier were just mixed, the major proportion of VA (~90) retained in the aqueous phase was mostly associated with Tween micelles. In the homogenised emulsified system, though, VA migrated to the o/w interface of the oil droplet. Another study [49] focused on FA and iFA (isoferulic acid) trying to determine the parameters that affect their partitioning behaviour in the emulsion phases. According to the research findings, an increase in the concentration of the emulsifiers resulted in enhanced solubilisation and migration of both FA and iFA at the interface, which subsequently resulted in clear antioxidant activities of both tested phenolic acids. Furthermore, Mitrus et al. [14] concluded that a higher surfactant volume fraction increased the percentage of antioxidants distributed in the interface. Other researchers attempted to elucidate the protective action of a few proteins commonly used in food applications (e.g., whey proteins, soy proteins, etc.) against the oxidative destabilisation of emulsion systems. Such an antioxidant effect is highly dependent not only on the type, structure, and concentration of the protein but also on their synergistic action with other antioxidants (e.g., enhanced antioxidant activity of proteins mixtures with tocopherols and other phenolics) [4,92]. An overall conclusion of the relevant studies is that proteins may, indeed, affect the interfacial partitioning of phenolic antioxidants by modifying the interfacial charge [93]. In doing so, the proteins can alter the location of transition metal ions and their subsequent interaction with polyphenols thereby determining their antioxidant activity in the interfacial systems [94]. A picture of the microstructure for a protein stabilised emulsion (produced in NTUA/Laboratory of Food Chemistry and Technology) is given in Figure 2 (Confocal Laser Scanning Microscopy-CSLM image of a 10% cottonseed o/w emulsion stabilized by 1% sodium caseinate).

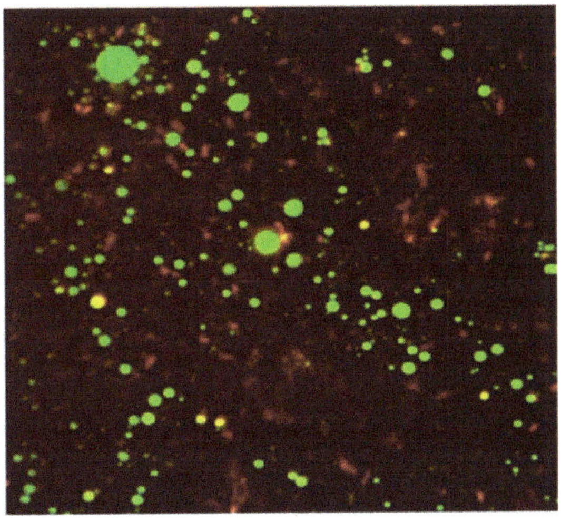

Figure 2. CSLM image of a 10% cottonseed o/w emulsion stabilized by 1% sodium caseinate. (Spherical oil droplets stabilized in the protein network, image size:131 × 131 μm) (produced in NTUA/Laboratory of Food Chemistry and Technology).

The effect of emulsifiers on phenolic antioxidant activities could be also seen from the "angle" of their reported impact on the oil droplet seize. According to Vilasaua et al. [95], the particle size decreases with increasing ionic/nonionic surfactant ratio, up to a certain ratio above which emulsions aggregate. Kiokias et al. [69] reported that in contrast to emulsions stabilized by low molecular weight emulsifiers (Tween 20, SDS), smaller oil droplets, obtained in protein-stabilized emulsions through the increase of protein concentration, were associated with enhanced protection from oxidative changes. Overall, we can claim that the stability of phenolic compounds that are located at the oil-water interface or inside the oil droplets of an o/w emulsion may be affected by droplet size [96]. A decrease in droplet size results in an increase of the oil and interfacial area, where the phenolic compounds can contact free radicals found in the water phase and thus oxidize faster. Additionally, the lipophilic phenolic compounds diffuse faster from the interior of an oil droplet to the exterior as the droplet size decreases [97].

Over the last few years, a number of studies have explored the distribution of phenolics in o/w emulsions by developing pseudophase kinetic models [74,84,98]. The advantage of this approach is that it works on intact emulsions, unlike other methods that require phase separation and cannot provide estimates of the amounts of antioxidants in the interfacial area. Via the use of partition constant values, such models determine the percentages of phenolic acids in the oil, aqueous and interfacial regions of the emulsions. The results of a few studies offer "libraries" of partition constant values for the distribution of various phenolic antioxidants (e.g., CA, 4-hydroxycinnamic acid, di-hydrocaffeic acid, and hydroxytyrosol derivatives) [53,75,99]. Lossada Barreiro et al. [98] investigated the phenolic distributions in intact emulsions by using the pseudo-phase kinetic model and estimated changes in the antioxidant concentration, by using a chemical probe and employing the Schaal oven test. They reported that at any given volume fraction of emulsifier, the concentrations of antioxidants in the interfacial region of stripped corn oil emulsions and their efficiency follow the order: propyl gallates > gallic acid > octylgallates > lauryl gallates. The authors highlighted that these results provide clear evidence that the antioxidant efficiency correlates with its concentration in the interfacial region. Furthermore, Meireles et al. [100] have employed the kinetic method to determine the distribution of a homologous series of chlorogenic acid in olive o/w emulsions. The authors concluded that the activity of chlorogenates in emulsions is greatly dependent on their interfacial concentrations.

The research evidence in o/w emulsions presented in this section (i) notes a clear effect of the type of emulsifier in the interfacial distribution of phenolics, which may vary depending on the mode of their binding interaction; (ii) highlights that the findings of recent kinetic studies (mainly by use of pseudo-phase models) tend to agree how important is the concentration of phenolic compounds in the interfacial region of o/w model emulsion systems in order to interpret their overall antioxidant potential.

3.3. Effect of Polarity on the Antioxidant Activity of Phenolic Compounds

A body of recent research evidence noted that the efficacy of antioxidants in emulsions is highly affected by their polarity [6]. A well-known theory in the field concerns the so-called "polar paradox" which illustrates the paradoxical behavior of antioxidants in different media and rationalizes the fact that polar antioxidants are more effective in less polar media, such as bulk oils, while nonpolar antioxidants are more effective in relatively more polar media, such as o/w emulsions or liposomes [8].

In line with the general concept of this theory, the more polar antioxidants (containing hydrophylic groups) are better distributed in the water phase or the oil-water interface where they may be easily oxidised (in particular under thermally accelerated oxidation conditions). On the other side, the less polar antioxidants (and normally hydrophobic in nature) are preferably located in the lipid core of the emulsion droplets, which is thereby better protected from oxidation [101]. Kanakidi et al. [87] arrived at similar results by highlighting that the more polar RA and carnosic acid, compared to the other phenolic diterpenes, are distributed in the aqueous phase and the oil-water interface where they are faster and promptly oxidized by trace metals or free radicals possibly present there. In the same mode of action, the antioxidant potential of phenolic compounds was reported to decrease with increasing hydrophobicity (e.g., gallic esters less protective than GA) during the oxidative deterioration of bulk oil systems [102]. Noon et al. [103] investigated the efficacy of four natural phenolic antioxidants with varying structures and polarities (quercetin, curcumin, rutin hydrate, and ascorbic acid) against the oxidative deterioration of different o/w emulsion environments. The authors concluded that the most non-polar compounds (curcumin and quercetin) yielded higher peroxide values (PV) reductions (of 65% and 74%, respectively) in 5% oil phase volume emulsions compared to 40% oil phase volume emulsions (just 28% and 43% PV reductions). Another study [74] further demonstrated the applicability of the polar paradox theory in the oxidative stability for n-3 high unsaturated fatty acid-rich water emulsions. While the hydrophilic ascorbic acid acted as a pro-oxidant, natural plant extracts rich in non-polar polyphenols and a-tocopherol presented a strong antioxidant character. Other research studies, though, came out with evidence that is not fully in line with the polar paradox concept. Poyato et al. [5] reported that a *Melissa officinalis* water extract rich in RA was more effective in o/w emulsions and inhibited lipid oxidation more effectively than the lipophilic antioxidant (BHA).

Di Mattia et al. [104] also came out with certain findings that are not in good agreement with the polar paradox theory. The authors reported that GA, although a phenolic compound with a high polarity, exerted a strong antioxidant effect upon its addition to olive o/w emulsions. GA was estimated to be present in the aqueous phase at a much higher (almost double) concentration than the more hydrophobic quercetin under the same experimental interfacial conditions.

Similarly, Bravo-Diaz et al. [105] used gallate and caffeate esters of variable alkyl chain length and hydrophobicity in o/w, Tween 20-based emulsions. The authors reported a decreasing activity with increasing hydrophobicity, which may be attributed to a distributional shifting of the phenolic antioxidants from the interfacial to oil region of the emulsion, based on their increasing oil solubility.

Moreover, Mitrus et al. [14] concluded that GA and its esters tend to accumulate in the interfacial region, and the percentage distributed in the interface increased with hydrophobicity up to a certain length of alkyl chain (called a cut-off effect), due to lack of water solubility of esters with a longer alkyl chain. Costa et al. [7], however, while evaluating the antioxidant efficiency of n-alkyl esters of phenolic compounds in olive o/w emulsions, concluded that the antioxidant distribution in the interface does not correlate directly with the length of the alkyl chain, thereby with the hydrophobicity.

Overall, we should note that various studies in the field do not fully align with polar paradox concepts concerning the antioxidant activity of phenolic compounds in emulsion systems. As a general conclusion, though, the polarity can be a critical factor in moderating the interfacial antioxidant activity as evidenced by recent studies that compare different types of phenolic acids (and other phenolic antioxidants) with varying structure and hydrophobicity in o/w emulsions. For this reason, we also included the next section (Section 3.4) which focuses on the interaction effects of phenolic antioxidants with varying polarities in o/w emulsion systems.

3.4. Synergistic Interfacial Activity of Phenolic Compounds against the Lipid Oxidation of o/w Emulsions

In Section 3.3, polarity was found to determine the phenolic antioxidant activity in various lipid-based systems. As a next step, the potential interaction effect of mixtures of phenolic antioxidants with different degrees of hydrophobicity against the lipid oxidation of o/w emulsions is further discussed. A number of research studies in this scientific field have widely reported over the last decade that more protection against oxidative deterioration of food products is offered through the synergistic action of antioxidants with varying structures, physical properties, and modes of action [106,107]. Filip et al. [108]

highlighted an enhanced antioxidant activity through the interaction of tocopherols and ascorbic acid in various oil model systems. They claimed that the synergistic effect occurs because of the reduction of tocopherol radicals by ascorbic acid. Zhou and Elias [76] observed a clear protective effect of CA, following its addition in o/w emulsions (stabilised by Tween 20 or Citrem emulsifiers), with tocopherols endogenously present in the oil phase. However, GA acted as a prooxidant accelerating lipid oxidation at the absence of any tocopherols in the emulsion. Wu et al. [108] reported a clear antioxidant character of GA and its alkyl esters when combined with α-tocopherol in o/w emulsions systems. The authors attributed this antioxidant synergy to the initial quick oxidation of α-tocopherol into α-tocopherol quinine, causing its transfer to the interfacial membrane and exposure to oxidation. The tocopherol radicals are reduced back by gallate esters regenerating α-tocopherol in the interfacial layer and enabling antioxidant activity in the emulsion system. Panya et al. [62] noted that the high polarity of RA, and consequently its distribution in the aqueous phase, was responsible for the observed low efficiency in emulsions. The authors reported, however, an enhanced antioxidant effect when RA was combined with α-tocopherol. This could be explained via an increased distribution of RA in the water phase allowing its interaction with the more hydrophobic α-tocopherol on the oil droplet's surface.

Noon et al. [103] found that combinations of ascorbic acid with quercetin or curcumin resulted in antioxidant synergism against the oxidation of o/w emulsions. This research builds on the understanding of the fundamental behaviour of natural antioxidants, such as plant extracts, within different emulsion formulations.

Similarly, Farooq et al. [11] reviewed how polarity and interactions of phenolic antioxidants may determine their antioxidant activity at the oil-water interface of food emulsions. The authors highlighted that synergistic interactions of polar with non-polar phenolics at the oil-water interface may be considered a promising strategy for improving the oxidative stability of emulsions. Further to synergistic activities, we cannot exclude the possibility of some "antagonistic effects between phenolic antioxidants. Liu et al. [109] also referred to the regeneration of α-tocopherol radicals, but this time to clarify antagonistic, rather than synergistic effects, in o/w systems. Following the addition of α-tocopherol at varying concentrations, the authors observed an increased partition of α-tocopherol in an aqueous phase when mixed with γ-oryzanol. They also proposed an inhibitory effect of γ-oryzanol against the regeneration of α-tocopherol. Such antagonism reduced the antioxidant potential of tocopherols in the emulsion system.

The Table 1 below presents an overview of several studies, the results of which have demonstrated a clear protective effect of phenolic compounds (either alone or in combinations) against the oxidative deterioration of o/w emulsions.

Table 1. Overview of literature reporting a clear antioxidant activity of certain phenolic compounds (alone or in combinations) against oxidative deterioration of various model oil-in-water emulsion systems.

Structure	Emulsion Systems/Reported Antioxidant Activities	Literature
α,δ Tocopherols 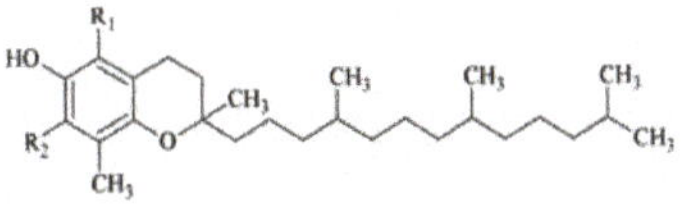	Flaxseed o/w emulsions/ Clear effect of tocopherols (δToc > α-Toc)	[30]
Flavonoids (general structure) 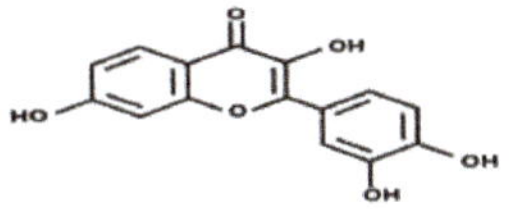	Clear effect of flavonoids in various o/w emulsions (with quercetin being the stronger antitoxidant)/	[34,35,37]
Caffeic acid (CA) 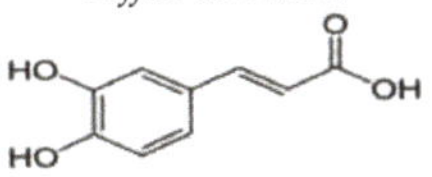	Tween-based linoleic acid o/w emulsions/clear CA effect	[46]
	Citrem- and Tween-based o/w emulsions/clear CA effect in the presence of endogenous tocopherols	[77]
Ferulic acid (FA) 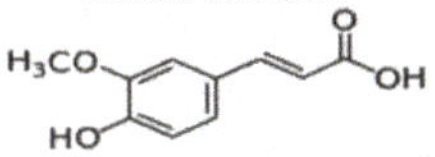	Corn oil-based o/w emulsions /clear FA effect	[49]
	Tween-linoleic acid-based emulsions/clear FA effect	[50]
Gallic acid (GA) 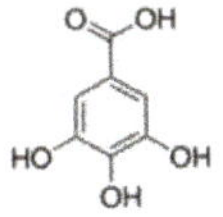	SDS stabilised rapeseed o/w nano-emulsions/clear GA effect	[53]
	Clear effect of GA or its alkyl esters added in combination with α-toc in o/w emulsions	[110]
Rosmarinic acid (RA) 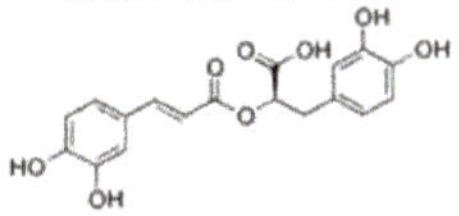	Tween-based o/w emulsions prepared with linoleic acid or soybean emulsions/clear RA effect	[62]
	Clear effect of RA-rich plant extracts in sunflower o/w emulsions	[19]
Vanillic acid (VA) 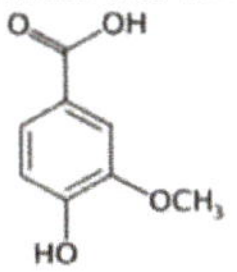	Tween 40-based o/w systems, at pH 3.5/clear effect of VA in the emulsified system	[64]
Ascorbic acid 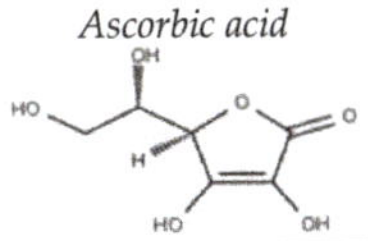	Mixtures of ascorbic acid with α-tocopherol in o/w emulsions/strong synergistic antioxidant effect	[108]
	Mixtures of ascorbic acid with quercetin in o/w emulsions/observed synergistic activity	[102]

4. Conclusions

The analysis of literature evidence undertaken in this work has concluded that the well-established antioxidant activity of several phenolic compounds in various o/w emulsion systems is highly dependent on their interfacial distribution. The partition of phenolics within the emulsion phases is a factor that could determine their antioxidant efficiencies in interfacial systems and is influenced by various parameters including pH, emulsifier, and their structure (hydrophilic/hydrophobic character). The highest activity is shown by the phenolic compounds located at the emulsion interface and correlates with their interfacial concentration. In addition, synergistic actions between phenolics have been reported to enhance interfacial antioxidant activity. By monitoring the interfacial distributions of phenolics in o/w emulsions, researchers could further promote their use as functional ingredients in novel food emulsion relevant products (such as various dressings, fresh cheese types, protein-stabilized drinks, etc.). Further work in this field may involve the design of new kinetic studies with the aim to validate and complete the current databases on the distribution constants of various phenolic compounds.

Author Contributions: Conceptualization, S.K.; resources, S.K. and V.O.; writing—original draft preparation, S.K.; review and editing, supervision, V.O. The authors have contributed substantially to the work reported. All authors have read and agreed to the published version of the manuscript.

Funding: This research received no external funding.

Institutional Review Board Statement: Not applicable.

Informed Consent Statement: Not applicable.

Data Availability Statement: Not applicable.

Conflicts of Interest: The authors declare no conflict of interest. The views expressed in this publication are purely those of the writers and may not in any circumstances be regarded as stating an official position of the European Commission.

References

1. Schroder, A.; Sprakel, J.; Boerkamp, W.; Schroen, K.; Berton-Carabin, C.C. Can we prevent lipid oxidation in emulsions by using fat-based pickering particles? *Food Res. Int.* **2019**, *120*, 352–363. [CrossRef] [PubMed]
2. Dimakou, C.; Oreopoulou, V. Antioxidant activity of carotenoids against the oxidative destabilization of sunflower oil-in-water emulsions. *LWT Food Sci. Technol.* **2012**, *46*, 393–400. [CrossRef]
3. Claire, C.; Berton-Carabin, C.-C.; Ropers, M.-H.; Genot, C. Lipid oxidation in Oil-in-Water Emulsions: Involvement of the Interfacial Layer. *Compreh. Rev. Food Sci. Food Saf.* **2019**, *13*, 945–977. [CrossRef]
4. Kiokias, S.; Gordon, M.; Oreopoulou, V. Compositional and processing factors that monitor oxidative deterioration of food relevant protein stabilized emulsions. *Crit. Rev. Food Sci. Nutr.* **2017**, *57*, 549–558. [CrossRef]
5. Poyato, C.; Navarro-Blasco, I.-E.; Calvo, M.-I.; Cavero, R.-Y.; Astiasarán, I.; Ansorena, D. Oxidative stability of O/W and W/O/W emulsions: Effect of lipid composition and antioxidant polarity. *Food Res. Int.* **2013**, *51*, 132–140. [CrossRef]
6. Shahidi, F.; Zhong, Y. Revisiting the polar paradox theory: A critical overview. *J. Agric. Food Chem.* **2011**, *59*, 3499–3504. [CrossRef]
7. Costa, M.; Barreiro, S.-L.; Magalhães, J.; Monteiro, L.-S.; Bravo-Díaz, C.; Paiva-Martins, F. Effects of the Reactive Moiety of Phenolipids on Their Antioxidant Efficiency in Model Emulsified Systems. *Foods* **2021**, *10*, 1028. [CrossRef]
8. Kiokias, S.; Varzakas, T. Innovative applications of food related emulsions. *Crit. Rev. Food Sci. Nutr.* **2017**, *57*, 3165–3172. [CrossRef]
9. Axmann, M.; Strobl, W.-M.; Plochberger, B.; Stangl, H. Cholesterol transfer at the plasma membrane. *Atherosclerosis* **2019**, *290*, 111–117. [CrossRef]
10. Kiokias, S. Antioxidant effects of vitamins C, E and provitamin A compounds as monitored by use of biochemical oxidative indicators linked to atherosclerosis and carcinogenesis. *Int. J. Nutr. Res.* **2019**, *1*, 1–13.
11. Farooq, S.; Hui, A.; Zhang, H.; Weiss, J. A comprehensive review on polarity, partitioning, and interactions of phenolic antioxidants at oil–water interface of food emulsions. *Compreh. Rev. Food Sci. Food Saf.* **2021**, *20*, 4250–4277. [CrossRef] [PubMed]
12. Phonsatta, N.; Deetae, P.; Luangpituksa, P.; Grajeda-Iglesias, C.; Figueroa-Espinoza, M.C.; Le Comte, J.; Villeneuve, P.; Decker, E.A.; Visessanguan, W.; Panya, A. Comparison of antioxidant evaluation assays for investigating antioxidative activity of gallic acid and its alkyl esters in different food matrices. *J. Agric. Food Chem.* **2017**, *65*, 7509–7518. [CrossRef] [PubMed]
13. Katsouli, M.; Polychniatou, V.; Tzia, C. Optimization of water in olive oil nano-emulsions composition with bioactive compounds by response surface methodology. *LWT Food Sci. Technol.* **2018**, *89*, 740–748. [CrossRef]
14. Mitrus, O.; Zuraw, M.; Losada-Barreiro, S.; Bravo-Díaz, C.; Paiva-Martins, F. Targeting Antioxidants to Interfaces: Control of the oxidative stability of lipid-based emulsions. *J. Agric. Food Chem.* **2019**, *67*, 3266–3274. [CrossRef]
15. Barouh, N.; Bourlieu-Lacanal, C.; Figueroa-Espinoza, M.C.; Durand, E.; Villeneuve, P. Tocopherols as antioxidants in lipid-based systems: The combination of chemical and physicochemical interactions determines their efficiency. *Compr. Rev. Food. Sci. Food Saf.* **2022**, *21*, 642–648. [CrossRef] [PubMed]

16. Arancibia, M.; Giménez, B.; López-Caballero, M.-E.; Gómez-Guillén, M.-C.; Montero, P. Release of cinnamon essential oil from polysaccharide bilayer films and its use for microbial growth inhibition in chilled shrimps. *LWT Food Sci. Technol.* **2014**, *59*, 989–995. [CrossRef]
17. Liudvinaviciute, D.; Rutkaite, R.; Bendoraitiene, J.; Klimaviciute, R. Thermogravimetric analysis of caffeic and rosmarinic acid containing chitosan complexes. *Carb. Polym.* **2019**, *222*, 115003. [CrossRef]
18. Garavand, F.; Jalai-Jivan, M.; Assadpour, E.; Jafari, S.M. Encapsulation of phenolic compounds within nano/microemulsion systems: A review. *Food Chem.* **2021**, *3641*, 130376. [CrossRef]
19. Choulitoudi, E.; Xristou, M.; Tsimogiannis, D.; Oreopoulou, V. The effect of temperature on the phenolic content and oxidative stability of o/w emulsions enriched with natural extracts from *Satureja thymbra*. *Food Chem.* **2021**, *349*, 129206. [CrossRef]
20. Lisete-Torres, P.; Losada-Barreiro, S.; Albuquerque, H.; Sanchez-Paz, V.; Paiva-Martins, F.; Bravo-Diaz, C. Distribution of hydroxytyrosol and hydroxytyrosol acetate in olive oil emulsions and their antioxidant efficiency. *J. Agric. Food Chem.* **2012**, *60*, 7318–7325. [CrossRef]
21. Liu, Y.; Carver, J.-A.; Calabrese, A.-N.; Pukala, T.-L. Gallic acid interacts with α-synuclein to prevent the structural collapse necessary for its aggregation. *Biochim. Biophys. Acta (BBA) Prot. Proteo.* **2014**, *1844*, 1481–1485. [CrossRef] [PubMed]
22. Kiokias, S.; Oreopoulou, V. A review on the health protective effects of phenolic acids against a range of severe pathologic conditions (incl Coronovrus based infections). *Molecules* **2021**, *26*, 5405. [CrossRef] [PubMed]
23. Oreopoulou, A.; Papavassilopoulou, E.; Bardouki, H.; Vamvakias, M.; Bimpilas, A.; Oreopoulou, V. Antioxidant recovery from hydrodistillation residues of selected Lamiaceae species by alkaline extraction. *J. Appl. Res. Medic. Aromat. Plants* **2018**, *8*, 83–89. [CrossRef]
24. Tsimogiannis, D.; Oreopoulou, V. Classification of phenolic compounds in plants. In *Polyphenols in Plants Isolation Purification and Extract Preparation*, 2nd ed.; Watson, R.R., Ed.; Elsevier Inc.: London, UK, 2019; pp. 263–284.
25. Durand, E.; Zhao, Y.; Ruesgas-Ramón, M.; Figueroa-Espinoza, M.-C.; Lamy, S.; Coupland, J.N.; Elias, R.-J.; Villeneuve, P. Evaluation of antioxidant activity and interaction with radical species using the vesicle conjugated autoxidizable triene (VesiCAT) assay. *Europ. J. Lipid Sci. Technol.* **2019**, *121*, 1800419. [CrossRef]
26. Laguerre, M.; Bily, A.; Roller, M.; Birtić, S. Mass transport phenomena in lipidoxidation and antioxidation. *Ann. Rev. Food Sci. Technol.* **2017**, *8*, 391–411. [CrossRef]
27. Kiokias, S.; Proestos, C.; Oreopoulou, V. Phenolic acids of plant origin—A review on their antioxidant activity in vitro (o/w emulsion systems) along with their in vivo health biochemical properties. *Foods* **2020**, *9*, 534. [CrossRef]
28. Traber, G.-M.; Jeffrey, A. Vitamin E, antioxidant and nothing more. *Free Radic. Biol. Med.* **2007**, *43*, 4–15. [CrossRef]
29. Kim, S.-K.; Im, G.-J.; An, Y.-S.; Lee, S.-H.; Jung, H.-H.; Park, S.-Y. The effects of the antioxidant α-tocopherol succinate on cisplatin-induced ototoxicity in HEI-OC1 auditory cells. *Int. J. Pediatr. Otorhin.* **2016**, *86*, 9–14. [CrossRef]
30. Wang, L.; Yu, X.; Geng, F.; Cheng, C.; Yang, J.; Deng, Q. Effects of tocopherols on the stability of flaxseed oil-in-water emulsions stabilized by different emulsifiers: Interfacial partitioning and interaction. *Food Chem.* **2022**, *374*, 131691. [CrossRef]
31. Rajasekaran, B.; Singh, A.; Nagarajan, M.; Benjakul, S. Effect of chitooligosaccharide and α-tocopherol on physical properties and oxidative stability of shrimp oil-in water emulsion stabilized by bovine serum albumin-chitosan complex. *Food Contr.* **2022**, *137*, 108899. [CrossRef]
32. Tsimogiannis, D.; Oreopoulou, V. Defining the role of flavonoid structure on cottonseed oil stabilization: Study of A- and C-ring substitution. *J. Am. Oil Chem. Soc.* **2007**, *84*, 129–136. [CrossRef]
33. Oh, W.Y.; Ambigaipalan, P.; Shahidi, F. Quercetin and its ester derivatives inhibit oxidation of food, LDL and DNA. *Food Chem.* **2021**, *364*, 130394. [CrossRef] [PubMed]
34. Yang, D.; Wang, X.Y.; Lee, J.H. Effects of flavonoids on physical and oxidative stability of soybean oil O/W emulsions. *Food Sci. Biotechnol.* **2015**, *24*, 851–858. [CrossRef]
35. Valerga, J.; Reta, M.; Lanari, M.C. Polyphenol input to the antioxidant activity of yerba mate (*Ilex paraguariensis*) extracts. *LWT Food Sci. Technol.* **2012**, *45*, 28–35. [CrossRef]
36. Skerget, M.; Kotnik, P.; Hadolin, M.; Hras, A.; Simonic, M.; Knez, Z. Phenols, pronahtocyaninds, flavones, and flavonols in some plant materials and their antioxidant activities. *Food Chem.* **2005**, *89*, 191–198. [CrossRef]
37. Beker, B.-Y.; Bakir, T.; Filiz, I.; Apak, R. Antioxidant protective effect of flavonoids on linoleic acid peroxidation induced by copper (II)/ascorbic acid. *Chem. Phys. Lipids* **2011**, *164*, 732–739. [CrossRef] [PubMed]
38. Roedig-Penman, A.; Gordon, M.H. Antioxidant properties of myricetin and quercetin in oil and emulsions. *J. Am. Oil Chem. Soc.* **1998**, *75*, 169–180. [CrossRef]
39. Duffus, L.J.; Norton, J.E.; Smith, P.; Norton, I.T.; Spyropoulos, F. A comparative study on the capacity of a range of food-grade particles to form stable O/W and W/O Pickering emulsions. *J. Coll. Interfac. Sci.* **2016**, *473*, 9–21. [CrossRef]
40. Tong, Q.; Yi, Z.; Ran, Y.; Chen, X.; Chen, G.; Li, X. Green tea polyphenol-stabilized gel-like high internal phase pickering emulsions. *ACS Sustain. Chem. Eng.* **2021**, *9*, 4076–4090. [CrossRef]
41. Skowyra, M.; Gallego, M.G.; Segovia, F.; Almajano, M.P. Antioxidant properties of Artemisia annua extracts in model food emulsions. *Antioxidants* **2014**, *3*, 116–128. [CrossRef]
42. García-Iñiguez de Ciriano, M.; Rehecho, S.; Calvo, M.I.; Cavero, R.Y.; Navarro, I.; Astiasarán, I.; Ansorena, D. Effect of lyophilized water extracts of Melissa officinalis on the stability of algae and linseed oil-in-water emulsion to be used as a functional ingredient in meat products. *Meat Sci.* **2010**, *85*, 373–377. [CrossRef] [PubMed]

43. Becker, E.M.; Ntouma, G.; Skibsted, L.H. Synergism and antagonism between quercetin and other chain-breaking antioxidants in lipid systems of increasing structural organisation. *Food Chem.* **2007**, *103*, 1288–1296. [CrossRef]
44. Kiokias, S.; Proestos, C.; Oreopoulou, V. Effect of natural Food Antioxidants against LDL and DNA Oxidative damages. *Antioxidants* **2018**, *7*, 133. [CrossRef]
45. Sugahara, S.; Chiyo, A.; Fukuoka, K.; Ueda, Y.; Tokunaga, Y.; Nishida, Y.; Kinoshita, H.; Matsuda, Y.; Igoshi, K.; Ono, M.; et al. Unique antioxidant effects of herbal leaf tea and stem tea from *Moringa oleifera* L. especially on superoxide anion radical generation systems. *Biosci. Biotechnol. Biochem.* **2018**, *82*, 1973–1984. [CrossRef] [PubMed]
46. Terpinc, P.; Polak, T.; Šegatin, N.; Hanzlowsky, A. Antioxidant properties of 4-vinyl derivatives of hydroxycinnamic acid. *Food Chem.* **2011**, *128*, 62–69. [CrossRef] [PubMed]
47. Conde, E.; Gordon, M.-H.; Moure, A.; Dominguez, H. Effects of caffeic acid and bovine serum albumin in reducing the rate of development of rancidity in oil-in-water and water-in-oil emulsions. *Food Chem.* **2011**, *129*, 1652–1659. [CrossRef]
48. Mojica, L.; Meyer, A.; Berhow, M.; González, E. Bean cultivars (*Phaseolus vulgaris* L.) have similar high antioxidant capacity, *in vitro* inhibition of α-amylase and α-glucosidase while diverse phenolic composition and concentration. *Food Res. Int.* **2015**, *69*, 38–48. [CrossRef]
49. Oehlke, K.; Heins, A.; Stöckmann, H.; Schwarz, K. Impact of emulsifier microenvironments on acid–base equilibrium and activity of antioxidants. *Food Chem.* **2010**, *118*, 48–55. [CrossRef]
50. Maqsood, S.; Benjakul, S. Comparative studies of four different phenolic compounds on in vitro antioxidative activity and the preventive effect on lipid oxidation of fish oil emulsion and fish mince. *Food Chem.* **2010**, *119*, 123–132. [CrossRef]
51. Shin, J.-A.; Jeong, S.-H.; Jia, C.-H.; Hong, S.-T.; Lee, K.-T. Comparison of antioxidant capacity of 4-vinylguaiacol with catechin and ferulic acid in oil-in-water emulsion. *Food Sci. Biotechnol.* **2019**, *28*, 35–41. [CrossRef]
52. Pandurangan, A.-K.; Mohebali, N.; Norhaizan, M.-E.; Looi, C.-Y. Gallic acid attenuates dextran sulfate sodium-induced experimental colitis in BALB/c mice. *Drug Des. Dev. Ther.* **2015**, *9*, 3923–3934. [CrossRef] [PubMed]
53. Da Silveira, T.F.; Laguerre, M.; Bourlieu-Lacanal, C.; Lecomte, J.; Durand, E.; Figueroa-Espinoza, M.C.; Barea, B.; Barouh, N.; Castro, I.A.; Villeneuve, P. Impact of surfactant concentration and antioxidant mode of incorporation on the oxidative stability of oil-in-water nanoemulsions. *LWT Food Sci. Technol.* **2021**, *141*, 110982. [CrossRef]
54. Evangeliou, V.; Panagopoulou, E.; Mandala, I. Encapsulation of EGCG and esterified EGCG derivatives in double emulsions containing Whey Protein Isolate, Bacterial Cellulose and salt. *Food Chem.* **2019**, *281*, 171–177. [CrossRef] [PubMed]
55. Zhu, M.; Wu, C.; Chen, Y.; Xie, M. Antioxidant effects of different polar gallic acid and its alkyl esters in oil-in-water emulsions. *J. Chin. Instit. Food Sci. Technol.* **2019**, *19*, 13–22.
56. Bou, R.; Boo, C.; Kwek, A.; Hidalgo, D.; Decke, E.A. Research ArticleEffect of different antioxidants on lycopene degradation inoil-in-water emulsions. *Eur. J. Lipid Sci. Technol.* **2011**, *113*, 724–729. [CrossRef]
57. Alavi Rafiee, S.; Farhoosh, R.; Sharif, A. Antioxidant activity of gallic acid as affected by an extra carboxyl group than pyrogallol in various oxidative environments. *Eur. J. Lipid Sci. Technol.* **2018**, *120*, 1800319. [CrossRef]
58. Trisha, S. Role of hesperdin, luteolin and coumaric acid in arthritis management: A Review. *Inter. J. Phys. Nutr. Phys. Educ.* **2018**, *3*, 1183–1186.
59. Park, J.; Gim, S.-Y.; Jeon, J.-Y.; Kim, M.-J.; Choi, H.-K.; Lee, J. Chemical profiles and antioxidant properties of roasted rice hull extracts in bulk oil and oil-in-water emulsion. *Food Chem.* **2019**, *272*, 242–250. [CrossRef]
60. Yashin, A.; Yashin, Y.; Xia, X.; Nemzer, B. Antioxidant activity of spices and their impact on human health: A review. *Antioxidants* **2017**, *6*, 70. [CrossRef]
61. Alamed, J.; Chaiyasit, W.; McClements, D.-J.; Decker, E.-A. Relationships between free radical scavenging and antioxidant activity in foods. *J. Agric. Food Chem.* **2009**, *257*, 2969–2976. [CrossRef]
62. Panya, A.; Kittipongpittaya, K.; Laguerre, M.; Bayrasy, C.; Lecomte, J.; Villeneuve, P.; Decker, E.-A. Interactions between α-tocopherol and rosmarinic acid and its alkyl esters in emulsions: Synergistic, additive, or antagonistic effect? *J. Agric. Food Chem.* **2012**, *60*, 10320–10330. [CrossRef] [PubMed]
63. Bakota, E.-L.; Winkler-Moser, J.-K.; Berhow, M.-A.; Eller, F.-J.; Vaughn, S.-F. Antioxidant activity and sensory evaluation of a rosmarinic acid-enriched extract of *Salvia officinalis*. *J. Food Sci.* **2015**, *80*, 711–717. [CrossRef] [PubMed]
64. Keller, S.; Locquet, N.; Cuvelier, M.E. Partitioning of vanillic acid in oil-in-water emulsions: Impact of the Tween®40 emulsifier. *Food Res. Int.* **2016**, *88*, 61–69. [CrossRef] [PubMed]
65. Li, R.; Dai, T.; Zhou, W.; Fu, G.; Wan, Y.; McClements, D.J.; Li, J. Impact of pH, ferrous ions, and tannic acid on lipid oxidation in plant-based emulsions containing saponin-coated flaxseed oil droplets. *Food Res. Int.* **2020**, *136*, 109618. [CrossRef]
66. Jayasinghe, C.; Gotoh, N.; Wada, S. Pro-oxidant/antioxidant behaviours of ascorbic acid, tocopherol, and plant extracts in n-3 highly unsaturated fatty acid rich oil-in-water emulsions. *Food Chem.* **2013**, *141*, 3077–3084. [CrossRef] [PubMed]
67. Narra, M.-R.; Rajendar, K.; Rudra, R.; Rao, J.-V.; Begum, G. The role of vitamin C as antioxidant in protection of biochemical and haematological stress induced by chlorpyrifos in freshwater fish Clarias batrachus. *Chemosphere* **2015**, *132*, 172–178. [CrossRef] [PubMed]
68. Dimakou, C.; Kiokias, S.; Tsaprouni, I.; Oreopoulou, V. Effect of processing and storage parameters on oxidative deterioration of oil-in-water emulsions. *Food Biophys.* **2007**, *2*, 38–45. [CrossRef]
69. Kiokias, S.; Dimakou, C.; Tsaprouni, I.; Oreopoulou, V. Effect of compositional factors against the thermal oxidation of novel food emulsions. *Food Biophys.* **2006**, *1*, 115–123. [CrossRef]

70. Raikos, V.; Neacsu, M.; Morrice, P.; Duthie, G. Physicochemical stability of egg protein-stabilized oil-in-water emulsions supplemented with vegetable powders. *Int. J. Food Sci. Technol.* **2014**, *49*, 2433–2440. [CrossRef]
71. Cheng, C.; Yu, X.; McClements, D.-J.; Huang, Q.-D.; Tang, H.; Yu, K.; Xiang, X.; Chen, P.; Wang, X.-T.; Deng, Q.-C. Effect of flaxseed polyphenols on physical stability and oxidative stability of flaxseed oil-in-water nanoemulsions. *Food Chem.* **2019**, *301*, 125207. [CrossRef]
72. Kiokias, S.; Oreopoulou, V. Antioxidant properties of natural carotenoid preparations against the AAPH-oxidation of food emulsions. *Innov. Food Sci. Emerg. Tech.* **2006**, *7*, 132–139. [CrossRef]
73. Seo, S.-R.; Lee, H.-Y.; Kim, J.-C. Thermo- and pH-responsiveness of emulsions stabilized with acidic thermosensitive polymers. *J. Food Eng.* **2016**, *111*, 449–457.
74. Branco, G.F.; Rodrigues, M.I.; Gioielli, L.A.; Castro, I.A. Effect of the simultaneous interaction among ascorbic acid, iron and ph on the oxidative stability of oil-in-water emulsions. *J. Agric. Food Chem.* **2011**, *59*, 12183–12192. [CrossRef] [PubMed]
75. Costa, M.; Losada-Barreiro, S.; Paiva-Martins, F.; Bravo-Dıaz, C.; Romsted, L.S. A direct correlation between the antioxidant efficiencies of caffeic acid and its alkyl esters and their concentrations in the interfacial region of olive oil emulsions. The pseudophase model interpretation of the "cut-off" effect. *Food Chem.* **2015**, *175*, 233–242. [CrossRef]
76. Zhou, L.; Elias, R.-J. *Antioxidant* and pro-oxidant activity of (−)-epigallocatechin-3-gallate in food emulsions: Influence of pH and phenolic concentration. *Food Chem.* **2013**, *138*, 1503–1509. [CrossRef] [PubMed]
77. Sørensen, A.-D.; Villeneuve, P.; Jacobsen, C. Alkyl caffeates as antioxidants in O/W emulsions: Impact of emulsifier type and endogenous tocopherols. *Europ. J. Lipid Sci. Technol.* **2017**, *119*, 6. [CrossRef]
78. Zhou, L.; Elias, R.-J. Factors influencing the antioxidant and pro-oxidant activity of polyphenols in oil-in-water emulsions. *J. Agric. Food Chem.* **2012**, *60*, 2906–2915. [CrossRef]
79. Roedig-Penman, A.; Gordon, M.-H. Antioxidant properties of catechins and green tea extracts in model food emulsions. *J. Agric. Food Chem.* **1997**, *45*, 4267–4270. [CrossRef]
80. Kim, J.; Choe, E. Effect of the pH on the lipid oxidation and polyphenols of soybean oil-in-water emulsion with added peppermint (*Mentha piperita*) extract in the presence and absence of iron. *Food Sci. Biotechnol.* **2018**, *27*, 1285–1292. [CrossRef]
81. Tian, L.; Zhang, S.; Yi, Z.; Cui, L.; Decker, E.-A.; McClements, D.-E. Antioxidant and prooxidant activities of tea polyphenols in oil-in-water emulsions depend on the level used and the location of proteins. *Food Chem.* **2022**, *1*, 375. [CrossRef]
82. Kittipongpittaya, K.; Panya, A.; Phonsatta, N.; Decker, E.A. Effects of Environmental pH on Antioxidant Interactions between Rosmarinic Acid and α-Tocopherol in Oil-in-Water (O/W) Emulsions. *J. Agric. Food Chem.* **2016**, *64*, 6575–6583. [CrossRef] [PubMed]
83. Losada-Barreiro, S.; Dıaz, C.-B.; Romsted, L.-S. Distributions of phenolic acid antioxidants between the interfacial and aqueous regions of corn oil emulsions: Effects of pH and emulsifier concentration. *Eur. J. Lipid Sci. Technol.* **2015**, *117*, 1801–1813. [CrossRef]
84. Costa, M.; Losada-Barreiro, S.; Paiva-Martins, F.; Bravo-Dıaz, C.; Romsted, L.S. Physical evidence that the variations in the efficiency of homologous series of antioxidants in emulsions are due to differences in their distribution. *J. Sci. Food Agric.* **2017**, *97*, 564–571. [CrossRef] [PubMed]
85. Zorić, Z.; Markić, J.; Pedisić, S.; Bučević-Popović, V.; Generalić-Mekinić, I.; Grebenar, K.; Kulišić-Bilušić, T. Stability of rosmarinic acid in aqueous extracts from different Lamiaceae species after in vitro digestion with human gastrointestinal enzymes. *Food Technol. Biotechnol.* **2016**, *54*, 97–102. [CrossRef] [PubMed]
86. Villeneuve, P.; Bourlieu-Lacanal, C.; Durand, E.; Lecomte, J.; McClements, D.-J.; Decker, E.-A. Lipid oxidation in emulsions and bulk oils: A review of the importance of micelles. *Crit. Rev. Food Sci. Nutr.* **2021**, *29*, 1–41. [CrossRef]
87. Kanikidi, L.-D.; Tsimogiannis, D.; Kiokias, S.; Oreopoulou, V. Formulation of Rosemary Extracts through Spray-Drying Encapsulation or Emulsification). *Neutraceuticals* **2022**, *2*, 1–21. [CrossRef]
88. Chaiyasit, W.; Elias, R.; McClements, D.-E.; Decker, E.-A. Role of physical structures in bulk oils on lipid oxidation. *Crit. Rev. Food Sci. Nutr.* **2007**, *47*, 299–317. [CrossRef]
89. Waraho, T.; McClements, D.-J.; Decker, E.-A. Mechanisms of lipid oxidation in food dispersions. *Trends Food Sci. Technol.* **2011**, *22*, 3–13. [CrossRef]
90. McClements, D.J.; Decker, E. Interfacial antioxidants: A review of natural and synthetic emulsifiers and co-emulsifiers that can inhibit lipid oxidation. *J. Agric. Food Chem.* **2018**, *66*, 20–35. [CrossRef]
91. Kiralan, S.-S.; Doğu-Baykut, E.; Kittipongpittaya, K.; McClements, D.-J.; Decker, E.-A. Increased antioxidant efficacy of tocopherols by surfactant solubilization in oil-in water emulsions. *J. Agric. Food Chem.* **2014**, *62*, 10561–10566. [CrossRef]
92. Almajano, M.-P.; Delgado, E.; Gordon, M.-H. Albumin causes a synergistic increase in the antioxidant activity of green tea catechins in oil-in-water emulsions. *Food Chem.* **2007**, *102*, 1375–1382. [CrossRef]
93. Sabouri, S.; Geng, J.-H.; Corredig, M. Tea polyphenols association to caseinate-stabilized oil-water interfaces. *Food Hydrocol.* **2015**, *51*, 95–100. [CrossRef]
94. Tian, B.; Wang, Y.-X.; Wang, T.-J.; Mao, L.-J.; Lu, Y.-N.; Wang, H.-T.; Feng, Z.-B. Structure and functional properties of antioxidant nano-emulsions prepared with tea polyphenols and soybean protein isolate. *J. Oleo Sci.* **2019**, *68*, 689–697. [CrossRef] [PubMed]
95. Vilasaua, J.; Solansa, C.; Gomezb, M.J.; Dabriob, J.; Mujika-Garaib, R.; Esquenaa, J. Influence of a mixed ionic/nonionic surfactant system and the emulsification process on the properties of paraffin emulsions. *Coll. Surf. A Physicochem. Eng. Aspects* **2011**, *392*, 38–44. [CrossRef]

96. Fu, Y.; McClements, D.J.; Luo, S.; Ye, J.; Chengmei, L. Degradation kinetic of rutin encapsulated in oil-in-water emulsions: Impact of particle size. *J. Sci. Food Agric.* **2022**, *103*, 770–778. [CrossRef]
97. Kharat, M.; Aberg, J.; Dai, T.; McClements, D.J. Comparison of Emulsion and Nanoemulsion Delivery Systems: The Chemical Stability of Curcumin Decreases as Oil Droplet Size Decreases. *J. Agric. Food Chem.* **2020**, *68*, 9205–9212. [CrossRef] [PubMed]
98. Losada-Barreiro, S.; Dıaz, C.-B.; Martins, F.-P.; Romsted, L.-S. Maxima in antioxidant distributions and efficiencies with increasing hydrophobicity of gallic acid and its alkyl esters. The pseudophase model interpretation of the "Cut off Effect". *J. Agric. Food Chem.* **2013**, *61*, 6533–6543. [CrossRef]
99. Almeida, J.; Losada-Barreiro, S.; Costa, M.; Paiva-Martins, F.; Bravo-Díaz, C.; Romsted, L.S. Interfacial Concentrations of Hydroxytyrosol and Its Lipophilic Esters in Intact Olive Oil-in-Water Emulsions: Effects of Antioxidant Hydrophobicity, Surfactant Concentration, and the Oil-to-Water Ratio on the Oxidative Stability of the Emulsions. *J. Agric. Food Chem.* **2016**, *64*, 5274–5283. [CrossRef]
100. Meireles, M.; Losada-Barreiro, S.; Costa, M.; Paiva-Martins, F.; Bravo-Díaz, C.; Monteiro, L.-S. Control of antioxidant efficiency of chlorogenates in emulsions: Modulation of antioxidant interfacial concentrations. *J. Sci. Food. Agric.* **2019**, *99*, 3917–3925. [CrossRef]
101. Huang, S.W.; Frankel, E.N.; Schwarz, K.; Aeschbach, R.; German, J.B. Antioxidant activity of carnosic acid and methyl carnosate in bulk oils and oil-in-water emulsions. *J. Agric. Food Chem.* **1996**, *44*, 2951–2956. [CrossRef]
102. Li, A.; Zhao, M.T.; Yin, F.W.; Zhang, M.; Liu, H.L.; Zhou, D.Y.; Shahidi, F. Antioxidant effects of gallic acid alkyl esters of various chain lengths in oyster during frying process. *Food Sci. Technol.* **2021**, *56*, 2938–2945. [CrossRef]
103. Noon, J.; Mills, T.B.; Norton, I.T. The use of natural antioxidants to combat lipid oxidation in O/W emulsions. *J. Food Eng.* **2020**, *281*, 110006. [CrossRef]
104. Di Mattia, C.D.; Sacchetti, G.; Mastrocola, D.; Sarker, D.K.; Pittia, P. Surface properties of phenolic compounds and their influence on the dispersion and oxidative stability of olive oil O/W emulsions. *Food Hydrocol.* **2010**, *24*, 652–658. [CrossRef]
105. Bravo-Dıaz, C.; Romsted, L.-S.; Liu, S.; Losada-Barreiro, S.; Gallego, M.-J.; Xiang, G. To model chemical reactivity in heterogeneous emulsions, think homogeneous microemulsions. *Langmuir* **2015**, *31*, 8961–8979. [CrossRef] [PubMed]
106. Kiokias, S.; Varzakas, T.; Oreopoulou, V. In vitro activity of vitamins, flavonoids, and natural phenolic antioxidants against the oxidative deterioration of oil-based systems. *Critic. Rev. Food Sci. Nutr.* **2008**, *48*, 78–93. [CrossRef] [PubMed]
107. Bast, A.; Haanen, G.-R.; den Berg, V. Antioxidant effects of carotenoids. *Int. J. Vit. Nutr. Res.* **2007**, *68*, 399–403.
108. Filip, V.; Hradkova, I.; Smidrkal, J. Antioxidants in margarine emulsions. *Czech J. Food Sci.* **2009**, *27*, 9–11. [CrossRef]
109. Liu, R.; Chang, M.; Liu, R.; Xu, Y.; Wang, X. Interactions between α-tocopherol and γ-oryzanol in oil-in-water emulsions. *Food Chem.* **2021**, *356*, 129648. [CrossRef]
110. Wu, C.; Zhou, X.Y.; Zhang, M.R.; Chen, Y.; Nie, S.P.; Xie, M.Y. Combined application of gallate ester and α-tocopherol in oil-in-water emulsion: Their distribution and antioxidant efficiency. *J. Disp. Sci. Technol.* **2019**, *41*, 909–917.

colloids and interfaces

MDPI

Article

Addition of Trans-Resveratrol-Loaded, Highly Concentrated Double Emulsion to Moisturizing Cream: Effect on Physicochemical Properties

Rocío Díaz-Ruiz [1,2], Amanda Laca [2], Ismael Marcet [2], Lemuel Martínez-Rey [2], María Matos [1,2] and Gemma Gutiérrez [1,2,*]

1 Instituto Universitario de Biotecnología de Asturias, University of Oviedo, 33006 Oviedo, Spain
2 Department of Chemical and Environmental Engineering, University of Oviedo, 33006 Oviedo, Spain
* Correspondence: gutierrezgemma@uniovi.es; Tel.: +34-98-510-3510; Fax: +34-98-510-3434

Abstract: Resveratrol is a compound increasingly studied for its many beneficial properties for health. However, it is a highly unstable photosensitive compound, and therefore it is necessary to encapsulate it to protect it if you want to use it in a commercial product. Emulsions are systems that allow the encapsulation of active ingredients, protecting them and allowing their release in a controlled manner. They are highly used systems in the pharmaceutical, cosmetic and food industries. The main objectives of this work are to study the feasibility of encapsulating resveratrol in concentrated water-in-oil-in-water double emulsions and the effect produced by adding the double emulsion with optimal formulation to a commercial cream for cosmetic applications. The effect of the selected optimal double emulsion on a commercial cream was studied, analyzing droplet size distribution, morphology, stability and rheology. The main conclusion of this work is that incorporating 1/3 of concentrated double emulsion $W_1/O/W_2$ into a commercial moisturizing cream had a positive physical effect and produced cream with a resveratrol concentration of up to 0.0042 mg/g.

Keywords: trans-resveratrol; encapsulation; high internal phase double emulsions; moisturizing cream

Citation: Díaz-Ruiz, R.; Laca, A.; Marcet, I.; Martínez-Rey, L.; Matos, M.; Gutiérrez, G. Addition of Trans-Resveratrol-Loaded, Highly Concentrated Double Emulsion to Moisturizing Cream: Effect on Physicochemical Properties. *Colloids Interfaces* **2022**, *6*, 70. https://doi.org/10.3390/colloids6040070

Academic Editors: César Burgos-Díaz, Mauricio Opazo-Navarrete and Eduardo Morales

Received: 6 October 2022
Accepted: 10 November 2022
Published: 16 November 2022

1. Introduction

Resveratrol (3,5,4′-trihydroxy-trans-stilbene), or RSV, is a polyphenol of the stilbene family that, due to its antioxidant character, seems to produce beneficial effects on human health, which has greatly increased its applications in different types of industries such as pharmaceuticals or cosmetics [1,2]. Several therapeutic properties of resveratrol are currently being studied, such as its ability to prevent cardiovascular or neurodegenerative diseases [3–5]. Although probably the most striking are its anticancer properties, hindering the spread of cancer as well as its initiation [6–10], being especially effective against possible skin cancer [11]. However, the property that has aroused the interest of the cosmetic industry in this compound is its antiaging effect, a consequence of its antioxidant properties [12,13]. According to previous studies, one of the properties of resveratrol is that it helps to retard skin aging since it helps protect cellular DNA from free radicals [14]. That is why the cosmetic industry is interested in the incorporation of resveratrol into creams or gels.

However, regarding resveratrol's physicochemical properties, it has a low solubility in water (0.03 g/L) compared to its high solubility in ethanol (50 g/L) [15], and its great UV, pH or thermal instability [15–21] are especially noteworthy.

Due to its low stability, it is necessary to protect it in order to slow down or even prevent its loss while increasing its bioavailability [22]. This can be achieved by encapsulating it with one of the currently available methods. One of the most prominent is using emulsions.

An emulsion is a heterogeneous liquid system of at least two phases in which one called the dispersed or internal phase is dispersed in the form of drops into another called the continuous or external phase. The liquid phases that compose the emulsions (typically water and oil) are immiscible with each other, making them highly unstable systems. To prevent droplet coalescence, stabilizing agents (surfactants, particles or proteins) are added at the interface.

The main two types of emulsions are oil-in-water (O/W) and water-in-oil (W/O). These types of systems possess the capacity to encapsulate antioxidant compounds of other bioactive compounds in their inner phase, which can be transported to desired points. Moreover, more complex emulsions systems can be found, as in the case of double emulsions, which are ternary systems where the dispersed droplets contain smaller droplets. Two main types to distinguish are water-in-oil-in-water ($W_1/O/W_2$) or oil-in-water-in-oil ($O_1/W/O_2$).

Emulsions are widely used to encapsulate compounds of interest for several bioapplications, such as polyphenols, lutein or vitamins [23–30]. Emulsions protect active compounds to maintain their desired properties and offer a controlled release of the encapsulated compounds. Emulsions are widely used in the cosmetic industry, and they are a good mechanism for encapsulating different active ingredients since they are found in the dispersed phase and the continuous phase would act as a barrier to protect them from the outside environment, thus increasing their stability and consequently their useful life. They also allow a controlled release. Double emulsions allow the encapsulating of aqueous soluble compounds in an aqueous matrix. Moreover, the organic intermediate phase offers an additional barrier for biocompound release, which enhances encapsulation and controlled release.

Incorporating active ingredients in cosmetic products was widely performed in the 1990s, and as a consequence, the term 'cosmeceutical' appeared, combining the terms 'cosmetic' and 'pharmaceutical' [31]. While a conventional cosmetic product is intended for beauty, a cosmeceutical product is intended to influence the biological functions of the skin by incorporating active ingredients. It is important in cosmeceutical products, as in the case of pharmaceuticals, that the encapsulated active ingredient should have a concentration above a certain threshold concentration, which is different for each compound, hence the importance of achieving highly concentrated double emulsions to arise a large biocompound concentration in the final product.

For this reason, the feasibility of encapsulating resveratrol in concentrated water in oil-in-water double emulsions formulated with Polyglycerol polyricinoleate (PGPR) and Tween 20 stabilizers has been studied. Moreover, the effect produced by the addition of the double emulsion to a commercial cream for cosmetic applications in order to take advantage of the benefits of the biocompound through its dermal application has been investigated.

2. Materials and Methods

2.1. Materials

Absolute ethanol, RSV, and Tween® 20 (polyoxyethylenesorbitan monolaurate) were purchased from Sigma Aldrich (San Louis, MO, USA). Miglyol® 812 (density 945 kg/m^3 at 20 °C), which is a neutral oil formed by esters of caprylic and capric fatty acids and glycerol, was supplied by Sasol GmbH (Hambuerg, Germany). Polyglycerol polyricinoleate (PGPR) was obtained from Brenntag AG (Frankfurt, Germany). Sodium chloride was supplied by Panreac (Barcelona, Spain), and the moisturizing cream was a balancing cream for normal-oily skin from the Deliplus brand acquired from the Mercadona supermarket chain.

2.2. Methods

2.2.1. Water-in-Oil (W_1/O) Inner Emulsion Preparation

The internal emulsion (W_1/O) was prepared with 30 vol.% of internal aqueous phase (W_1) and 70 vol.% of oily phase (O). As oily phase, Miglyol 812 containing 5 wt.% of the hydrophobic emulsifier (PGPR) was used. PGPR was previously dissolved in the oily phase by magnetic stirring at room temperature for 30 min. PGPR was selected since it is a common stabilizer in food formulations with highly stabilizing effect [32–34].

RSV has low water solubility, and its solubility in alcohol decreases as the carbon number of the alcohol increases [35]. For this reason, 20% ethanol (*v*/*v*) was used in the W_1. Moreover, a concentration of 50 mg/L of RSV was incorporated into the W_1 solution.

Additionally, 0.1M NaCl was added to W_1 to ensure aqueous droplet stability since it has been found that the addition of electrolytes to aqueous phases increases emulsion stability [32,36,37].

Both phases were placed together in glass vessels and emulsified by high-shear mixing at 15,000 rpm for 5 min using Silentcruser M Homogenizer (Heidolph, Germany) with a 6 mm dispersing tool.

2.2.2. Water-in-Oil-in-Water ($W_1/O/W_2$) Double Emulsions Preparation

W_2 was composed of a 2% (*w*/*v*) Tween 20 solution, which was ensured to have more than 300 times the CMC value [38,39]. In order to equilibrate the osmotic pressure of both aqueous phases, 0.1 M NaCl was also added to the W_2 [40].

The $W_1/O/W_2$ double emulsions were formulated, dispersing the W_1/O primary emulsion, and incorporated into W_2 at the volumetric ratios of W_1/O in W_2: 80/20.

This second emulsification step was carried out by mixing the inner emulsion and external aqueous phase at 5000 rpm for 2 min with the aforementioned Silentcruser M Homogenizer. Milder agitation conditions were used in order to avoid the rupture of the inner emulsion W_1/O, producing a final simple O/W emulsion.

Emulsification conditions and emulsion formulation parameters were selected according to the optimal results obtained in previous studies [24,41,42].

2.2.3. Addition of Trans-Resveratrol-Loaded, Highly Concentrated Double Emulsion ($W_1/O/W_2$) to Moisturizing Cream

Once the optimal formulation for the double emulsions had been chosen by considering the results of previous studies [42], its effect was studied by mixing it with a commercial cream with proportions of 1/4, 1/3 and 1/2 emulsion (*w*/*w*). The mixture between the cream and the emulsion was carried out by manual agitation.

2.2.4. Characterization

The influence of incorporating a double emulsion in its optimal formulation to a commercial cream was studied.

The same parameters were studied for a commercial cream as well as for a mixture between the commercial cream and the incorporation of the double emulsion in proportions of 1/4, 1/3 and 1/2 emulsion.

The double emulsions were previously characterized in terms of droplet size distribution, stability, encapsulation efficiency and rheology [42].

Droplet Size Distribution and Morphology

The droplet size distribution was measured with the long-bench Mastersizer S (Malvern Instruments Ltd., Malvern, UK) based on diffraction of laser radiation. A representation of the different sizes and volume occupied by them was obtained. The size results were expressed in equivalent sphere diameter, that is, the diameter of the sphere that has the same volume as the measured particle. Although for emulsions, spherical droplets can be assumed. The refractive index of the Miglyol 812 (1.54) was used for double emulsions droplet size measurements.

Micrographs of the emulsions were obtained with an Olympus BX50 light microscope (Olympus, Tokyo, Japan) with 10–100× magnification using UV–vis and fluorescence lamps. Micrographs were used for visual inspection of emulsions and to confirm the droplet size obtained by laser light scattering.

Colloidal Stability

Dispersed phase droplets could produce emulsion destabilization due to the migration of the low-density drops to the top part of the container, producing a top layer rich in droplets and a top layer formed mainly by external continuous phase. This destabilization phenomenon is known as creaming.

The stability of the moisturizing cream and its mixture with double emulsions was determined with the TurbiscanLabExpert (Formulaction, Toulouse, France). This instrument emits a light beam that passes through the sample and measures transmission and backscattering of the light beam.

Samples without dilution were placed in the cells. Transmitted and backscattered light was monitored as a function of time and cell height. The optical reading head scans the sample in the cell, providing TS and BS data every 40 μm as a function of the sample height (in mm). The obtained profiles built up a macroscopic fingerprint of the cream or mixture, providing useful information about sample changes with time [27].

Rheology

The rheological tests were carried out with a Haake MARS II rotational rheometer with a Haake UTC Peltier temperature control unit. All the analyses were developed at 25 ± 0.1 °C, employing a parallel-plate sensor system (PP60Ti) with a gap of 1 mm. Before starting any measurement, the sample rested for at least 5 min, allowing the stresses induced during sample load to relax.

Viscosity measurements were conducted in control rate (CR) mode from 0 to 1000 s^{-1} for 300 s. Herschel–Bulkley model, commonly used to characterize different hydrocolloid dispersions and food emulsions [43], has been selected to fit the flow curves obtained

$$\tau = \tau_0 + K\dot{\gamma}^n \tag{1}$$

where τ is the shear stress (Pa), τ_0 is the yield point (Pa), $\dot{\gamma}$ is the shear rate (s^{-1}), K is the consistency coefficient (Pa s^n) and n is the flow behavior index (-).

The frequency sweeps were carried out from 10 to 0.1 Hz at a constant shear stress of 0.1 Pa.

All measurements were developed at least in triplicate.

Texturometry

Texturometry measurements were performed with a TA.XT.plus Texture Analyzer (Stable Micro Systems, Godalming, UK). A load cell of 5000 g was used as a penetration test, a 5 mm penetration was made with a 0.5 in. diameter probe (S/0.5) at a speed of 2 mm/s. Force versus time was recorded. Maximum force was taken as a measurement of hardness, while the maximum negative force was taken as an indication of stickiness. Test was made at room temperature and measurements were developed in duplicate.

3. Results and Discussion

In light of the results of the previous studies [42], the optimal formulation chosen was the double emulsion $W_1/O/W_2$ with a proportion of 80/20 with an internal phase W_1/O with a proportion of 30/70. How the addition of this emulsion to a commercial cream affected the physicochemical properties was studied.

3.1. Droplet Size Distribution and Morphology

Figure 1 shows the results of the droplet size distribution for the double emulsion and commercial cream as well as the mixtures of the commercial cream and optimal formulation chosen in proportions of 1/2, 1/3 and 1/4 emulsion.

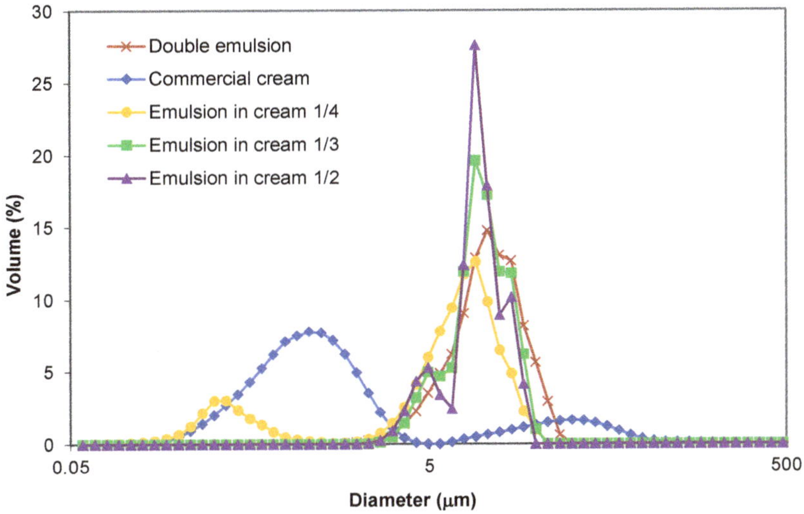

Figure 1. Droplet size distribution of double emulsion, commercial cream and cream-emulsion mixtures.

The results of Figure 1 show that mixing the emulsion has a great influence on the droplet size, considerably displacing the main peak to the right and producing an increase in the average droplet size. It can be observed that the main peaks of the mixtures are similar to the main peak obtained for the double emulsion droplet size distribution.

It can also be seen how, as a greater amount of emulsion is incorporated, the size distribution of the commercial cream (the peak at low sizes) is transformed into the typical size distribution of double emulsions with an internal emulsion ratio of 30/70 [42,44]. Thus, it can be observed how the mixture that contains the lowest proportion of emulsion shows its first peak close to the main peak of the size distribution of the cream, and a second typical peak of the double emulsions of the same internal concentration appears. As the amount of emulsion in the mixture increased, the size distribution obtained became similar to that of the double emulsion itself [44].

A similar trend was observed in previous studies when the emulsion was incorporated into yogurt. It was seen that in all samples, the presence of emulsion affects the size distribution of the matrix [45]. In the present case, the addition of the emulsion caused an increase in the average droplet size and produced a bimodal distribution as was observed in the case when the emulsion was incorporated into the food matrix [45]. Moreover, the addition of RSV has been found to have some interactions with the matrix where it is incorporated since its presence produces physical changes in the system [46].

It is also important to point out that mixed systems with 1/3 and 1/2 of emulsion on cream present a narrower size distribution than double emulsion itself, which is more noticeable as the emulsion ratio in cream increases. It can be also observed that the presence of the smaller droplets (around 5 μm) disappears when the emulsion portion on the cream increases. This behavior indicates a clear interaction of both systems.

The microscope images obtained for the different samples are shown below in Figure 2.

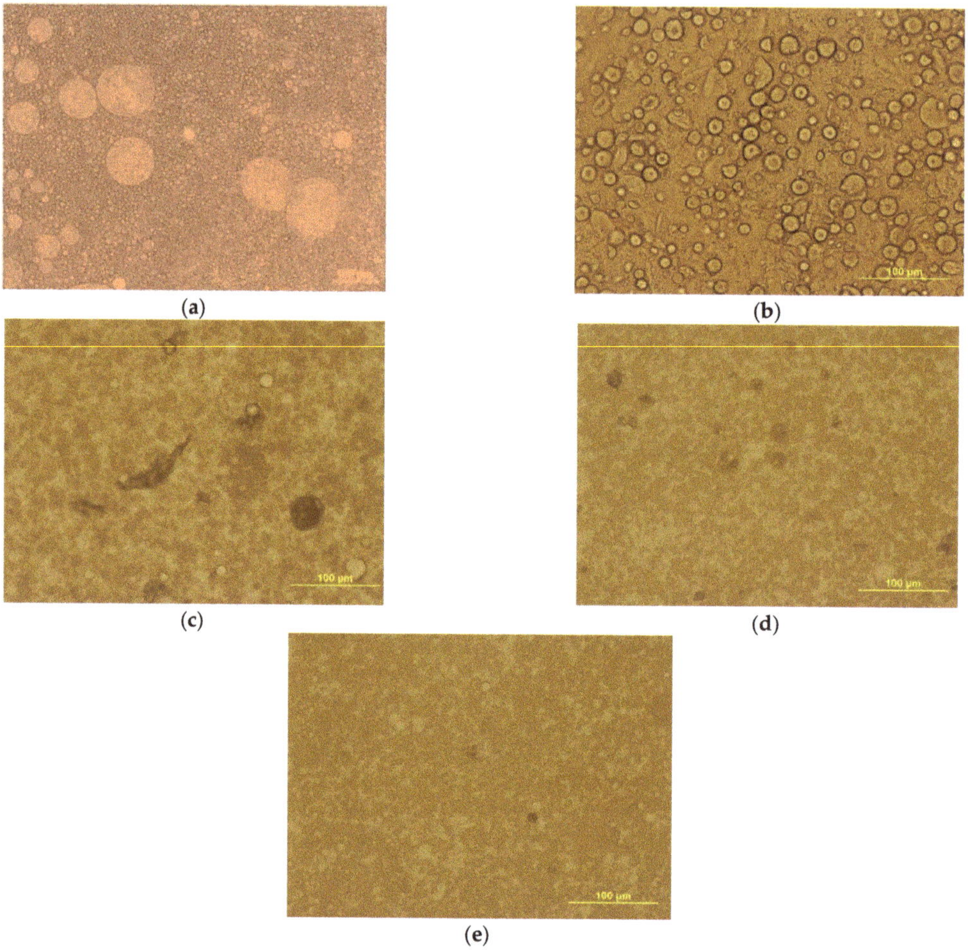

Figure 2. Micrographs of the commercial cream and mixtures of the commercial cream with the double emulsion $W_1/O/W_2$ of concentration 80/20 with an internal phase ratio W_1/O of 30/70: (**a**) double emulsion; (**b**) commercial cream; (**c**) commercial cream with 1/4 emulsion; (**d**) commercial cream with 1/3 emulsion and (**e**) commercial cream with 1/2 emulsion.

It can be observed the individual and large particle sizes registered on the double emulsion (Figure 2a). Moreover, it can be seen how the commercial cream is highly structured but with a lower droplet size (Figure 2b). It is difficult to quantify and compare the concentration of dispersed phases in both individual systems. However, it can be noticed that double emulsion seems to have a larger internal fraction than the commercial cream, which could be because of the large droplet size measured when the mixtures are characterized.

In the mixtures, it can be observed that the structure observed in the commercial cream is broken, even with just a quarter (Figure 2c), and it is even more noticeable when a larger proportion of emulsion is added. This confirms the interaction observed in Figure 1 between both systems, in which clear interaction of both systems is observed in the mixtures. It is difficult to appreciate individual large drops in the mixed systems as in the cases of plain emulsion or plain cream which could indicate that the measurements made by the laser diffraction technique could be responsible for the measurements of agglomerates produced by the interaction of emulsion droplets that could be acting as flocculants of individual cream droplets.

3.2. *Colloidal Stability*

The results of the stability measurements are shown in Figure 3.

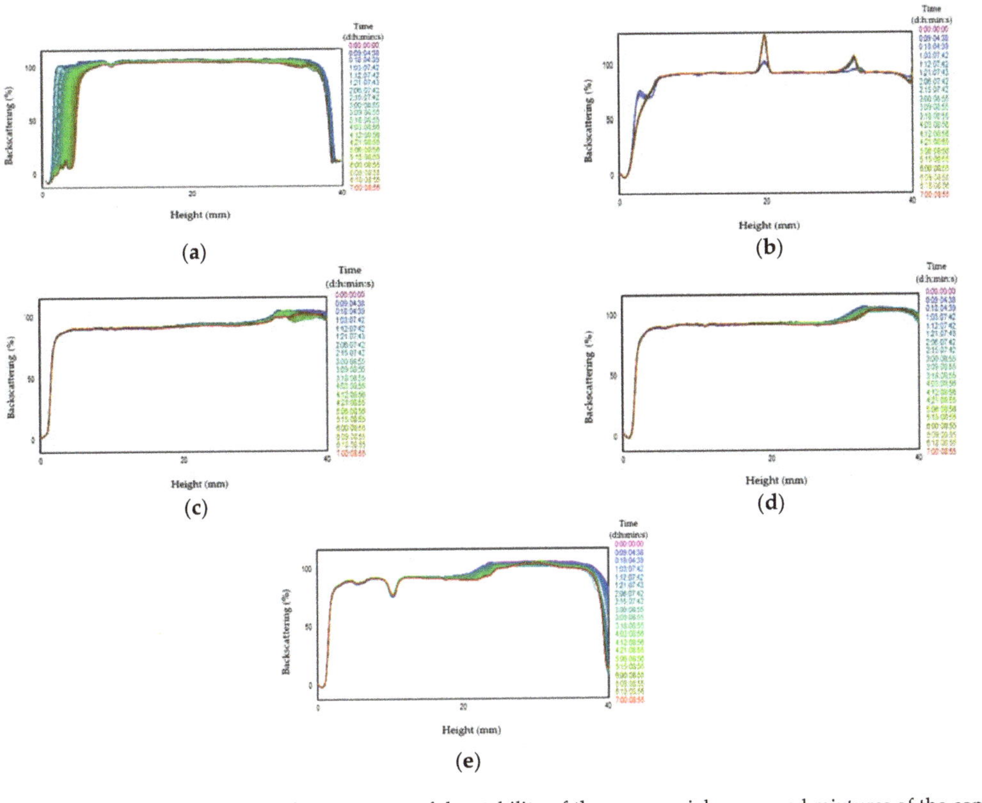

Figure 3. Measurements of the stability of the commercial cream and mixtures of the commercial cream with the double emulsion $W_1/O/W_2$ of concentration 80/20 with an internal phase ratio W_1/O of 30/70: (**a**) double emulsion; (**b**) commercial cream; (**c**) commercial cream with 1/4 emulsion; (**d**) commercial cream with 1/3 emulsion and (**e**) commercial cream with 1/2 emulsion.

It can be observed that double emulsion presents a clarification on the bottom part of the cell (Figure 3a), which is not appreciable when this emulsion was mixed with the commercial cream. This is probably due to the large viscosity of the sample, which increases the stability of the system versus clarification or creaming phenomena.

On the other hand, plain commercial cream presented high stability (Figure 3b). However, when adding the emulsion, it can be seen how the backscattering decreases slightly, especially in the upper part of the cell. Particularly in the case when 1/2 of emulsion was included in the commercial cream, a change on the top part of the system was observed, indicating an instability process related to creaming, probably due to drop coalescence, which seems to be promoted by the interaction between both systems. This creaming effect is less noticeable when 1/4 and 1/3 of emulsion were added to the commercial cream.

3.3. *Rheology*

Viscosity measurements are shown in Figure 4. It can be observed that the commercial cream presents a typical pseudoplastic behavior, while the behavior registered for the double emulsion is closer to a Newtonian fluid [47]. In the case of the mixtures, it can be observed that all of them present a pseudoplastic character. The case when 1/2 of emulsion was added to the mixture is less pronounced, which indicates a high influence of the emulsion on the system viscosity.

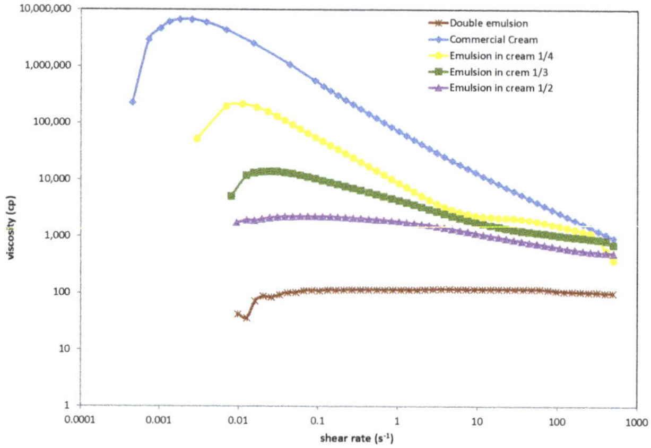

Figure 4. Viscosity measurements of the commercial cream and mixtures of the commercial cream with the double emulsion $W_1/O/W_2$ of concentration 80/20 with an internal phase ratio W_1/O of 30/70.

It is also important to consider that commercial cream viscosity is up to 10^5 times higher than that showed by the emulsion at low shear rates (value around 0.0015 s^{-1}), arising intermediate values for the mixed systems. However, these differences became lower as shear rate increased, showing close values for the commercial cream and mixtures at shear rates of 500 s^{-1}, the point at which the viscosity of the emulsion is 10 times lower than the viscosity of the samples that contain commercial cream.

The parameter values obtained from the Herschel–Bulkley model fitting to the flow curve data for all the samples are shown in Table 1.

Table 1. Parameter values obtained fitting the Herschel–Bulkley rheological model to flow curve data of measured samples.

Sample	Herschel–Bulkley		
	τ_0	k	n
Double emulsion	0.020 ± 0.002	0.149 ± 0.003	0.940 ± 0.003
Commercial cream	18.860 ± 2.202	47.220 ± 5.128	0.356 ± 0.019
Mixture 1/4	28.940 ± 5.165	51.070 ± 6.304	0.344 ± 0.047
Mixture 1/3	5.643 ± 0.412	46.000 ± 5.402	0.220 ± 0.052
Mixture 1/2	−3.103 ± 0.107	5.985 ± 0.602	0.710 ± 0.171

When considering the Herschel–Bulkley model, a relatively high value of the yield point, the stress at which a material begins to deform plastically, is observed in the commercial cream [48], whereas for the double emulsion, the value is close to zero. The yield point increases when mixing a quarter of emulsion but notably decreases in the other mixtures. The n-value indicates the almost Newtonian character (n = 0.94) observed for the double emulsion. The n-value is lower in the case of the mixture of commercial cream and 1/2 of double emulsion up to a value of 0.71. However, the commercial cream presented an n-value of 0.36, close to that registered for the mixtures with 1/4 of double emulsion, which indicates the presence of 1/4; of emulsion does not affect the flow behavior of the sample.

In Figure 5, results obtained from the frequency sweep analysis are shown. The elastic (G′) and the viscous (G″) moduli are represented vs. the frequency. These tests provide information about the viscoelastic character of the samples.

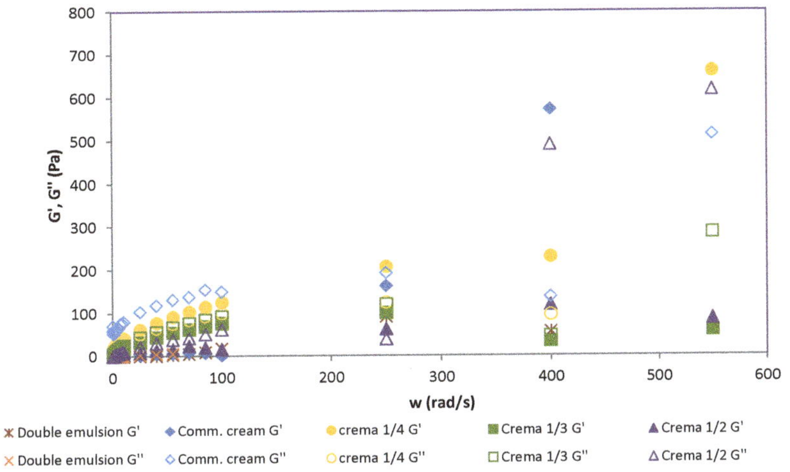

Figure 5. Frequency sweeps of the commercial cream and mixtures of the double emulsion, commercial cream and mixtures cream and double emulsion.

It can be seen that the double emulsion presents a larger G″ than G′, which indicates a character more viscous than elastic. On the contrary, the commercial cream shows a behavior notably more elastic than viscous (G′ > G″) in all the frequency ranges studied.

Observing the mixtures, it can be appreciated that when just 1/4 of emulsion is added to the cream, the samples still have a larger elastic modulus than viscous. However, when 1/3 of the emulsion is added, the values of both moduli were closer, except for the high frequency (550 rad/s) point at which G″ > G′. This fact is even more pronounced when 1/2 of emulsion is added, showing a behavior similar to that exhibited by the double emulsion.

3.4. Texturometry

Texturometry results are shown in Table 2.

Table 2. Results of the texturometry tests for the commercial cream and mixtures of the commercial cream with the double emulsion $W_1/O/W_2$ of concentration 80/20 with an internal phase ratio W_1/O of 30/70.

Sample	Hardness (N)	Stickiness (N)
Commercial cream	0.182 ± 0.002	−0.140 ± 0.002
Mixture 1/4	0.212 ± 0.004	−0.160 ± 0.004
Mixture 1/3	0.268 ± 0.004	−0.175 ± 0.006
Mixture 1/2	0.056 ± 0.001	−0.037 ± 0.002

According to hardness values, it can be seen that the addition of the emulsion barely affects the firmness of the commercial sample until a high proportion of emulsion is achieved, i.e., the mixture 1/2 showed a notably lower value of hardness in comparison with the rest of the samples. A similar trend is observed regarding stickiness (which has a negative value simply because it is a force exerted in the opposite direction) since the 1/2 sample exhibited the lowest stickiness value. Again, this is due to the fact that this sample has a higher proportion of emulsion.

According to the literature, hardness differences below 13% are not perceptible by touch. Moreover, the change in the perception of stickiness seems to be insignificant when the differences are below 6–8% [49–51]. The registered values indicate that even the addition of 1/4 of emulsion to the mixture has acceptable changes in hardness, whereas the presence of double emulsion could produce noticeable changes in stickiness in comparison to cream.

However, it is important to point out the wide range of creams and cosmetics that are commercially available, which entails a wide range of different textural properties [51,52].

4. Conclusions

The presence of emulsion in commercial cream considerably changed the values of the parameters studied. However, its appearance is still acceptable. By increasing the amount of emulsion in the mixture above a third of the total, the cream structure suffered breakage and partial destabilization of the mixture.

The predominant droplet size increased considerably when adding emulsion to commercial cream.

The stability of the commercial cream remained high when it was mixed with the emulsion. Mixing the cream with the emulsion increased its yield resistance. However, this resistance decreased considerably if the amount of emulsion in the mixture was increased by more than 1/3 of emulsion, probably due to a breakage of the structure.

The behavior of commercial cream was predominantly elastic, while double emulsion presented a predominantly viscous character. The mixture with just 1/4 of emulsion presented a behavior similar to cream, while the addition of 1/2 of emulsion produced a change in behavior closer to the one registered for double emulsion alone, with the behavioral intermediate being when 1/3 of emulsion was added.

It has been demonstrated the feasibility to incorporate concentrated double emulsions containing resveratrol into a commercial cream, obtaining a cream with resveratrol concentration up to 0.0042 mg/g and suitable physical properties for satisfactory consumer use.

Author Contributions: R.D.-R., investigation, data curation, and writing—original draft preparation; L.M.-R., investigation and data curation; A.L., data curation, contextualization, supervision, and writing—review and editing; I.M., investigation, data curation and contextualization, M.M., data curation, conceptualization, visualization, writing—review and editing, and funding acquisition; G.G., conceptualization, supervision, writing—review and editing, and funding acquisition. All authors have read and agreed to the published version of the manuscript.

Funding: This research was funded by the Consejería de Economía y Empleo del Principado de Asturias through the 'Plan de Ciencia, Tecnología e Innovación, 2013e2017' [Grants Refs. GRUPIN14-022 and IDI/2018/000185]. Support from the European Regional Development Fund is also gratefully acknowledged.

Data Availability Statement: Not applicable.

Conflicts of Interest: The authors declare no conflict of interest.

References

1. Vang, O.; Ahmad, N.; Baile, C.A.; Baur, J.A.; Brown, K.; Csiszar, A.; Das, D.K.; Delmas, D.; Gottfried, C.; Lin, H.Y.; et al. What Is New for an Old Molecule? Systematic Review and Recommendations on the Use of Resveratrol. *PLoS ONE* **2011**, *6*, e19881. [CrossRef] [PubMed]
2. Yang, T.; Wang, L.; Zhu, M.; Zhang, L.; Yan, L. Properties and Molecular Mechanisms of Resveratrol: A Review. *Pharm. Int. J. Pharm. Sci.* **2015**, *70*, 501–506. [CrossRef]
3. Neves, A.R.; Lucio, M.; Lima, J.L.C.; Reis, S. Resveratrol in Medicinal Chemistry: A Critical Review of Its Pharmacokinetics, Drug-Delivery, and Membrane Interactions. *Curr. Med. Chem.* **2012**, *19*, 1663–1681. [CrossRef] [PubMed]
4. Smoliga, J.M.; Baur, J.A.; Hausenblas, H.A. Resveratrol and Health—A Comprehensive Review of Human Clinical Trials. *Mol. Nutr. Food Res.* **2011**, *55*, 1129–1141. [CrossRef] [PubMed]
5. Pangeni, R.; Sahni, J.K.; Ali, J.; Sharma, S.; Baboota, S. Resveratrol: Review on Therapeutic Potential and Recent Advances in Drug Delivery. *Expert Opin. Drug Deliv.* **2014**, *11*, 1285–1298. [CrossRef] [PubMed]
6. Aluyen, J.K.; Ton, Q.N.; Tran, T.; Yang, A.E.; Gottlieb, H.B.; Bellanger, R.A. Resveratrol: Potential as Anticancer Agent Resveratrol. *J. Diet. Suppl.* **2012**, *9*, 45–46. [CrossRef] [PubMed]
7. Yang, X.; Li, X.; Ren, J. From French Paradox to Cancer Treatment: Anti-Cancer Activities and Mechanisms of Resveratrol. *Anticancer. Agents Med. Chem.* **2014**, *14*, 806–825. [CrossRef]
8. Athar, M.; Back, J.H.; Tang, X.; Kim, K.H.; Kopelovich, L.; Bickers, D.R.; Kim, A.L. Resveratrol: A Review of Preclinical Studies for Human Cancer Prevention. *Toxicol. Appl. Pharmacol.* **2007**, *224*, 274–283. [CrossRef]
9. Kraft, T.E.; Parisotto, D.; Schempp, C.; Efferth, T. Fighting Cancer with Red Wine? Molecular Mechanisms of Resveratrol. *Crit. Rev. Food Sci. Nutr.* **2009**, *49*, 782–799. [CrossRef]

10. Jang, M.; Cai, L.; Udeani, G.O.; Slowing, K.V.; Thomas, C.F.; Beecher, C.W.; Fong, H.H.; Farnsworth, N.R.; Kinghorn, A.D.; Mehta, R.G.; et al. Cancer Chemopreventive Activity of Resveratrol, a Natural Product Derived from Grapes. *Science* **1997**, *275*, 218–220. [CrossRef]
11. Ruivo, J.; Francisco, C.; Oliveira, R.; Figueiras, A. The Main Potentialities of Resveratrol for Drug Delivery Systems. *Braz. J. Pharm. Sci.* **2015**, *51*, 499–514. [CrossRef]
12. Baxter, R.A. Anti-Aging Properties of Resveratrol: Review and Report of a Potent New Antioxidant Skin Care Formulation. *J. Cosmet. Dermatol.* **2008**, *7*, 2–7. [CrossRef]
13. Serravallo, M.; Jagdeo, J.; Glick, S.A.; Siegel, D.M.; Brody, N.I. Sirtuins in Dermatology: Applications for Future Research and Therapeutics. *Arch. Dermatol. Res.* **2013**, *305*, 269–282. [CrossRef] [PubMed]
14. Betz, J. *Resveratrol and Its Effects on Human Health and Longevity—Myth or Miracle?* Natural News: Taichung, Taiwan, 2011.
15. Robinson, K.; Mock, C.; Liang, D. Pre-Formulation Studies of Resveratrol. *Drug Dev. Ind. Pharm.* **2015**, *41*, 1464–1469. [CrossRef] [PubMed]
16. Lafarge, E.; Villette, S.; Cario-André, M.; Lecomte, S.; Faure, C. Transdermal Diffusion of Resveratrol by Multilamellar Liposomes: Effect of Encapsulation on Its Stability. *J. Drug Deliv. Sci. Technol.* **2022**, *76*, 103742. [CrossRef]
17. Sessa, M.; Tsao, R.; Liu, R.; Ferrari, G.; Donsì, F. Evaluation of the Stability and Antioxidant Activity of Nanoencapsulated Resveratrol during in Vitro Digestion. *J. Agric. Food Chem.* **2011**, *59*, 12352–12360. [CrossRef]
18. Liu, Y.; Fan, Y.; Gao, L.; Zhang, Y.; Yi, J. Enhanced PH and Thermal Stability, Solubility and Antioxidant Activity of Resveratrol by Nanocomplexation with α-Lactalbumin. *Food Funct.* **2018**, *9*, 4781–4790. [CrossRef]
19. Wu, Y.; He, D.; Zong, M.; Wu, H.; Li, L.; Zhang, X.; Xing, X.; Li, B. Improvement in the Stability and Bioavailability of Trans-Resveratrol with Hydrolyzed Wheat Starch Complexation: A Theoretical and Experimental Study. *Food Struct.* **2022**, *32*, 100267. [CrossRef]
20. Vian, M.; Tomao, V.; Gallet, S.; Coulomb, P.; Lacombe, J. Simple and Rapid Method for *cis*- and *trans*-Resveratrol and Piceid Isomers Determination in Wine by High-Performance Liquid Chromatography Using Chromolith Columns. *J. Chromatogr. A* **2005**, *1085*, 224–229. [CrossRef]
21. Silva, C.G.; Monteiro, J.; Marques, R.R.N.; Silva, A.M.T.; Martínez, C.; Canle, L.M.; Faria, J.L. Photochemical and Photocatalytic Degradation of Trans-Resveratrol. *Photochem. Photobiol. Sci.* **2013**, *12*, 638–644. [CrossRef]
22. Francioso, A.; Mastromarino, P.; Masci, A.; D'Erme, M.; Mosca, L. Chemistry, Stability and Bioavailability of Resveratrol. *Med. Chem.* **2014**, *10*, 237–245. [CrossRef] [PubMed]
23. Hemar, Y.; Cheng, L.J.; Oliver, C.M.; Sanguansri, L.; Augustin, M. Encapsulation of Resveratrol Using Water-in-Oil-in-Water Double Emulsions. *Food Biophys.* **2010**, *5*, 120–127. [CrossRef]
24. Matos, M.; Gutiérrez, G.; Coca, J.; Pazos, C. Preparation of Water-in-Oil-in-Water ($W_1/O/W_2$) Double Emulsions Containing *trans*-Resveratrol. *Colloids Surf. A Physicochem. Eng. Asp.* **2014**, *442*, 69–79. [CrossRef]
25. Lu, W.; Kelly, A.L.; Miao, S. Emulsion-Based Encapsulation and Delivery Systems for Polyphenols. *Trends Food Sci. Technol.* **2016**, *47*, 1–9. [CrossRef]
26. Gutiérrez, G.; Matos, M.; Benito, J.M.; Coca, J.; Pazos, C. Preparation of HIPEs with Controlled Droplet Size Containing Lutein. *Colloids Surf. A Physicochem. Eng. Asp.* **2014**, *442*, 111–122. [CrossRef]
27. Matos, M.; Gutiérrez, G.; Iglesias, O.; Coca, J.; Pazos, C. Enhancing Encapsulation Efficiency of Food-Grade Double Emulsions Containing Resveratrol or Vitamin B12 by Membrane Emulsification. *J. Food Eng.* **2015**, *166*, 212–220. [CrossRef]
28. Matos, M.; Gutiérrez, G.; Iglesias, O.; Coca, J.; Pazos, C. Characterization, Stability and Rheology of Highly Concentrated Monodisperse Emulsions Containing Lutein. *Food Hydrocoll.* **2015**, *49*, 156–163. [CrossRef]
29. Juškaitė, V.; Ramanauskienė, K.; Briedis, V. Design and Formulation of Optimized Microemulsions for Dermal Delivery of Resveratrol. *Evid.-Based Complement. Altern. Med.* **2015**, *2015*, 540916. [CrossRef]
30. Yang, Y.; McClements, D.J. Encapsulation of Vitamin E in Edible Emulsions Fabricated Using a Natural Surfactant. *Food Hydrocoll.* **2013**, *30*, 712–720. [CrossRef]
31. Reszko, A.E.; Berson, D.; Lupo, M.P. Cosmeceuticals: Practical Applications. *Dermatol. Clin.* **2009**, *27*, 401–416. [CrossRef]
32. Márquez, A.L.; Medrano, A.; Panizzolo, L.A.; Wagner, J.R. Effect of Calcium Salts and Surfactant Concentration on the Stability of Water-in-Oil (w/o) Emulsions Prepared with Polyglycerol Polyricinoleate. *J. Colloid Interface Sci.* **2010**, *341*, 101–108. [CrossRef]
33. Wilson, R.; Van Schie, B.; Howes, D. Overview of the Preparation, Use and Biological Studies on Polyglycerol Polyricinoleate (PGPR). *Food Chem. Toxicol.* **1998**, *36*, 711–718. [CrossRef]
34. Wolf, F.; Koehler, K.; Schuchmann, H. Stabilization of Water Droplets in Oil with PGPR for Use in Oral and Dermal Applications. *J. Food Process Eng.* **2013**, *36*, 276–283. [CrossRef]
35. Sun, X.; Peng, B.; Yan, W. Measurement and Correlation of Solubility of Trans-Resveratrol in 11 Solvents at T = (278.2, 288.2, 298.2, 308.2, and 318.2) K. *J. Chem. Thermodyn.* **2008**, *40*, 735–738. [CrossRef]
36. Frasch-Melnik, S.; Spyropoulos, F.; Norton, I.T. $W_1/O/W_2$ Double Emulsions Stabilised by Fat Crystals—Formulation, Stability and Salt Release. *J. Colloid Interface Sci.* **2010**, *350*, 178–185. [CrossRef] [PubMed]
37. Jiang, J.; Mei, Z.; Xu, J.; Sun, D. Effect of Inorganic Electrolytes on the Formation and the Stability of Water-in-Oil (W/O) Emulsions. *Colloids Surf. A Physicochem. Eng. Asp.* **2013**, *429*, 82–90. [CrossRef]
38. Szymczyk, K.; Szaniawska, M.; Taraba, A. Micellar Parameters of Aqueous Solutions of Tween 20 and 60 at Different Temperatures: Volumetric and Viscometric Study. *Colloids Interfaces* **2018**, *2*, 34. [CrossRef]

39. Mittal, K.L. Determination of CMC of Polysorbate 20 in Aqueous Solution by Surface Tension Method. *J. Pharm. Sci.* **1972**, *61*, 1334–1335. [CrossRef]
40. Miyagishi, S.; Okada, K.; Asakawa, T. Salt Effect on Critical Micelle Concentrations of Nonionic Surfactants, N-Acyl-N-Methylglucamides (MEGA-N). *J. Colloid Interface Sci.* **2001**, *238*, 91–95. [CrossRef]
41. Matos, M.; Gutiérrez, G.; Martínez-Rey, L.; Iglesias, O.; Pazos, C. Encapsulation of Resveratrol Using Food-Grade Concentrated Double Emulsions: Emulsion Characterization and Rheological Behaviour. *J. Food Eng.* **2018**, *226*, 73–81. [CrossRef]
42. Díaz-Ruiz, R.; Martínez-Rey, L.; Laca, A.; Álvarez, J.R.; Gutiérrez, G.; Matos, M. Enhancing Trans-Resveratrol Loading Capacity by Forcing $W_1/O/W_2$ Emulsions up to Its Colloidal Stability Limit. *Colloids Surf. B Biointerfaces* **2020**, *193*, 111130. [CrossRef] [PubMed]
43. García, V.; Laca, A.; Martínez, L.A.; Paredes, B.; Rendueles, M.; Díaz, M. Development and Characterization of a New Sweet Egg-Based Dessert Formulation. *Int. J. Gastron. Food Sci.* **2015**, *2*, 72–82. [CrossRef]
44. Díaz-Ruiz, R.; Valdeón, I.; Álvarez, J.R.; Matos, M.; Gutiérrez, G. Simultaneously Encapsulation of Trans-Resveratrol and Vitamin D3 in Highly Concentrated Double Emulsions. *J. Sci. Food Agric.* **2020**, *101*, 3654–3664. [CrossRef] [PubMed]
45. Díaz-Ruiz, R.; Laca, A.; Sánchez, M.; Fernández, M.R.; Matos, M.; Gutiérrez, G. Addition of Trans-Resveratrol-Loaded Highly Concentrated Double Emulsion to Yoghurts: Effect on Physicochemical Properties. *Int. J. Mol. Sci.* **2022**, *23*, 85. [CrossRef]
46. Davidov-Pardo, G.; McClements, D.J. Resveratrol Encapsulation: Designing Delivery Systems to Overcome Solubility, Stability and Bioavailability Issues. *Trends Food Sci. Technol.* **2014**, *38*, 88–103. [CrossRef]
47. Żołek-Tryznowska, Z. 6—*Rheology of Printing Inks*; Izdebska, J., Thomas, S.B.T.-P.P., Eds.; William Andrew Publishing: Norwich, NY, USA, 2016; pp. 87–99. ISBN 978-0-323-37468-2.
48. Abdewi, E.F. *Mechanical Properties of Reinforcing Steel Rods Produced by Zliten Steel Factory*; Elsevier: Amsterdam, The Netherlands, 2017; ISBN 978-0-12-803581-8.
49. Aktar, T.; Chen, J.; Ettelaie, R.; Holmes, M. Tactile Sensitivity and Capability of Soft-Solid Texture Discrimination. *J. Texture Stud.* **2015**, *46*, 429–439. [CrossRef]
50. Yeon, J.; Kim, J.; Ryu, J.; Park, J.-Y.; Chung, S.-C.; Kim, S.-P. Human Brain Activity Related to the Tactile Perception of Stickiness. *Front. Hum. Neurosci.* **2017**, *11*, 8. [CrossRef]
51. Gilbert, L.; Savary, G.; Grisel, M.; Picard, C. Predicting Sensory Texture Properties of Cosmetic Emulsions by Physical Measurements. *Chemom. Intell. Lab. Syst.* **2013**, *124*, 21–31. [CrossRef]
52. Cisneros Estevez, A.; Toro-Vazquez, J.F.; Hartel, R.W. Effects of Processing and Composition on the Crystallization and Mechanical Properties of Water-in-Oil Emulsions. *J. Am. Oil Chem. Soc.* **2013**, *90*, 1195–1201. [CrossRef]

colloids and interfaces

MDPI

Article

Stability Studies and the In Vitro Leishmanicidal Activity of Hyaluronic Acid-Based Nanoemulsion Containing *Pterodon pubescens* Benth. Oil

Sirlene Adriana Kleinubing [1], Priscila Miyuki Outuki [1], Éverton da Silva Santos [1], Jaqueline Hoscheid [2,3,*,†], Getulio Capello Tominc [2], Mariana Dalmagro [3], Edson Antônio da Silva [4], Marli Miriam de Souza Lima [1], Celso Vataru Nakamura [1] and Mara Lane Carvalho Cardoso [1]

1 Programa de Pós-Graduação em Ciências Farmacêuticas, Universidade Estadual de Maringá, Maringá 87020-900, Brazil
2 Programa de Mestrado Profissional em Plantas Medicinais e Fitoterápicos na Atenção Básica, Universidade Paranaense, Umuarama 87502-210, Brazil
3 Programa de Pós-Graduação em Biotecnologia Aplicada à Agricultura, Universidade Paranaense, Umuarama 87502-210, Brazil
4 Programa de Pós-Graduação em Engenharia Química, Centro de Engenharias e Ciências Exatas, Universidade Estadual do Oeste do Paraná, Toledo 85903-000, Brazil
* Correspondence: jaqueline.hoscheid@gmail.com
† Current address: Universidade Paranaense, Umuarama 87502-210, Brazil.

Citation: Kleinubing, S.A.; Outuki, P.M.; Santos, É.d.S.; Hoscheid, J.; Tominc, G.C.; Dalmagro, M.; Silva, E.A.d.; Lima, M.M.d.S.; Nakamura, C.V.; Cardoso, M.L.C. Stability Studies and the In Vitro Leishmanicidal Activity of Hyaluronic Acid-Based Nanoemulsion Containing *Pterodon pubescens* Benth. Oil. *Colloids Interfaces* **2022**, *6*, 64. https://doi.org/10.3390/colloids6040064

Academic Editors: César Burgos-Díaz, Mauricio Opazo-Navarrete, Eduardo Morales and Georgi G. Gochev

Received: 14 September 2022
Accepted: 1 November 2022
Published: 3 November 2022

Abstract: The physicochemical and microbiological stability of a hyaluronic acid-based nanostructured topical delivery system containing *P. pubescens* fruit oil was evaluated, and the in vitro antileishmanial activity of the nanoemulsion against *Leishmania amazonensis* and the cytotoxicity on macrophages was investigated. The formulation stored at 5 ± 2 °C, compared with the formulation stored at 30 and 40 ± 2 °C, showed a higher chemical and physical stability during the period analyzed and in the accelerated physical stability study. The formulation stored at 40 °C presented a significant change in droplet diameter, polydispersity index, zeta potential, pH, active compound, and consistency index and was considered unstable. The microbiological stability of the formulations was confirmed. The leishmanicidal activity of the selected system against intracellular amastigotes was significantly superior to that observed for the free oil. However, further research is needed to explore the use of the hyaluronic acid-based nanostructured system containing *P. pubescens* fruit oil for the treatment of cutaneous leishmaniasis.

Keywords: sucupira; nanoemulsion; accelerated physical stability; hyaluronic acid; *Leishmania amazonensis*

1. Introduction

Leishmaniasis is one of the most important parasitic diseases, caused by the protozoa of the genus *Leishmania*, with a great impact on global public health [1,2]. This infection manifests itself in several clinical forms, mainly visceral, mucocutaneous, and cutaneous leishmaniasis (CL). The cutaneous form is worldwide the most prevalent clinical form of leishmaniasis, characterized by chronic lesions and permanent scars on the skin, with deformations in the infected areas [3–5]; it is caused by several species of *Leishmania*, including *Leishmania tropica*, *Leishmania major*, *Leishmania amazonensis*, and *Leishmania braziliensis*. The chemotherapy of leishmaniasis includes pentavalent antimonials, miltefosine, amphotericin B, and paromomycin [2]. However, the high cost, the side effects, and the development of resistance are the main disadvantages of these drugs that compromise the efficacy of the treatment [6]. Therefore, many efforts have been made to develop new drug therapies [7,8].

Recent studies focused on antileishmanial activities of medicinal plants showed their potential to inhibit the growth of several species of *Leishmania* [6,8,9]. *Pterodon pubescens* Benth. is a native plant of the central region of Brazil, popularly known as "sucupira branca". The *P. pubescens* fruit oil is used in popular medicine to treat various diseases for its anti-rheumatic, analgesic, anti-inflammatory, and antileishmanial properties [10–17]. According to some studies, the fruit oil is rich in derivatives of geranylgeraniol and vouacapan diterpenes, which are related to the leishmanicidal activity of this plant [7,18,19].

Nanoemulsions have attracted great attention as vehicles for the delivery of hydrophobic active substances, due to the stability conferred by these systems and to the reduced droplet size, which provides a greater surface area and a high bioavailability of the active substance [20,21]. Targeted drug delivery systems have been considered an effective strategy for the prevention and treatment of leishmaniasis [22]. Since macrophages are the main phagocytic cells involved in leishmaniasis infection, drug delivery systems that can target the antileishmanial agent to these cells can represent a promising approach to increasing the therapeutic efficacy for this disease, and to reducing the toxic effects in the normal cells [2,23].

The potential of hyaluronic acid (HA) for use in drug delivery systems is related to its biological, rheological, and physicochemical properties, in addition to it being biocompatible, non-toxic, and completely biodegradable [24–27]. Macrophages are known to express CD44 receptors (HA receptors). HA is considered the main ligand of the CD44 receptor. The CD44-HA binding allows for the targeting of nanocarriers with HA to cells that contain this receptor. Furthermore, HA can cross the cell membrane, an extremely useful strategy for intracellular therapeutic delivery [28]. Thus, HA-based nanocarriers may be used for targeting active substances into macrophages, aiming to treat diseases associated with this cell [28–30].

In a previous study, we reported the development of HA-based nanoemulsion containing *P. pubescens* fruit oil [31]. In the present study, the physicochemical stability of the previously selected system was investigated under different storage conditions. In addition, considering the therapeutic properties of *P. pubescens* oil, the in vitro leishmanicidal activity of the nanoemulsion against the intracellular amastigotes forms of *Leishmania amazonensis* and its cytotoxicity in macrophages was also investigated.

2. Materials and Methods

2.1. Materials

Soybean phospholipids (Lipoid S100—soybean lecithin, ≥94% phosphatidylcholine) and polyethylene glycol hydrogenated castor oil/sorbitan oleate (PEG-40H) were kindly provided by Lipoid GMBH (Ludwigshaffen, Germany) and Oxiteno (São Paulo, Brazil), respectively. Hyaluronic acid (HA) was obtained from Via Farma (São Paulo, Brazil). Ultra-purified water was obtained using a Milli-Q Plus (Millipore Corporation, Billerica, MA, USA) system. All other chemicals and reagents were of analytical grade.

2.2. Oil Extraction of *P. pubescens* Fruit

The *P. pubescens* fruit was collected in the city of Nossa Senhora do Livramento, Mato Grosso state, Brazil (15°89′ S; longitude 56°41′ W). The taxonomic identification was performed, and voucher specimens were deposited at the herbarium of the Federal University of Mato Grosso (number 39551) and State University of Maringá (number 20502). The oil extraction from the *P. pubescens* fruit was performed as previously reported [32], by turbo extraction (Ultra-Turrax UTC115KT, IKA® Works, Wilmington, NC, USA).

2.3. Nanoemulsion Preparation

The oil phase consisting of the *P. pubescens* oil (3.0%, *w*/*w*) and the lipoid S100 surfactant (1.0%, *w*/*w*) was heated to 70 °C and added to the aqueous phase constituted of PEG-40H (10.0%, *w*/*w*) and water (at 70 °C) and under constant stirring, using an Ultra-Turrax T-25 (IKA® Works, Wilmington, NC, USA), at 18,000 rpm for 15 min. Afterward, HA (0.2%, *w*/*w*) was added to the formulation and stirred for 2 h at room temperature, using a magnetic stirrer. The formulation was kept at rest for 24 h at 25 ± 1 °C before further characterization [31].

2.4. *Accelerated Physical Stability*

The physical stability of the nanoemulsion was determined by centrifugal force using a multisampling analytical centrifuge (LUMiSizer®; LUM GmbH, Berlin, Germany). This study was performed 24 h after preparation of the formulations. Samples were placed in rectangular cuvettes with an optical path of 10.0 mm, and exposed to a rotational speed of 4000 rpm for 8 h, at temperatures of 5, 30, and 40 °C (the same temperatures tested during the stability study for 180 d). The physical stability of the formulations was analyzed by the transmission profiles and the instability index, calculated using the SepView 6.0 software (LUM, Berlin, Germany).

2.5. *Physicochemical Stability Study*

The study of the physicochemical properties of a nanostructured system can reveal valuable information about its stability [21]. This stability study was performed according to a guide for stability studies [33].

The nanoemulsion was added to glass containers (5 mL) with a bung and screw cap and incubated at different storage conditions: 5 ± 2 °C, 30 ± 2 °C, and 40 ± 2 °C (75% relative humidity). After pre-determined intervals (24 h and 30, 60, 90, and 180 d), the samples were evaluated with respect to macroscopic analysis, droplet diameter, PDI, zeta potential, pH, vouacapan content, and consistency index. Free *P. pubescens* oil samples were also kept under the same storage conditions, to compare the chemical stability of the free oil with the oil present in the nanoemulsion.

2.5.1. Macroscopic Analysis

The samples were visually evaluated in relation to color, appearance, and phase separation.

2.5.2. Droplet Diameter and Polydispersity Index (PDI)

The determination of average droplet diameter and PDI formulation were performed using a particle analyzer (NanoPlus-3 zeta/nano particle analyzer, Micromeritics Instrument Corporation, Norcross, GA, USA) by dynamic light scattering (DLS). All measurements were taken at a fixed angle of 90° and at a temperature of 25 ± 0.1 °C. The samples were diluted 1:10 in ultra-purified water before measurement, to avoid multiple scattering effects [34].

2.5.3. Zeta Potential

The zeta potential was determined by electrostatic mobility using a NanoPlus/zeta particle analyzer (Micromeritics Instrument Corporation, GA, USA) at 25 ± 0.1 °C. Before measurements, the samples were diluted 1:10 in ultra-purified water, pH 6.8.

2.5.4. pH

The pH of the samples was measured at 25 ± 1 °C, using a calibrated pH meter (TECNAL, São Paulo, Brazil).

2.5.5. Chemical Analysis of *P. pubescens* Oil by GC-MS

The chromatographic profiles of *P. pubescens* oil present in the formulation were obtained by gas chromatography coupled with mass spectrometry (GC-MS) (Thermo Electron Corporation DSQ II; TLC, Thermo Fisher Scientific Inc., Waltham, MA, USA), equipped with an HP-5 capillary column (30 m × 0.25 mm). The vouacapan, used as a marker, was quantified by the monitoring of selected ions using a single ion monitoring system [34]. Lyophilized aliquots of the formulation were resuspended in 1 mL of chloroform and injected for analysis. The percentage of recovery was calculated as the ratio of the experimental concentration of vouacapan present in the samples to the theoretical concentration multiplied by 100.

2.5.6. Consistency Index

The consistency index of the formulation was determined by continuous shear, using a MARS II (Haake®) controlled stress rheometer (Thermo Fisher Scientific Inc., Newington, Germany), inflow mode, at 25 ± 0.1 °C, with a controlled shear rate (CR), equipped with a cone-plate geometry of 35 mm and a 2° angle, separated by a fixed distance of 0.052 mm [31]. The Ostwald–de Waele equation (*Power Law*) was used to obtain the consistency index.

2.6. Microbiological Stability

The microbiological stability of nanoemulsion was determined according to the Brazilian Pharmacopoeia [35]. The test was carried out with fresh samples (after 24 h of preparation) and with the samples kept at 30 °C for 180 d. The analysis was performed using the pour plate technique for the following microorganisms: heterotrophic bacteria, fungi, and molds. Aliquots of nanoemulsion diluted in sterile saline were placed in Petri dishes containing 20 mL of a TSA medium (for heterotrophic bacteria) and Sabouraud dextrose agar (for fungi and molds), previously melted and stabilized at 45 °C. After homogenization and solidification of the medium, the samples were incubated in BOD incubators (Biogenic Oxygen Demand, TE-390 model, TECNAL, Piracicaba, Brazil) under the following conditions: 72 h at 35 °C to determine the presence of total heterotrophic bacteria, and 7 d at 28 °C to determine the presence of fungi and molds. After this period, the number of colony-forming units (CFU)/mL was determined. All analyses were performed in triplicate.

2.7. In Vitro Antileishmanial Activity

Peritoneal macrophages were obtained from BALB/c mice (age 3–8 weeks). Syringes were filled aseptically with 0.01 M cold phosphate buffer (PBS), and the content was injected into the peritoneal cavity of mice previously euthanized with lidocaine (10 mg/kg) and thiopental (200 mg/kg). The contents were then removed from the cavity, and maintained at 4 °C. This suspension was centrifuged for 10 min at 1500 rpm and 4 °C, promoting the formation of a cell pellet. This cell pellet was resuspended in an RPMI 1640 medium supplemented with 10% inactivated fetal bovine serum (FBS, Gibco), to form a suspension with a concentration of 5×10^5 macrophages/mL. These cells were added to glass coverslips in 24-well plates. The plates were incubated at 37 °C and 5% CO_2 for 2 h, to promote the adhesion of macrophages to the coverslips. The cells that did not adhere were removed by rinsing the well with a culture medium. Afterward, macrophages were infected with *L. amazonensis* promastigotes cultured for 5–6 d at a 7:1 parasite/macrophage ratio, and then incubated at 34 °C and 5% CO_2 for 4 h. The wells were again washed to remove non-internalized protozoa.

Afterwards, the free *P. pubescens* oil and the nanoemulsion, at different concentrations, were diluted in an RPMI medium supplemented with 10% FBS, and added to the wells. The plates were then incubated for 48 h at 34 °C and 5% CO_2. At the end of this period, the supernatant was removed, and the coverslips were fixed with methanol for 10 min and stained with 10% Giemsa in 0.01 M PBS for 40 min. Slides were analyzed under an optical microscope by counting the total number of macrophages, the number of infected macrophages, and the number of amastigotes per cell. The concentration capable of inhibiting 50% of the parasite growth (IC_{50}) was calculated by linear regression analysis. The results were compared with the results for positive (Miltefosine) and negative (culture medium alone) controls.

2.8. Cytotoxicity Assays

Macrophages (J774A1) were cultured in an RPMI 1640 medium (Sigma-Aldrich Corporation, St. Louis, MO, USA) at pH 7.6 supplemented with 10% FBS, penicillin (5000 U/mL), and streptomycin (5 mg/mL). Cytotoxic assays were performed in sterile 96-well plates, with an initial inoculum of 5×10^5 cells/mL in an RPMI 1640 medium incubated at 37 °C and 5% CO_2 for 24 h. Afterwards, the free *P. pubescens* oil and the nanoemulsion were added to the cell monolayers and incubated for 48 h. The culture medium was then removed, the cells were washed with PBS, and 50 μL of MTT (3-[4,5-dimethylthiazol-2-yl]-2,5-diphenyltetrazolium bromide formazan; 2 mg/mL in PBS) were added. The cells were incubated while protected from light for 4 h at 37 °C and 5% CO_2. To solubilize the formazan crystals, 150 μL of dimethylsulfoxide DMSO were added to the plates. The absorbance of the resulting solutions was read at 570 nm in spectrophotometer plates (Bio-Tek—Power Wave XS) to measure cell metabolic viability [36]. The concentration capable of inhibiting 50% of the cell growth was expressed as the cytotoxic concentration (CC_{50}). The selectivity index was determined by the ratio of cytotoxicity (CC_{50}) on J774A1 cells to leishmanicidal activity (IC_{50}) against *L. amazonensis*.

2.9. Statistical Analysis

All experiments were performed in triplicate, and were presented as mean ± standard deviation. ANOVA and a post hoc Tukey's test were used for statistical comparison of the results. The results were considered statistically significant for values of $p \leq 0.05$.

3. Results

3.1. Physical Stability Accelerated

The transmission profile result obtained at different temperatures is shown in Figure 1. Overlapped transmission profiles indicate a stable system, and the best result was found at 5 °C, indicating a greater stability of the formulation.

Table 1 shows the results of the instability indexes. It was observed that the instability index increased by around 65% when centrifuged at 40 °C, as compared with 30 °C, demonstrating the instability of the nanoemulsion at this temperature. Comparing the results obtained at 5 and 30 °C, the instability index was approximately 10% lower at 5 °C, suggesting that the formulation would be more stable when stored at this temperature.

Table 1. Instability index of HA-based nanoemulsion containing *P. pubescens* oil at temperatures of 5, 30, and 40 °C.

	Temperature		
	5 °C	**30 °C**	**40 °C**
Instability index	0.13 ± 0.06^{b}	0.15 ± 0.00^{b}	0.42 ± 0.12^{a}

Notes: The instability index was calculated using the SepView 6.0 software (LUM, Berlin, Germany). Different letters are considered to be statistically different ($p < 0.05$).

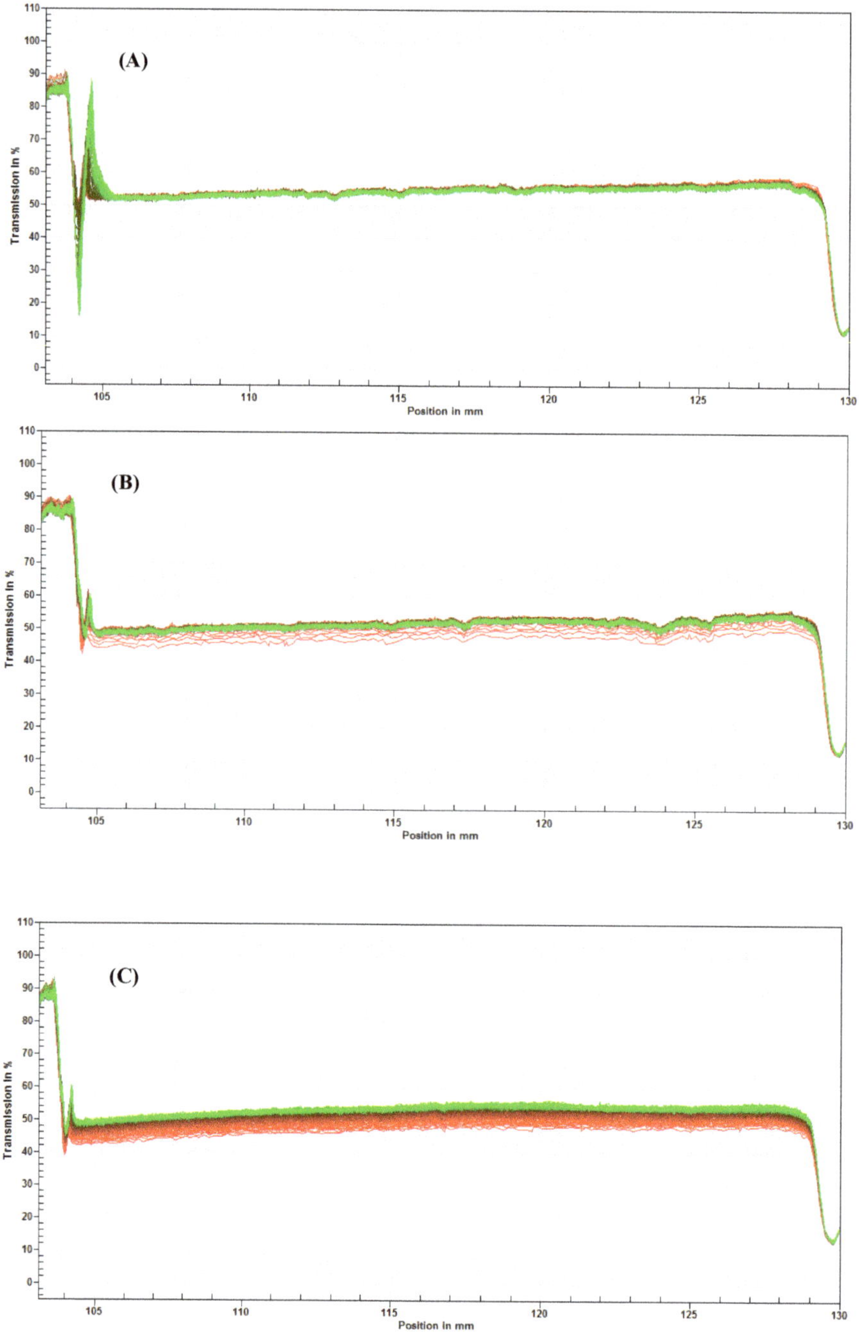

Figure 1. Evolution of the transmission profiles of the HA-based nanoemulsion containing *P. pubescens* oil at (**A**) 5 °C, (**B**) 30 °C, and (**C**) 40 °C.

3.2. Physicochemical Stability Study

This study was carried out to test the best storage conditions for the developed system. The formulation developed for the stability study was characterized 24 h after preparation. It was translucent, homogeneous, and without any signal of phase separation.

The pictures in Figure 2 reveal no change in the macroscopic aspects of the formulation stored for 180 d, at the different temperatures. The formulation presented no signal of phase separation, and had the same transparent and homogeneous appearance as the freshly prepared formulation.

Figure 2. Pictures of the HA-based nanoemulsion containing *P. pubescens* oil stored for 180 d at temperatures of 5, 30, and 40 °C.

The average droplet diameter of the formulation at temperatures of 5, 30, and 40 °C, at intervals of 24 h and 30, 60, 90, and 180 d, are presented in Table 2. When stored at 5 °C, there was an increase in diameter from 60 to 180 d, which was statistically significant ($p < 0.05$), in comparison with 24 h and 30 d.

At 30 °C, an increase in droplet diameter occurred at 90 d and remained constant until the end of the study period. Regarding the temperature of 40 °C, the droplet diameter did not change until 90 d, but a decrease was observed at 180 d.

In general, a slight increase in the PDI as a function of time was observed when stored at 5 and 30 °C. However, the formulation had a homogeneous droplet distribution, since the PDI of the samples remained at <0.3. The zeta potential values remained stable for the formulation stored at 5 °C and 30 °C until 180 d.

On the other hand, the formulation stored at 40 °C showed a decrease in zeta potential, suggesting that this system may present less physical stability compared with the formulation stored at 5 and 30 °C. In the accelerated stability study, the instability index was higher for the formulation evaluated at 40 °C. Thus, both studies suggest that this temperature may decrease the stability of the formulation.

It is important to monitor the pH values of the formulation, since pH change indicates chemical reactions, and may compromise system quality. It was observed that the pH of the nanoemulsion at temperatures of 5 and 30 °C decreased during storage, but remained around 6.0 at the end of the study. However, when the formulation was stored at 40 °C, there was a more pronounced decrease in the pH at 180 d, with a value slightly above 4.0.

Table 2. Droplet diameter, PDI, zeta potential, and pH of the HA-based nanoemulsion containing *P. pubescens* oil stored at temperatures of 5, 30, and 40 °C for 180 d.

		Time				
		24 h	30 d	60 d	90 d	180 d
Droplet diameter (nm)	5 °C	24.86 ± 0.64 [b]	24.66 ± 0.85 [b]	29.50 ± 1.20 [a]	31.33 ± 1.50 [a]	32.36 ± 1.45 [a]
	30 °C	24.86 ± 0.64 [b]	24.83 ± 0.68 [b]	25.06 ± 0.49 [b]	27.30 ± 0.46 [a]	27.13 ± 1.10 [a]
	40 °C	24.86 ± 0.64 [a]	23.70 ± 0.75 [a]	23.90 ± 0.11 [a]	23.66 ± 0.75 [a]	18.83 ± 0.64 [b]
PDI	5 °C	0.24 ± 0.04 [c]	0.26 ± 0.07 [b]	0.26 ± 0.09 [b]	0.26 ± 0.01 [b]	0.29 ± 0.06 [a]
	30 °C	0.24 ± 0.04 [c]	0.24 ± 0.04 [cd]	0.23 ± 0.03 [d]	0.29 ± 0.09 [b]	0.30 ± 0.04 [a]
	40 °C	0.24 ± 0.04 [a]	0.23 ± 0.05 [b]	0.23 ± 0.02 [b]	0.22 ± 0.06 [c]	0.19 ± 0.03 [d]
Zeta potential (mV)	5 °C	−31.45 ± 0.26 [a]	ND	ND	ND	−32.44 ± 0.20 [a]
	30 °C	−31.45 ± 0.26 [a]	ND	ND	ND	−29.67 ± 0.09 [a]
	40 °C	−31.45 ± 0.26 [a]	ND	ND	ND	−18.47 ± 1.50 [b]
pH	5 °C	6.80 ± 0.03 [a]	6.70 ± 0.05 [b]	6.56 ± 0.03 [ab]	6.33 ± 0.05 [bc]	6.00 ± 0.01 [c]
	30 °C	6.80 ± 0.03 [a]	6.50 ± 0.01 [b]	6.47 ± 0.04 [b]	6.15 ± 0.07 [c]	6.09 ± 0.14 [c]
	40 °C	6.80 ± 0.03 [a]	5.22 ± 0.07 [b]	4.79 ± 0.04 [c]	4.55 ± 0.01 [d]	4.22 ± 0.00 [e]

Notes: Non-determined (ND). Mean ± SD (n = 3). One-way ANOVA followed by post hoc Tukey's test; different letters (a to e) are statistically different ($p < 0.05$).

The total content of the *P. pubescens* oil was determined by the quantification of the derivatives of the vouacapans. Values close to 100% were found, demonstrating the efficacy of the method used to prepare the nanoemulsion. Figure 3 presents the percentage recovery of vouacapans (used as markers) from the nanoemulsion (Figure 3A) and from the free *P. pubescens* oil (Figure 3B), stored at temperatures at 5, 30, and 40 °C for 180 d.

The percentage of recovery of the oil present in the nanoemulsion showed a small, non-significant change during the analyzed period, when maintained at 5 °C. However, when stored at 30 °C, a small decrease occurred in the content of vouacapans from 90 d. For the free oil stored at temperatures of 5 and 30 °C, it can be seen that the content of the active compound was much lower, compared with the formulation.

The oil content in the nanoemulsion stored at 40 °C had a pronounced decrease from the second month of analysis (approximately 50%). For the free oil, at 40 °C, a marked reduction of the active compound was observed in the first month of analysis.

Figure 4 presents the chromatographic profile of the *P. pubescens* oil present in the nanoemulsion stored at 5 °C for 180 d. It is possible to observe the presence of characteristic peaks of the derivatives of geranylgeraniol (between 11 and 15 min) and vouacapan diterpenes (between 22 and 25 min) in the *P. pubescens* oil.

The consistency index obtained for the nanoemulsion during the stability study is presented in Figure 5.

The initial consistency index was 210.37 ± 1.34 mPa·s. At all temperatures analyzed, the consistency index decreased over time. However, comparing the formulation maintained at 5 and 30 °C, it was observed that the consistency index of the formulation maintained at 5 °C showed a less pronounced decrease (around 18%) than that of the formulation stored at 30 °C, which showed a decrease of approximately 50%.

For the formulation maintained at 40 °C, a marked decrease in the consistency index was observed in the first month of study, and this decrease was almost 100% until 180 d, demonstrating the physical instability of the system at this temperature. Thus, based on these results, it is suggested that the formulation is more stable when stored at 5 °C. These results are in agreement with the results obtained in the accelerated stability study using centrifugal force, in which the formulation evaluated at 5 °C was also more stable.

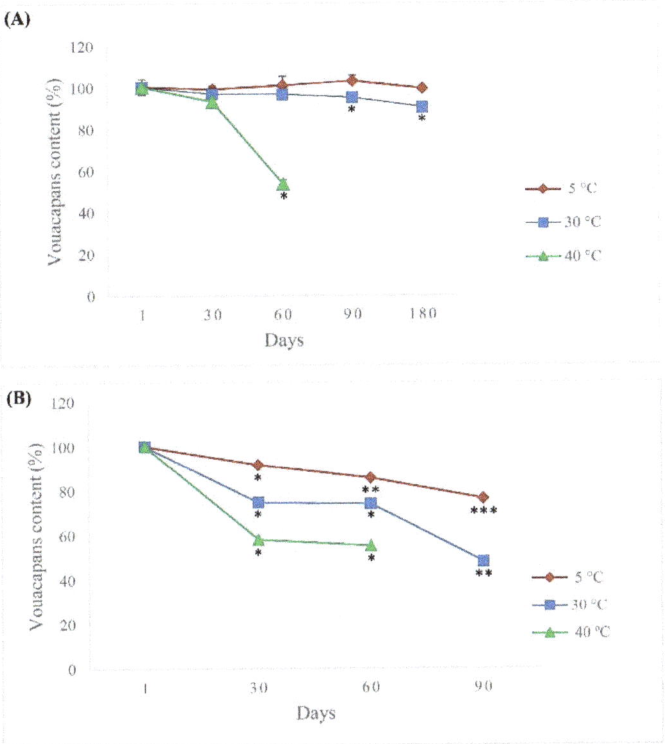

Figure 3. Percentage of recovery of vouacapans of the nanoemulsion (**A**) and free *P. pubescens* oil (**B**), stored at temperatures of 5, 30, and 40 °C for 180 d. Mean ± SD ($n = 3$). One-way ANOVA followed by post hoc Tukey's test; different specimens (*, **, and ***) are considered statistically different at $p < 0.05$.

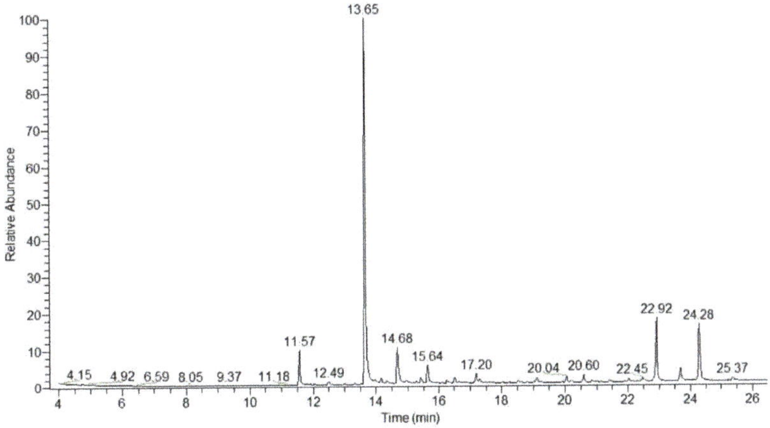

Figure 4. Chromatographic profile of *P. pubescens* oil present in the nanoemulsion stored at 5 °C for 180 d.

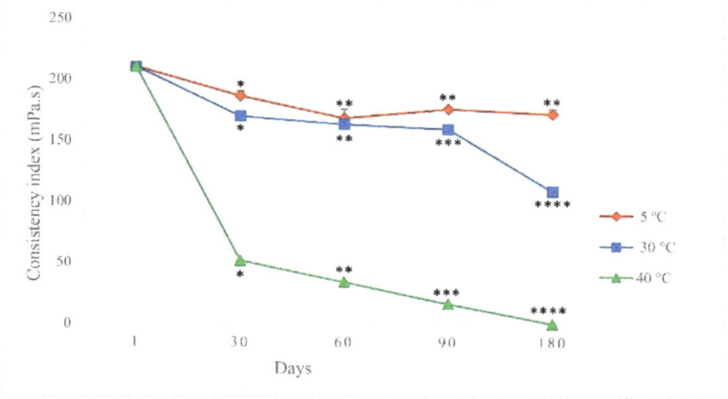

Figure 5. Consistency index of HA-based nanoemulsion containing *P. pubescens* oil stored at temperatures of 5, 30, and 40 °C for 180 d. Mean ± SD ($n = 3$). One-way ANOVA followed by post hoc Tukey's test; different specimens (*, **, ***, and ****) are considered statistically different at $p < 0.05$.

3.3. Microbiological Stability

The results showed that there was no growth of the microorganisms tested.

3.4. In Vitro Antileishmanial Activity and Cytotoxicity Assays

The leishmanicidal activity (IC_{50}) for the free *P. pubescens* oil and the formulation is shown in Table 3. The IC_{50} values demonstrated an expressive leishmanicidal activity of the nanoemulsion, compared with the free *P. pubescens* oil, which represents a decrease in IC_{50} of approximately 95%, suggesting the great potential of the formulation in the treatment of CL. The assays of cytotoxicity showed that the free oil was cytotoxic to macrophages with CC_{50} values of approximately 36 μg/mL.

Table 3. Leishmanicidal activity (IC_{50}), cytotoxicity (CC_{50}) and selectivity index.

Samples	IC_{50} (μg/mL)	CC_{50} (μg/mL)	Selectivity Index
Nanoemulsion	2.00 ± 0.04	3.50 ± 1.00	1.75
Free *P. pubescens* oil	41.50 ± 3.50	36.00 ± 1.40	0.87
Miltefosine	0.70 ± 0.02	22.40 ± 0.80	32.00

Notes: IC_{50}: concentration capable of inhibiting 50% of the parasite growth; CC_{50}: concentration capable of inhibiting 50% of the cell growth.

4. Discussion

The droplet diameter and the PDI are representative parameters that indicate the stability of nanoemulsions [37]. The small initial droplet diameter contributed to the transparency of the nanoemulsion [38]. Despite the increase in the diameter of the droplets for the system at 5 and 30 °C, there was no evidence of creaming or coalescence of the droplets during storage, since the diameters remained around 30 nm, contributing to the kinetic stability of the formulation [38].

The PDI is a representation of the distribution of droplet diameter within a given sample, which can range from 0 to 1. PDI values < 0.2 indicate the existence of a homogeneous particle size, while values close to 1 indicate heterogeneity and physical instability [39,40]. In drug delivery applications using lipid-based carriers such as nanoemulsions, a PDI < 0.3 is considered acceptable [39,41,42], indicating a homogenous population of droplets and a high physical stability during 180 d of storage [43–45].

Additionally, the zeta potential may be an indicator, but it is not a unique criterion to predict the stability of a nanoemulsion [46]. Zeta potential values above 30 mV (absolute value) provide a high energy barrier that causes repulsion of the adjacent droplets, resulting in the formation of stabilized emulsions [47] and the high physical stability of the system [48]. The negative charge observed in this study is related to the anionic nature of HA, which is absorbed in the oil–water interface [47].

Furthermore, high temperatures can change the solubility of nonionic surfactant (PEG-40H), due to the breakdown of hydrogen bonds at the surfactant surface, facilitating aggregation and coalescence by reducing the mechanical barrier, and resulting in a loss of stability [45]. The formulation stored at 40 °C showed a decrease in zeta potential, compared with those stored at 5 and 30 °C, and in the accelerated stability study, the instability index was higher for the formulation evaluated at 40 °C; thus, it is suggested that this temperature may decrease the stability of the formulation, by a reduction of electrical and mechanical barrier energies [44,45].

The pH values close to 6.0 observed for the formulation stored at temperatures of 5 and 30 °C are acceptable, since these values are within the favorable pH range for an emulsified system that remains stable, and are safe values for topical application. On the other hand, the pH values close to 4.0 found for the formulation stored at 40 °C could be the reason for the decrease of zeta potential. pH change indicates the occurrence of chemical reactions that may compromise the quality of the final product [45]. A pH decrease may have occurred due to oxidation of the oil phase, leading to hydroperoxide formations, or due to hydrolysis of triglycerides with the formation of free fatty acids [49]. Additionally, the chromatographic profile of the *P. pubescens* oil present in the nanoemulsion stored at 40 °C for 180 d showed a high chemical degradation of the constituents of the oil present in the formulation, demonstrating the instability of the *P. pubescens* oil at this temperature.

However, when stored at 5 °C, the developed system presented a good recovery content. Comparing the formulation with the free oil, it was observed that the degradation of *P. pubescens* oil present in the nanoemulsion was lower than that of the free oil, indicating that the nanoemulsion protected the vegetable oil from degradation, mainly at temperatures of 5 and 30 °C. These results confirm the results obtained in the thermal analysis [31,50], which demonstrated the ability of the nanoemulsion to protect the *P. pubescens* oil from thermal degradation. According to Bajerski et al. [51], one of the advantages of using nanoemulsion to encapsulate vegetable oils is the ability of these systems to protect the vegetable oil from photo-, thermal, and volatilization instability, improving the chemical stability of the oil.

The viscosity of the system is of great importance for the formation and stability of emulsions. A decrease in the viscosity over time may indicate a kinetic instability of the system [52]. Instabilities arising from variation in droplet size, particle number, and emulsifier orientation or migration over time, can be detected by changes in the viscosity of the product [49].

In this study, we observed that the PDI increased during storage, which may be attributed, in part, to the reduction in consistency index [45]. The maintenance of the consistency index is an important factor to consider, since viscous emulsions are more stable compared with less viscous emulsions, due to the delay of the phenomena of physical instability. According to Ali et al. [52], the factors that govern creaming are the dispersed-phase globule size and the viscosity of the external phase. The viscosity reduction in a nanoemulsified system increases the probability of droplets moving freely and colliding with each other, favoring the phenomenon of coalescence. In addition, for a topical formulation, maintenance of the viscosity is necessary [49].

The stability of the nanoemulsion is associated with physical integrity, chemical stability, and protection against microbial contamination [53]. Direct sources of contamination (raw materials, packaging, and production environment) and indirect sources (resulting from cleaning equipment and/or training operators) can affect the microbial quality of the formulation [54]. Microbial contamination is associated with the loss of therapeutic efficacy due to the chemical degradation of the constituents of the oil or changes in chemical and physical parameters [45]. The microbiological stability showed that there was no growth of the microorganisms tested. According to some studies [11,55], *P. pubescens* extracts present antimicrobial activity, which may favor the maintenance of the microbiological stability of the developed system.

The results obtained in the biological assay against the intracellular amastigotes of *L. amazonensis* indicate that the nanoscale droplets ($\approx$25 nm) and the presence of HA in the formulation may have contributed to the internalization of *P. pubescens* oil in the infected macrophages, increasing leishmanicidal activity in approximately 95%, compared with the free oil. Typically, delivery using nanocarriers to the target cell increases the bioavailability of active components and decreases toxicity in normal cells [2].

These results suggest that the *P. pubescens* oil was successfully internalized in the infected macrophages using the formulation, suggesting that a site-specific drug delivery system is promising for the treatment of CL. However, further investigations regarding the internalization of the *P. pubescens* oil present in the nanoemulsion by HA receptors, are necessary.

An important aspect of active compounds with potential therapeutic application is the absence of toxic effects on the host cells, which can be evaluated via the selectivity index. This index reveals whether the compounds act preferentially on the parasite or on the host defense cells. The greater the selectivity index, the more selective the formulation is in inhibiting the parasite [56].

Evaluation of the formulation showed that the CC_{50} value decreased, but the selectivity index was approximately double that of the free *P. pubescens* oil, demonstrating the potential of the nanoemulsion to improve the selectivity of the active substance over cells containing the parasite. According to the literature, furanoditerpenes with a vouacapan skeleton are the main compounds in *Pterodon* species capable of causing cytotoxicity in cancer cells [57,58]. Considering the potential of *P. pubescens* oil in the treatment of leishmaniasis [18,19], and the results obtained in this study, the development of a drug release system capable of increasing the therapeutic efficacy of *P. pubescens* oil is encouraged; however, additional studies on the compound responsible for its cytotoxicity are needed.

5. Conclusions

The HA-based nanoemulsion containing *P. pubescens* oil was more stable when stored at 5 °C, in which the parameters evaluated remained practically the same during the study period. The degradation of the *P. pubescens* oil present in the nanoemulsion was lower than that of the free oil, showing that the chemical stability of the oil increased when it was encapsulated in the nanoemulsion. The microbiological stability test confirmed the absence of bacteria or fungi in the formulation.

The biological assay against the intracellular amastigotes of *L. amazonensis* indicates that the *P. pubescens* oil was successfully internalized in the infected macrophages using the formulation, suggesting that a site-specific drug delivery system is promising for the treatment of CL. The development of a drug release system capable of increasing the therapeutic efficacy of *P. pubescens* oil is encouraged, but further investigations on the internalization of the oil present in the nanoemulsion by HA receptors, and on the increase in selectivity index, together with in vivo studies, are needed.

Author Contributions: We declare that this work was carried out by the author(s) named in this article and that all liabilities pertaining to claims relating to the content of this article will be borne by the authors. S.A.K., P.M.O. and J.H. contributed to laboratory work management, data analysis, and manuscript drafting. G.C.T. and M.D. contributed to microbiological stability and to critical reading of the manuscript. M.M.d.S.L. contributed to rheometry analysis, accelerated physical stability, and manuscript drafting. É.d.S.S. and C.V.N. contributed to biological studies (in vitro antileishmanial activity and cytotoxicity assays) and to manuscript drafting. E.A.d.S. and M.L.C.C. designed the study, supervised the laboratory work, and contributed to critical reading of the manuscript. All authors have read and agreed to the published version of the manuscript.

Funding: This research received no external funding.

Data Availability Statement: Not applicable.

Acknowledgments: The authors are grateful to Financiadora de Estudos e Projetos (FINEP), Fundação Araucária, Coordenadoria de Aperfeiçoamento de Pessoal do Ensino Superior (CAPES), and Conselho Nacional de Desenvolvimento Científico e Tecnológico (CNPq) for their financial support.

Conflicts of Interest: The authors declare that they have no conflict of interest.

References

1. Souza, T.M.; Morais-Braga, M.F.B.; Saraiva, A.A.F.; Rolón, M.; Veja, C.; Arias, A.R.; Costa, J.G.M.; Menezes, I.R.A. Evaluation of the anti-Leishmania activity of ethanol extract and fractions of the leaves from *Pityrogramma calomelanos* (L.) link. *Nat. Prod. Res.* **2013**, *27*, 992–996. [CrossRef] [PubMed]
2. Akbari, M.; Oryan, A.; Hatam, G. Application of nanotechnology in treatment of leishmaniasis: A Review. *Acta Trop.* **2017**, *172*, 86–90. [CrossRef] [PubMed]
3. Kheirandish, F.; Delfan, B.; Mahmoudvand, H.; Moradi, N.; Ezatpour, B.; Ebrahimzadeh, F.; Rashidipour, M. Antileishmanial, antioxidant, and cytotoxic activities of *Quercus infectoria* Olivier extract. *Biomed. Pharm.* **2016**, *82*, 208–215. [CrossRef]
4. Ammar, A.A.; Nasereddin, A.; Ereqat, S.; Dan-Goor, M.; Jaffe, C.L.; Zussman, E.; Abdeen, Z. Amphotericin B-loaded nanoparticles for local treatment of cutaneous leishmaniasis. *Drug Deliv. Transl. Res.* **2019**, *9*, 76–84. [CrossRef]
5. Galvão, E.L.; Pedras, M.J.; Cota, G.F.; Rabello, A.; Simões, T.C. How cutaneous leishmaniasis and treatment impacts in the patients' lives: A cross-sectional study. *PLoS ONE* **2019**, *14*, e0211374. [CrossRef] [PubMed]
6. Essid, R.; Rahali, F.Z.; Msaada, K.; Sghair, I.; Hammami, M.; Bouratbine, A.; Aoun, K.; Limam, F. Antileishmanial and cytotoxic potential of essential oils from medicinal plants in Northern Tunisia. *Ind. Crop. Prod.* **2015**, *77*, 795–802. [CrossRef]
7. Oliveira, L.A.R.; Oliveira, G.A.R.; Borges, L.L.; Bara, M.T.F.; Silveira, D. Vouacapane diterpenoids isolated from *Pterodon* and their biological activities. *Braz. J. Pharm.* **2017**, *27*, 663–672. [CrossRef]
8. Moraes, A.R.D.P.; Tavares, G.D.; Rocha, F.J.S.; Paula, E.; Giorgio, S. Effects of nanoemulsions prepared with essential oils of copaiba- and andiroba against *Leishmania infantum* and *Leishmania amazonensis* infections. *Exp. Parasitol.* **2018**, *187*, 12–21. [CrossRef]
9. Bouyahya, A.; Et-Touys, A.; Dakka, N.; Fellah, H.; Abrini, J.; Bakri, Y. Antileishmanial potential of medicinal plant extracts from the North-West of Morocco. *Beni.-Suef. Univ. J. Basic. Appl. Sci.* **2018**, *7*, 50–54. [CrossRef]
10. Coelho, L.P.; Reis, P.A.; Castro, F.L.; Gayer, C.R.M.; Lopes, C.S.; e Silva, M.C.C.; Sabino, K.C.C.; Todeschini, A.R.; Coelho, M.G.P. Antinociceptive properties of ethanolic extract and fractions of *Pterodon pubescens* Benth. Seeds. *J. Ethnopharmacol.* **2005**, *98*, 109–116. [CrossRef]
11. Bustamante, K.G.L.; Lima, A.D.F.; Soares, M.L.; Fiuza, T.S.; Tresvenzol, L.M.F.; Bara, M.T.F.; Pimenta, F.C.; Paula, J.R. Avaliação da atividade antimicrobiana do extrato etanólico bruto da casca da sucupira branca (*Pterodon emarginatus* Vogel)—Fabaceae. *Rev. Bras. Plantas Med.* **2010**, *12*, 341–345. [CrossRef]
12. Spindola, H.M.; Servat, L.; Rodrigues, R.A.F.; Sousa, I.M.O.; Carvalho, J.E.; Foglio, M.A. Geranylgeraniol and 6α,7β-dihydroxyvouacapan-17β-oate methyl ester isolated from *Pterodon pubescens* Benth: Further investigation on the antinociceptive mechanisms of action. *Eur. J. Pharm.* **2011**, *656*, 45–51. [CrossRef]
13. Nucci, C.; Mazzardo-Martins, L.; Stramosk, J.; Brethanha, L.C.; Pizzolatti, M.G.; Santos, A.R.S.; Martins, D.F. Oleaginous extract from the fruits *Pterodon pubescens* Benth induces antinociceptionin animal models of acute and chronic pain. *J. Ethnopharmacol.* **2012**, *143*, 170–178. [CrossRef]
14. Alves, S.F.; Borges, L.L.; de Paula, J.A.M.; Vieira, R.F.; Ferri, P.H.; de Couto, R.O.; de Paula, J.R.; Bara, M.T.F. Chemical variability of the essential oils from fruits of *Pterodon emarginatus* in the Brazilian Cerrado. *Braz. J. Pharm.* **2013**, *23*, 224–229. [CrossRef]
15. Hoscheid, J.; Bersani-Amado, C.A.; da Rocha, B.A.; Outuki, P.M.; da Silva, M.A.R.C.P.; Froehlich, D.L.; Cardoso, M.L.C. Inhibitory Effect of the Hexane Fraction of the Ethanolic Extract of the Fruits of *Pterodon pubescens* Benth. *Acute Chronic Inflamm. ECAM* **2013**, *2013*, 272795. [CrossRef]
16. Hoscheid, J.; Cardoso, M.L.C. Sucupira as a potential plant for arthritis treatment and other diseases. *Arthritis* **2015**, *2015*, 379459. [CrossRef]
17. Goes, P.R.N.; Hoscheid, J.; Silva-Filho, S.E.; Froehlich, D.L.; Pelegrini, B.L.; Canoff, J.R.A.; Lima, M.M.S.; Cuman, R.K.N.; Cardoso, M.L.C. Rheological behavior and antiarthritic activity of *Pterodon pubescens* nanoemulsion. *Res. Soc. Dev.* **2020**, *9*, e179108119. [CrossRef]
18. Santos, E.S.; Garcia, F.P.; Outuki, P.M.; Hoscheid, J.; de Goes, P.R.N.; Cardozo-Filho, L.; Nakamura, C.V.; Cardoso, M.L.C. Optimization of extraction method and evaluation of antileishmanial activity of oil and nanoemulsions of *Pterodon pubescens* Benth. *Fruit Extracts. Exp. Parasitol.* **2016**, *170*, 252–260. [CrossRef]
19. Oliveira, L.A.R.; Oliveira, G.A.R.; Lemes, G.F.; Romão, W.; Vaz, B.G.; Albuquerque, S.; Gonçalez, C.; Lião, L.M.; Bara, M.T.F. Isolation and Structural Characterization of Two New Furanoditerpenes from *Pterodon emarginatus* (*Fabaceae*). *J. Braz. Chem. Soc.* **2017**, *28*, 1911–1916. [CrossRef]
20. Wulff-Pérez, M.; Torcello-Gómez, A.; Gálvez-Ruíz, M.J.; Martín-Rodríguez, A. Stability of emulsions for parenteral feeding: Preparation and characterization of o/w nanoemulsions with natural oils and Pluronic f68 as surfactant. *Food Hydrocoll.* **2009**, *23*, 1096–1102. [CrossRef]

21. Bangia, J.K.; Singh, M.; Om, H.; Behera, K. A comparative study on the effect of temperature on density, sound velocity and refractive index of nanoemulsions formed by castor, olive and linseed oils in aqueous cellulose acetate propionate and butyrate and Tween80. *Thermochim. Acta* **2016**, *641*, 43–48. [CrossRef]
22. Want, M.Y.; Islamuddin, M.; Chouhan, G.; Dasgupta, A.K.; Chattopadhyay, A.P.; Afrin, F. A new approach for the delivery of artemisinin: Formulation, characterization, and ex-vivo antileishmanial studies. *J. Colloid Interface Sci.* **2014**, *432*, 258–269. [CrossRef] [PubMed]
23. Souza, A.; Martins, D.S.S.; Mathias, S.L.; Monteiro, L.M.; Yukuyama, M.N.; Scarim, C.B.; Löberberg, R. Promising nanotherapy in treating leishmaniasis. *Int. J. Pharm.* **2018**, *547*, 421–431. [CrossRef]
24. Vasi, A.M.; Popa, M.I.; Butnaru, M.; Dodi, G.; Verestiuc, L. Chemical functionalization of hyaluronic acid for drug delivery applications. *Mater. Sci. Eng. C* **2014**, *38*, 177–185. [CrossRef]
25. Kutlusoy, T.; Oktay, B.; Apohan, N.K.; Süleymanoglu, M.; Kuruca, S.E. Chitosan-co-Hyaluronic acid porous cryogels and their application intissue engineering. *Int. J. Biol. Macromol.* **2017**, *103*, 366–378. [CrossRef] [PubMed]
26. Pedrosa, S.S.; Pereira, P.; Correia, A.; Gama, F.M. Targetability of hyaluronic acid nanogel to cancer cells: In Vitro and in vivo studies. *Eur. J. Pharm. Sci.* **2017**, *104*, 102–113. [CrossRef]
27. Quinones, J.P.; Jokinen, J.; Keinänen, S.; Covas, C.P.; Brüggemann, O.; Ossipov, D. Self-assembled hyaluronic acid-testosterone nanocarriers for delivery of anticancer drugs. *Eur. Pol. J.* **2018**, *99*, 384–393. [CrossRef]
28. Montanari, E.; Di Meo, C.; Oates, A.; Coviello, T.; Matricardi, P. Pursuing Intracellular Pathogens with Hyaluronan. From a 'Pro-Infection' Polymer to a Biomaterial for 'Trojan Horse' Systems. *Molecules* **2018**, *23*, 939. [CrossRef]
29. Stefanello, T.F.; Szarpak-Jankowska, A.; Appaix, F.; Louage, B.; Hamard, L.; De Geest, B.G.; Sander, B.V.D.; Nakamura, C.V.; Auzély-Velty, R. Thermoresponsive hyaluronic acid nanogels as hydrophobic drug carrier to macrophages. *Acta Biomater.* **2014**, *10*, 750–4758. [CrossRef]
30. Micale, N.; Piperno, A.; Mahfoudh, N.; Schurigt, U.; Schultheis, M.; Mineo, P.G.; Schirmeister, T.; Scala, A.; Grassi, G. A hyaluronic acid–pentamidine bioconjugate as a macrophage mediated drug targeting delivery system for the treatment of leishmaniasis. *RSC Adv.* **2015**, *5*, 95545–95550. [CrossRef]
31. Kleinubing, S.A.; Outuki, P.M.; Hoscheid, J.; Pelegrini, B.L.; da Silva, E.A.; Canoff, J.R.A.; Lima, M.M.S.; Cardoso, M.L.C. Hyaluronic acid incorporation into nanoemulsions containing *Pterodon pubescens* Benth. fruit oil for topical drug delivery. *Biocatal. Agric. Biotechnol.* **2021**, *32*, 101939. [CrossRef]
32. Hoscheid, J.; Reinas, A.; Cortez, D.A.G.; da Costa, W.F.; Cardoso, M.L.C. Determination by GC–MS-SIM of furanoditerpenes in *Pterodon pubescens* Benth.: Development and validation. *Talanta* **2012**, *100*, 372–376. [CrossRef] [PubMed]
33. Brasil. Resolução No 01, de 29 de Julho de 2005. Guia Para a Realização de Estudos de Estabilidade. Brasília: Agência Nacional de Vigilância Sanitária. 2005. Available online: http://bvsms.saude.gov.br/bvs/saudelegis/anvisa/2005/res0001_29_07_2005.html (accessed on 7 July 2022).
34. Hoscheid, J.; Outuki, P.M.; Kleinubing, S.A.; Silva, M.F.; Bruschi, M.L.; Cardoso, M.L.C. Development and characterization of *Pterodon pubescens* oil nanoemulsions as a possible delivery system for the treatment of rheumatoid arthritis. *Colloids Surf. A Physicochem. Eng. Asp.* **2015**, *484*, 19–27. [CrossRef]
35. Brasil. Farmacopeia Brasileira. *Diário Of. União* **2010**, *1*, 236–253.
36. Mosmann, T. Rapid colorimetric assay for cellular growth and survival: Application to proliferation and cytotoxicity assays. *J. Immunol. Methods* **1983**, *65*, 55–63. [CrossRef]
37. Kelmann, R.G.; Kuminek, G.; Teixeira, H.F.; Koester, L.S. Carbamazepine parenteral nanoemulsions prepared by spontaneous emulsification process. *Int. J. Pharm.* **2007**, *342*, 231–239. [CrossRef]
38. Teixeira, M.C.; Severino, P.; Andreani, T.; Boonme, P.; Santini, A.; Silva, A.M.; Souto, E.B. D-α-tocopherol nanoemulsions: Size properties, rheological behavior, surface tension, osmolarity and cytotoxicity. *Saudi Pharm. J.* **2017**, *25*, 231–235. [CrossRef]
39. Danaei, M.; Dehghankhold, M.; Ataei, S.; Hasanzadeh Davarani, F.; Javanmard, R.; Dokhani, A.; Khorasani, S.; Mozafari, M.R. Impact of Particle Size and Polydispersity Index on the Clinical Applications of Lipidic Nanocarrier Systems. *Pharmaceutics* **2018**, *10*, 57. [CrossRef]
40. Sato, T.S.; De Medeiros, T.M.; Hoscheid, J.; Prochnau, I.S. Proposta de formulação contendo extrato de folhas de Eugenia involucrata e análise da atividade antimicrobiana. Proposal of a formulation containing leaves extract of *Eugenia involucrata*. *Rev. Fitos* **2018**, *12*, 68–82. [CrossRef]
41. Vighi, E.; Ruozi, B.; Montanari, M.; Battini, R.; Leo, E. pDNA condensation capacity and in vitro gene delivery properties of cationic solid lipid nanoparticles. *Int. J. Pharm.* **2010**, *389*, 254–261. [CrossRef] [PubMed]
42. Outuki, P.M.; Kleinubing, S.A.; Hoscheid, J.; Montanha, M.C.; da Silva, E.A.; do Couto, R.O.; Kimura, E.; Cardoso, M.L.C. The incorporation of *Pterodon pubescens* fruit oil into optimized nanostructured lipid carriers improves its effectiveness in colorectal cancer. *Ind. Crops Prod.* **2018**, *123*, 719–730. [CrossRef]
43. Galindo-Alvarez, J.; Le, K.-A.; Sadtler, V.; Marchal, P.; Perrin, P.; Tribet, C.; Marie, E.; Durand, A. Enhanced stability of nanoemulsions using mixtures of nonionic surfactant and amphiphilic polyelectrolyte. *Colloids Surf. A Physicochem. Eng. Asp.* **2011**, *389*, 237–245. [CrossRef]
44. Li, B.; Ge, Z.Q. Nanostructured lipid carriers improve skin permeation and chemical stability of idebenone. *AAPS PharmSciTech* **2012**, *13*, 276–283. [CrossRef]

45. Hoscheid, J.; Outuki, P.M.; Kleinubing, S.A.; de Goes, P.R.N.; Lima, M.M.S.; Cuman, R.K.N.; Cardoso, M.L.C. *Pterodon pubescens* oil nanoemulsions: Physiochemical and microbiological characterization and in vivo anti-inflammatory efficacy studies. *Braz. J. Pharmacog.* **2017**, *27*, 375–383. [CrossRef]
46. Kong, M.; Park, H.J. Stability investigation of hyaluronic acid based nanoemulsion and its potential as transdermal carrier. *Carbohydr. Polym.* **2011**, *83*, 1303–1310. [CrossRef]
47. Yang, S.C.; Benita, S. Enhanced Absorption and Drug Targeting by Positively Charged Submicron Emulsions. *Drug Dev. Res.* **2000**, *50*, 476–486. [CrossRef]
48. Luo, X.; Zhou, Y.; Bai, L.; Liu, F.; Deng, Y.; McClements, D.J. Fabrication of b-carotene nanoemulsion-based delivery systems using dual-channel microfluidization: Physical and chemical stability. *J. Colloid Interf. Sci.* **2017**, *490*, 328–335. [CrossRef]
49. Barradas, T.N.; Senna, J.P.; Cardoso, S.A.; Nicoli, A.; Padula, C.; Santi, P.; Rossi, F.; e Silva, K.G.H.; Mansur, C.R.E. Hydrogel-thickened nanoemulsions based on essential oils for topical delivery of psoralen: Permeation and stability studies. *Eur. J. Pharm. Biopharm.* **2017**, *116*, 38–50. [CrossRef]
50. Reinas, A.E.; Hoscheid, J.; Outuki, P.M.; Cardoso, M.L.C. Preparation and characterization of microcapsules of *Pterodon pubescens* Benth. by using natural polymers. *Braz. J. Pharm. Sci.* **2014**, *50*, 920–930. [CrossRef]
51. Bajerski, L.; Michels, L.R.; Colomé, L.M.; Bender, E.A.; Freddo, R.J.; Bruxel, F.; Haas, S.E. The use of Brazilian vegetable oils in nanoemulsions: An update on preparation and biological applications. *Braz. J. Pharm. Sci.* **2016**, *52*, 348–363. [CrossRef]
52. Ali, M.S.; Alam, M.S.; Alam, N.; Anwer, T.; Safhi, M.M.A. Accelerated Stability Testing of a Clobetasol Propionate-Loaded Nanoemulsion as per ICH Guidelines. *Sci. Pharm.* **2013**, *81*, 1089–1100. [CrossRef] [PubMed]
53. Alam, M.S.; Ali, M.S.; Alam, M.I.; Anwer, T.; Safhi, M.M.A. Stability Testing of Beclomethasone Dipropionate Nanoemulsion. *Trop. J. Pharm. Res.* **2015**, *14*, 15–20. [CrossRef]
54. Ferreira, B.C.A.; Novais, E.B.; Ribeiro, R.B.C.; Fernandes, C.K.C. Estudo de estabilidade físico-química e microbiológica de dipirona em gotas armazenadas em residências do municipio de São Luis de Montes Belos-GO. *Rev. Facul Montes Belos.* **2014**, *7*, 109–120.
55. Dutra, R.C.; Braga, F.G.; Coimbra, E.S.; Silva, A.D.; Barbosa, N.R. Antimicrobial and leishmanicidal activities of seeds of *Pterodon emarginatus*. *Braz. J. Pharm.* **2009**, *19*, 429–435. [CrossRef]
56. Caldeira, L.R.; Fernandes, F.R.; Costa, D.F.; Frézard, F.; Afonso, L.C.C.; Ferreira, L.A.M. Nanoemulsions loaded with amphotericin B: A new approach for the treatment of leishmaniasis. *Eur. J. Pharm. Sci.* **2015**, *70*, 125–131. [CrossRef] [PubMed]
57. Vieira, C.R.; Marques, M.F.; Soares, P.R.; Matuda, L.; Oliveira, C.M.A.; Kato, L.; Silva, C.C.; Grillo, L.A. Antiproliferative activity of *Pterodon pubescens* Benth. seed oil and its active principle on human melanoma cells. *Phytomedicine* **2008**, *15*, 528–532. [CrossRef]
58. Spindola, H.M.; Carvalho, J.E.; Ruiz, A.L.T.G.; Rodrigues, R.A.F.; Denny, C.; Sousa, I.M.O.; Tamashiro, J.Y.; Foglio, M.A. Furanoditerpenes from *Pterodon pubescens* Benth. with selective in vitro anticancer activity for prostate cell line. *J. Braz. Chem. Soc.* **2009**, *20*, 569–575. [CrossRef]

colloids and interfaces

MDPI

Article

Structural Modification of Jackfruit Leaf Protein Concentrate by Enzymatic Hydrolysis and Their Effect on the Emulsifier Properties

Carolina Calderón-Chiu [1], Montserrat Calderón-Santoyo [1], Julio César Barros-Castillo [1], José Alfredo Díaz [2] and Juan Arturo Ragazzo-Sánchez [1,*]

[1] Laboratorio Integral de Investigación en Alimentos, Tecnológico Nacional de México/Instituto Tecnológico de Tepic, Av. Tecnológico #2595, Colonia Lagos del Country, Tepic C.P. 63175, Nayarit, Mexico

[2] Departamento de Química Física e Inorgánica, Facultad de Ciencias Naturales y Matemáticas, Ciudad Universitaria, Escuela de Química, Universidad de El Salvador, San Salvador C.P. 01101, El Salvador

* Correspondence: jragazzo@ittepic.edu.mx

Abstract: Jackfruit leaf protein concentrate (LPC) was hydrolyzed by pepsin (H–Pep) and pancreatin (H–Pan) at different hydrolysis times (30–240 min). The effect of the enzyme type and hydrolysis time of the LPC on the amino acid composition, structure, and thermal properties and its relationship with the formation of O/W emulsions were investigated. The highest release of amino acids (AA) occurred at 240 min for both enzymes. H–Pan showed the greatest content of essential and hydrophobic amino acids. Low β-sheet fractions and high β-turn contents had a greater influence on the emulsifier properties. In H–Pep, the β-sheet fraction increased, while in H–Pan it decreased as a function of hydrolysis time. The temperatures of glass transition and decomposition were highest in H–Pep due to the high content of β-sheets. The stabilized emulsions with H–Pan (180 min of hydrolysis) showed homogeneous distributions and smaller particle sizes. The changes in the secondary structure and AA composition of the protein hydrolysates by the effect of enzyme type and hydrolysis time influenced the emulsifying properties. However, further research is needed to explore the use of H–Pan as an alternative to conventional emulsifiers or ingredients in functional foods.

Keywords: jackfruit; leaf protein concentrate; leaf protein hydrolysate; enzymatic hydrolysis; amino acid profile; secondary structure; thermal properties; O/W emulsions

Citation: Calderón-Chiu, C.; Calderón-Santoyo, M.; Barros-Castillo, J.C.; Díaz, J.A.; Ragazzo-Sánchez, J.A. Structural Modification of Jackfruit Leaf Protein Concentrate by Enzymatic Hydrolysis and Their Effect on the Emulsifier Properties. *Colloids Interfaces* **2022**, *6*, 52. https://doi.org/10.3390/colloids6040052

Academic Editors: César Burgos-Díaz, Mauricio Opazo-Navarrete and Eduardo Morales

Received: 24 August 2022
Accepted: 21 September 2022
Published: 28 September 2022

1. Introduction

Oil-in-water (O/W) emulsions are widely used in the food, pharmaceutical, and cosmetic industries [1]. They are commonly stabilized by proteins such as whey protein, casein, soy, egg white proteins, gelatin, bovine serum albumin, collagen, etc., which are of animal origin [2,3]. However, the high cost and the negative environmental impacts of animal protein production are opening new opportunities for the exploration of alternative proteins from nonconventional sources [4,5]. Plant proteins are a green trend that could be applied in the pharmaceutical, cosmetic, and food industries [6]. The partial or total replacement of animal proteins for new plant proteins obtained from food by-products has been driven by the assurance of sustainability in food production, which is due to the lower CO_2 emission associated with their primary production [4,7,8].

In that sense, the exploration of the emulsifying properties of proteins is one of the most interesting areas of current research, as it focuses on the identification of proteins of plant origin and their hydrolysates as possible emulsifiers [9]. For this reason, studies on protein hydrolysates with emulsifying properties from plant proteins such as chickpea [10,11], fava bean [12], rice bran [13], and navy bean [14], among others, have increased. These protein hydrolysates have shown a higher rate of diffusion into the oil/water interface and greater coverage at the interface than the native protein. Hence, they could be a suitable alternative to animal proteins and be used as emulsifiers.

Recently, the LPC (leaf protein concentrate) obtained from the leaves (generated during the pruning) of jackfruit (*Artocarpus heterophyllus* L.) trees in Nayarit, México, was considered an alternative source of protein (65.82% in d.b.) with potential for food applications. However, the low solubility of LPC (~15% at pH 8.0) limits its application as an emulsifier in food formulations. Thus, to expand the application of LPC as a functional ingredient, LPC was modified by enzymatic hydrolysis with pepsin and pancreatin [15,16].

Similarly, protein hydrolysates of LPCs have been obtained with Flavourzyme, Protamex, Neutrase, Alcalase, Trypsin, and Chymotrypsin [16,17]. However, these protein hydrolysates have been evaluated for their biological activities such as their antioxidant, antidiabetic, antihypertensive, and antimicrobial activities. Nonetheless, the study of the functional properties of leaf protein hydrolysates such as emulsifying properties has been little explored. In a previous study [15], the ability to produce functional protein hydrolysates from jackfruit LPC with pepsin (H–Pep) and pancreatin (H–Pan) at different hydrolysis times was evaluated. The H–Pep and H–Pan showed functional properties such as solubility, foaming, and emulsifying properties, and antioxidant properties significantly improved concerning the LPC, which demonstrates its dual functionality.

The improvement in functional and antioxidant properties is due to enzymatic hydrolysis. The enzymatic hydrolysis has been used to improve the functionality of plant protein in a safe, simple, and relatively inexpensive way [12]. The breakdown of peptide bonds during the hydrolysis of LPC causes an increase in the number of free amino and carboxyl groups, exposes hydrophobic patches hidden inside the protein structure, and releases short-chain peptides [5]. This results in the release of soluble peptides and the removal of antinutritional factors while improving the antioxidant activity, solubility, and adsorption at the O/W interface [18]. Besides the intrinsic changes in the primary structure of the protein, the reduction in the protein chain due to enzymatic hydrolysis also affects the secondary structure. The modification of the secondary structure produces changes in the hydrophilicity, hydrophobicity, and structural stability of the protein, which are related to functional properties such as emulsifying properties [19]. The above could indicate that the changes in amino acid composition, secondary structure, and thermal properties of jackfruit leaf protein hydrolysates by the effect of enzymatic hydrolysis influence the emulsifying properties.

Although the possible potential of jackfruit leaf protein hydrolysates as alternative emulsifiers has been demonstrated [15], the utilization of new protein sources from plants as emulsifiers requires a thorough investigation for application in foods. The elucidation of the structure–function relationship is a prerequisite for developing new food materials, which is critical for replacing conventional animal-based proteins with plant protein [20].Consequently, it is essential to know the structural and thermal properties of these hydrolysates, since this would allow them to be handled properly in such a way that they function correctly in a food system and improve their applicability [21].

Nevertheless, knowledge of the structure and thermal properties of LPC and protein hydrolysates from jackfruit leaf is still unidentified. Thus, this study aimed to evaluate the effect of the enzyme type and hydrolysis time of LPC on the amino acid profile, structural and thermal properties, and its relationship with the formation of oil-in-water (O/W) emulsions. Therefore, the determination of the amino acid profile by GC-MS and the characterization by FTIR, TGA, and DSC techniques of LPC and protein hydrolysates were carried out. Subsequently, the use of protein hydrolysates (0.5% *w*/*v*) from jackfruit leaf as an emulsifier was explored, to select the one with the best emulsifying capacity, which in successive studies can be used to formulate an emulsion with the desirable characteristics in the food industry.

2. Materials and Methods

2.1. Materials

The leaves of jackfruit were obtained from the "Tierras Grandes" orchard (Zacualpan, Compostela, Nayarit, Mexico) in February 2019. The leaves were washed and dried at 60 °C in a convective drying oven (Novatech, HS60-AID, Guadalajara, Jalisco, Mexico) for 24 h. Subsequently, they were ground (NutriBullet® SERIE 900, Los Angeles, CA, USA) and sieved (No. 100 mesh, 150 μm diameter). The flour (protein content of 24.06%) was packed in vacuum-sealed bags and stored at room temperature until use [15].

2.2. Chemical Substances

Pepsin (EC 3.4.23.1), pancreatin (EC 232-468-9), amino acid standards (AA-S-18), L-Norleucine (Nor), *N-tert*-Butyldimethylsilyl-*N*-methyltrifluoroacetamide (MTBSTFA), and other chemicals used in the experiments of GS-MS (highest purity available) were purchased from Sigma-Aldrich (St. Louis, MO, USA). Sodium hydroxide (NaOH) and hydrochloric acid (HCl) were of analytical grade purchased from Thermo Fisher Scientific Inc. (Waltham, MA, USA).

2.3. Extraction and Hydrolysis of the LPC of Jackfruit

The extraction and hydrolysis of the LPC were carried out following that described by Calderón-Chiu et al. [15]. For extraction, the leaf flour (30 g) was mixed with distilled water (563 mL) and 0.2 M NaOH (188 mL). The mixture was placed in the high hydrostatic pressure (HHP) equipment (Avure Technologies AB, Middletown, OH, USA) at 300 MPa for 20 min (25 °C). The mixture was centrifuged at 15,000× g for 20 min at 4 °C (Hermle Z 326 K, DEU). The supernatant was recovered and adjusted to pH 4.0 with 1 N HCl to precipitate proteins, which were recuperated by centrifugation. Finally, the precipitate was diafiltered through a 1 kDa membrane and lyophilized at −50 °C and 0.12 mbar in a freeze-dried (Labconco FreeZone 4.5, USA).

Subsequently, the LPC was hydrolyzed with pepsin and pancreatin for 30, 60, 120, 180, and 240 min. Briefly, the LPC was suspended in distilled water (1% w/v) and incubated in a shaking water bath (Shaking Hot Tubs 290200, Boekel Scientific, Feasterville-Trevose, PA, USA) at 37 °C, 115 rpm (30 min). Then, the pH solutions were adjusted to 2.0 (for pepsin) and 7.0 (for pancreatin) with 1 N HCl and NaOH, respectively. The enzyme: substrate ratio was 1:100 (w/w) and the pH was maintained with HCl or NaOH if necessary. The reaction was terminated by heating at 95 °C (15 min), followed by centrifugation at 10,000× g for 15 min at 4 °C (Hermle Z, 326 K, DEU), and the pH adjustment to 7.0 for both enzymes. The protein hydrolysates were filtered (0.45 μm) and freeze dried. Protein hydrolysates of pepsin (H–Pep) and pancreatin (H–Pan) at different hydrolysis times were obtained with DHs (degree of hydrolysis) from 1.78–3.44% and 6.12–9.21%, respectively [15].

2.4. Amino Acid Profile

The determination of the amino acids was carried out with the methodology described by Brion-Espinoza et al. [16]. Samples of flour, LPC, and hydrolysates were subjected to acid hydrolysis with 6 M HCl for 24 h at 110 °C. Then, the hydrolyzed samples were derivatized with MTBSTFA. Briefly, 100 μL of hydrolysate and 10 μL of Nor (internal standard, concentration 0.2 mg/mL) were evaporated under nitrogen gas to dry residue. The resulting precipitate was dissolved in 100 μL of acetonitrile and 100 μL of MTBSTFA. This solution was incubated at 100 °C for 2.5 h in a glycerol bath. The derivatization reaction was carried out into a 2 mL PTFE-lined screw-capped vial. For the L-amino acids standards mixture, the same procedure mentioned above was followed. Then, 1 μL of the test solution was injected into a gas chromatograph. The analysis of CG-MS was carried out using GC equipment 7890A (Agilent Technologies; Palo Alto, CA, USA) coupled with mass spectrometry (MS) 240 Ion Trap (Agilent Technologies; Palo Alto, CA, USA). The amino acid profile was reported as g of amino acid/100 g of protein. Cysteine, arginine, and tryptophan were not quantified.

2.5. Fourier Transform Infrared Spectroscopy

Infrared spectra were recorded using a spectrometer (Thermo Scientific, Nicolet iS5 FT-IR, USA) equipped with an iD7 ATR accessory with ZnSe crystal (4000–400 cm^{-1}) at 25 °C. Automatic signals were collected in 16 scans at a resolution of 4 cm^{-1}. Then, all spectra were corrected with a linear baseline using spectroscopy software Knowitall (v.8.2 Bio-Rad Inc., USA). Subsequently, interpretations of the changes in the overlapping amide I band (1600–1700 cm^{-1}) and curve-fitting procedure (second-derivative analysis) to quantify the protein secondary structure was carried out using Peak-Fit software (v4.12, SPSS Inc., Chicago, IL, USA) Then, the relative contents of α-helix, β-Sheet, β-turn, and random coil were determined from the fitted peak areas [22].

2.6. Thermal Properties

Thermogravimetric analysis (TGA) and differential scanning calorimetry (DSC) of LPC and protein hydrolysates were performed [23]. For TGA (TGA 550 equipment, TA Instruments, New Castle, DE, USA), the lyophilized samples (3–5 mg) were heated from 25 to 800 °C at a heating rate of 10 °C/min under a nitrogen flow. The first decomposition stage in the TGA curve was assigned to moisture loss and the midpoint temperature of the second decomposition stage was assumed as the decomposition temperature (T_{dec}, [24]). As for the DSC (DSC 250 equipment, TA Instruments, New Castle, DE, USA), lyophilized samples (3–5 mg) were hermetically sealed in aluminum pans. Subsequently, they were heated from 0 to 300 °C at a heating rate of 10 °C/min under a nitrogen flow. An empty pan was used as a reference. Thermograms obtained by DSC and TGA were analyzed by the software TRIOS 5.0.0.44616. Then, the derivative thermogravimetric (DTG, obtained from the TGA curve), the glass transition (Tg) and decomposition (T_{dec}) temperatures, and denaturation enthalpy (Δh) were obtained.

2.7. Preparation of Emulsions with LPC and Protein Hydrolysates

To demonstrate the effect of the hydrolysis time and type of enzyme on the emulsifying properties of the hydrolysates, emulsions were formulated with the different hydrolysates and LPC at 0.5% (w/v). Then, particle size distribution was determined and related to the emulsifying activity index (EAI, m^2/g) and emulsion stability index (ESI, min) of the protein hydrolysates previously evaluated [15]. Briefly, the emulsions were formulated as described by Zang et al. [13] with modifications. The aqueous phase was prepared by dissolving LPC or protein hydrolysates (0.5%, w/v) in distilled water at pH 7.0. The solution was stirred in a magnetic plate stirrer at 700 rpm for 2 h at room temperature, followed by centrifugation (Hermle Z, 326 K, Wehingen, Germany) at 2350× g for 30 min (25 °C). A homogenizer basic ultra turrax (IKA T10, IKA, Staufen, Germany) was used to obtain the pre-emulsion, which consisted of 10 wt% olive oil (oil phase) and 90 wt% LPC or protein hydrolysate solutions (aqueous phase). After mixing at 16,000 rpm for 2 min (ultra turrax), the pre-emulsion was processed by a Digital Sonifier® Unit (Model S-150D, Branson Ultrasonics Corporation, Danbury, CT, USA) at 24 kHz for 5 min and the particle size distribution was determined.

Subsequently, to investigate in more detail the emulsifying capacity of H–Pan (hydrolyzed with better emulsifying properties), emulsions were prepared and adjusted to pH 5.0, 7.0, and 9.0 to evaluate the effect of pH on the particle size distribution as an indication of the stability of the emulsion.

2.8. Particle Size Distribution of Emulsions

The emulsion particle size distribution was measured in a Mastersizer 3000 (Hydro EV, Malvern, Worcestershire, UK). In brief, the emulsion was added dropwise in the Hydro EV unit until a laser obscuration of 8–12% was obtained [23]. The emulsions were analyzed five successive times in the diffractometer at 25 °C. Then, the volume-weighted mean particle diameter (D [4,3]) and polydispersity index (PDI) were calculated automatically with the

Mastersizer software (version 3.60, Worcestershire, UK). The refractive index was set as 1.46 for the dispersed phase (olive oil) and 1.33 for the dispersant (water).

2.9. Statistical Analysis

All the samples were analyzed by triplicate. The data were analyzed with a one-way analysis of variance (ANOVA), followed by the post-hoc LSD (Least Significant Difference) for the mean comparison ($p < 0.05$) with the Statistica v10.0 software (StatSoft, Inc., Tulsa, OK, USA).

3. Results and Discussion

3.1. Amino Acid Composition

The total amino acid (TAA), essential amino acids (EAA), and hydrophobic amino acids (HAA) of the LPC were higher ($p < 0.05$) than in the flour (Table 1). This behavior means that the extraction by HHP allowed an appropriate recovery of amino acids and could constitute a suitable method for obtaining the LPC of jackfruit. The LPC presented a high content of leucine, valine, alanine, isoleucine, aspartic acid, and glutamic acid. For protein hydrolysates, a gradual release of amino acids during the enzymatic process was observed. The protein hydrolysates showed a lower content of TAA, EAA, and HAA than the LPC ($p < 0.05$).H–Pan showed high levels of TAA, EAA, and HAA ($p < 0.05$). Previous studies by Calderón-Chiu et al. [15] showed that H–Pan (6.12–9.21%) presented a higher degree of hydrolysis (DH) than H–Pep (1.78–3.44%), which would explain the higher content of amino acids in H–Pan.

Table 1. Amino acids profile (g/100 g protein) of LPC and protein hydrolysates obtained after treatment with pepsin and pancreatin.

			H–Pep at Different Hydrolysis Times (min)				
Amino Acid	**Leaf Flour**	**LPC**	**30**	**60**	**120**	**180**	**240**
Alanine	1.68 ± 0.17 ^{b}	6.55 ± 0.19 ^{e}	3.1 ± 0.12 ^{a}	2.57 ± 0.1 ^{ac}	2.29 ± 0.68 ^{bc}	3.19 ± 0.06 ^{a}	4.22 ± 0.42 ^{d}
Glycine	0.32 ± 0.15 ^{c}	0.97 ± 0.05 ^{d}	0 ± 0 ^{a}	0.13 ± 0.01 ^{ab}	0.13 ± 0.04 ^{ab}	0.17 ± 0 ^{bc}	0 ± 0 ^{a}
Valine	1.77 ± 0.2 ^{d}	8.62 ± 0.42 ^{e}	5.58 ± 0.8 ^{c}	3.77 ± 0.11 ^{a}	3.53 ± 0.72 ^{a}	4.25 ± 0.27 ^{ab}	5.17 ± 0.84 ^{bc}
Leucine	2.52 ± 0.14 ^{c}	11.26 ± 0.13 ^{e}	6.9 ± 0.65 ^{d}	4.44 ± 0.05 ^{ab}	3.87 ± 0.29 ^{ac}	4.8 ± 0.15 ^{ab}	5.77 ± 1.33 ^{bd}
Isoleucine	1.36 ± 0.12 ^{d}	6.02 ± 0.01 ^{e}	3.91 ± 0.3 ^{c}	2.56 ± 0.08 ^{a}	2.38 ± 0.41 ^{a}	2.83 ± 0.22 ^{ab}	3.67 ± 0.93 ^{bc}
Proline	1.46 ± 0.08 ^{a}	2.68 ± 0.06 ^{b}	1.11 ± 0.15 ^{a}	1.28 ± 0.12 ^{a}	1.04 ± 0.79 ^{a}	1.01 ± 0.01 ^{a}	0.89 ± 0.03 ^{a}
Serine	0.28 ± 0.03 ^{e}	2.87 ± 0.11 ^{d}	1.49 ± 0.01b ^{c}	1.27 ± 0.05 ^{bc}	1.06 ± 0.05 ^{a}	1.73 ± 0.05 ^{c}	2.72 ± 0.23 ^{d}
Threonine	0.76 ± 0.05 ^{a}	5.02 ± 0.05 ^{b}	2.66 ± 0.11 ^{c}	2.14 ± 0.23 ^{d}	1.7 ± 0.42 ^{e}	3.22 ± 0.01 ^{f}	4.05 ± 0.04 ^{g}
Phenylalanine	1.27 ± 0.06 ^{c}	5.27 ± 0.11 ^{d}	3.43 ± 0.03 ^{a}	2.62 ± 0.91 ^{ab}	1.95 ± 0.13 ^{bc}	2.77 ± 0.01 ^{ab}	2.89 ± 0.07 ^{a}
Aspartic acid	1.65 ± 0.11 ^{a}	5.87 ± 0.08 ^{f}	2.85 ± 0.15 ^{c}	2.55 ± 0.01 ^{bc}	2.05 ± 0.45 ^{ab}	3.81 ± 0.27 ^{d}	4.68 ± 0.08 ^{e}
Glutamic acid	1.06 ± 0.12 ^{a}	5.66 ± 0.07 ^{d}	1.94 ± 0.12 ^{ac}	2.71 ± 0.04 ^{bc}	1.43 ± 1.27 ^{a}	3.38 ± 0.02 ^{b}	3.4 ± 0.06 ^{b}
Tyrosine	3.34 ± 0.36 ^{a}	3.41 ± 0.08 ^{a}	0 ± 0 ^{b}	0 ± 0 ^{b}	0 ± 0 ^{b}	0 ± 0 ^{b}	0 ± 0 ^{b}
HAA	13.41 ± 0.43 ^{a}	43.81 ± 0.16 ^{b}	24.02 ± 0.13 ^{c,A}	17.24 ± 0.5 ^{d,A}	15.07 ± 0.33 ^{e,A}	18.84 ± 0.11 ^{f,A}	22.62 ± 0.28 ^{g,A}
AAA	4.62 ± 0.86 ^{c}	8.68 ± 0.23 ^{d}	3.43 ± 0.15 ^{b,A}	2.62 ± 0.31 ^{ab}	1.95 ± 0.45 ^{a,A}	2.77 ± 0.27 ^{ab,A}	2.89 ± 0.13 ^{ab,A}
EAA	7.69 ± 0.11 ^{b}	36.19 ± 0.17 ^{f}	22.47 ± 0.2 ^{a,A}	15.54 ± 0.23 ^{d,A}	13.44 ± 1.8 ^{c,A}	17.87 ± 0.41 ^{e,A}	21.56 ± 1.27 ^{a,A}
TAA	17.48 ±1.03 ^{b}	64.19± 0.1 ^{f}	32.96 ± 0.23 ^{a,A}	26.04 ± 0.82 ^{d,A}	21.45 ± 1.41 ^{c,A}	31.15 ± 0.44 ^{a,A}	37.47 ± 1.24 ^{e,A}
Alanine	1.68 ± 0.17 ^{a}	6.55 ± 0.19 ^{e}	2.91 ± 0.26 ^{bc}	1.92 ± 0.17 ^{a}	3.04 ± 0.06 ^{c}	3.61 ± 0.19 ^{d}	2.54 ± 0.12 ^{b}
Glycine	0.32 ± 0.15 ^{b}	0.97 ± 0.05 ^{c}	3.13 ± 0.27 ^{d}	0.79 ± 0.15 ^{ac}	0.65 ± 0.03 ^{ac}	0.59 ± 0.05 ^{ab}	0.48 ± 0.08 ^{ab}
Valine	1.77 ± 0.2 ^{c}	8.62 ± 0.42 ^{e}	3.07 ± 0.3 ^{a}	3.07 ± 0.2 ^{a}	4.75 ± 0.65 ^{b}	4.99 ± 0.71 ^{b}	6.73 ± 0.8 ^{d}
Leucine	2.52 ± 0.14 ^{c}	11.26 ± 0.13 ^{e}	3.84 ± 0.33 ^{a}	3.59 ± 0.14 ^{a}	5.89 ± 0.86 ^{b}	5.86 ± 0.13 ^{b}	7.18 ± 0.65 ^{d}
Isoleucine	1.36 ± 0.12 ^{c}	6.02 ± 0.01 ^{e}	2.09 ± 0.18 ^{a}	2.19 ± 0.12 ^{a}	3.41 ± 0.47 ^{b}	3.61 ± 0.01 ^{b}	4.31 ± 0.3 ^{d}
Proline	1.46 ± 0.08 ^{b}	2.68 ± 0.06 ^{c}	4.62 ± 0.24 ^{d}	3.29 ± 0.08 ^{a}	3.26 ± 0.01 ^{a}	3.31 ± 0.06 ^{a}	7.57 ± 0.15 ^{e}
Serine	0.28 ± 0.03 ^{b}	2.87 ± 0.11 ^{f}	2.23 ± 0.08 ^{e}	0.62 ± 0.03 ^{c}	1.04 ± 0.11 ^{a}	1.62 ± 0.11 ^{d}	1.03 ± 0 ^{a}
Threonine	0.76 ± 0.05 ^{b}	5.02 ± 0.05 ^{f}	2.65 ± 0.12 ^{a}	1.39 ± 0.05 ^{c}	2.49 ± 0.03 ^{a}	2.95 ± 0.2 ^{d}	3.3 ± 0.11 ^{e}
Phenylalanine	1.27 ± 0.06 ^{b}	5.27 ± 0.11 ^{d}	1.97 ± 1.48 ^{ab}	1.67 ± 0.06 ^{ab}	3.15 ± 0.37 ^{ac}	2.9 ± 0.74 ^{ac}	4.21 ± 0.03 ^{cd}
Aspartic acid	1.65 ± 0.11 ^{c}	5.87 ± 0.08 ^{e}	3.73 ± 0.45 ^{a}	0.71 ± 0.11 ^{b}	2.54 ± 0.57 ^{d}	3.65 ± 0.4 ^{a}	3.64 ± 0.15 ^{a}
Glutamic acid	1.06 ± 0.12 ^{b}	5.66 ± 0.07 ^{c}	3.28 ± 0.71 ^{a}	0.64 ± 0.12 ^{b}	2.83 ± 1.19 ^{a}	3.99 ± 0.08 ^{a}	2.8 ± 0.69 ^{a}
Tyrosine	3.34 ± 0.36 ^{a}	3.41 ± 0.08 ^{a}	0 ± 0 ^{b}	0 ± 0 ^{b}	0 ± 0 ^{b}	0 ± 0 ^{b}	0 ± 0.73 ^{b}
HAA	13.41 ±0.43 ^{b}	43.81 ± 0.16 ^{f}	18.51 ± 0.1 ^{d,B}	15.73 ± 0.43 ^{c,B}	23.5 ± 0.43 ^{a,B}	24.28 ± 0.25 ^{a,B}	32.54 ± 1.34 ^{e,B}
AAA	4.62 ± 0.86 ^{b}	8.68 ± 0.23 ^{a}	1.97 ± 0.08 ^{b,B}	1.67 ± 0.06 ^{b,B}	3.15 ± 0.09 ^{c,B}	2.9 ± 0.02 ^{c,A}	4.21 ± 0.15 ^{b,B}

Table 1. *Cont.*

Amino Acid	Leaf Flour	LPC	H–Pep at Different Hydrolysis Times (min)				
			30	60	120	180	240
EAA	7.69 ± 0.11 [b]	36.19 ± 0.17 [f]	13.63 ± 0.27 [d,B]	11.9 ± 0.11 [c,B]	19.69 ± 0.57 [a,B]	20.31 ± 0.01 [a,B]	25.73 ± 1.61 [e,B]
TAA	17.48 ±1.03 [b]	64.19 ± 0.1 [f]	33.53 ± 0.33 [a,A]	19.88 ± 1.03 [c,B]	33.04 ± 0.1 [a,B]	37.08 ± 0.23 [d,B]	43.8 ± 0.17 [e,B]

LPC, leaf protein concentrate; H–Pep, hydrolysate of pepsin; H–Pan, hydrolysate of pancreatin; HAA, hydrophobic amino acids; AAA, aromatic amino acids; EAA, essential amino acids; TAA; total amino acid. Different lowercase letters in the same row indicate significant differences ($p < 0.05$) between treatments. Different capital letters in the same column indicate significant differences ($p < 0.05$) between H–Pep and H–Pan at a given time.

On the other hand, the highest release of amino acids occurred after 240 min of hydrolysis with both enzymes. The H–Pep was rich in leucine, valine, aspartic acid, alanine, and threonine, whereas H–Pan presented high amounts of proline, leucine, valine, isoleucine, and phenylalanine. Tyrosine was not detected in any protein hydrolysates. The absence of tyrosine in the hydrolysates may be due to its degradation during hydrolysis with 6 M HCL [25], which is necessary for amino acid determination. This behavior could be due to the low levels of tyrosine released during the enzymatic hydrolysis by both enzymes, which caused its loss in the samples during the treatment with HCL. Glycine was in very low concentrations or was not detected in some H–Pep. The differences in amino acid composition can be attributed to the type of enzyme used.

The content of the EAA of LPC, H–Pep, and H–Pan was lower than reported for the LPCs of amaranth, eggplant, and pumpkin [21], as well as for amaranth leaf protein hydrolysate [26]. These differences in amino acid content are attributed to the type of plant (species and age), and the environmental and cultivation conditions [27]. Moreover, amaranth leaf protein hydrolysates presented a higher DH than those reported in this study. However, the EAA such as isoleucine, leucine, threonine, valine, and phenylalanine of the LPC and protein hydrolysates obtained at the extended hydrolysis are at levels that are recommended in adults by the WHO/FAO [28]. This suggests that jackfruit LPC and those protein hydrolysates are good sources of high-quality protein and could be used to supplement foods with deficiencies in these amino acids. The composition of AA in protein hydrolysates could contribute to the functionality of LPCs and protein hydrolysates [21].

3.2. FTIR Spectra Analysis

The FTIR spectra of protein hydrolysates showed bands of greater intensity than the LPC (Figure 1A,B), which suggests that the LPC structure changed with the enzymatic hydrolysis. The IR spectral bands were assigned and described in Table 2. In general, the spectrum of LPC and hydrolysates showed bands between 3323–3241 cm^{-1} and 2917–2061 cm^{-1} corresponding to band amide A (stretching vibration of the NH bond coupled to the O–H stretching vibration of water during the hydrolysis [29]) and amide B (asymmetric stretching vibration of C−H that was found in the aliphatic side chain of proteins, [30]), respectively.

Compared with the LPC, the protein hydrolysates showed shifting in the peak position of the amide A and B bands (Table 2). The intensity of amide A and B were dependent on the hydrolysis time and enzyme used. The H–Pan showed a lower intensity in these bands (Figure 1) and the extended hydrolysis time caused the absence of signals in band amide B (Table 2). This behavior has been attributed to the destruction of hydrogen bonds of hydroxyl (–OH) groups during enzymatic hydrolysis, which reduced the density and strength of those bands [22]. The shifting to a higher wavenumber in amide B after hydrolysis by pancreatin (from 2923 cm^{-1} in LPC to 2925–2936 cm^{-1} in H–Pan, Table 2) can be due to the exposure of buried hydrophobic patches with more aliphatic side chains. This suggests that there was a greater exposure of buried hydrophobic patches in H–Pan as a function of time, which also coincides with the high level of HAA in these samples (Table 1). Similarly, the above could also explain the high DHs reported for these samples [15]. H–Pan (6.12–9.21%) showed higher DHs than H–Pep (1.78–3.44%), which would also explicate the apparent low intensity in the bands of H–Pan.

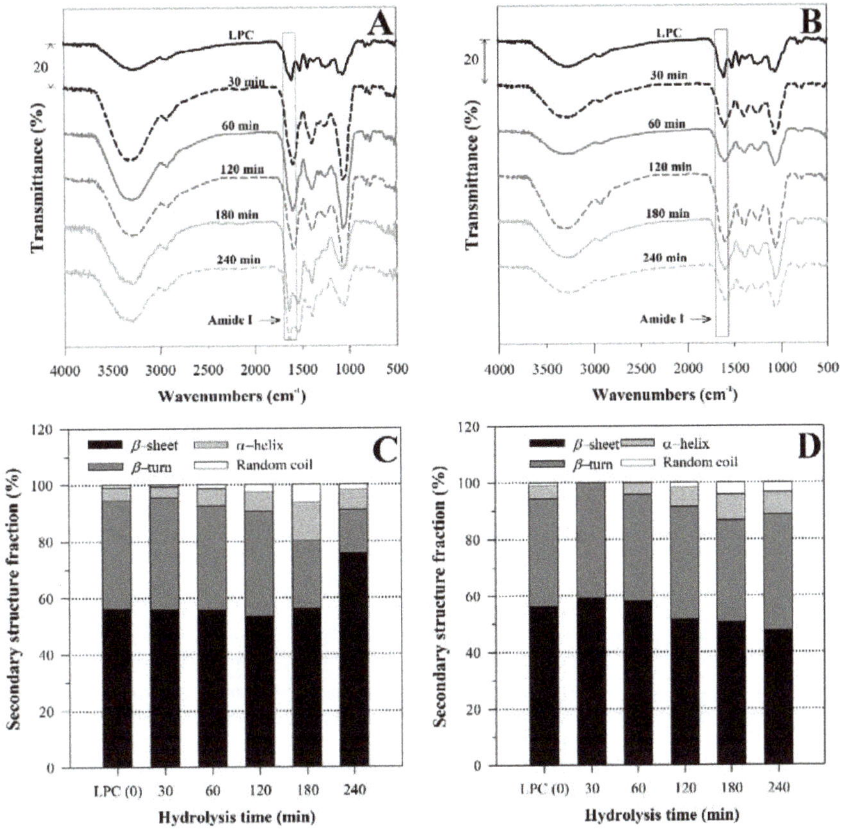

Figure 1. FTIR spectra (**A**,**B**) and secondary structure fractions (**C**,**D**) of leaf protein concentrate (LPC) and protein hydrolysates obtained with pepsin (**A**,**C**) and pancreatin (**B**,**D**) at different hydrolysis times.

Table 2. FTIR analysis of LPC and its hydrolysates.

		IR Spectral Bands (cm^{-1})						
Sample	**Hydrolysis Time (min)**	**Amide A**	**Amide B**	**Amide I**	**Amide II**	**CH_3 Bending**	**Amide III**	**C-O**
LPC	0	3280.3	2923.0	1605.0	1517.2	nd	1235.2	1062.1
H–Pep	30	3305.9	2918.0	1636.0	1594.8	1398.6	1257.4	1065.0
	60	3289.0	2919.0	1635.0	1598.7	1395.3	1258.8	1065.0
	120	nd	nd	1636.0	1596.3	1399.1	1253.5	1064.0
	180	3283.2	2917.0	1636.0	1596.3	1395.3	1255.9	1062.1
	240	3289.0	2961.4	1636.8	1540.8	1399.6	1260.2	1078.0
H–Pan	30	3247.1	2936.0	1635.0	1599.2	1387.1	1242.4	1069.3
	60	3288.0	2925.0	1636.3	1599.7	1388.5	1235.2	1069.3
	120	3323.2	2925.0	1636.0	1598.7	1389.9	1242.9	1068.4
	180	3239.0	nd	1636.0	1596.3	1388.1	1238.1	1068.9
	240	3262.0	nd	1636.0	1598.2	1388.0	1243.9	1067.4

LPC, leaf protein concentrate; H–Pep, hydrolysate of pepsin; H–Pan, hydrolysate of pancreatin. nd: not detected.

On the other hand, three characteristic protein bands at 1605.0, 1517.2, and 1235.2 cm^{-1} were observed in the LPC and are attributed to amide I (1600–1700 cm^{-1}), amide II (1600–1500 cm^{-1}), and amide III (1220–1300 cm^{-1}), respectively [18]. After enzymatic hydrolysis, the amide I, II, and III regions of the protein hydrolysates shifted the peak positions to higher wavenumbers than the LPC (Table 2) and with less intensity in H–Pan. This could indicate greater structural changes due to hydrolysis with pancreatin. Additionally, at 1438.2 cm^{-1} a band was detected in the LPC and was assigned to CH_2 bending vibration [31]. However, that band shifted the peak position to lower wavenumbers in the protein hydrolysates at 1395.3–1399.1 cm^{-1} for H–Pep and 1387.1–1388.9 cm^{-1} for H–Pan, which corresponds to the CH_3 bending vibration. Additionally, signals at 1062.1–1069.3 cm^{-1}, both for the LPC and hydrolysates, could correspond to the C−O group of α-anomer [31].

Subsequently, the amide I band region was divided into β-sheet (1610–1640 cm^{-1} and 1680-1695 cm^{-1}), random coil (1640–1650 cm^{-1}), α-helix (1650–1660 cm^{-1}), and β-turn (1660–1700 cm^{-1}) structures [32]. The results of the secondary structure fractions revealed that the LPC showed 4.49% of α-helix, 56.24% of β-sheet, 38.24% of β-turn, and 1.02% random coil (Figure 1). These fractions were different from those reported in the LPC of *Diplazium esculentum*, which presented a high content of α-helix (69.16%) and β-turn (35.68%) and a low β-sheet (17.02%, [31]). The differences between these two LPCs could be attributed to the composition of the matrices studied and to the conditions of extraction.

The protein hydrolysates showed a increase in the contents of α-helix (from 4.76–13%) and random coil (from 1.02–6.20%) in comparison to the LPC. In H–Pep (Figure 1C), the β-sheet fraction showed a considerable increase during the enzymatic hydrolysis (from 56.24–75.74). However, the β-turn decreased (from 38.24–15.38%) considerably with a possible rearrangement to α–helical and random coil structures as a function of hydrolysis time. Meanwhile, in H–Pan (Figure 1D), the decrease in the β-sheet from 56.24–47.55% was observed with an apparent rearrangement to α-helical (0.14–9.13%), β-turn (38.24–40.94%), and random coil (0–4.27%) fractions. The increase in these structures at the extended hydrolysis time was also a trend found in the protein hydrolysates of rice glutelin [33] and *Cinnamomum camphora* seed kernel [34].

The β-sheet is a structure present in the internal parts of the folded protein, maintaining hydrophobic amino acids. Then, the unfolding of protein molecules during hydrolysis with pancreatin exposed buried hydrophobic residues, which caused the β-sheet structure to loosen and its proportions to reduce [22], releasing greater amounts of HAA as shown in Table 1. This might suggest that the β-sheet fraction was more opened by hydrolysis with pancreatin, whereas the β-turn fraction was more susceptible to hydrolysis with pepsin, which caused the reduction in this fraction in H–Pep. On the contrary, the β-turn fraction was increased in H–Pan, which agrees with the high contents of glycine and proline in these hydrolysates (Table 1), since it has been established that glycine and proline amino acid residues contribute to the formation of these fractions [35]. On the other side, the low content of β-sheets in H–Pan could describe the high solubility and DHs that were reported in previous studies [15]. This suggests that the changes in the secondary structure of the protein hydrolysates could affect the functionality of these hydrolysates.

3.3. *Thermal Properties*

3.3.1. Thermogravimetric Analysis

Curves of TGA and DTG of the LPC and protein hydrolysates obtained at different hydrolysis times showed that the thermal decomposition profiles of H–Pep (Figure 2A) and H–Pan (Figure 2B) were different from the LPC. The LPC showed two weight losses, while H–Pep (Figure 2C) showed three weight losses up to 120 min of hydrolysis. The extension of hydrolysis at 180 and 240 min showed four weight losses. On the contrary, in H–Pan (Figure 2D), only three weight losses were observed. In general, the LPC and hydrolysates showed a first weight loss (from 25–100 °C) attributed to the release of free water or water weakly bound to protein molecules [36].

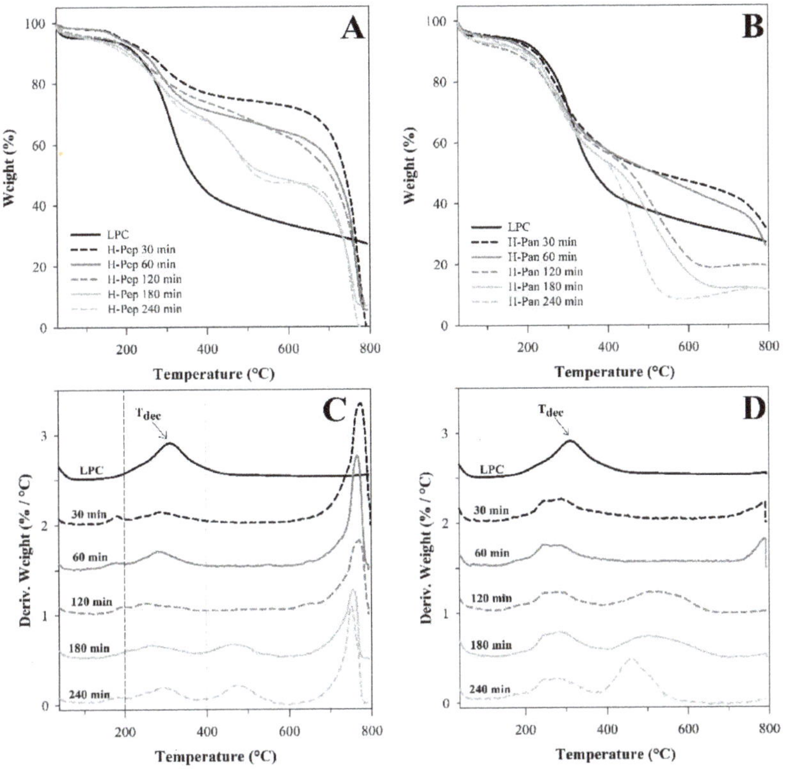

Figure 2. Curves of TGA (**A**,**B**) and DTG (**C**,**D**) of leaf protein concentrate (LPC) and protein hydrolysates obtained with pepsin (**A**,**C**) and pancreatin (**B**,**D**) at different hydrolysis times.

The analysis of the decomposition stages of the LPC and its hydrolysates (Table 3) showed that the moisture content (stage I) in all samples ranged between 1.93–7.94%, with H–Pep being the samples with the ($p < 0.05$) lowest moisture content. The low moisture content indicates a lower hydrophilic character [24], which is consistent with the low solubility reported for H–Pep (19–41%) compared to H–Pan (60–98%, [15]). The second weight loss (stage II) of LPC was 61.6% and occurred between 252.4 and 370.4 °C, which coincides with the content of protein in LPC (65.8%) previously reported [15], indicating that this stage corresponds to the decomposition of protein in LPC. Meanwhile, in hydrolysates, the second weight loss started around 230.7 °C for H–Pep and at 218.5 °C for H–Pan, and these temperatures decreased significantly ($p < 0.05$) with the progress of hydrolysis (Table 3). The weight loss at this stage could be attributed to the volatilization of protein fragments that were generated during enzymatic hydrolysis [24].

Table 3. Thermogravimetric analysis of the stages of decomposition of LPC and its hydrolysates.

Sample	Hydrolysis Time (min)	Stages								
		I		II		III		IV		Final Residue (%)
		Temperature Range (°C)	Weight Loss (%)	Temperature Range (°C)	Weight Loss (%)	Temperature Range (°C)	Weight Loss (%)	Temperature Range (°C)	Weight Loss (%)	
LPC	0	28.3–39.2 ± 0.23 ^{a}	3.93 ± 0.03 ^{a}	252.4–370.4 ± 0.18 ^{a}	61.65 ± 0.80 ^{a}	***	***	***	***	34.42 ± 0.12 ^{a}
H–Pep	30	34.5–47.0 ± 0.13 ^{b,A}	1.97 ± 0.02 ^{b,A}	230.7–357.2 ± 0.07 ^{b,A}	23.60 ± 0.03 ^{b,A}	736.5–774.6 ± 0.26 ^{a,A}	67.58 ± 0.21 ^{a,A}	***	***	6.85 ± 0.07 ^{b,A}
	60	35.3–36.9 ± 0.22 ^{c,A}	1.93 ± 0.13 ^{b,A}	228.2–334.8 ± 0.43 ^{c,A}	28.92 ± 0.12 ^{c,A}	736.6–764.8 ± 0.15 ^{b,A}	57.33 ± 0.41 ^{b,A}	***	***	11.82 ± 0.18 ^{c,A}
	120	30.2–47.6 ± 0.16 ^{d,A}	4.33 ± 0.02 ^{c,A}	192.2–295.8 ± 0.32 ^{d,A}	27.50 ± 0.14 ^{d,A}	725.2–767.7 ± 0.08 ^{c,A}	64.27 ± 0.26 ^{c,A}	***	***	3.90 ± 0.23 ^{d,A}
	180	30.8–42.4 ± 0.25 ^{e,A}	5.11 ± 0.10 ^{d,A}	202.0–318.4 ± 0.13 ^{e,A}	24.61 ± 0.22 ^{e,A}	426.4–515.5 ± 0.23 ^{d,A}	21.79 ± 0.05 ^{d,A}	719.7–762.6 ± 0.33 ^{a}	42.32 ± 0.21 ^{a}	6.18 ± 0.17 ^{e,A}
	240	30.3–37.1 ± 0.06 ^{c,A}	4.94 ± 0.07 ^{e,A}	227.2–328.6 ± 0.26 ^{f,A}	25.54 ± 0.32 ^{f,A}	439.1–524.1 ± 0.29 ^{e,A}	21.36 ± 0.39 ^{d,A}	727.4–761.6 ± 0.27 ^{a}	47.60 ± 0.32 ^{b}	0.56 ± 0.05 ^{f,A}
H–Pan	30	35.2–45.2 ± 0.02 ^{b,B}	5.13 ± 0.12 ^{b,B}	218.5–353.8 ± 0.11 ^{b,B}	44.27 ± 0.07 ^{b,B}	746.6–785.3 ± 0.39 ^{a,B}	18.61 ± 0.12 ^{a,B}	***	***	31.99 ± 0.12 ^{b,B}
	60	35.0–45.8 ± 0.17 ^{b,B}	5.55 ± 0.15 ^{b,B}	210.4–334.1 ± 0.34 ^{c,A}	44.13 ± 0.14 ^{b,B}	763.5–784.9 ± 0.23 ^{a,B}	23.82 ± 0.19 ^{b,B}	***	***	26.51 ± 0.18 ^{c,B}
	120	31.6–47.1 ± 0.09 ^{c,B}	7.94 ± 0.13 ^{c,B}	211.3–322.9 ± 0.18 ^{d,B}	34.74 ± 0.21 ^{c,B}	461.8–605.0 ± 0.36 ^{c,B}	37.50 ± 0.07 ^{c,B}	***	***	19.82 ± 0.11 ^{d,B}
	180	32.0–46.8 ± 0.18 ^{d,B}	7.32 ± 0.28 ^{d,B}	220.8–327.0 ± 0.25 ^{e,B}	38.94 ± 0.56 ^{d,B}	456.1–595.4 ± 0.19 ^{d,B}	41.92 ± 0.11 ^{d,B}	***	***	11.83 ± 0.15 ^{e,B}
	240	32.8–41.9 ± 0.21 ^{e,B}	6.54 ± 0.02 ^{e,B}	213.2–321.3 ± 0.71 ^{f,B}	37.73 ± 0.97 ^{e,B}	429.4–518.8 ± 0.22 ^{e,B}	44.27 ± 0.16 ^{e,B}	***	***	11.46 ± 0.45 ^{e,B}

LPC, leaf protein concentrate; H–Pep, hydrolysate of pepsin; H–Pan, hydrolysate of pancreatin. Different lowercase letters in the same column indicate significant differences ($p < 0.05$) between hydrolysis time for the same enzyme. Different capital letters indicate significant differences ($p < 0.05$) between H–Pep and H–Pan at a given time. ***: not detected.

The third weight loss started (stage III) around 736.5 °C for H–Pep-30 and decreased at 725.2 °C in H–Pep-120 ($p < 0.05$). Similarly in H–Pan-30, this stage of decomposition started at 746.53 °C and reduced at 763.5 °C ($p < 0.05$) in H–Pan-60 (Table 3). Subsequently, the increase in the hydrolysis time for both enzymes reduced ($p < 0.05$) the onset temperatures to around 439.1 for H–Pep-240 and 429.4 °C for H–Pan-240. Additionally, the H–Pep at 180 and 240 min of hydrolysis showed a fourth weight loss (stage IV), which began at 719.7 and 727.4 °C, respectively (Table 3). The additional weight loss steps could be the result of the slow decomposition of peptides (with different thermal stability) from the previous step since with increasing temperature complex decomposition reactions can occur [36]. Thus, this suggests that H–Pep presented larger protein fragments than H–Pan that were degraded more slowly, causing more decomposition stages. The reduction in the onset temperatures of the different stages as a function of time could indicate the reduction in the stability of the hydrolysates due to the increase in smaller peptides in the long hydrolysis time.

On the other hand, the decomposition temperature (T_{dec}) of the LPC (Figure 3F) was 311.9 °C. After hydrolysis, a decrease in T_{dec} was observed. Although the results of the T_{dec} of the protein hydrolysates showed certain variations, it could be said that apparently, H–Pep presents higher T_{dec}, which is probably caused by the high content of β-sheets in these samples, which gives it greater stability. The differences in the decomposition profiles and thermogravimetric curves of the protein hydrolysates were affected by enzymatic treatment, which caused changes in the protein structure and led to marked differences in the loss of absorbed water, stages of decomposition, weight loss, and residual materials.

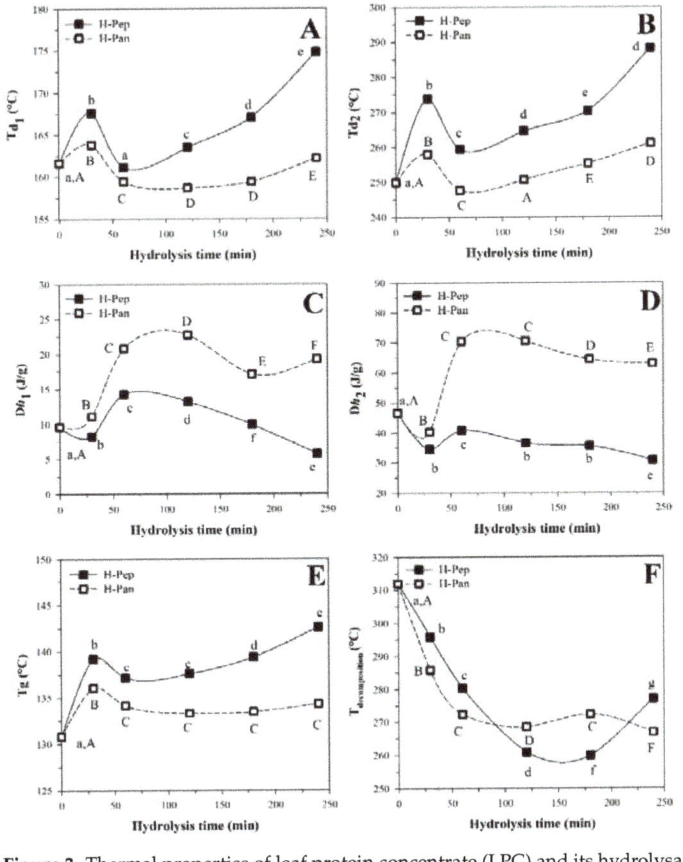

Figure 3. Thermal properties of leaf protein concentrate (LPC) and its hydrolysates. (**A**) and (**B**) show

the denaturation temperature (T_d), while (**C**) and (**D**) the denaturation enthalpy (Δh) corresponding to the first (Td_1 and Δh_1) and second (Td_2 and Δh_2) endothermic peak, respectively. (**E**) and (**F**) are the glass transition temperature (Tg) and decomposition temperature of LPC and protein hydrolysates as a function of time, respectively. Different letters in the same treatment indicate a significant difference ($p < 0.05$) with respect to time.

3.3.2. Differential Scanning Calorimetry Analysis

DSC provides information on the structural stability of a protein based on the endothermic and exothermic events [37]. The Δh represents the enthalpy changes that occur during the denaturation process and reflects the extent of the ordered secondary structure of a protein [38], while T_d is a measurement of its thermal stability. After this temperature, irreversible changes occur in the structure of the protein. A higher T_d value is associated with higher thermal stability [5]. Thermographs of LPC, H–Pep (Figure 4A), and H–Pan (Figure 4B) displayed two endothermic peaks. The presence of two peaks in the DSC profiles of proteins has been associated with the denaturation of major proteins that have different stability [39]. LPC of jackfruit shows majority fractions of glutenin (69.33%) and prolamin (17.34%). Therefore, these endothermic peaks could be attributed to the denaturation of these protein fractions, since when analyzed by DSC, they presented a profile similar to LPC [40].

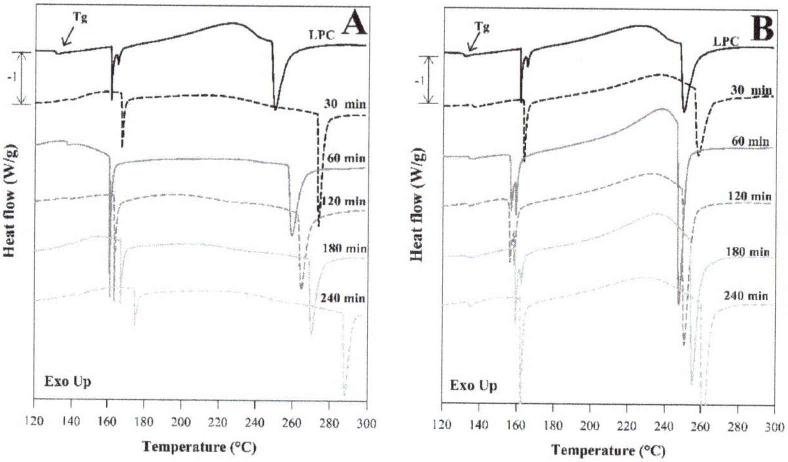

Figure 4. Curves of DSC of leaf protein concentrate (LPC) and protein hydrolysates obtained with pepsin (**A**) and pancreatin (**B**) at different hydrolysis times.

The first endothermic peak in LPC had a T_{d1} and Δh_1 of 161.6 °C and 9.6 J/g, respectively. The T_{d1} of protein hydrolysates increased considerably ($p < 0.05$) as a function of time (Figure 3A), in H–Pep from 167.6–174.8 °C and H–Pan from 163.8–162.2 °C. Regarding their Δh_1, H–Pep (from 14.3–5.7 J/g) and H–Pan (from 20.8–19.3 J/g) showed a considerable increase in Δh_1 in the first 60 min of hydrolysis concerning LPC ($p < 0.05$), which decreased as the enzymatic process progressed (Figure 4C). Concerning to second endothermic peak, the T_{d2} and Δh_2 of the LPC were 249.9 °C and 46.7 J/g, respectively. Similar behavior of T_{d1} was observed in T_{d2} since an increased markedly as a function of hydrolysis time ($p < 0.05$) in H–Pep (from 273.8–288.2 °C) and H–Pan (from 258.0–261.0 °C) was observed (Figure 4B). In what refers to the Δh_2 (Figure 4D), different behaviors were observed depending on the enzyme used. The Δh_2 decreased considerably in H–Pep as a function of hydrolysis ($p < 0.05$), while in H–Pan an increase in Δh_2 was observed in the first 60 min but decreased with an increase in the hydrolysis time.

The increase in Δh in some protein hydrolysates is attributed to the unfolding of the protein which produced polypeptides with stronger protein-protein interactions concerning the LPC [38]. During the hydrolysis process, there could be a rearrangement or reassociation of the released protein fragments, which form rigid particles and more compact conformations, especially at the beginning of hydrolysis [41]. However, as hydrolysis progresses, the reduction in Δh as a function of time indicates the decrease in the ordered structure and aggregation [39], which indicate that the protein aggregates were held together mainly by the noncovalent hydrophobic bonds [38]. On the other hand, the high Δh_1 and Δh_2 in H–Pan (regarding H–Pep) could be caused by the high contents of β-turn in this protein hydrolysates. The β-turn is a structure mainly formed by proline and glycine residues. These amino acids reduce the steric strain of protein hydrolysates, which decreases the entropy of the denatured state [42].

As for T_d, both T_{d1} and T_{d2} of H–Pep were notably higher ($p < 0.05$) than in H–Pan (Figure 3A,B). This indicates that these hydrolysates had more hydrophobic interactions, which are the main interactive forces responsible for the formation of the protein aggregates [38]. Additionally, the high denaturation temperatures have also been attributed to the high percentages of β-sheet in these hydrolysates (Figure 1C). Proteins with large fractions of β-sheet structure usually exhibit high denaturation temperatures [43]. This behavior would explain why these hydrolysates had more decomposition stages in DTG (Figure 2C). Similar trends have been reported in protein hydrolysates of glutelin [44] peanut [39], and chickpea [10].

On the other hand, the Tg of the LPC was 130.8 °C. In protein hydrolysates, this temperature was dependent on the hydrolysis time and the enzyme used (Figure 3E). H–Pep (from 139.2–142.0 °C) showed an increase in Tg as a function of hydrolysis time ($p < 0.05$), while in H–Pan (from 136.1–134.0 °C) contrary comportment was observed. The H–Pep showed higher Tg and T_{dec} ($p < 0.05$), indicating that the β-sheet fraction significantly influences the thermal stability of these hydrolysates. Then, the results clearly show that the thermal events and the shape of the endothermic peaks in the different DSC curves depend on the secondary structure.

3.4. Particle Size Distribution of Emulsions Stabilized with LPC and Hydrolysates

The LPC stabilized emulsion exhibited a trimodal size distribution (Figure 5) with high PDI (2.88 ± 0.63, Figure 5C) and D [4,3] (122.33 ± 22.23, Figure 5D). These results indicate the poor capacity of LPC to form a stable emulsion, which coincides with the low emulsifying activity index (EAI, 32.38 m^2/g) and emulsion stability index (ESI, 30.24 min) for the LPC reported in a previous study [15]. After enzymatic hydrolysis, considerable changes in particle size distributions, PDI, and D [4,3] were observed. Emulsions stabilized with H–Pep showed very similar distributions to LPC but with a shift in the distribution towards smaller particles (first peak) than LPC and with greater magnitude (Figure 5A). In the case of emulsions stabilized with H–Pan, monomodal distributions were observed after 60 min of hydrolysis. These emulsions showed narrower distributions (Figure 5B) and lowered PDI (Figure 5C) and D [4,3] (Figure 5D) than emulsions stabilized by LPC and H–Pep.

Both the emulsions stabilized with H–Pep and H–Pan showed a reduction in PDI (Figure 5C) and D [4,3] (Figure 5D) with increasing hydrolysis time. In emulsions stabilized with H–Pep, the PDI decreases from 4.75–1.93 and the D [4,3] from 62.6–35.9 μm. The lowest values of PDI and D [4,3] for H–Pep were obtained in 240 min of hydrolysis.

Whereas in emulsions stabilized with H Pan the PDI decreased from 2.38–1.49 and the D [4,3] from 20.7–2.78 μm up to 180 min of hydrolysis. H–Pan with 180 min of hydrolysis showed the lowest PDI and D [4,3]. However, the increase in time of hydrolysis caused a rise in PDI and D [4,3] in these emulsions, although the increase in D [4,3] was not significant ($p > 0.05$). This behavior coincides with the results of EAI and ESI reported for these hydrolysates. The best EAI (6.25 m^2/g) and ESI (88.79 min) were determined after

180 min of hydrolysis for H–Pan and the subsequent increase in time caused the decrease in these properties [15].

In general, the decrease in particle size in emulsions stabilized with protein hydrolysates could be attributed to the release of shorter protein chains during enzymatic hydrolysis and the exposure of hidden hydrophobic residues that improved the solubility. Enzymatic hydrolysis reduced the HAA content in LPC (Table 1), which improved the hydrophobic-hydrophilic balance of the protein hydrolysates. LPC showed a high HAA content, which explains its low solubility in water (15% at pH 8.0, [15]). The low solubility reduced the adsorption of protein at the interface and minimized protein-oil interactions, causing the aggregation of oil droplets and leading to nonhomogeneous emulsions [22].

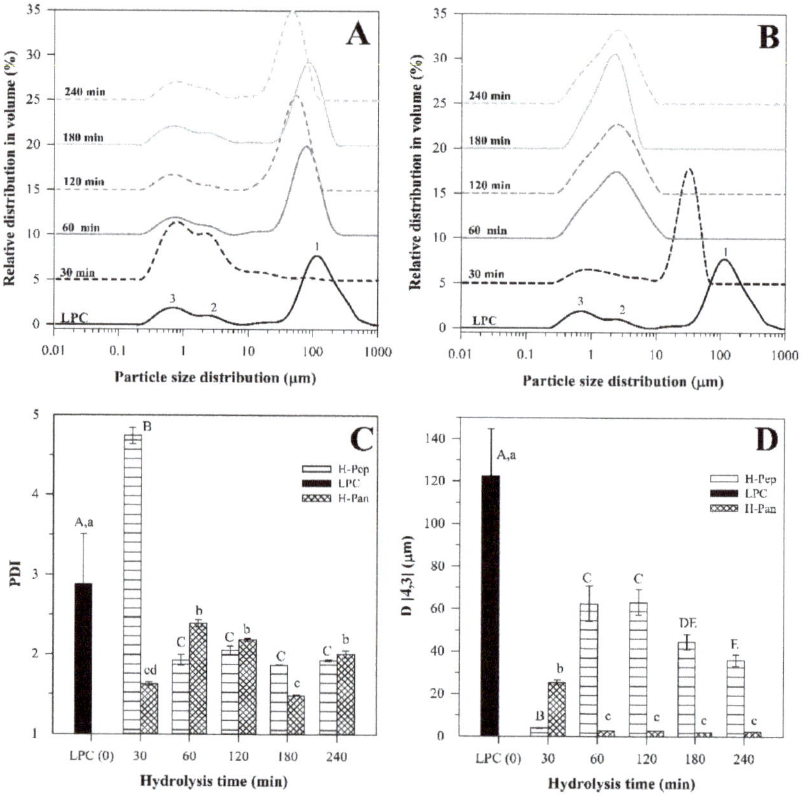

Figure 5. Effect of hydrolysis time and type of enzyme on the particle size distribution (**A**,**B**), polydispersity index (PDI, **C**), and volume-weighted mean particle diameter (D [4,3], **D**) of emulsions stabilized (pH = 7.0) with H-Pep (**A**) and H-Pan (**B**) obtained at different hydrolysis time. LPC, Leaf protein concentrate. Different letters indicate significant differences ($p < 0.05$) between the hydrolysis time in a specified sample.

Nevertheless, the best results based on smaller particle sizes and homogeneous distributions were obtained in emulsions stabilized with H–Pan, which was attributed to the low T_d and β-sheet contents of these hydrolysates compared to H–Pep. The high T_d of H–Pep indicates a greater amount of protein aggregates that reduced the adsorption of protein at the interface. As for the H–Pan, the decrease in the β-sheet caused the increase in the β-turn structures (Figure 1D), revealing buried hydrophobic patches that led to a greater hydrophobicity of the surface. This has been associated with the improvement of emulsifying properties [22]. Moreover, the α-helix, β-turn, and random coil are structures that are relatively flexible and open, while β-sheet structures are more stable [33]. Hence, the decreasing of the β-sheets and the increasing of the β-turns in H–Pan indicate that the hydrolysis by pancreatin produced more flexible protein fragments that can spread rapidly at oil–water interfaces [45]. The same trend was reported in protein hydrolysates of rice glutelin [33] and potatoes [22].

Effect of pH on the Particle Size Distribution of Emulsions Stabilized with H–Pan

The effect of pH on the PDI (Figure 6A) and D [4,3] (Figure 6B) of emulsions stabilized with protein hydrolysates obtained at different hydrolysis times was evaluated only with H–Pan. This is because emulsions stabilized with H–Pep showed significantly higher PDI and D [4,3], as well as nonhomogeneous particle size distributions. In all the pH values tested, high values of PDI and D [4,3] in emulsions stabilized with H–Pan obtained in a short hydrolysis time (60–120 min) were observed. However, the progress of hydrolysis at 180 min led to the reduction in these values, followed by an increase in PDI and D [4,3] at an extended hydrolysis time (240 min), indicating that excessive hydrolysis reduced the ability of hydrolysate to stabilize the emulsion [46]. This would explain the reduction in EAI (from 56.24–51.18) and ESI (from 88.79–67.18 min) of the protein hydrolysates obtained from 180–240 min of hydrolysis, respectively [15].

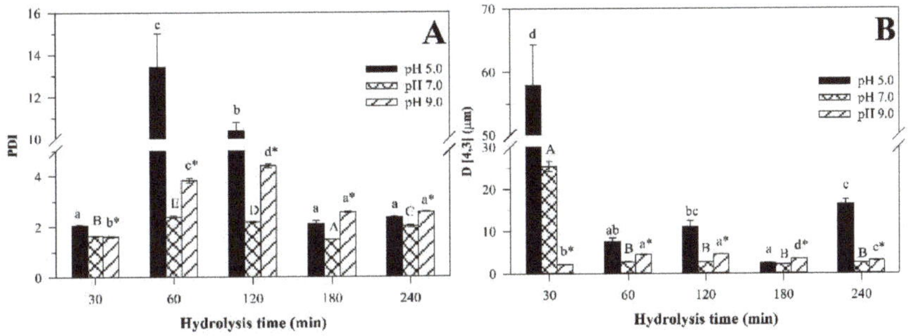

Figure 6. Effect of pH on polydispersity index (PDI, **A**) and volume-weighted mean particle diameter (D [4,3], **B**) of emulsions stabilized with H–Pan obtained at different hydrolysis times. Different letters indicate significant differences ($p < 0.05$) between the hydrolysis time at a specified pH. *: Visual guide for the reader in treatments at pH 9.

Hence, moderate hydrolysis could produce flexible peptides with greater hydrophobicity, which could facilitate anchoring to the O/W interface and improve their emulsifying properties. Conversely, protein hydrolysates obtained at an extended hydrolysis time could release smaller peptides and this increase in low molecular weight peptides could increment peptide–peptide interactions instead of peptide–oil interactions. Then, the excess of the unabsorbed peptides can lead to depletion flocculation, thereby increasing particle size and PDI [22]. Similar trends were reported in protein hydrolysates of chickpea [10], fava bean [12], and oat [46].

Regarding the pH, the highest PDI and D [4,3] values were obtained in emulsions at pH 5.0, indicating extensive droplet aggregation. This is because at pHs close to the isoelectric point (acidic pH), the net charge of the droplets is very low. Then, the electrostatic repulsion is not enough to overcome the Van der Waals forces, which promote the flocculation of the droplets and the consequent increase in particle size.

The low pH led to a reduction in the solubility of hydrolysate; therefore, it could not move quickly to the interface [47,48]. However, increasing the pH to 7.0–9.0 significatively reduced ($p < 0.05$) the PDI and the particle size. This reduction in particle size coincides with the solubility increase in protein hydrolysates as a function of pH reported in a previous study [15]. The increase in the solubility improved the absorption and reorientation of peptides on the surfaces of the oil droplets [47,48]. During the hydrolysis of proteins, carboxyl and amino groups are released due to the cleavage of peptide bonds, which increase as the hydrolysis time increases. These carboxyl groups acquire a negative charge when the pH of the medium is increasing. Therefore, the increase in pH in the emulsions led to a greater negative charge on the surface of the emulsion droplets, which has been related to strong electrostatic repulsions between the oil droplets, thus preventing the physical destabilization of the emulsions by coagulation, flocculation, and the increase in particle size [22,49].

Finally, the emulsions stabilized with H–Pep and H–Pan showed a PDI greater than 1.0, indicating that current protein hydrolysates are not suitable for stabilizing emulsions. However, the results indicate that H–Pan at 180 min presented a lower PDI (near 1.0) and D [4,3], and monomodal distributions (compared with H–Pep). These results suggest that H–Pan could be used as an alternative emulsifier. However, additional studies are required to confirm its potential use in the food industry.

4. Conclusions

This study showed that the hydrolysis of LPC with pepsin and pancreatin modified the amino acid composition, secondary structure, and thermal properties of leaf protein hydrolysates, which influence the emulsifying capacity. This modification was dependent on the hydrolysis time and the type of enzyme. Both H–Pep and H–Pan showed slightly different amino acid patterns, with higher amounts of HAA and EAA at 240 min. The lower β-sheet fractions and high β-turn contents had a greater influence on the emulsifier properties of H–Pan, which improved the flexibility of these hydrolysates. The high content of β-sheets in H–Pep caused high decomposition and glass transition temperatures, as well as a low emulsifying capacity. These results showed that H–Pan at 180 min of hydrolysis produced emulsions with monomodal distributions, a low PDI (near to 1.0), and D [4,3], resulting in more-stable emulsions. Then, the hydrolysis with pancreatin can be used to improve the emulsifier properties of LPC.

However, this study explored the ability of H–Pan (at 180 min of hydrolysis) of jackfruit leaf as a possible emulsifier using a concentration of 0.5% (w/v). Therefore, a further study exploring different concentrations of H–Pan (at 180 min of hydrolysis) and blends with other polymers is recommended. In addition, the protein hydrolysate is a mixture of peptides with different molecular weights, so the influence of molecular weight on the emulsifying properties will continue to be investigated to improve the stability of the emulsions in future research.

Author Contributions: Conceptualization, C.C.-C., M.C.-S. and J.A.R.-S.; data curation, C.C.-C. and J.C.B.-C.; formal analysis, C.C.-C.; funding acquisition, J.A.R.-S.; investigation, C.C.-C.; methodology, C.C.-C., M.C.-S., J.C.B.-C., J.A.D. and J.A.R.-S.; project administration, J.A.R.-S.; resources, J.A.R.-S.; supervision, J.A.R.-S.; visualization, C.C.-C., M.C.-S., J.C.B.-C., J.A.D. and J.A.R.-S.; writing—original draft, C.C.-C. and J.A.R.-S.; writing—review and editing, C.C.-C., M.C.-S., J.C.B.-C., J.A.D. and J.A.R.-S. All authors have read and agreed to the published version of the manuscript.

Funding: Thanks to project code 316948 of the CYTED thematic network code 319RT0576.

Institutional Review Board Statement: Not applicable.

Informed Consent Statement: Not applicable.

Data Availability Statement: Not applicable.

Acknowledgments: The authors thank CONACYT (Consejo Nacional de Ciencia y Tecnología-Mexico) for their support through scholarship number 713740 granted to Carolina Calderón-Chiu.

Conflicts of Interest: The authors declare no conflict of interest.

References

1. Drapala, K.P.; Mulvihill, D.M.; O'Mahony, J.A. A Review of the Analytical Approaches Used for Studying the Structure, Interactions and Stability of Emulsions in Nutritional Beverage Systems. *Food Struct.* **2018**, *16*, 27–42. [CrossRef]
2. Dapčević-Hadnađev, T.; Dizdar, M.; Pojić, M.; Krstonošić, V.; Zychowski, L.M.; Hadnađev, M. Emulsifying Properties of Hemp Proteins: Effect of Isolation Technique. *Food Hydrocoll.* **2019**, *89*, 912–920. [CrossRef]
3. Burger, T.G.; Zhang, Y. Recent Progress in the Utilization of Pea Protein as an Emulsifier for Food Applications. *Trends Food Sci. Technol.* **2019**, *86*, 25–33. [CrossRef]
4. Pojić, M.; Mišan, A.; Tiwari, B. Eco-Innovative Technologies for Extraction of Proteins for Human Consumption from Renewable Protein Sources of Plant Origin. *Trends Food Sci. Technol.* **2018**, *75*, 93–104. [CrossRef]
5. Galves, C.; Galli, G.; Miranda, C.G.; Kurozawa, L.E. Improving the Emulsifying Property of Potato Protein by Hydrolysis: An Application as Encapsulating Agent with Maltodextrin. *Innov. Food Sci. Emerg. Technol.* **2021**, *70*, 102696. [CrossRef]
6. Chang, C.; Nickerson, M.T. Encapsulation of Omega 3-6-9 Fatty Acids-Rich Oils Using Protein-Based Emulsions with Spray Drying. *J. Food Sci. Technol.* **2018**, *55*, 2850–2861. [CrossRef]
7. Grasberger, K.F.; Gregersen, S.B.; Jensen, H.B.; Sanggaard, K.W.; Corredig, M. Plant-Dairy Protein Blends: Gelation Behaviour in a Filled Particle Matrix. *Food Struct.* **2021**, *29*, 100198. [CrossRef]
8. Fernández-Sosa, E.I.; Chaves, M.G.; Henao Ossa, J.S.; Quiroga, A.V.; Avanza, M.V. Protein Isolates from *Cajanus cajan* L. as Surfactant for o:W Emulsions: PH and Ionic Strength Influence on Protein Structure and Emulsion Stability. *Food Biosci.* **2021**, *42*, 101159. [CrossRef]
9. Ozturk, B.; McClements, D.J. Progress in Natural Emulsifiers for Utilization in Food Emulsions. *Curr. Opin. Food Sci.* **2016**, *7*, 1–6. [CrossRef]
10. Ghribi, A.M.; Gafsi, I.M.; Sila, A.; Blecker, C.; Danthine, S.; Attia, H.; Bougatef, A.; Besbes, S. Effects of Enzymatic Hydrolysis on Conformational and Functional Properties of Chickpea Protein Isolate. *Food Chem.* **2015**, *187*, 322–330. [CrossRef]
11. Tamm, F.; Herbst, S.; Brodkorb, A.; Drusch, S. Functional Properties of Pea Protein Hydrolysates in Emulsions and Spray-Dried Microcapsules. *Food Hydrocoll.* **2016**, *58*, 204–214. [CrossRef]
12. Liu, C.; Bhattarai, M.; Mikkonen, K.S.; Heinonen, M. Effects of Enzymatic Hydrolysis of Fava Bean Protein Isolate by Alcalase on the Physical and Oxidative Stability of Oil-in-Water Emulsions. *J. Agric. Food Chem.* **2019**, *67*, 6625–6632. [CrossRef] [PubMed]
13. Zang, X.; Yue, C.; Wang, Y.; Shao, M.; Yu, G. Effect of Limited Enzymatic Hydrolysis on the Structure and Emulsifying Properties of Rice Bran Protein. *J. Cereal Sci.* **2019**, *85*, 168–174. [CrossRef]
14. Zhang, Y.; Romero, H.M. Exploring the Structure-Function Relationship of Great Northern and Navy Bean (*Phaseolus vulgaris* L.) Protein Hydrolysates: A Study on the Effect of Enzymatic Hydrolysis. *Int. J. Biol. Macromol.* **2020**, *162*, 1516–1525. [CrossRef] [PubMed]
15. Calderón-Chiu, C.; Calderón-Santoyo, M.; Herman-Lara, E.; Ragazzo-Sánchez, J.A. Jackfruit (*Artocarpus heterophyllus* Lam) Leaf as a New Source to Obtain Protein Hydrolysates: Physicochemical Characterization, Techno-Functional Properties and Antioxidant Capacity. *Food Hydrocoll.* **2021**, *112*, 106319. [CrossRef]
16. Brion-Espinoza, I.A.; Iñiguez-Moreno, M.; Ragazzo-Sánchez, J.A.; Barros-Castillo, J.C.; Calderón-Chiu, C.; Calderón-Santoyo, M. Edible Pectin Film Added with Peptides from Jackfruit Leaves Obtained by High-Hydrostatic Pressure and Pepsin Hydrolysis. *Food Chem. X* **2021**, *12*, 100170. [CrossRef]
17. Zhang, Y.; Shen, Y.; Zhang, H.; Wang, L.; Zhang, H.; Qian, H.; Qi, X. Isolation, Purification and Identification of Two Antioxidant Peptides from Water Hyacinth Leaf Protein Hydrolysates (WHLPH). *Eur. Food Res. Technol.* **2018**, *244*, 83–96. [CrossRef]
18. Yan, X.; Liang, S.; Peng, T.; Zhang, G.; Zeng, Z.; Yu, P.; Gong, D.; Deng, S. Influence of Phenolic Compounds on Physicochemical and Functional Properties of Protein Isolate from *Cinnamomum camphora* Seed Kernel. *Food Hydrocoll.* **2020**, *102*, 105612. [CrossRef]
19. Böcker, U.; Wubshet, S.G.; Lindberg, D.; Afseth, N.K. Fourier-Transform Infrared Spectroscopy for Characterization of Protein Chain Reductions in Enzymatic Reactions. *Analyst* **2017**, *142*, 2812–2818. [CrossRef]
20. del Mar Contreras, M.; Lama-Muñoz, A.; Manuel Gutiérrez-Pérez, J.; Espínola, F.; Moya, M.; Castro, E. Protein Extraction from Agri-Food Residues for Integration in Biorefinery: Potential Techniques and Current Status. *Bioresour. Technol.* **2019**, *280*, 459–477. [CrossRef]
21. Famuwagun, A.A.; Alashi, A.M.; Gbadamosi, S.O.; Taiwo, K.A.; Oyedele, D.J.; Adebooye, O.C.; Aluko, R.E. Comparative Study of the Structural and Functional Properties of Protein Isolates Prepared from Edible Vegetable Leaves. *Int. J. Food Prop.* **2020**, *23*, 955–970. [CrossRef]
22. Akbari, N.; Mohammadzadeh Milani, J.; Biparva, P. Functional and Conformational Properties of Proteolytic Enzyme-modified Potato Protein Isolate. *J. Sci. Food Agric.* **2020**, *100*, 1320–1327. [CrossRef] [PubMed]
23. Miss-Zacarías, D.M.; Iñiguez-Moreno, M.; Calderón-Santoyo, M.; Ragazzo-Sánchez, J.A. Optimization of Ultrasound-Assisted Microemulsions of Citral Using Biopolymers: Characterization and Antifungal Activity. *J. Dispers. Sci. Technol.* **2022**, *43*, 1373–1382. [CrossRef]
24. Ricci, L.; Umiltà, E.; Righetti, M.C.; Messina, T.; Zurlini, C.; Montanari, A.; Bronco, S.; Bertoldo, M. On the Thermal Behavior of Protein Isolated from Different Legumes Investigated by DSC and TGA. *J. Sci. Food Agric.* **2018**, *98*, 5368–5377. [CrossRef] [PubMed]
25. Sanger, F.; Thompson, E.O.P. Halogenation of Tyrosine during Acid Hydrolysis. *Biochim. Biophys. Acta* **1963**, *71*, 468–471. [CrossRef]
26. Famuwagun, A.A.; Alashi, A.M.; Gbadamosi, O.S.; Taiwo, K.A.; Oyedele, D.; Adebooye, O.C.; Aluko, R.E. Antioxidant and Enzymes Inhibitory Properties of Amaranth Leaf Protein Hydrolyzates and Ultrafiltration Peptide Fractions. *J. Food Biochem.* **2021**, *45*, e13396. [CrossRef]

27. Tenorio, A.T.; Kyriakopoulou, K.E.; Suarez-Garcia, E.; van den Berg, C.; van der Goot, A.J. Understanding Differences in Protein Fractionation from Conventional Crops, and Herbaceous and Aquatic Biomass—Consequences for Industrial Use. *Trends Food Sci. Technol.* **2018**, *71*, 235–245. [CrossRef]
28. WHO/FAO. *Report of a Joint WHO/FAO/UNU Expert Consultation 2007*; World Health Organization/Food and Agricultural Organization: Geneva, Switzerland, 2007.
29. Kchaou, H.; Jridi, M.; Benbettaieb, N.; Debeaufort, F.; Nasri, M. Bioactive Films Based on Cuttlefish (Sepia Officinalis) Skin Gelatin Incorporated with Cuttlefish Protein Hydrolysates: Physicochemical Characterization and Antioxidant Properties. *Food Packag. Shelf Life* **2020**, *24*, 100477. [CrossRef]
30. Trigui, I.; Yaich, H.; Zouari, A.; Cheikh-Rouhou, S.; Blecker, C.; Attia, H.; Ayadi, M.A. Structure-Function Relationship of Black Cumin Seeds Protein Isolates: Amino-Acid Profiling, Surface Characteristics, and Thermal Properties. *Food Struct.* **2021**, *29*, 100203. [CrossRef]
31. Saha, J.; Deka, S.C. Functional Properties of Sonicated and Non-Sonicated Extracted Leaf Protein Concentrate from *Diplazium esculentum*. *Int. J. Food Prop.* **2017**, *20*, 1051–1061. [CrossRef]
32. Gómez, A.; Gay, C.; Tironi, V.; Avanza, M.V. Structural and Antioxidant Properties of Cowpea Protein Hydrolysates. *Food Biosci.* **2021**, *41*, 101074. [CrossRef]
33. Xu, X.; Liu, W.; Liu, C.; Luo, L.; Chen, J.; Luo, S.; McClements, D.J.; Wu, L. Effect of Limited Enzymatic Hydrolysis on Structure and Emulsifying Properties of Rice Glutelin. *Food Hydrocoll.* **2016**, *61*, 251–260. [CrossRef]
34. Yan, X.; Zhang, G.; Zhao, J.; Ma, M.; Bao, X.; Zeng, Z.; Gong, X.; Yu, P.; Wen, X.; Gong, D. Influence of Phenolic Compounds on the Structural Characteristics, Functional Properties and Antioxidant Activities of Alcalase-Hydrolyzed Protein Isolate from *Cinnamomum camphora* Seed Kernel. *LWT* **2021**, *148*, 111799. [CrossRef]
35. Marcelino, A.M.C.; Gierasch, L.M. Roles of β-Turns in Protein Folding: From Peptide Models to Protein Engineering. *Biopolymers* **2008**, *89*, 380–391. [CrossRef] [PubMed]
36. López, D.N.; Ingrassia, R.; Busti, P.; Bonino, J.; Delgado, J.F.; Wagner, J.; Boeris, V.; Spelzini, D. Structural Characterization of Protein Isolates Obtained from Chia (*Salvia hispanica* L.) Seeds. *LWT* **2018**, *90*, 396–402. [CrossRef]
37. Li, C.; Yang, J.; Yao, L.; Qin, F.; Hou, G.; Chen, B.; Jin, L.; Deng, J.; Shen, Y. Characterisation, Physicochemical and Functional Properties of Protein Isolates from *Amygdalus pedunculata* Pall Seeds. *Food Chem.* **2020**, *311*, 125888. [CrossRef]
38. Asen, N.D.; Aluko, R.E. Physicochemical and Functional Properties of Membrane-Fractionated Heat-Induced Pea Protein Aggregates. *Front. Nutr.* **2022**, *9*, 852225. [CrossRef] [PubMed]
39. Zhao, G.; Liu, Y.; Zhao, M.; Ren, J.; Yang, B. Enzymatic Hydrolysis and Their Effects on Conformational and Functional Properties of Peanut Protein Isolate. *Food Chem.* **2011**, *127*, 1438–1443. [CrossRef]
40. Calderón-Chiu, C.; Instituto Tecnológico de Tepic, Tepic, Mexico; Ragazzo-Sánchez, J.A.; Instituto Tecnológico de Tepic, Tepic, Mexico; Calderón-santoyo, M.; Instituto Tecnológico de Tepic, Tepic, Mexico. Personal Communication, Non-Published Material. 2022.
41. Tang, C.-H.; Ma, C.-Y. Heat-Induced Modifications in the Functional and Structural Properties of Vicilin-Rich Protein Isolate from Kidney (*Phaseolus vulgaris* L.) Bean. *Food Chem.* **2009**, *115*, 859–866. [CrossRef]
42. Fu, H.; Grimsley, G.R.; Razvi, A.; Scholtz, J.M.; Pace, C.N. Increasing Protein Stability by Improving Beta-Turns. *Proteins Struct. Funct. Bioinforma* **2009**, *77*, 491–498. [CrossRef]
43. Damodaran, S. Amino Acids, Peptides, and Protein. In *Fennema's Food Chemistry*; CRC Press: Boca Raton, FL, USA, 2017; pp. 235–356. ISBN 9781315372914.
44. Zheng, X.; Wang, J.; Liu, X.; Sun, Y.; Zheng, Y.; Wang, X.; Liu, Y. Effect of Hydrolysis Time on the Physicochemical and Functional Properties of Corn Glutelin by Protamex Hydrolysis. *Food Chem.* **2015**, *172*, 407–415. [CrossRef] [PubMed]
45. Fu, X.; Huang, X.; Jin, Y.; Zhang, S.; Ma, M. Characterization of Enzymatically Modified Liquid Egg Yolk: Structural, Interfacial and Emulsifying Properties. *Food Hydrocoll.* **2020**, *105*, 105763. [CrossRef]
46. Zheng, Z.; Li, J.; Liu, Y. Effects of Partial Hydrolysis on the Structural, Functional and Antioxidant Properties of Oat Protein Isolate. *Food Funct.* **2020**, *11*, 3144–3155. [CrossRef] [PubMed]
47. Chen, C.; Chi, Y.-J.; Zhao, M.-Y.; Xu, W. Influence of Degree of Hydrolysis on Functional Properties, Antioxidant and ACE Inhibitory Activities of Egg White Protein Hydrolysate. *Food Sci. Biotechnol.* **2012**, *21*, 27–34. [CrossRef]
48. Chang, C.; Tu, S.; Ghosh, S.; Nickerson, M.T. Effect of PH on the Inter-Relationships between the Physicochemical, Interfacial and Emulsifying Properties for Pea, Soy, Lentil and Canola Protein Isolates. *Food Res. Int.* **2015**, *77*, 360–367. [CrossRef]
49. Ruiz-Álvarez, J.M.; del Castillo-Santaella, T.; Maldonado-Valderrama, J.; Guadix, A.; Guadix, E.M.; García-Moreno, P.J. PH Influences the Interfacial Properties of Blue Whiting (*M. Poutassou*) and Whey Protein Hydrolysates Determining the Physical Stability of Fish Oil-in-Water Emulsions. *Food Hydrocoll.* **2022**, *122*, 107075. [CrossRef]

colloids and interfaces

MDPI

Article

Ultrasound-Assisted Extraction of *Artocarpus heterophyllus* L. Leaf Protein Concentrate: Solubility, Foaming, Emulsifying, and Antioxidant Properties of Protein Hydrolysates

Julián Vera-Salgado, Carolina Calderón-Chiu, Montserrat Calderón-Santoyo, Julio César Barros-Castillo, Ulises Miguel López-García and Juan Arturo Ragazzo-Sánchez *

Laboratorio Integral de Investigación en Alimentos, Tecnológico Nacional de México/Instituto Tecnológico de Tepic, Av. Tecnológico #2595, Col. Lagos del Country, Tepic 63175, Nayarit, Mexico
* Correspondence: jragazzo@ittepic.edu.mx

Citation: Vera-Salgado, J.; Calderón-Chiu, C.; Calderón-Santoyo, M.; Barros-Castillo, J.C.; López-García, U.M.; Ragazzo-Sánchez, J.A. Ultrasound-Assisted Extraction of *Artocarpus heterophyllus* L. Leaf Protein Concentrate: Solubility, Foaming, Emulsifying, and Antioxidant Properties of Protein Hydrolysates. *Colloids Interfaces* **2022**, *6*, 50. https://doi.org/10.3390/colloids6040050

Academic Editors: César Burgos-Díaz, Mauricio Opazo-Navarrete and Eduardo Morales

Received: 10 September 2022
Accepted: 20 September 2022
Published: 26 September 2022

Abstract: The impact of ultrasound-assisted extraction (UAE) was evaluated on the functionality of jackfruit leaf protein hydrolysates. Leaf protein concentrate was obtained by ultrasound (LPCU) and conventional extractions by maceration (LPCM). LPCM and LPCU were hydrolyzed with pancreatin (180 min), and hydrolysates by maceration (HM) and ultrasound (HU) were obtained. The composition of amino acids, techno-functional (solubility, foaming, and emulsifying properties), and antioxidant properties of the hydrolysates were evaluated. A higher amount of essential amino acids was found in HU, while HM showed a higher content of hydrophobic amino acids. LPCs exhibited low solubility (0.97–2.89%). However, HM (67.8 ± 0.98) and HU (77.39 ± 0.43) reached maximum solubility at pH 6.0. The foaming and emulsifying properties of the hydrolysates were improved when LPC was obtained by UAE. The IC_{50} of LPCs could not be quantified. However, HU (0.29 ± 0.01 mg/mL) showed lower IC_{50} than HM (0.32 ± 0.01 mg/mL). The results reflect that the extraction method had a significant ($p < 0.05$) effect on the functionality of protein hydrolysates. The UAE is a suitable method for enhancing of quality, techno-functionality, and antioxidant properties of LPC.

Keywords: ultrasound-assisted extraction; enzymatic hydrolysis; protein hydrolysates; antioxidant capacity; techno-functional properties

1. Introduction

Proteins consist of vital amino acids for human beings due to their nutritional value. In addition, they are widely used in the food industry because of their excellent techno-functional properties [1]. Currently, the main sources for obtaining proteins are of animal origin. Moreover, the production processes, as well as the waste generated, have a high environmental impact [2]. Contrarily, the obtention of proteins from vegetal sources represents a lower environmental impact [3].

The supply and consumption of plant-based proteins and dietary transition to plant-based protein consumption patterns are framed among the top global food trends [4]. Additionally, in recent years, efforts have been made to valorize the waste generated by the agroindustry to produce many value-added products [5]. Therefore, becoming the agricultural sector the leading supplier of raw materials for different processes is essential. The circular economy emerges as an agricultural alternative to counteract the production of agro-industrial waste. Thus, by-products gain added value to be used in further processes. Additionally, agro-waste is the biological and techno-functional compounds in agro-waste that could produce suitable profits [6]. Alike, proteins have been used to obtain protein hydrolysates with enzymes to improve the properties of native proteins [7]. Hydrolysates have been associated with multiple benefits to human health, such as antihypertensives [8] and antioxidants [9]. The functionality of hydrolysates is related to their amino acid composition with diverse applications in the cosmetic and food industry [10]. Although the circular economy has been developed in many fields, new technologies are required to maintain the balance between economic, industrial development, and ecosystem protection with effective resource use [5].

The primary agricultural waste sources are associated with fruit production and crop residues, such as jackfruit. Under conventional harvesting conditions of this crop, it is estimated that leaves produced by tree pruning are around 10,378 tons/ha per year [11]. Recently, phenolic compounds in jackfruit leaves were reported, and extracts showed suitable antioxidant and antimicrobial activity [12]. In another study, the techno-functional properties of protein hydrolysates from jackfruit leaves obtained by high hydrostatic pressures were evaluated. The results evidenced the suitability of peptides for their application as carrier material. As well, the peptides showed suitable foaming and emulsifying properties and high solubility [11]. The quality of proteins and peptides is associated with their physicochemical (color, texture, solubility) and techno-functional (foaming, emulsifying, clarifying, thickening) properties [13]. These features define the possible areas of application, such as the cosmetic, pharmaceutical, and food industries.

Proteins are generally obtained by conventional methods such as maceration. However, this methodology has several disadvantages since it is time-consuming, requires high amounts of solvent, and has high energy consumption [14]. To maximize the protein isolation from plant matrixes, an efficient diffusion of the extraction solvent is crucial to break intramolecular bonds and weaken cellular structures. In addition, alternative extraction methodologies in agreement with principles of green chemistry have been implemented, such as microwave, supercritical fluid, and ultrasound-assisted extractions. Recently, the ultrasound-assisted process was reported as an effective treatment to increase the yield, enzymatic efficiency, amino acid composition, and bio-functionality of hydrolysates [15,16].

The use of ultrasound-assisted extraction (UAE) to obtain proteins is aimed at overcoming the issues of maceration, reducing extraction time, energy costs, and solvent amounts. Moreover, it yields a more homogeneous mixture, higher energy transfer rate, reduced temperature gradients, selective extraction, reduced equipment size, and greater process control [17]. Protein extraction yields more significantly than 30% have been reported from plant sources, including pumpkin seeds [18], bitter melon seeds [19], and soybeans [20]. Until now, reports do not exist about the use of UAE to obtain jackfruit leaves protein and their relationship with the techno-functionality properties of hydrolysates. The research aimed to extract jackfruit leaf concentrate protein from UAE and evaluate the techno-functional and antioxidant properties of hydrolysates.

2. Materials and Methods

2.1. Vegetal Material

Jackfruit leaves were handpicked after trees pruning in the "Tierras Grandes" orchard in Zacualpan, Compostela, Nayarit, Mexico, in May 2022. Then, they were transferred to polyethylene bags and transported to the laboratory. After washing, the leaves were dehydrated in a convective drying oven (Novatech, HS60-AID, Guadalajara, Mexico) for 24 h at 60 °C. Subsequently, they were ground using a high-speed blender (NutriBullet® SERIE 900, Los Angeles, CA, USA) and sieved (#100 mesh, 150 μm diameter). The flour was vacuum-packed and stored at room temperature for subsequent experiments [11].

2.2. Chemical Substances

Pancreatin (EC 232-468-9), amino acids standard (AAS-18), L-norleucine (Nor, ≥98%), *N-tert*-butyldimethylsilyl-*N*-methyltrifluoroacetamide (MTBSTFA, >97%), acetonitrile (ACN), potassium persulfate, 2,2′-azinobis-(3-ethylbenzothiazoline-6-sulfonic acid (ABTS+), 6-hydroxy-2,5,7,8-tetramethylchroman-2-carboxylic acid (Trolox), Bradford reagent, and sodium dodecyl sulfate (SDS) were purchased from Sigma-Aldrich (St. Louis, MO, USA). Analytical-grade chemicals such as sodium hydroxide (NaOH) and hydrochloric acid (HCl) were provided from Thermo Fisher Scientific Inc. (Waltham, MA, USA).

2.3. Preparation of Flour for Extraction Procedures

A depigmentation procedure was developed prior to extraction. Briefly, the flour was mixed with acetone in the liquid-to-solid ratio of 1:10 (*w/v*), then the sample was stirred on a magnetic plate for 24 h. Subsequently, the sediment was separated by decantation, and the depigmented flour was dried at room temperature until dry. Samples of leaf flour (30 g) were placed in a 1 L beaker containing 563 mL of distilled water and 188 mL of a solution (0.2 M NaOH), then the mixture was stirred for 10 min on a magnetic stirring plate. The same procedure was conducted for the following extraction approaches.

2.3.1. Maceration Extraction (M)

The homogenized solution was transferred to conical centrifuge tubes (50 mL) and centrifuged at 1500× *g* for 20 min at 10 °C (HERMLE, Z 326K, Waseerburg, Germany). The supernatant was collected and adjusted to pH 4.0 with 1 N HCl until the isoelectric precipitation. The samples were left to stand for 2 h and then centrifuged at 1500× *g* for 20 min at 10 °C [11]. Finally, the precipitate obtained by maceration (LPCM) was collected and used for further analysis.

2.3.2. Ultrasound-Assisted Extraction (UAE)

The homogenized solution was subjected to UAE using an ultrasonic bath (Digital Ultrasonic Cleaner, CD-4820, Guangdong, CHN) at 42 kHz for 20 min. The mixture was centrifuged at 1500× *g* for 20 min at 10 °C (HERMLE, Z 326K, Waseerburg, Germany). The supernatant was recovered, pH was adjusted, proteins were precipitated, and recovered by centrifugation in the same way mentioned above. The pellet of LPC obtained by UAE (LPCU) was used for comparison with the LPCM.

2.4. LPC Hydrolysis

LPCM and LPCU were hydrolyzed with pancreatin enzyme for 180 min [21]. A solution of LPC (1%, *w*/*v*) was prepared with distilled water and incubated at 37 °C, 115 rpm in a shaking water bath (Shaking Hot Tubs 290200, Boekel Scientific, Feasterville-Trevose, PA, USA). The solution was adjusted to pH 7.0 (1 N NaOH), and the enzyme was added in an enzyme-substrate ratio of 1:100 (*w*/*w*). The pH was maintained by adding NaOH if necessary. The hydrolysis process was stopped by enzyme inactivation; thus, the mixture was heated at 95 °C for 15 min. The pH 7.0 was adjusted, and the solution was centrifuged at 10,000× *g* for 15 min at 10 °C (HERMLE, Z 326K, Waseerburg, Germany). The supernatant was filtered (0.45 µm) and evaporated in a convective oven at 50 °C until the required concentration for further analysis. The hydrolysates by maceration (HM) and UAE (HU) were obtained. The yield was calculated (Equation (1)).

$$\text{Yield}(\%) = \frac{m_h}{m_{LPC}} \cdot 100 \quad (1)$$

where:

HM: mass of hydrolysate, g;
LPCM: mass of LPC, g.

2.5. Analysis of Amino Acids by Gas-Chromatography Mass-Spectrometry (GC-MS)

The amino acid determination was carried out following the protocol proposed by Brion-Espinoza et al. [7]. Samples of flour, LPC, and hydrolysates were subjected to acid hydrolysis with 6 M HCl for 24 h at 110 °C. Then, they were derivatized with MTBSTFA (*N*-*tert*-Butyldimethylsilyl-*N*-methyltrifluoroacetamide), a reactant for GC derivatization. Briefly, 100 µL of hydrolysate and 10 µL of L-norleucine (internal standard, 0.2 mg/mL in HCL 0.1 M) were evaporated under nitrogen gas to dryness. The resulting precipitate was dissolved in 200 µL of acetonitrile and 200 µL of MTBSTFA. This solution was incubated at 100 °C for 2.5 h in a glycerol bath. The derivatization reaction was performed in a 2 mL PTFE-lined screw-capped vial. For the L-amino acids standards mixture, the same procedure was followed. The GC-MS was performed using a GC 7890A coupled to MS 240 Ion Trap (Agilent Technologies; Palo Alto, CA, USA). A capillary column Agilent J&W VF-5ms (30 m × 0.25 mm, i.d., 0.25 µm film thickness) was used for the separation. The carrier gas was helium (99.99%) at a flow rate of 2 mL/min. The oven temperature program was set at 150 °C for 2 min, increased at 3 °C/min to 280 °C. A total of 2 µL were injected with autosampler in split mode (20:1) in the GC injector port at 260 °C. MS parameters were as follows: energy of ionization (70 eV), full scan mode (35–650 m/z), ion trap (150 °C), manifold (80 °C), and transfer line (130 °C). Linear retention indexes were calculated using a mixture of straight-chain alkanes (C_7–C_{30}), injected under the same analysis conditions. The amino acid profile was reported as g of amino acid/100 g of sample.

2.6. Techno-Functional Properties

2.6.1. Solubility

The solubility was carried out following the protocol used by Calderón-Chiu et al. [11]. Samples of LPCs and hydrolysates (10 mg) were placed inside a 3 mL conical tube, and 1 mL of distilled water was added. Then, tubes were vortexed for 30 s to dissolve the sample. Each sample was adjusted at pHs 2.0, 4.0, 6.0, 8.0, and 10.0 with 1 N HCl or NaOH. The solutions were vortexed for 30 min and centrifuged (7500× g, 15 min) (Hettich MIKRO 220R, Tuttlingen, Germany). The protein content of the supernatant recovered was determined by the Bradford method [22]. After the solubilization of the sample in 1 mL of 0.5 N NaOH solution, the total protein content was determined, and solubility (%) was calculated (Equation (2)).

$$\text{Solubility } (\%) = \frac{P_{Snat}}{P_{Total}} \cdot 100 \quad (2)$$

where:

P_{Snat}: protein content in the supernatant, g;

P_{Total}: total protein content in the sample, g.

2.6.2. Foaming Properties

The foaming capacity (FC) and foaming stability (FS) of LPCs, and the HM and HU hydrolysates were determined according to the methodology used by Calderón-Chiu et al. [11]. A protein solution at 4.6 mg/mL in distilled water was prepared for each test. Then, an aliquot of 6 mL was placed in a 15 mL conical tube. The sample was homogenized using Ultra-Turrax (IKA T10, Staufen, Germany) at 16,000 rpm for 2 min to incorporate air bubbles, and the test was developed at room temperature. The total solution was immediately transferred to a 15 mL glass graduated cylinder; after 30 s, the total volume was recorded, and FC (%) was calculated (Equation (3)).

$$\text{FC}(\%) = \frac{A_0 - B}{B} \cdot 100 \quad (3)$$

where:

A_0: volume after homogenization, mL;

B: volume before homogenization, mL.

For the FS (%) determination, the same homogenized sample was used, and after 10 min allowed to stand, the volume was recorded and calculated (Equation (4)).

$$\text{FS } (\%) = \frac{At - B}{B} \cdot 100 \quad (4)$$

where:

At: volume after rest, mL;

B: volume before homogenization, mL.

2.6.3. Emulsifying Properties

The turbidimetric method was used for the emulsifying properties. Briefly, in a 15 mL test tube, 2 mL of olive oil were mixed with 6 mL of solution 4.6 mg/mL of LPC or hydrolysates in distilled water. Initially, the samples were homogenized using Ultra-Turrax (IKA T10, Staufen, Germany) at 10,000 rpm for 1 min. Then, aliquots of 50 µL were taken from the bottom of the tube at 0 and 10 min and diluted 100 times separately in 5 mL of 0.1% SDS solution. The sample was stirred for 10 s on a magnetic stirring plate. A spectrophotometer (Cary 50 Bio UV–Visible, Varian, Mulgrave, Australia) was used to measure the absorbance (500 nm) at the time (t) of 0 min (A_0) and 10 min (A_{10}) after emulsion formation. The emulsifying activity index (EAI) and emulsion stability index (ESI) were calculated (Equation (5)) and (Equation (6)) respectively [23].

$$\text{EAI} = \frac{2{\cdot}2.303{\cdot}DF{\cdot}A_0}{c{\cdot}f{\cdot}10{,}000} \tag{5}$$

$$\text{ESI} = \frac{A_0}{A_0 - A_{10}} \cdot t \tag{6}$$

where:

DF: dilution factor, 100;

c: mass of the sample, g;

f: mass fraction of olive oil in the emulsion, 0.25.

2.7. Antioxidant Properties

To evaluate the radical-scavenging activity (RSA), an $ABTS^+$ stock solution at a concentration of 7 mM $ABTS^+$ in 2.45 mM potassium persulfate was prepared and maintained in total darkness at 25 °C for 15 h. A dilution with distilled water of an aliquot of the stock solution was adjusted to an absorbance of 0.70 ± 0.02 at 734 nm. Then, 50 µL of LPC or hydrolysate solutions (0.3–1 mg/mL) were taken and mixed with 950 µL of $ABTS^+$ radical, shaking vigorously for 10 s. The absorbance was recorded at 734 nm on a spectrophotometer (Cary 50 Bio UV–Visible, Varian, Mulgrave, Australia) after 7 min. $ABTS^+$ RSA was calculated with Equation (7).

$$\text{ABTS}^{+}\text{RSA } (\%) \; \frac{Acontrol - Asample}{Acontrol} \times 100 \tag{7}$$

where:

$A_{control}$: absorbance of the $ABTS^+$ solution;

A_{Sample}: absorbance of the reaction ($ABTS^+$ with sample).

Subsequently, curves representing radical-scavenging activity (RSA, %) y-axis versus sample concentration (mg/mL) x-axis were plotted for each sample. The corresponding point at 50% of antioxidant activity with the x-intercept was defined as the IC_{50} value. A linear regression equation of curves was used for this purpose [11].

2.8. Statistical Analysis

Data obtained in triplicate were analyzed with a one-way analysis of variance (ANOVA), followed by the post hoc Tukey test for the mean comparison ($p < 0.05$) with the STATISTICA software (version 12.0, StatSoft, Inc., 2011, Caty, CN, USA).

3. Results and Discussion

3.1. Yield and Amino Acid Composition

Table 1 showed no significant difference ($p > 0.05$) in the yield between LPCM and LPCU. However, LPCU showed a slightly higher yield than LPCM, which can be attributed to ultrasound. The UAE facilitates the disruption and degradation of the leaf. Then, the extraction solvent penetrates the internal structure of vegetal material, enhancing the mass transfer and increasing yield [24]. These results differ from those reported by Moreno-

Nájera et al. [25], who reported a yield of 9.74% of jackfruit leaf protein extracted by ultrasound. The differences are attributed to the extraction solvent since these authors used a 1 M NaCl solution for the extraction.

Table 1. Amino acid composition and yield of leaf protein concentrates obtained by maceration and ultrasound and their hydrolysates.

Amino Acid (mg/100 g Protein)	Leaf Protein Concentrate (LPC)		Hydrolysate (H)		Suggested Intake (%) [1]
	Maceration (M)	Ultrasound (U)	Maceration (M)	Ultrasound (U)	
Yield (%)	6.76 ± 0.60 [a]	7.04 ± 0.29 [a]	38.80 ± 0.92 [b]	41.38 ± 4.14 [b]	NA
Alanine	2.06 ± 0.12 [a]	1.89 ± 0.45 [a]	3.78 ± 0.48 [b]	3.16 ± 0.07 [ab]	NA
Glycine	2.10 ± 0.11 [a]	1.67 ± 0.53 [a]	3.70 ± 0.39 [b]	3.79 ± 0.19 [b]	NA
Valine	2.15 ± 0.14 [a]	2.26 ± 0.46 [a]	4.72 ± 0.0 [b]	3.67 ± 0.02 [c]	3.9
Leucine	2.81 ± 0.10 [a]	2.89 ± 0.61 [a]	6.76 ± 0.23 [c]	4.48 ± 0.04 [b]	5.9
Isoleucine	1.60 ± 0.08 [a]	1.80 ± 0.33 [a]	3.52 ± 0.13 [c]	2.71 ± 0.07 [b]	3.0
Proline	1.54 ± 0.06 [a]	2.23 ± 0.36 [ab]	4.40 ± 0.67 [c]	3.96 ± 0.59 [bc]	NA
Methionine	0.37 ± 0.02 [a]	0.55 ± 0.08 [ab]	0.71 ± 0.06 [b]	0.61 ± 0.08 [ab]	2.2
Serine	1.15 ± 0.03 [a]	1.46 ± 0.28 [ab]	1.54 ± 0.48 [ab]	2.35 ± 0.10 [c]	NA
Threonine	1.12 ± 0.04 [a]	1.44 ± 0.26 [a]	1.88 ± 0.05 [a]	2.16 ± 0.23 [b]	2.3
Phenylalanine	1.56 ± 0.08 [a]	2.01 ± 0.41 [ab]	3.20 ± 0.32 [c]	2.83 ± 0.22 [bc]	3.8
Aspartic acid	2.99 ± 0.08 [a]	3.11 ± 0.84 [a]	3.18 ± 0.58 [a]	5.68 ± 0.40 [b]	NA
Glutamic acid	3.19 ± 0.26 [a]	4.45 ± 1.01 [a]	3.19 ± 0.56 [a]	6.69 ± 0.77 [b]	NA
Lysine	2.02 ± 0.31 [ab]	9.03 ± 0.30 [b]	nd	7.02 ± 1.12 [c]	4.5
HAA	12.09 ± 0.61 [a]	13.63 ± 1.69 [a]	27.10 ± 1.11 [b]	21.42 ± 0.22 [c]	NA
AAA	1.56 ± 0.08 [a]	2.01 ± 0.41 [ab]	3.20 ± 0.32 [c]	2.83 ± 0.22 [bc]	NA
EAA	10.51 ± 0.73 [a]	18.54 ± 1.18 [ab]	18.91 ± 0.96 [ab]	21.32 ± 2.56 [b]	NA
NCAA	6.18 ± 0.34 [a]	7.56 ± 0.86 [a]	6.36 ± 0.14 [a]	12.37 ± 1.17 [b]	NA
TAA	24.65 ± 1.44 [a]	34.78 ± 2.93 [ab]	40.58 ± 2.16 [b]	49.11 ± 2.55 [c]	NA

Hydrophobic amino acids: HAA; aromatic amino acids: AAA; essential amino acids: EAA; negatively charged amino acids: NCAA; total amino acid: TAA; nd: not detected. [a–c] Different small letters in the same row indicate significant differences among the treatment ($p < 0.05$). Suggested profile of EAA requirements for an adult human by FAO/WHO [26] [1].

Likewise, the extraction of the LPC by UAE and NaOH solution could lead to protein structural damage, which produces aggregates that do not solubilize with the extraction solvent. These protein aggregates possibly remain in the centrifugation residue, leading to low yield [27]. Then, the results could suggest that less extraction time is required to improve yield. Therefore, in subsequent studies, it is recommended to evaluate shorter extraction times. On the other hand, the hydrolysates showed a higher yield than the concentrates, which suggests high cleavage of peptide bonds by the enzyme [11]. However, no significant differences ($p > 0.05$) were shown between HM and HU. The above indicates that the extraction method did not influence the yield.

Regarding the amino acid profile (Table 1), although the LPCU presents a higher content of the total amino acid (TAA), essential amino acids (EAA), and hydrophobic amino acids (HAA) than LPCM, no significant differences were observed between both treatments ($p > 0.05$). These concentrates showed high contents of lysine, glutamic acid, aspartic acid, leucine, valine, and proline. Concerning the hydrolysates, HM and HU showed significantly ($p < 0.05$) higher content of TAA, EAA, and HAA than LPCs, which indicates that the release of amino acids during the enzymatic process with pancreatin was successful. HU showed higher content of lysine, glutamic acid, aspartic acid, glycine, serine, and threonine, whereas HM showed high content of leucine, valine, proline, alanine, isoleucine, and phenylalanine. Notwithstanding, HM showed EAA such as valine, leucine, isoleucine, and methionine, which are found in the requirements suggested by FAO/WHO [26]. For its part, HU presented methionine and lysine at levels required too. Lysine is a vital amino acid from a nutritional point of view since its deficiency in children is responsible for retarded growth. Likewise, it is important to note that the content of valine, leucine, isoleucine, and lysine of HU was higher than those reported for *Spirulina platensis* protein hydrolysates obtained by enzymatic hydrolysis with pancreatin [28].

Hence, HU could be used as an alternative source for plant-based foods that are low in lysine [29]. This suggested that UAE changed the molecular structure of LPC. This trend has been reported by Sun et al. [30] for peanut protein isolate extracted by ultrasound. The results showed that the extraction method affected the amino acid profile of the samples. Therefore, this confirms that UAE could maintain the quality of the protein concentrate and hydrolysates. Thus, UAE could be considered an alternative technique to obtain LPC with better quality, which would affect the functionality of the protein hydrolysates.

3.2. Techno-Functional Properties

3.2.1. Solubility

The functional properties of food proteins are fundamental in food processing, which can influence food texture and organoleptic characteristics [31]. In general, the solubility of the samples was dependent on the pH (Table 2). The LPCM and LPCU did not present solubility at acidic pH (2.0–4.0), which is attributed to proximity to the isoelectric point (pI) of the samples [28]. Subsequently, the increase in pH to 6.0–8.0 improved the solubility in both treatments. The LPCU exhibited slightly higher solubility than LPCM; however, this was not significant ($p > 0.05$).

Table 2. Solubility of protein concentrates and hydrolysates at different pH.

pH	Leaf Protein Concentrate		Hydrolysate	
	Maceration	**Ultrasound**	**Maceration**	**Ultrasound**
2	0 ± 0 ᵃ	0 ± 0 ᵃ	1.00 ± 0.02 ᵇ,ᴬ	3.83 ± 0.23 ᶜ,ᴮ
4	0 ± 0 ᵃ	0 ± 0 ᵃ	60.14 ± 0.10 ᵈ,ᴬ	72.12 ± 3.05 ᵃᵇ,ᴮ
6	0.97 ± 0.18 ᵃᵇ,ᴬ	2.59 ± 0.1 ᵃ,ᴬ	67.8 ± 0.98 ᵃ,ᴮ	77.39 ± 0.43 ᵇ,ᶜ
8	2.95 ± 0.18 ᵇ,ᴬ	2.89 ± 0.91 ᵃ,ᴬ	65.27 ± 1.35 ᵃ,ᴮ	66.26 ± 1.70 ᵃ,ᴮ
10	0 ± 0 ᵃ	0 ± 0 ᵃ	52.74 ± 0.10 ᶜ,ᴬ	66.16 ± 0.14 ᵃ,ᴮ

A–C Different capital letters in the same row indicate significant differences among the treatment ($p < 0.05$). a–d Different small letters in the same column indicate significant differences in the concentration of the sample ($p < 0.05$). nd, not detected.

On the other hand, the protein hydrolysates showed a significant ($p < 0.05$) increase in solubility from pH 2 (concerning LPCs), reaching the maximum solubility at pH 6.0 for both hydrolysates. The increase in solubility at acid pH could be due to a shift in the isoelectric point (pI) [32]. The shift in the pI after hydrolysis has also been attributed to differences in the types and numbers of charged groups on the proteins after hydrolysis. This trend was reported by Xu et al. [33] in protein hydrolysates of rice glutelin.

The HU showed significantly ($p < 0.05$) higher solubility than HM. This indicates that the ultrasonic cavitation derived from the collapse of gas bubbles, high-intensity shock waves, shear forces, and turbulence caused structural changes and denaturation of the protein substrates, decreasing the particle size of substrates and consequently exposing more enzymatic cleavage sites [34].

Therefore, in enzymatic hydrolysis of LPCU, there were more digested proteins than short-chain peptides. This meant that the molecular weight of polypeptide chains decreased, and the hydrophilic property was enhanced by increasing the number of polar functional groups ($-NH_2^+$ and $-COO^-$). These groups played a key role in developing the overall hydration of proteins [34,35]. The negatively charged amino acids (NCAA), such as glutamic and aspartic acid in HU ($12.37 \pm 1.17\%$), were higher than HM ($6.36 \pm 3.14\%$). These NCAA, in an alkaline environment, provide a strong net negative charge leading to more interaction with an aqueous environment; hence the solubility is increased [36,37]. This trend was similar to that reported by Chen et al. [15], who observed that UAE improved the enzymatic accessibility of soy protein isolate, allowing the protein to be easily hydrolyzed and rendered soluble.

3.2.2. Foaming Properties

The property of proteins to form stable foams is important in producing various foods [38]. The FC and FS of LPCs and hydrolysates were dependent on the extraction method (Table 3). The LPCs did not exhibit desirable foaming properties. This behavior is due to the low solubility that the LPCs presented (Table 2) since the stable foams are formed with soluble proteins, which can interact and form thick viscous films [38]. The aggregation of LPC interfered with interactions between the protein and water, which is needed to form foam [39].

Table 3. Functional properties of leaf protein concentrate obtained by maceration and ultrasound and their hydrolysates.

Functionals Properties	Leaf Protein Concentrate		Hydrolysate	
	Maceration	Ultrasound	Maceration	Ultrasound
Foaming capacity (FC, %)	0 ± 0 [a]	0 ± 0 [a]	15.56 ± 2.55 [b]	35 ± 1.67 [c]
Foaming stability (FS, %)	0 ± 0 [a]	0 ± 0 [a]	0 ± 0 [a]	21.11 ± 4.19 [b]
Emulsifying activity index (EAI, m^2/g)	9.59 ± 1.10 [a]	33.99 ± 0.91 [b]	67.57 ± 1.31 [c]	78.28 ± 0.03 [d]
Emulsion stability index (ESI, min)	32.74 ± 3.7 [a]	46.19 ± 4.81 [b]	200.34 ± 11.44 [c]	480.89 ± 10.77 [d]

[a–d] Different small letters in the same row indicate significative differences ($p < 0.05$) between treatments.

Conversely, the hydrolysates significantly ($p < 0.05$) increased the FC concerning the LPCs. Nevertheless, the FS only improved in the HU. This behavior is because several molecular properties influence the FC and FS. The FC is affected by the adsorption rate, flexibility, and hydrophobicity. In contrast, the FS depends on the rheological properties of films, such as hydration, thickness, protein concentration, and favorable intermolecular interactions [11,40]. As mentioned above, HU showed high content of NCAA, which are responsible for the hydration properties, an essential requirement for foam stability. Hence, these findings indicate that although the extraction method did not improve foaming properties in LPCs, the UAE contributed to the partially unfolded structures of proteins. This behavior improved the enzymatic process, releasing protein hydrolysates with better molecular flexibility and solubility [39], which increased the FC and FS, as reported in previous studies [39,41,42].

3.2.3. Emulsifying Properties

The EAI and ESI were higher ($p < 0.05$) in the LPCU than LPCM. On the other hand, the hydrolysates showed significantly ($p < 0.05$) better emulsifying properties than LPCs (Table 3). HU showed better ($p < 0.05$) EAI and ESI than HM, this indicates that the extraction method influences functionality. This observation is mainly attributed to the cavitation and mechanical effects of UAE on LPC. These mechanisms play an essential role in changing the molecular structure of the substrate, reducing the substrate particle size, and making it more sensitive to enzymolysis [43–45], which allows the release of peptides with different characteristics from those obtained by maceration.

According to Table 1, LPCU hydrolysis allowed the release of protein hydrolysates with lower HAA content and higher NCAA levels than LPCM. This behavior could indicate that the HU presents a better hydrophobic-hydrophilic balance, given that the high content of HAA in HM could limit protein-lipid interactions [45]. Consequently, protein molecules could not be more effectively adsorbed at the interface O/W, decreasing the emulsifying properties [44]. Similar trends were observed by Chen et al. [15] for protein hydrolysates of soy protein.

However, the results of the emulsifying properties of HU are better than those obtained for hydrolysates of jackfruit leaf protein (under the same hydrolysis conditions). However, the LPC was extracted by hydrostatic pressure [11]. These hydrolysates exhibited EAI and ESI values of ~56.25 m^2/g and ~88.79 min, respectively. Furthermore, it must be emphasized that the concentrations used to evaluate the emulsifying properties of HU were lower than those used by the authors mentioned above. The preceding indicates that the choice of a suitable method for extraction is essential before enzyme hydrolysis. Because the functionality of the protein hydrolysates depends on their structure unfold generated during the UAE.

3.3. Antioxidant Properties

In general, the $ABTS^+$ radical-scavenging activity of LPCs and hydrolysates depends on the extraction method and sample concentration of the sample (Table 4). LPCs showed significantly lower antioxidant capacity than protein hydrolysates. However, HU showed significantly better $ABTS^+$ radical-scavenging activity than HM at 0.5–0.8 mg/mL concentrations, reaching the highest antioxidant capacity (99%) at a concentration of 1 mg/mL for both treatments.

Table 4. $ABTS^+$ radical-scavenging activity (%) at different concentrations and IC_{50} (mg/mL) of leaf protein concentrates obtained by maceration and ultrasound and their hydrolysates.

Concentration (mg/mL)	Leaf Protein Concentrate		Hydrolysate	
	Maceration	Ultrasound	Maceration	Ultrasound
0.3	11.52 ± 0.59 ^{a,A}	18.09 ± 2.44 ^{a,B}	13.88 ± 0.42 ^{a,AB}	12.21 ± 0.83 ^{a,A}
0.5	11.97 ± 0.14 ^{a,B}	23.38 ± 1.06 ^{a,A}	54.58 ± 1.02 ^{b,C}	59.41 ± 3.04 ^{b,C}
0.8	21.25 ± 2.61 ^{b,B}	30.91 ± 0.17 ^{b,A}	75.28 ± 3.62 ^{c,D}	85.59 ± 1.37 ^{c,C}
1.0	20.37 ± 0.69 ^{b,B}	38.15 ± 0.39 ^{c,A}	99.68 ± 0.23 ^{d,C}	99.63 ± 0.26 ^{d,C}
IC_{50}	nd	nd	0.32 ± 0.01 ^{B}	0.29 ± 0.01 ^{A}

$^{A–D}$ Different large letters in the same row indicate significant differences among the treatment ($p < 0.05$). $^{a–d}$ Different small letters in the same column indicate significant differences in the concentration of the sample ($p < 0.05$). IC_{50}, concentration of sample (mg/mL) required to achieve 50% of antioxidant activity.

The antioxidant capacity of ABTS can be reported in IC_{50} values. Low IC_{50} values mean better activity. The LPCs did not reach 50% antioxidant capacity by ABTS; therefore, their IC_{50} could not be quantified. On the contrary, HU showed lower IC_{50} than HM, reflecting that the extraction method had a significant ($p < 0.05$) effect. The UAE breaks Van der Waals forces, hydrogen bonds, and other non-covalent bonds of LPC [46]. The above-mentioned changes the structural configuration of proteins, increasing the accessibility of the proteases, which influences the chain length of peptides and amino acids released in the enzymatic process [47].

The amino acid composition showed that HU had higher contents of the acidic amino acids (glutamic and aspartic) and lysine (Table 1). The acidic amino acids can donate electrons and act as metal chelating agents. Likewise, positively charged amino acids, such as lysine, can bind and neutralize negatively charged free radicals [48], which would explain the high antioxidant properties of HU. The same trend was observed by Fadimu et al. [49] in lupin protein hydrolysates. The results demonstrated the potential of the UAE as a suitable method for enhancing the release of novel bioactive peptides with better antioxidative properties.

4. Conclusions

The UAE of LPC was evaluated on the functionality of protein hydrolysates of jackfruit compared to conventionally extracted LPC. LPCs presented did not show desirable techno-functional and antioxidant properties. Enzymatic hydrolysis with pancreatin improved the techno-functional and antioxidant properties, but it was dependent on the extraction method. The results indicated that UAE of LPC improves the enzymatic hydrolysis process. This was decisive during the hydrolysis since protein hydrolysates were obtained with a different amino acid composition concerning the HM, which evidences changes in the structure of LPC by ultrasound. The above resulted in better solubility, foaming, and emulsifying properties and antioxidant capacity of HU. Ultrasound-assisted extraction could be an excellent method for obtaining plant proteins, such as leaf protein, since it leads to desirable modifications that improve enzymatic hydrolysis. This has a significant impact on the functionality of the hydrolysates since multiple properties, such as emulsifying, foaming, and antioxidant, are improved. Obtaining these multifunctional ingredients is of great interest in the food industry. However, the influence of concentration, molecular weight, and functional group charges on the functionality of protein hydrolysates will continue to be investigated in future research.

Author Contributions: Conceptualization, J.V.-S., C.C.-C., M.C.-S., J.C.B.-C., U.M.L.-G. and J.A.R.-S.; Data curation, J.V.-S., C.C.-C. and J.C.B.-C.; Formal analysis, J.V.-S. and C.C.-C.; Funding acquisition, J.A.R.-S.; Investigation, J.V.-S., C.C.-C., M.C.-S. and J.C.B.-C.; Methodology, J.V.-S., C.C.-C., J.C.B.-C. and U.M.L.-G.; Project administration, M.C.-S. and J.A.R.-S.; Resources, J.A.R.-S.; Supervision, M.C.-S. and J.A.R.-S.; Validation, C.C.-C. and J.A.R.-S.; Visualization, C.C.-C., M.C.-S., U.M.L.-G. and J.A.R.-S.; Writing—original draft, C.C.-C., M.C.-S., J.C.B.-C. and J.A.R.-S.; Writing—review and editing, C.C.-C., M.C.-S., J.C.B.-C., U.M.L.-G. and J.A.R.-S. All authors have read and agreed to the published version of the manuscript.

Funding: Thanks to project code 316948 of the CYTED thematic network code 319RT0576 and to "Frutos Tropicales de la Bahía" S.P.R. of R.L., located in the ejido of Ixtapa de la Concepción, municipality of Compostela Nayarit.

Data Availability Statement: Not applicable.

Acknowledgments: The authors thank CONACYT (Consejo Nacional de Ciencia y Tecnología-Mexico) for their support through scholarship number 805853 granted to Julián Vera-Salgado.

Conflicts of Interest: The authors declare no conflict of interest.

References

1. Tahergorabi, R.; Hosseini, S.V. Proteins, Peptides, and Amino Acids. In *Nutraceutical and Functional Food Components: Effects of Innovative Processing Techniques*; Elsevier Science: Amsterdam, The Netherlands, 2017; ISBN 9780128052570.
2. Deprá, M.C.; Dias, R.R.; Sartori, R.B.; de Menezes, C.R.; Zepka, L.Q.; Jacob-Lopes, E. Nexus on Animal Proteins and the Climate Change: The Plant-Based Proteins Are Part of the Solution? *Food Bioprod. Process.* **2022**, *133*, 119–131. [CrossRef]
3. Detzel, A.; Krüger, M.; Busch, M.; Blanco-Gutiérrez, I.; Varela, C.; Manners, R.; Bez, J.; Zannini, E. Life Cycle Assessment of Animal-Based Foods and Plant-Based Protein-Rich Alternatives: An Environmental Perspective. *J. Sci. Food Agric.* **2021**, *102*, 5098–5110. [CrossRef] [PubMed]
4. Estell, M.; Hughes, J.; Grafenauer, S. Plant Protein and Plant-Based Meat Alternatives: Consumer and Nutrition Professional Attitudes and Perceptions. *Sustainability* **2021**, *13*, 1478. [CrossRef]
5. Yaashikaa, P.R.; Senthil Kumar, P.; Varjani, S. Valorization of Agro-Industrial Wastes for Biorefinery Process and Circular Bioeconomy: A Critical Review. *Bioresour. Technol.* **2022**, *343*, 126126. [CrossRef] [PubMed]
6. Yaashikaa, P.R.; Senthil Kumar, P.; Varjani, S.J.; Saravanan, A. Advances in Production and Application of Biochar from Lignocellulosic Feedstocks for Remediation of Environmental Pollutants. *Bioresour. Technol.* **2019**, *292*, 122030. [CrossRef] [PubMed]
7. Brion-Espinoza, I.A.; Iñiguez-Moreno, M.; Ragazzo-Sánchez, J.A.; Barros-Castillo, J.C.; Calderón-Chiu, C.; Calderón-Santoyo, M. Edible Pectin Film Added with Peptides from Jackfruit Leaves Obtained by High-Hydrostatic Pressure and Pepsin Hydrolysis. *Food Chem. X* **2021**, *12*, 100170. [CrossRef]
8. Fadimu, G.J.; Gill, H.; Farahnaky, A.; Truong, T. Improving the Enzymolysis Efficiency of Lupin Protein by Ultrasound Pretreatment: Effect on Antihypertensive, Antidiabetic and Antioxidant Activities of the Hydrolysates. *Food Chem.* **2022**, *383*, 132457. [CrossRef]
9. Karkouch, I.; Tabbene, O.; Gharbi, D.; Ben Mlouka, M.A.; Elkahoui, S.; Rihouey, C.; Coquet, L.; Cosette, P.; Jouenne, T.; Limam, F. Antioxidant, Antityrosinase and Antibiofilm Activities of Synthesized Peptides Derived from Vicia Faba Protein Hydrolysate: A Powerful Agents in Cosmetic Application. *Ind. Crops Prod.* **2017**, *109*, 310–319. [CrossRef]
10. Halim, N.R.A.; Azlan, A.; Yusof, H.M.; Sarbon, N.M. Antioxidant and Anticancer Activities of Enzymatic Eel (*Monopterus* Sp) Protein Hydrolysate as Influenced by Different Molecular Weight. *Biocatal. Agric. Biotechnol.* **2018**, *16*, 10–16. [CrossRef]
11. Calderón-Chiu, C.; Calderón-Santoyo, M.; Herman-Lara, E.; Ragazzo-Sánchez, J.A. Jackfruit (*Artocarpus heterophyllus* Lam) Leaf as a New Source to Obtain Protein Hydrolysates: Physicochemical Characterization, Techno-Functional Properties and Antioxidant Capacity. *Food Hydrocoll.* **2021**, *112*, 106319. [CrossRef]
12. Vázquez-González, Y.; Ragazzo-Sánchez, J.A.; Calderón-Santoyo, M. Characterization and Antifungal Activity of Jackfruit (*Artocarpus heterophyllus* Lam.) Leaf Extract Obtained Using Conventional and Emerging Technologies. *Food Chem.* **2020**, *330*, 127211. [CrossRef] [PubMed]

13. Gigliotti, J.C.; Jaczynski, J.; Tou, J.C. Determination of the Nutritional Value, Protein Quality and Safety of Krill Protein Concentrate Isolated Using an Isoelectric Solubilization/Precipitation Technique. *Food Chem.* **2008**, *111*, 209–214. [CrossRef]
14. Kumar, M.; Tomar, M.; Potkule, J.; Verma, R.; Punia, S.; Mahapatra, A.; Belwal, T.; Dahuja, A.; Joshi, S.; Berwal, M.K.; et al. Advances in the Plant Protein Extraction: Mechanism and Recommendations. *Food Hydrocoll.* **2021**, *115*, 106595. [CrossRef]
15. Chen, L.; Chen, J.; Ren, J.; Zhao, M. Effects of Ultrasound Pretreatment on the Enzymatic Hydrolysis of Soy Protein Isolates and on the Emulsifying Properties of Hydrolysates. *J. Agric. Food Chem.* **2011**, *59*, 2600–2609. [CrossRef]
16. Wu, Q.; Zhang, X.; Jia, J.; Kuang, C.; Yang, H. Effect of Ultrasonic Pretreatment on Whey Protein Hydrolysis by Alcalase: Thermodynamic Parameters, Physicochemical Properties and Bioactivities. *Process Biochem.* **2018**, *67*, 46–54. [CrossRef]
17. Chemat, F.; Rombaut, N.; Sicaire, A.G.; Meullemiestre, A.; Fabiano-Tixier, A.S.; Abert-Vian, M. Ultrasound Assisted Extraction of Food and Natural Products. Mechanisms, Techniques, Combinations, Protocols and Applications. A Review. *Ultrason. Sonochem.* **2017**, *34*, 540–560. [CrossRef]
18. Das, M.; Devi, L.M.; Badwaik, L.S. Ultrasound-Assisted Extraction of Pumpkin Seeds Protein and Its Physicochemical and Functional Characterization. *Appl. Food Res.* **2022**, *2*, 100121. [CrossRef]
19. Naik, M.; Natarajan, V.; Modupalli, N.; Thangaraj, S.; Rawson, A. Pulsed Ultrasound Assisted Extraction of Protein from Defatted Bitter Melon Seeds (*Momardica charantia* L.) Meal: Kinetics and Quality Measurements. *LWT* **2022**, *115*, 112997. [CrossRef]
20. Ding, Y.; Ma, H.; Wang, K.; Azam, S.M.R.; Wang, Y.; Zhou, J.; Qu, W. Ultrasound Frequency Effect on Soybean Protein: Acoustic Field Simulation, Extraction Rate and Structure. *LWT* **2021**, *145*, 111320. [CrossRef]
21. Calderón-chiu, C.; Calderón-santoyo, M.; Damasceno-gomes, S.; Ragazzo-Sánchez, J.A. Use of Jackfruit Leaf (*Artocarpus heterophyllus* L.) Protein Hydrolysates as a Stabilizer of the Nanoemulsions Loaded with Extract-Rich in Pentacyclic Triterpenes Obtained from *Coccoloba uvifera* L. Leaf. *Food Chem. X* **2021**, *12*, 100138. [CrossRef]
22. Bradford, M. A Rapid and Sensitive Method for the Quantitation of Microgram Quantities of Protein Utilizing the Principle of Protein-Dye Binding. *Anal. Biochem.* **1976**, *72*, 248–254. [CrossRef]
23. Pearce, K.N.; Kinsella, J.E. Emulsifying Properties of Proteins: Evaluation of a Turbidimetric Technique. *J. Agric. Food Chem.* **1978**, *26*, 716–723. [CrossRef]
24. Kingwascharapong, P.; Chaijan, M.; Karnjanapratum, S. Ultrasound-Assisted Extraction of Protein from Bombay Locusts and Its Impact on Functional and Antioxidative Properties. *Sci. Rep.* **2021**, *11*, 17320. [CrossRef] [PubMed]
25. Moreno-Nájera, L.C.; Ragazzo-Sánchez, J.A.; Gastón-Peña, C.R.; Calderón-Santoyo, M. Green Technologies for the Extraction of Proteins from Jackfruit Leaves (*Artocarpus heterophyllus* Lam). *Food Sci. Biotechnol.* **2020**, *29*, 1675–1684. [CrossRef] [PubMed]
26. WHO/FAO World Health Organization/Food and Agricultural Organization. *Protein and Amino Acid Requirement in Human Nutrition: Report of a Joint WHO/FAO/UNU Expert Consultation*; World Health Organization: Geneva, Switzerland, 2007.
27. Aguilar-Acosta, L.; Serna-Saldivar, S.; Rodríguez-Rodríguez, J.; Escalante-Aburto, A.; Chuck-Hernández, C. Effect of Ultrasound Application on Protein Yield and Fate of Alkaloids during Lupin Alkaline Extraction Process. *Biomolecules* **2020**, *10*, 292. [CrossRef]
28. Mohammadi, M.; Soltanzadeh, M.; Ebrahimi, A.R.; Hamishehkar, H. Spirulina Platensis Protein Hydrolysates: Techno-Functional, Nutritional and Antioxidant Properties. *Algal Res.* **2022**, *65*, 102739. [CrossRef]
29. Zou, Y.; Li, P.P.; Zhang, K.; Wang, L.; Zhang, M.H.; Sun, Z.L.; Sun, C.; Geng, Z.M.; Xu, W.M.; Wang, D.Y. Effects of Ultrasound-Assisted Alkaline Extraction on the Physiochemical and Functional Characteristics of Chicken Liver Protein Isolate. *Poult. Sci.* **2017**, *96*, 2975–2985. [CrossRef]
30. Sun, X.; Zhang, W.; Zhang, L.; Tian, S.; Chen, F. Effect of Ultrasound-assisted Extraction on the Structure and Emulsifying Properties of Peanut Protein Isolate. *J. Sci. Food Agric.* **2021**, *101*, 1150–1160. [CrossRef]
31. Zhao, Y.; Wen, C.; Feng, Y.; Zhang, J.; He, Y.; Duan, Y.; Zhang, H.; Ma, H. Effects of Ultrasound-Assisted Extraction on the Structural, Functional and Antioxidant Properties of *Dolichos lablab* L. Protein. *Process Biochem.* **2021**, *101*, 274–284. [CrossRef]
32. Jamdar, S.N.; Rajalakshmi, V.; Pednekar, M.D.; Juan, F.; Yardi, V.; Sharma, A. Influence of Degree of Hydrolysis on Functional Properties, Antioxidant Activity and ACE Inhibitory Activity of Peanut Protein Hydrolysate. *Food Chem.* **2010**, *121*, 178–184. [CrossRef]
33. Xu, X.; Liu, W.; Liu, C.; Luo, L.; Chen, J.; Luo, S.; McClements, D.J.; Wu, L. Effect of Limited Enzymatic Hydrolysis on Structure and Emulsifying Properties of Rice Glutelin. *Food Hydrocoll.* **2016**, *61*, 251–260. [CrossRef]
34. O'Sullivan, J.; Murray, B.; Flynn, C.; Norton, I. The Effect of Ultrasound Treatment on the Structural, Physical and Emulsifying Properties of Animal and Vegetable Proteins. *Food Hydrocoll.* **2016**, *53*, 141–154. [CrossRef]
35. Hu, Y.; Wu, Z.; Sun, Y.; Cao, J.; He, J.; Dang, Y.; Pan, D.; Zhou, C. Insight into Ultrasound-Assisted Phosphorylation on the Structural and Emulsifying Properties of Goose Liver Protein. *Food Chem.* **2022**, *373*, 131598. [CrossRef] [PubMed]
36. Wu, D.; Tu, M.; Wang, Z.; Wu, C.; Yu, C.; Battino, M.; El-Seedi, H.R.; Du, M. Biological and Conventional Food Processing Modifications on Food Proteins: Structure, Functionality, and Bioactivity. *Biotechnol. Adv.* **2020**, *40*, 107491. [CrossRef] [PubMed]
37. Xu, X.; Qiao, Y.; Shi, B.; Dia, V.P. Alcalase and Bromelain Hydrolysis Affected Physicochemical and Functional Properties and Biological Activities of Legume Proteins. *Food Struct.* **2021**, *27*, 100178. [CrossRef]
38. Zayas, J.F. Foaming Properties of Proteins. In *Functionality of Proteins in Food*; Springer: Berlin/Heidelberg, Germany, 1997; pp. 260–309.
39. Zou, Y.; Wang, L.; Cai, P.; Li, P.; Zhang, M.; Sun, Z.; Sun, C.; Xu, W.; Wang, D. Effect of Ultrasound Assisted Extraction on the Physicochemical and Functional Properties of Collagen from Soft-Shelled Turtle Calipash. *Int. J. Biol. Macromol.* **2017**, *105*, 1602–1610. [CrossRef]

40. Damodaran, S. Amino Acids, Peptides, and Protein. In *Fennema's Food Chemistry*; CRC Press: Boca Raton, FL, USA, 2017; pp. 235–356. ISBN 9781315372914.
41. Dong, Z.Y.; Li, M.Y.; Tian, G.; Zhang, T.H.; Ren, H.; Quek, S.Y. Effects of Ultrasonic Pretreatment on the Structure and Functionality of Chicken Bone Protein Prepared by Enzymatic Method. *Food Chem.* **2019**, *299*, 125103. [CrossRef]
42. Lv, S.; Taha, A.; Hu, H.; Lu, Q.; Pan, S. Effects of Ultrasonic-Assisted Extraction on the Physicochemical Properties of Different Walnut Proteins. *Molecules* **2019**, *24*, 4260. [CrossRef]
43. Li, S.; Yang, X.; Zhang, Y.; Ma, H.; Liang, Q.; Qu, W.; He, R.; Zhou, C.; Mahunu, G.K. Effects of Ultrasound and Ultrasound Assisted Alkaline Pretreatments on the Enzymolysis and Structural Characteristics of Rice Protein. *Ultrason. Sonochem.* **2016**, *31*, 20–28. [CrossRef]
44. Wang, Q.; Wang, Y.; Huang, M.; Hayat, K.; Kurtz, N.C.; Wu, X.; Ahmad, M.; Zheng, F. Ultrasound-Assisted Alkaline Proteinase Extraction Enhances the Yield of Pecan Protein and Modifies Its Functional Properties. *Ultrason. Sonochem.* **2021**, *80*, 105789. [CrossRef] [PubMed]
45. Akbari, N.; Mohammadzadeh Milani, J.; Biparva, P. Functional and Conformational Properties of Proteolytic Enzyme-modified Potato Protein Isolate. *J. Sci. Food Agric.* **2020**, *100*, 1320–1327. [CrossRef] [PubMed]
46. Wen, C.; Zhang, J.; Zhang, H.; Duan, Y.; Ma, H. Effects of Divergent Ultrasound Pretreatment on the Structure of Watermelon Seed Protein and the Antioxidant Activity of Its Hydrolysates. *Food Chem.* **2019**, *299*, 125165. [CrossRef] [PubMed]
47. Kadam, S.U.; Álvarez, C.; Tiwari, B.K.; O'Donnell, C.P. Extraction and Characterization of Protein from Irish Brown Seaweed Ascophyllum Nodosum. *Food Res. Int.* **2017**, *99*, 1021–1027. [CrossRef] [PubMed]
48. Aluko, R.E. *Amino Acids, Peptides, and Proteins as Antioxidants for Food Preservation*; Elsevier Ltd.: Amsterdam, The Netherlands, 2015; ISBN 9781782420972.
49. Fadimu, G.J.; Gill, H.; Farahnaky, A.; Truong, T. Investigating the Impact of Ultrasound Pretreatment on the Physicochemical, Structural, and Antioxidant Properties of Lupin Protein Hydrolysates. *Food Bioprocess Technol.* **2021**, *14*, 2004–2019. [CrossRef]

colloids and interfaces

MDPI

Article

Utilizing TPGS for Optimizing Quercetin Nanoemulsion for Colon Cancer Cells Inhibition

Hadel A. Abo Enin [1,*], Ahad Fahd Alquthami [2,†], Ahad Mohammed Alwagdani [2,†], Lujain Mahmoud Yousef [2,†], Majd Safar Albuqami [2,†], Miad Abdulaziz Alharthi [2,†] and Hashem O. Alsaab [2,*]

1 Pharmaceutics Department, National Organization of Drug Control and Research (NODCAR), Giza P.O. Box 12511, Egypt
2 Department of Pharmaceutics and Pharmaceutical Technology, College of Pharmacy, Taif University, P.O. Box 11099, Taif 21944, Saudi Arabia
* Correspondence: hadelaboenin@outlook.com (H.A.A.E.); h.alsaab@tu.edu.sa (H.O.A.); Tel.: +966-556-047-523 (H.O.A.)
† These authors contributed equally to this work.

Abstract: *Background:* Colorectal cancer is one of the most challenging cancers to treat. Exploring novel therapeutic strategies is necessary to overcome drug resistance and improve patient outcomes. Quercetin (QR) is a polyphenolic lipophilic compound that was chosen due to its colorectal anticancer activity. Nanoparticles could improve cancer therapy via tumor targeting by utilizing D-tocopheryl polyethylene glycol succinate (vitamin-E TPGS) as a surfactant in a nanoemulsion preparation, which is considered an efficient drug delivery system for enhancing lipophilic antineoplastic agents. Thus, this study aims to develop and optimize QR-loaded nanoemulsions (NE) using TPGS as a surfactant to enhance the QR antitumor activity. *Method:* The NE was prepared using a self-assembly technique using the chosen oils according to QR maximum solubility and TPGS as a surfactant. The prepared QR-NE was evaluated according to its particle morphology and pH. QR entrapment efficiency and QR in vitro drug release rate were determined from the selected QR-NE then we measured the QR-NE stability. The anticancer activity of the best-selected formula was studied on HT-29 and HCT-116 cell lines. *Results:* Oleic acid was chosen to prepare QR-NE as it has the best QR solubility. The prepared NE, which had particles size < 200 nm, maximum entrapment efficiency > 80%, and pH 3.688 + 0.102 was selected as the optimal formula. It was a physically stable formula. The prepared QR-NE enhanced the QR release rate (84.52 ± 0.71%) compared to the free drug. QR-NPs significantly improved the cellular killing efficiency in HCT-116 and HT-29 colon cancer cell lines (lower IC50, two folds more than free drug). *Conclusion:* The prepared QR-NE could be a promising stable formula for improving QR release rate and anticancer activity.

Keywords: colorectal cancer; nanoemulsion; quercetin; TPGS; drug delivery system

Citation: Enin, H.A.A.; Alquthami, A.F.; Alwagdani, A.M.; Yousef, L.M.; Albuqami, M.S.; Alharthi, M.A.; Alsaab, H.O. Utilizing TPGS for Optimizing Quercetin Nanoemulsion for Colon Cancer Cells Inhibition. *Colloids Interfaces* **2022**, *6*, 49. https://doi.org/10.3390/colloids6030049

Academic Editors: César Burgos-Díaz, Mauricio Opazo-Navarrete and Eduardo Morales

Received: 26 July 2022
Accepted: 2 September 2022
Published: 19 September 2022

1. Introduction

Colorectal cancer is one of the most common cancers in men and the third among women globally. More than 1.9 million new colorectal cancer cases and 935,000 deaths were estimated to have occurred in 2020 [1,2]. Although its mortality rates have decreased in all populations, the American Indians and Alaskan natives still suffer mostly at a high rate [3,4]. Nowadays, newer applications are used to diagnose and treat colorectal cancer due to their excellent enhancement of conventional methods and the development of novel approaches for detection and therapy, such as nanotechnology [5–7]. Recent nanotechnologies applications in colorectal cancer therapy have utilized nano-sized particle (NP)-based specific delivery systems for enhancing chemo and targeted therapy for tumors. These novel treatment approaches enhance drug permeability and drug retention effect [8,9]. The enhancing permeability and retention (EPR) effect arises from the leaky vasculature and impaired lymphatic drainage in the tumor cells [10,11].

Tumor targeting is one of the main nanotechnology advantages in cancer treatment. A NPs drug delivery system could improve drug delivery to malignant cells with less accumulation in nonmalignant cells [12]. This could be achieved by passive and/or active targeting, for targeting malignant and nonmalignant cells. In passive targeting, nano-drug particles could entrap the tumor cells, which could enhance the permeability and retention (EPR) [13]. To localize nano-drug particles to cancer cells, active targeting is promising, using selective molecular recognition as antigens or frequently expressed proteins on the surfaces of cancer cells. It also takes advantage of biochemical properties associated with cancer, such as matrix metalloproteinase secretion [14].

Nanoscale drug delivery systems, commonly referred to as nanocarriers, are nano-sized materials that can carry multiple drugs or imaging agents. Nanoemulsions (NE) as nanoscale delivery systems can deliver hydrophobic cytotoxic antineoplastic drugs with efficient pharmacokinetics and pharmacodynamics patterns. NE can also enhance the dose efficacy, reduce the drugs' side effects, increase the drug surface area, ease preparation, improve the thermodynamic stability, and help target and sustain controlled drug delivery [15–17].

Quercetin (QR) is a bioactive flavonoid compound (flavanol class) with strong anti-cancer activity besides its anti-inflammatory, anti-oxidant, and vasodilator effects [18,19]. QR anticancer activity works by adjusting cell-cycle progression, promoting apoptosis by reducing the expression stage of anti-apoptotic proteins, inhibiting angiogenesis and metastasis progression, affecting autophagy, and inhibiting cell proliferation [19,20]. According to its rapid clearance and lower water solubility and stability, the lower QR's oral bioavailability and high therapeutic dose (about 500 mg twice daily) are the primary oral QR limitations [21]. QR stability changes in a physiological medium as it is affected by pH, temperature, and oxidation [21]. In addition, using D-tocopheryl polyethylene glycol succinate (vitamin E TPGS) surfactant in nanocarrier is promising. TPGS is the water-soluble derivative of vitamin E. It contains a hydrophilic polar head and the lipophilic alkyl tail, which has been shown to have beneficial effects on the solubilization of QR, inhibiting P-gp mediated by hindering P-gp efflux on intestinal brush borders. TPGS has proved its essential role in chemotherapy by inducing cell-cycle arrest, promoting apoptosis, and enhancing permeation in the cancer cell [22]. Therefore, in this study, QR will be loaded in NPs as a NE formulation using TPGS as a surfactant to overcome the QR drawbacks and enhance its antitumor efficacy.

2. Materials and Methods

2.1. Materials and Cell Lines

The QR, medium-chain triglyceride, glyceryl monooleate (GMO), oleic acid, PEG 400, and surfactants (TPGS) were bought from Sigma-Aldrich Co (St. Louis, MO, USA). Colon cancer cell lines HCT-116 and HT-29 ATCC® were taken from our collaborator and were cultured in Gibco DMEM (Dulbecco's modified Eagle's medium) containing 4.5 g of glucose/liter (Thermofisher Scientific, Waltham, MA, USA) and including 10% fetal bovine serum (FBS; Thermo Fisher Scientific, Waltham, MA, USA). For MTT studies, the cell culture media were maintained at 37 °C and 5% CO_2 after being supplemented with 100 units/mL of penicillin and 100 μg/mL of streptomycin.

2.2. Selection of the Oil Phase

The appropriate oil for NE formulation is the oil improving the drug solubility to increase the drug-entrapping efficiency. An excess QR amount was separately added to 5 mL of certain oils such as MCT, glyceryl monooleate (GMO), and oleic acid (all of the above vehicles were in a liquid phase at 37 °C) for 72 h to achieve a dissolution equilibrium state. The samples were centrifuged at 10,000 rpm for 10 min. The QR concentrations in the supernatant were quantified by UV analysis at 370 nm using Synergy 2 UV Microplate Reader by BioTek Instruments, Inc. The solubility result was the mean of three experiments ± SD [23].

2.3. Preparation of Quercetin-Loaded NEs (QR-NEs)

QR-NE was prepared by the high-pressure homogenization (HPH) method, as described previously by Sessa et al. [24]. The emulsion components were selected based on several preliminary trials, and the chosen oil was the oil that had the maximum drug solubility. Oil phase (the selected oil) 10 $w/w\%$ was heated at 70 °C, then 0.1 $w/w\%$ QR was added and magnetically stirred for 15 min, till complete drug solubilization. The aqueous phase consists of distilled water containing 8 $w/w\%$ surfactants (TPGS) heated to the same temperature as the oil phase. Coarse emulsions were prepared by adding the oil phase into the aqueous phase under magnetic stirring at 500 rpm for 10 min. The coarse emulsions were passed through a high-pressure homogenizer to obtain final NE (AH100D, ATS Engineering, BVI, Canada), further dispersed by high-speed stirring using Ultra-Turrax (FM200, FLUKO Technology, Saarbrücken, Germany) at different homogenization rates (10,000, 12,000, 15,000) rpm for different times (10, 12, 15 min). To investigate the impacts of each independent variable (homogenizer speed and rotation time) and their combined effect on dependent variables (particle size and entrapment efficiency) one-way ANOVA was used with consideration of significant difference at $p < 0.05$ using Design Expert 7.0.0 Stat Ease. Inc. Minneapolis, MN, software [25].

2.4. Evaluation of the Prepared QR-NE

2.4.1. Nanoemulsion Particle Morphology

NE droplet size and zeta potential was determined by a Zeta-sizer (Zeta-sizer Ver. 7.01, Malvern Instruments). Separately, 0.1 mL of the prepared QR-NE formulation was dispersed in 50 mL of water and mixed well, and monitored at 25 °C ± 1.

2.4.2. Determination of Entrapment Efficiency

Entrapment efficiency (EE%) and loading capacity were determined by difference according to the following equations, respectively:

$$\text{EE (\%)} = \text{Weight of total drug in the formulation} - \text{Weight of drug in the aqueous phase (un-entrapped)} \times 100/\text{Weight of total drug in the formulation.} \quad (1)$$

$$\text{Drug loading (\%)} = (\text{Total weight of the drug}/\text{Total weight of sample}) \times 100\%. \quad (2)$$

The free drug concentration (un-entrapped) was determined in the supernatant after centrifugation of the prepared QR-NE formulations at 13,300 rpm and 4 °C for 60 min (Eppendorf, Hamburg, Germany). The QR amount in the supernatant was assessed against a blank (free QR formulation we prepared and treated under the same conditions) using a UV spectrophotometer (Pharmacia/Amersham Ultrospec 4000 UV/VIS Spectrophotometer) at 370 nm [26].

2.4.3. Determination of pH

The pH of the formulation was measured using a digital pH meter (Metler Toledo, OH, USA). Results were taken as the mean ± SD of three measures to reduce the error. pH is an important parameter as mainly the used excipients in the formulation decide the pH of the final preparation and hence the route of administration.

2.5. Studying the Physical Stability of QR-NE

2.5.1. Centrifugation Method

The selected formula was centrifuged at 5000 rpm for 10 min to check its physical stability. The nanoemulsion system was observed visually for the appearance of any creaming or phase separation [27].

2.5.2. Agitation Test

Three grams of QR-NE selected formula were accurately weighed and placed in a petri dish on a platform shaker at 50 rpm for 24 h at room temperature using an orbital shaker. The sample was observed for signs of any cream or oil droplets (phase separation) [28].

2.5.3. Heating Cooling Cycle

The formulae were stored between refrigerator temperatures 4 °C and 45 °C for six cycles with storage at each temperature and not less than 48 h. The formulations, stable at these temperatures, were subjected to centrifugation [29].

2.6. In Vitro Drug Release Study of QR from QR-NEs

The in vitro release study was conducted separately on free drug (QR) suspension and the selected formula (QR-NE). Two grams from each, separately, were put in a dialysis bag (molecular weight cut-off 3500). The dialysis bags were put into a beaker containing 50 mL of the release medium at 37 °C and mechanically stirred at 100 rpm. At specific time intervals, 2 mL release medium (10% alcoholic water; to maintain a sink condition) was withdrawn and replaced with an equivalent fresh medium. QR-released concentration was determined using a UV-spectrophotometer. Three independent experiments were conducted, and the data were expressed as mean ± SD [30].

2.7. Transmission Electron Microscopy (TEM)

For morphology, transmission electron microscopy (TEM) characterization of the tested selected formula was tested using the JEOL JEM-1000 instrument (JEOL Ltd., Tokyo, Japan). Fifty microliters of the selected sample were placed on a film-coated 200-mesh copper specimen grid for 10 min, and the fluid excess was eliminated using filter paper. The prepared grid was stained using 3% phosphotungstic acid (one drop) and dried for 3 min. The dried sample was examined using the TEM microscope (Philips, CM 12). The sample was observed by operating at 120 kV.

2.8. Evaluation of the Anticancer Activity against HT-29 and HCT-116 Cells

The cytotoxicity of the selected QR-NE formula was studied on HT-29 and HCT-116 cells using MTT-colorimetric method. An MTT assay was used to evaluate the viability of cancer cell lines following 48 h of treatment with quercetin at 5, 10, 20, 50, and 100 µM. Free QR was used as a control. Both cells were seeded in 96-well plates at a density of 5×10^3 cells and then incubated for 24 h. Subsequently, the cells were treated with series concentrations of free QR and the selected formula of QR-NE (containing an equivalent concentration of QR) separately for 24 h. The cell viability was evaluated with MTT on a Synergy 2 Multi-Detection Microplate Reader by BioTek Instruments, Inc at 570 nm. The inhibitory concentration (50%) was determined from 3 independent experiments conducted, and the result was expressed as mean ± standard deviation compared to full proliferation (100%), which was obtained from untreated cells and was considered a negative control [31].

2.9. Animals' Treatment and Histological Analysis of Tissues after QR-NE Treatment

To evaluate the safety of QR-NE, male Wistar rats (200 ± 20 g) were used. The study was conducted according to the guidelines of the Declaration of Helsinki and approved by the Institutional Animal Care and Use Committee at Taif University, Taif, Saudi Arabia. The protocol number is (42-0112). The rats were kept and housed in the animal facility within optimal conditions (a quiet, stress-free, temperature-controlled environment, on a 12-h light/dark cycle). After housing for about a week in the laboratory conditions, the rats were randomly distributed into 3 groups (4 rats per group). Group (I): vehicle group (the rats were intraperitoneally treated with 1 mL/kg of normal saline); Group (II): free drug (the rats were intraperitoneally injected with 50 mg/kg of QR treatment); Groups (III): QR-NE treatment (the rats were intraperitoneally injected with 50 mg/kg of QR-NE treatment). All the treatments were continued for 5 consecutive days. Then, for evaluating the effects of the compounds, the 12 h fasted rats were sacrificed by anesthetizing and both whole blood and the tissues were collected. For histopathological examination, a part of the collected tissues was fixed in 10% neutral buffered formalin. The histopathological alterations were assessed using the extracted specimens from liver, kidney, and spleen. The specimens were instantaneously fixed after being extracted in 10% formaldehyde and embedded in paraffin wax. Sections from specimens were slided (3–4 μm) after being deparaffinized and hydrated in distilled water. Then the sections were stained with hematoxylin and eosin (H&E) according to standard protocols to evaluate tissue architecture. Finally, histological images were taken using an inverted fluorescence microscope (Leica DMI8, Leica, Wetzlar, Germany).

2.10. Statistical Analysis

The results of in vitro and anticancer activity studies were expressed as the mean of three replicates ± SD. The paired t-test was used to compare two variables, while one-way ANOVA was used to assess the difference between groups using Design Expert 7.0.0 Stat Ease. Inc. software at a probability level $p < 0.05$ for significant differences. Sigmoidal concentration–response curve-fitting models and best-fit straight lines were used by Sigma plot software to calculate cell viability and cellular apoptosis, respectively. Cell viability was expressed as a percentage of survival compared to untreated cells, whereas cellular apoptosis was represented in folds compared to untreated cells (negative control).

3. Results and Discussion

3.1. The Selection of Oils

Different oils (GMO, medium-chain triglyceride, and oleic acid) were chosen as excipients due to their biocompatibility and low toxicity. Moreover, they have also been reported to form stable nanoemulsions without precipitation. As shown in Figure 1a, the solubility of QR was found to be highest in the oleic acid (18 carbon chain) [32]. Hence, the oleic was chosen as an oil phase to prepare nanoemulsion containing QR in this study ($p < 0.05$).

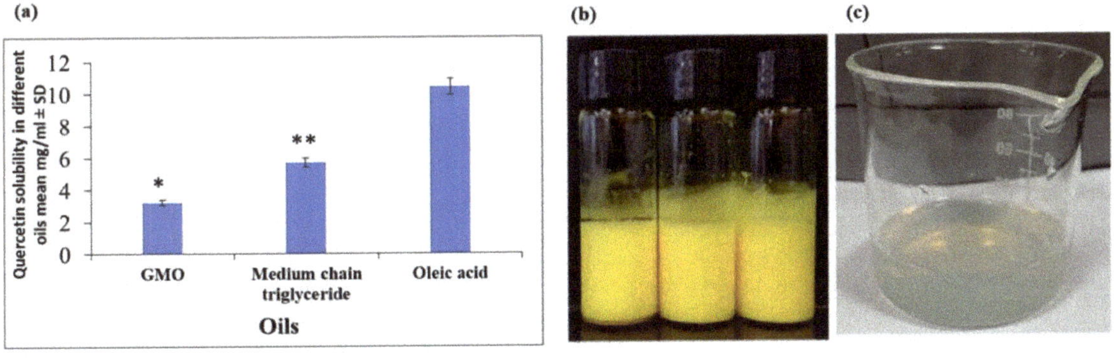

Figure 1. (**a**) Selection of oil for QR-NE preparation (there is a significant difference between GMO and MCT solubility results from oleic acid's results at $p = 0.068$ * and 0.041 **, respectively). (**b**) Quercetin emulsion using the selected oil (the composition as in the methods section). (**c**) Quercetin nanoemulsion after homogenization.

The NE preparations were mainly affected by their components. A preliminary study was conducted; the accepted formula was only the official formula considering the NEs have a homogenous yellow milky-like consistency. After homogenization treatments, stable, clear, and homogenous with no oil droplets solutions were obtained as shown in Figure 1b,c. Table 1 summarizes the nine experiment runs by studying the independent variables on the accepted particle size, DL, and EE%. As presented in Table 1, all formulations have an average particle size in the nano-range ranging from 9.522–273 nm. The drug-loading content and encapsulation efficiency of the QR-NE can be measured compared to a standard curve. In this study, we obtained enough drug-loading ranging from 46.87% to 53.65%. As observed, increasing the homogenization speed led to decreasing the particle size as observed from the formulae from one to six, and a further increase in the homogenization speed of 15,000 rpm led to a further increase in the particle size as observed in the formulae from 7 to 9. This may be because either the droplet breakup was not continuous due to turbulent–inertial forces appearing in the used homogenizer, or there was retardation of droplet breakup or significant re-coalescence at higher pressures. These results agreed with previously reported results by Jafari et al. [33].

Table 1. Formulation parameters affect the particle size and entrapment efficiency.

PN	Speed (rpm)	Time (Min)	EE%	DL	PS (nm)
1	10,000	10	87.3 ± 2.5	46.87%	273.9 ± 2.22
2	10,000	12	91.2 ± 2.12	48.97%	127.7 ± 1.14
3	10,000	15	93.2 ± 1.87	51.84%	203.2 ± 1.98
4	12,000	10	83.5 ± 1.09	53.65%	50.59 ± 0.56
5	12,000	12	85.5 ± 1.11	50.98%	9.522 ± 0.11
6	12,000	15	88.9 ± 2.01	52.115	62.37 ± 0.87
7	15,000	10	79.1 ± 1.76	47.87%	155.3 ± 1.53
8	15,000	12	78.8 ± 1.65	49.97%	85.6 ± 0.16
9	15,000	15	77.1 ± 1.23	50.84%	105.2 ± 0.54

Abbreviations: PN: Patch number; EE%: Encapsulation efficiency; DL: Drug loading; PS: Particle size.

Also, Figure 2 illustrates the influence of homogenization speed and time on particle size. The following equation shows the effect of the different variables on QR-NEs particle size:

$$PS = 357.543362573099 - 0.0139235087719298 \times Speed - 5.39640350877193 \times Time. \quad (3)$$

The expressed equation (Equation (3)) revealed that the negative sign of the speed and time coefficients indicated a significant inverse relationship between the speed and time of homogenization and the particle size. The p-value for homogenization speed was 0.0368 ($p < 0.05$) which indicates that increasing the homogenization speed had a significant effect on particle size reduction while homogenization time has no significant effect on decreasing particle size; p-value = 0.0714 ($p > 0.05$). This could be due to increasing homogenization speed, which could amplify mechanical and hydraulic shear that breaks the emulsion gel structure into NE vesicles with a smaller particle size [34]. To be more precise, increasing the homogenization speed (from 10,000 rpm to 12,000 rpm) and homogenization time (from 10 min to 12 min) led to a decrease in the PS of the nanoemulsion formulations. Further increases resulted in a significant increase in particle size.

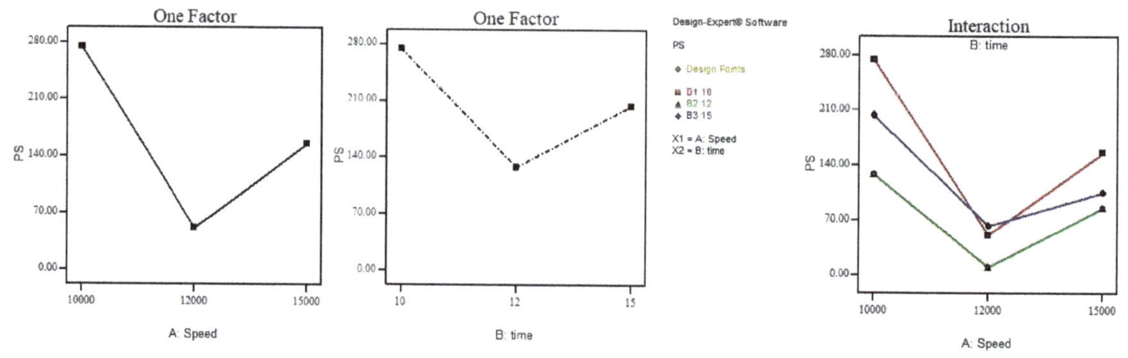

Figure 2. Effect of homogenization speed and time on the particle size (PS).

The effect of different formulation variables on EE% is noted in Table 1 and Figure 3. As noted in the following equation:

$$EE = 96.166081871345 - 0.0023140350877193 \times \text{Speed} + 0.612280701754387 \times \text{Time}. \quad (4)$$

The negative sign of homogenization speed indicates a significant antagonistic effect on EE% ($p < 0.05$). Increasing homogenization speed decreases the entrapping of QR into NE particles by reducing their contact. In contrast, a significant direct relation between homogenization time and EE% could be observed ($p < 0.05$). By increasing the contact time between drug and NE components, therefore EE% could be improved. This result agreed with previously reported works [35].

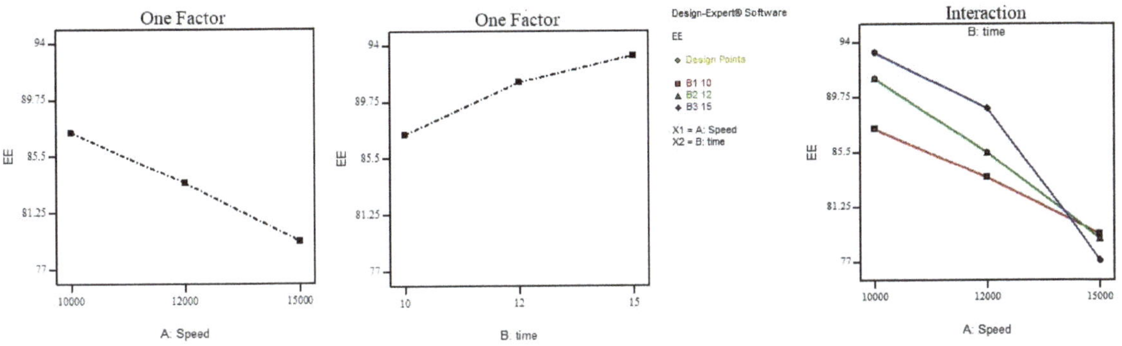

Figure 3. Effect of homogenization speed and time on the entrapping efficiency (EE).

3.2. Optimized QR-NE Formula Selection: Morphology and pH Determination

For formula optimization, the formulae with PS < 200 nm, higher DL > 50%, and maximum EE > 80% had higher priority and were chosen for further study. Therefore, formulae (F4, F5, F6) were selected. A paired t-test was applied to these formulae's PS and EE in pairs, and it was found that there were no significant differences between these formulae in the EE% results, while there was a substantial difference in the PS results. Therefore, F5 was chosen as it had a high EE% value, 85.5%, and the uniformly smallest PS with a high intensity of 96%, as represented in Figure 4a.

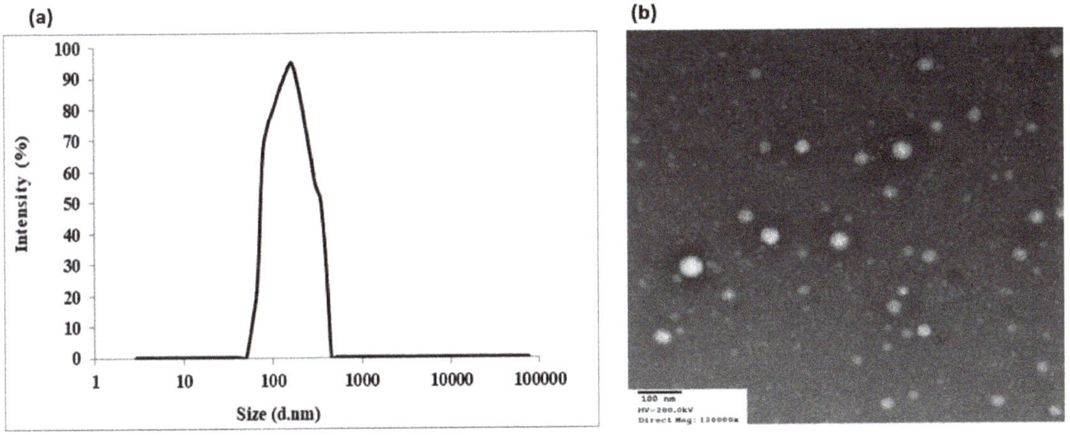

Figure 4. (**a**) Particle size distribution of QR-NE formula (F5). (**b**) QR-NEs (F5) Morphology under transmission electron microscopy (TEM) (magnification × 120,000).

Transmission electron microscopy characterization of F5 in Figure 4b shows that the nanoemulsion particles are uniformly nano-sized and have a spherical shape with a smooth and flexible boundary. No aggregation appears between the nanoparticles, indicating their stability against Ostwald ripening due to globular collapse [2]. This result agreed with previously reported results that TPGS could form a smooth particle layer, protecting the NE particles from severe structural changes.

The pH of the selected formula (F5) was 3.688 ± 0.102. The result indicates the suitability for oral administration of the prepared emulsion as it is closer to the stomach pH range (1.5–4). In addition, it is also more comparable to the large intestine pH (4–7); therefore, QR-NEs are suitable for colon administration [36].

3.3. *Studying the Physical Stability of QR-NEs*

After agitation and centrifugation on QR-NE (F5), there was no phase separation or creamy appearance in the prepared QR-NE formula. Nanoemulsion is a physically stable form of the prepared emulsion. In general, the nanoemulsion particles exhibit Brownian movement; therefore, no coalescence of droplets could occur except in the absence of this Brownian movement [37]. Centrifugation and agitation could contribute to the energy with which the NE droplets impinge upon each other. The lack of phase separation and creamy appearance after agitation and centrifugation indicates that QR-NE has good stability and can withstand the gravitational and mechanical forces during transportation and handling.

3.4. *In Vitro Drug Release Study of QR from QR-NPs*

The release results of the best-selected sample (QR-NE-F5) were compared in this respect with the release of free drug (QR). The release values were calculated as the percentage of QR dissolute according to the predetermined yield value. As noted in Figure 5, the QR-NPs showed an excellent release rate compared to free QR release. At the end of the release time (4 h), 25.63 ± 1.12% of free QR was released, which was considerably lower ($p < 0.05$) than released from QR-NE-F5 (84.52 ± 0.71%). The nanoemulsion formula's higher release rate can be attributed to its smaller particle size, which increases the surface area for diffusion. In general, NE is an excellent tool for enhancing the solubility of hydrophobic drugs such as QR; thereby, the bioavailability of the drug owing to small-scale globule size is improved. The presence of TPGS in NE can be attributed to the surfactant nature of TPGS, which enhances the QR release rate via increasing QR solubility [38,39]. TPGS as a surface active agent could improve drug wettability in contact with drug release media by adsorption on a larger surface area of nanoparticles and rapid drug partitioning into diluted dissolution medium, primarily from small droplets [40]. The TPGS layer around the QR NPs could protect the drug against gastrointestinal degradation, following which intestinal absorption would be further improved.

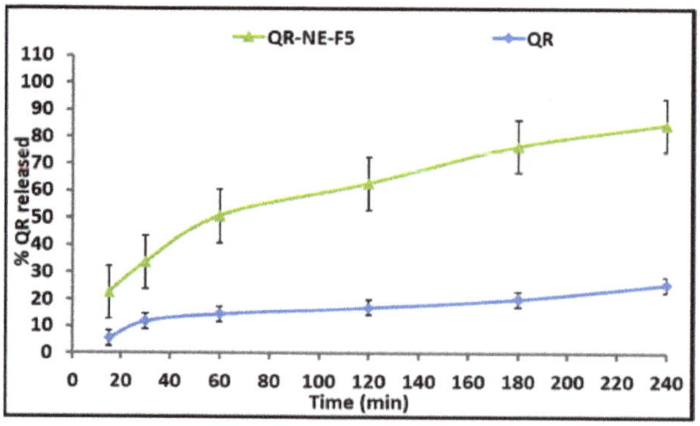

Figure 5. The cumulative released amount of quercetin (QR) and the selected formula (QR-NE-F5).

Many studies have concluded results in agreement with our results, which show that NE formulations usually result in an enhanced hydrophobic drugs release rate due to the effect of oil and interfacial film barriers [40,41].

3.5. Anticancer Activity of QR-NE; Cytotoxicity Study against Colorectal Cancer Cells

An MTT assay was used to evaluate the viability of HCT-116 and HT-29 colon cancer cell lines following 48 h of treatment with quercetin at 5, 10, 20, 50, and 100 µM. QR-NE inhibited CRC cell viability in a dose-dependent manner, which was more effective than the free drug. An increase in the concentration enhanced the viability of all cancer cell lines. More than twofold lower IC50 value of QR-NE (around 18 µM) indicates the superior anticancer effect of nanoparticle-based formulations on HCT-116 colon cancer cell lines, as shown in Figure 6a. We also found that the NE significantly improved the cellular killing in HT-29 colon cancer cell lines (around 15 µM) compared with the free drug, as shown in Figure 6b. Furthermore, the higher release rate of the QR combined with the lipid structure effect of the nanoemulsion increases drug concentration in colon cancer cells, improving the anticancer activity of the QR-NE formulations compared to free drug [41].

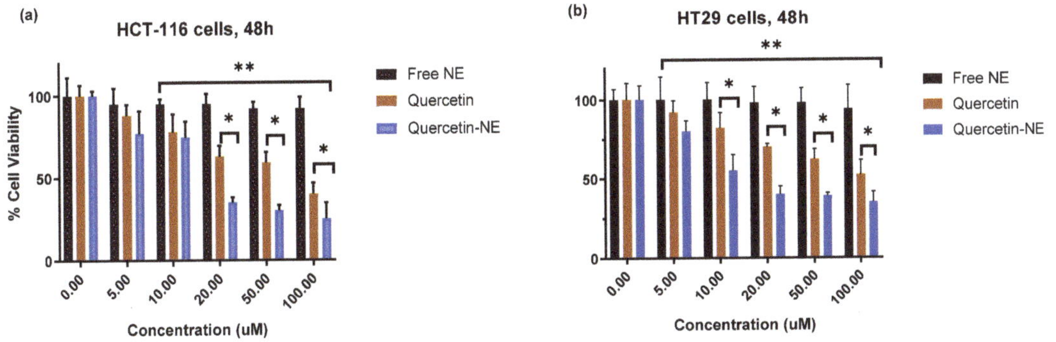

Figure 6. Cell viability of QR anticancer drug treated with 5–100 µM QR, QR-NE, and free NE up to 48 h for (**a**) HCT-116 and (**b**) HT-29 colon cancer cell lines. Results indicated that formulation inhibited CRC cell viability in a dose-dependent manner of QR-NE compared to free drugs, while empty NPs are nontoxic. The twofold lower IC50 value of QR-NE indicates the superior anticancer effect of nanoparticle-based formulations. Results are expressed as mean ± SD (n = 3). * $p < 0.01$, ** $p < 0.001$ based on two-way ANOVA followed by Sidak's multiple comparisons test.

3.6. QR-NE Safety Evaluation on the Animal Model

Macroscopic examination of all organs, including the liver, spleen, kidney, heart, lung, and brain tissue, revealed no differences 5 days after QR-NE (50 mg/kg) administration compared to organs of control rats which were treated with free NE and vehicle. There were no signs of atrophy, hyperplasia, necrosis, or inflammation.

To investigate tissue abnormalities, histological examination was performed using hematoxylin–eosin staining. For liver tissues, when compared to the control group, rats treated with QR-NE experienced no adverse effects on the liver. The parenchymal architecture revealed normal hepatocytes with no evidence of steatosis, inflammation, or fibrosis (Figure 7). Similarly, as shown in Figure 7, microscopic examination of the kidney, spleen, and heart revealed no evidence of inflammation or fibrosis. The histological sections of rats' liver, kidney, spleen, and heart are normal. Based on all gathered data, indications point to the safety of utilizing this nanoformulation for future efficacy studies.

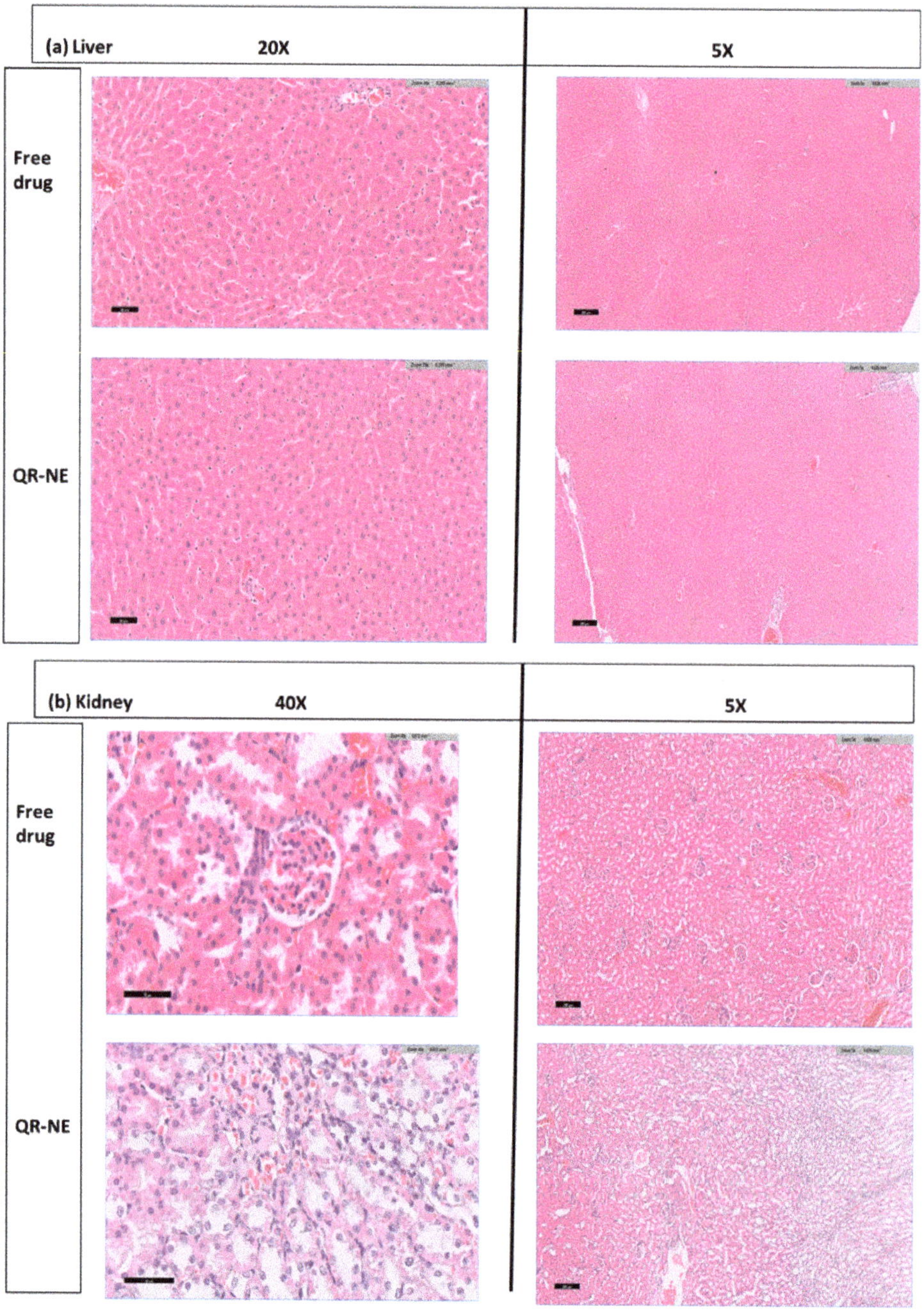

Figure 7. *Cont.*

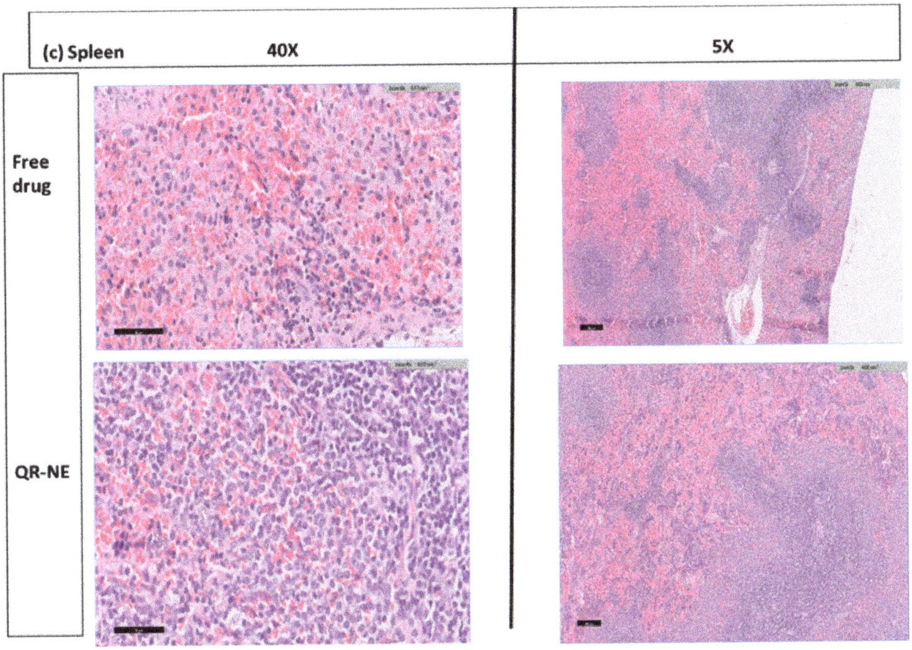

Figure 7. QR-NE safety evaluation on the animal model. Photomicrographs of (**a**) liver, (**b**) kidney, and (**c**) spleen sections after treatment with QR-NE and QR-free drug (H&E staining, magnification 40×, scale bar: 20 μm and 5×, scale bar: 200 μm). H&E hematoxylin and eosin sections were all treated with quercetin NE after 14 days. Histological evaluation was performed by ImageJ software version 1.52v.

4. Conclusions

The present study describes the development and safety of the QR-NE anticancer effect. NE prepared using oleic acid as oil and TPGS as a surfactant is an excellent tool for enhancing the solubility of hydrophobic drugs. Increasing the homogenization speed led to a significant effect of decreasing particle size. While increasing the homogenization time decreased NE particle size and increased the contact time between drug and NE components, EE% could be improved. QR-NE is a stable formula that can withstand gravitational and mechanical forces during transportation and handling. The QR-NE showed an excellent release rate compared to the free QR release. Using TPGS in NE produces uniform nanoparticles with no aggregation, and a smooth film around the particles improves the used drug stability and solubility. The latter could enhance the QR release rate and could protect the drug against gastrointestinal degradation, following which intestinal absorption would be further improved. QR-NE inhibited CRC cell viability in a dose-dependent manner, which was more effective than the drug alone. We also found that the NE significantly enhanced the cellular toxicity efficiency in HT-29 and HCT-116 cancer cell lines, resulting in efficient cell killing compared with free agents.

Author Contributions: Conceptualization, H.A.A.E., and H.O.A.; methodology, H.A.A.E., A.F.A., A.M.A., L.M.Y., M.S.A., M.A.A. and H.O.A.; software, H.A.A.E.; validation, H.A.A.E. and H.O.A.; formal analysis, H.A.A.E. and H.O.A.; investigation, H.O.A.; resources, H.O.A.; data curation, H.A.A.E. and H.O.A.; writing—original draft preparation, H.A.A.E., A.F.A., A.M.A., L.M.Y., M.S.A., M.A.A. and H.O.A.; writing—review and editing, H.A.A.E. and H.O.A.; visualization, H.A.A.E. and H.O.A.; supervision, H.A.A.E. and H.O.A.; project administration, H.A.A.E. and H.O.A.; funding acquisition, H.O.A. All authors have read and agreed to the published version of the manuscript.

Funding: His research received no external funding.

Institutional Review Board Statement: The study was conducted according to the guidelines of the Declaration of Helsinki and approved by the Institutional Animal Care and Use Committee at Taif University, Taif, Saudi Arabia. The protocol number is (42-0112).

Informed Consent Statement: Not applicable.

Data Availability Statement: No new data were created or analyzed in this study.

Acknowledgments: Hashem O. Alsaab would like to acknowledge Taif University Researchers. Supporting Project number (TURSP-2020/67), Taif University, Taif, Saudi Arabia.

Conflicts of Interest: The authors declare no conflict of interest.

References

1. Sung, H.; Ferlay, J.; Siegel, R.L.; Laversanne, M.; Soerjomataram, I.; Jemal, A.; Bray, F. Global cancer statistics 2020: GLOBOCAN estimates of incidence and mortality worldwide for 36 cancers in 185 countries. *CA Cancer J. Clin.* **2021**, *71*, 209–249. [CrossRef] [PubMed]
2. Gokhale, J.P.; Mahajan, H.S.; Surana, S.J. Quercetin loaded nanoemulsion-based gel for rheumatoid arthritis: In vivo and in vitro studies. *Biomed. Pharmacother.* **2019**, *112*, 108622. [CrossRef] [PubMed]
3. Fedewa, S.A.; Flanders, W.D.; Ward, K.C.; Lin, C.C.; Jemal, A.; Goding Sauer, A.; Doubeni, C.A.; Goodman, M. Racial and ethnic disparities in interval colorectal cancer incidence: A population-based cohort study. *Ann. Intern. Med.* **2017**, *166*, 857–866. [CrossRef] [PubMed]
4. US Cancer Statistics Working Group. *United States Cancer Statistics: 1999–2014 Incidence and Mortality Web-based Report*; US Department of Health and Human Services, Centers for Disease Control and Prevention and National Cancer Institute: Atlanta, GA, USA, 2017.
5. Bhise, K.; Sau, S.; Alsaab, H.; Kashaw, S.K.; Tekade, R.K.; Iyer, A.K. Nanomedicine for cancer diagnosis and therapy: Advancement, success and structure–activity relationship. *Ther. Deliv.* **2017**, *8*, 1003–1018. [CrossRef]
6. Hussain, Z.; Arooj, M.; Malik, A.; Hussain, F.; Safdar, H.; Khan, S.; Sohail, M.; Pandey, M.; Choudhury, H.; Ei Thu, H. Nanomedicines as emerging platform for simultaneous delivery of cancer therapeutics: New developments in overcoming drug resistance and optimizing anticancer efficacy. *Artif. Cells Nanomed. Biotechnol.* **2018**, *46*, 1015–1024. [CrossRef]
7. Sau, S.; Alsaab, H.O.; Bhise, K.; Alzhrani, R.; Nabil, G.; Iyer, A.K. Multifunctional nanoparticles for cancer immunotherapy: A groundbreaking approach for reprogramming malfunctioned tumor environment. *J. Control. Release* **2018**, *274*, 24–34. [CrossRef]
8. Anitha, A.; Maya, S.; Sivaram, A.J.; Mony, U.; Jayakumar, R. Combinatorial nanomedicines for colon cancer therapy. *Wiley Interdiscip. Rev. Nanomed. Nanobiotechnol.* **2016**, *8*, 151–159. [CrossRef]
9. Pellino, G.; Gallo, G.; Pallante, P.; Capasso, R.; De Stefano, A.; Maretto, I.; Malapelle, U.; Qiu, S.; Nikolaou, S.; Barina, A.; et al. Noninvasive biomarkers of colorectal cancer: Role in diagnosis and personalised treatment perspectives. *Gastroenterol. Res. Pract.* **2018**, *2018*, 2397863. [CrossRef]
10. Bennie, L.A.; McCarthy, H.O.; Coulter, J.A. Enhanced nanoparticle delivery exploiting tumour-responsive formulations. *Cancer Nanotechnol.* **2018**, *9*, 10. [CrossRef]
11. Iyer, A.K.; Khaled, G.; Fang, J.; Maeda, H. Exploiting the enhanced permeability and retention effect for tumor targeting. *Drug Discov. Today* **2006**, *11*, 812–818. [CrossRef]
12. Balasubramanian, V.; Liu, Z.; Hirvonen, J.; Santos, H.A. Bridging the knowledge of different worlds to understand the big picture of cancer nanomedicines. *Adv. Healthc. Mater.* **2018**, *7*, 1700432. [CrossRef] [PubMed]
13. Subhadarshini, S.; Merchant, N.; Raju, G.S.R. Nanomaterials: Diagnosis and Therapeutic Properties. In *Role of Tyrosine Kinases in Gastrointestinal Malignancies*; Springer: Berlin/Heidelberg, Germany, 2018; pp. 235–241.
14. Yoo, J.; Park, C.; Yi, G.; Lee, D.; Koo, H. Active targeting strategies using biological ligands for nanoparticle drug delivery systems. *Cancers* **2019**, *11*, 640. [CrossRef] [PubMed]
15. Gué, E.; Since, M.; Ropars, S.; Herbinet, R.; Le Pluart, L.; Malzert-Fréon, A. Evaluation of the versatile character of a nanoemulsion formulation. *Int. J. Pharm.* **2016**, *498*, 49–65. [CrossRef] [PubMed]
16. Kang, L.; Gao, Z.; Huang, W.; Jin, M.; Wang, Q. Nanocarrier-mediated co-delivery of chemotherapeutic drugs and gene agents for cancer treatment. *Acta Pharm. Sin. B* **2015**, *5*, 169–175. [CrossRef] [PubMed]
17. Sahu, P.; Das, D.; Mishra, V.K.; Kashaw, V.; Kashaw, S.K. Nanoemulsion: A novel eon in cancer chemotherapy. *Mini Rev. Med. Chem.* **2017**, *17*, 1778–1792. [CrossRef]
18. Park, J.B. Flavonoids are potential inhibitors of glucose uptake in U937 cells. *Biochem. Biophys. Res. Commun.* **1999**, *260*, 568–574. [CrossRef]
19. Tang, S.-M.; Deng, X.-T.; Zhou, J.; Li, Q.-P.; Ge, X.-X.; Miao, L. Pharmacological basis and new insights of quercetin action in respect to its anti-cancer effects. *Biomed. Pharmacother.* **2020**, *121*, 109604. [CrossRef]
20. Ghobrial, I.M.; Witzig, T.E.; Adjei, A.A. Targeting apoptosis pathways in cancer therapy. *CA A Cancer J. Clin.* **2005**, *55*, 178–194. [CrossRef]

21. Barbosa, A.I.; Costa Lima, S.A.; Reis, S. Application of pH-responsive fucoidan/chitosan nanoparticles to improve oral quercetin delivery. *Molecules* **2019**, *24*, 346. [CrossRef]
22. Bu, H.; He, X.; Zhang, Z.; Yin, Q.; Yu, H.; Li, Y. A TPGS-incorporating nanoemulsion of paclitaxel circumvents drug resistance in breast cancer. *Int. J. Pharm.* **2014**, *471*, 206–213. [CrossRef]
23. Sharma, S.; Sahni, J.K.; Ali, J.; Baboota, S. Effect of high-pressure homogenization on formulation of TPGS loaded nanoemulsion of rutin–pharmacodynamic and antioxidant studies. *Drug Deliv.* **2015**, *22*, 541–551. [CrossRef] [PubMed]
24. Sessa, M.; Balestrieri, M.L.; Ferrari, G.; Servillo, L.; Castaldo, D.; D'Onofrio, N.; Donsì, F.; Tsao, R. Bioavailability of encapsulated resveratrol into nanoemulsion-based delivery systems. *Food Chem.* **2014**, *147*, 42–50. [CrossRef] [PubMed]
25. El-Enin, H.A.; Al-Shanbari, A.H. Nanostructured liquid crystalline formulation as a remarkable new drug delivery system of anti-epileptic drugs for treating children patients. *Saudi Pharm. J.* **2018**, *26*, 790–800. [CrossRef] [PubMed]
26. Ren, Y.; Li, X.; Han, B.; Zhao, N.; Mu, M.; Wang, C.; Du, Y.; Wang, Y.; Tong, A.; Liu, Y.; et al. Improved anti-colorectal carcinomatosis effect of tannic acid co-loaded with oxaliplatin in nanoparticles encapsulated in thermosensitive hydrogel. *Eur. J. Pharm. Sci.* **2019**, *128*, 279–289. [CrossRef]
27. Riquelme, N.; Zúñiga, R.; Arancibia, C. Physical stability of nanoemulsions with emulsifier mixtures: Replacement of tween 80 with quillaja saponin. *LWT* **2019**, *111*, 760–766. [CrossRef]
28. Ha, J.-W.; Yang, S.-M. Rheological responses of oil-in-oil emulsions in an electric field. *J. Rheol.* **2000**, *44*, 235–256. [CrossRef]
29. Panapisal, V. Effects of surfactant mixture ratio and concentration on nanoemulsion physical stability. *Thai J. Pharm. Sci.* **2016**, *40*, 45–48.
30. Pangeni, R.; Choi, S.W.; Jeon, O.-C.; Byun, Y.; Park, J.W. Multiple nanoemulsion system for an oral combinational delivery of oxaliplatin and 5-fluorouracil: Preparation and in vivo evaluation. *Int. J. Nanomed.* **2016**, *11*, 6379–6399. [CrossRef]
31. Galaup, A.; Opolon, P.; Bouquet, C.; Li, H.; Opolon, D.; Bissery, M.-C.; Tursz, T.; Perricaudet, M.; Griscelli, F. Combined effects of docetaxel and angiostatin gene therapy in prostate tumor model. *Mol. Ther.* **2003**, *7*, 731–740. [CrossRef]
32. Nesamony, J.; Kalra, A.; Majrad, M.S.; Boddu, S.H.S.; Jung, R.; Williams, F.E.; Schnapp, A.M.; Nauli, S.M.; Kalinoski, A.L. Development and characterization of nanostructured mists with potential for actively targeting poorly water-soluble compounds into the lungs. *Pharm. Res.* **2013**, *30*, 2625–2639. [CrossRef]
33. Jafari, S.M.; Assadpoor, E.; He, Y.; Bhandari, B. Re-coalescence of emulsion droplets during high-energy emulsification. *Food Hydrocoll.* **2008**, *22*, 1191–1202. [CrossRef]
34. Bei, D.; Marszalek, J.; Youan, B.-B.C. Formulation of dacarbazine-loaded cubosomes—part I: Influence of formulation variables. *Aaps PharmScitech* **2009**, *10*, 1032–1039. [CrossRef] [PubMed]
35. Iqbal, R.; Mehmood, Z.; Baig, A.; Khalid, N. Formulation and characterization of food grade O/W nanoemulsions encapsulating quercetin and curcumin: Insights on enhancing solubility characteristics. *Food Bioprod. Process.* **2020**, *123*, 304–311. [CrossRef]
36. Pillay, V.; Fassihi, R. In vitro release modulation from crosslinked pellets for site-specific drug delivery to the gastrointestinal tract: I. Comparison of pH-responsive drug release and associated kinetics. *J. Control. Release* **1999**, *59*, 229–242. [CrossRef]
37. Jadhav, C.; Kate, V.; Payghan, S.A. Investigation of effect of non-ionic surfactant on preparation of griseofulvin non-aqueous nanoemulsion. *J. Nanostructure Chem.* **2015**, *5*, 107–113. [CrossRef]
38. Alsaab, H.O.; Sau, S.; Alzhrani, R.M.; Cheriyan, V.T.; Polin, L.A.; Vaishampayan, U.; Rishi, A.K.; Iyer, A.K. Tumor hypoxia directed multimodal nanotherapy for overcoming drug resistance in renal cell carcinoma and reprogramming macrophages. *Biomaterials* **2018**, *183*, 280–294. [CrossRef]
39. Cheriyan, V.T.; Alsaab, H.O.; Sekhar, S.; Stieber, C.; Kesharwani, P.; Sau, S.; Muthu, M.; Polin, L.A.; Levi, E.; Iyer, A.K.; et al. A CARP-1 functional mimetic loaded vitamin E-TPGS micellar nano-formulation for inhibition of renal cell carcinoma. *Oncotarget* **2017**, *8*, 104928–104945. [CrossRef]
40. Buyukozturk, F.; Benneyan, J.C.; Carrier, R.L. Impact of emulsion-based drug delivery systems on intestinal permeability and drug release kinetics. *J. Control. Release* **2010**, *142*, 22–30. [CrossRef]
41. Md, S.; Alhakamy, N.A.; Aldawsari, H.M.; Husain, M.; Kotta, S.; Abdullah, S.T.; A Fahmy, U.; Alfaleh, M.A.; Asfour, H.Z. Formulation design, statistical optimization, and in vitro evaluation of a naringenin nanoemulsion to enhance apoptotic activity in A549 lung cancer cells. *Pharmaceuticals* **2020**, *13*, 152. [CrossRef]

colloids and interfaces

MDPI

Article

Comparison between Quinoa and *Quillaja saponins* in the Formation, Stability and Digestibility of Astaxanthin-Canola Oil Emulsions

Daniela Sotomayor-Gerding [1,*], Eduardo Morales [2] and Mónica Rubilar [2,3,*]

1 Programa de Doctorado en Ciencias de Recursos Naturales, Universidad de La Frontera, Avenida Francisco Salazar 01145, Temuco 4811230, Chile
2 Scientific and Technological Bioresource Nucleus, BIOREN, Universidad de La Frontera, Avenida Francisco Salazar 01145, Temuco 4811230, Chile
3 Department of Chemical Engineering, Faculty of Engineering and Sciences, Universidad de La Frontera, Avenida Francisco Salazar 01145, Temuco 4811230, Chile
* Correspondence: d.sotomayor02@ufromail.cl (D.S.-G.); monica.rubilar@ufrontera.cl (M.R.); Tel.: +56-45-2744232 (M.R.)

Abstract: Saponins from *Quillaja saponaria* and *Chenopodium quinoa* were evaluated as natural emulsifiers in the formation of astaxanthin enriched canola oil emulsions. The aim of this study was to define the processing conditions for developing emulsions and to evaluate their physical stability against environmental conditions: pH (2–10), temperature (20–50 °C), ionic strength (0–500 mM NaCl), and storage (35 days at 25 °C), as well as their performance in an in vitro digestion model. The emulsions were characterized, evaluating their mean particle size, polydispersity index (PDI), and zeta potential. Oil-in-water (O/W) emulsions were effectively produced using 1% oil phase and 1% emulsifier (saponins). Emulsions were stable over a wide range of pH values (4–10), but exhibited particle aggregation at lower pH, salt conditions, and high temperatures. The emulsion stability index (ESI) remained above 80% after 35 days of storage. The results of our study suggest that saponins can be an effective alternative to synthetic emulsifiers.

Keywords: oil-in-water (O/W) emulsion; emulsifier; saponin; astaxanthin; *Quillaja saponaria*; *Chenopodium quinoa*; canola oil; zeta potential; particle size; in vitro digestion

Citation: Sotomayor-Gerding, D.; Morales, E.; Rubilar, M. Comparison between Quinoa and *Quillaja saponins* in the Formation, Stability and Digestibility of Astaxanthin-Canola Oil Emulsions. *Colloids Interfaces* **2022**, *6*, 43. https://doi.org/10.3390/colloids6030043

Academic Editor: Spencer Taylor

Received: 19 July 2022
Accepted: 25 August 2022
Published: 28 August 2022

1. Introduction

Emulsifiers are surface active substances that facilitate emulsion formation and promote emulsion stability, affecting the particle size and the electrical repulsion between the particles [1]. Emulsifiers can be classified as synthetic, natural, finely dispersed solids, and auxiliary agents based on their chemical structure [2]. In recent years, the demand for healthier food products, containing more natural and environmentally friendly ingredients has increased, for which the use of natural emulsifiers has been the focus of recent research [3].

Currently, proteins, phospholipids, polysaccharides, lipopolysaccharides, bioemulsifiers (e.g., saponins, sophorolipids, rhamnolipids, and mannoproteins), and bioemulsifiers isolated from plant materials or produced by fermentation using bacteria, yeasts, or fungi (e.g., glycolipids, lipoproteins, and lipopeptides) are used as natural emulsifiers [2].

Due to their natural foam-like quality, the application of saponins as natural biosurfactants to improve the surface properties of food has recently been the subject of intensive study. Saponins are secondary metabolites mainly derived from plant materials [4]. These biosurfactants commonly contain a mixture of different amphiphilic constituents that have demonstrated their ability to form micelles when dispersed in water and support the formation and stabilization of oil-in-water emulsions. Their amphiphilic nature is given by the presence of hydrophilic regions (e.g., sugar groups) and hydrophobic regions (e.g., phenolic groups) distributed within a single molecule [5]. Saponins have been found to have pharmaceutical properties of hemolytic, molluscicidal, anti-inflammatory, antioxidant, antifungal, antimicrobial, antiparasitic, antitumor, antiviral, and immune adjuvant activities [4,6]. Saponins may also be effective at inhibiting lipid oxidation in emulsions because of their radical-scavenging capacity [7].

Saponins from *Quillaja saponaria* have been used for the preparation of oil-in-water (O/W) emulsions in several studies, using medium chain triglycerides (MCT) [8,9], orange oil [10], and for the encapsulation of vitamin E [11]. Furthermore, the use of saponins from *Q. saponaria* mixed with other surfactants such as sodium caseinate, pea protein, rapeseed lecithin, egg lecithin [12], Tween 80 [13], β-lactoglobulin [14], or hydrolyzed rice glutelin [15] have been reported in the literature.

The presence of saponins has been also reported in quinoa (*Chenopodium quinoa)* [16,17]. Several studies have reported the use of *C. quinoa* extracts as emulsifier. Authors have reported the use of quinoa starch for the preparation of Pickering emulsions [18–20] and the use of protein isolated for the preparation of emulsion gels [21] and high internal phase emulsions [22–24]. However, the use of quinoa saponins in the preparation of emulsions has not been reported in the literature. On the other hand, studies have highlighted the health benefits of quinoa derived products [25] and a recent study reported a safety assessment for the oral use of saponins from *C. quinoa* in rats reporting no adverse effects under a dose of 50 mg/kg/day [26].

Although there are a wide variety of studies on the use of saponins as emulsifiers, the incorporation of bioactive ingredients has not been extensively studied. The incorporation of carotenoids such as astaxanthin in emulsions is of great interest in the food industry, as this is a pigmented compound with many health benefits [27]. Nevertheless, their utilization as nutraceutical ingredients within foods is currently limited because of their poor water-solubility, high melting point, chemical instability, and low bioavailability [28].

Consequently, the aim of this study was to evaluate saponins from *Quillaja saponaria* and *Chenopodium quinoa* as natural emulsifiers in the formation and stabilization of astaxanthin-enriched canola oil emulsions. The performance of these extracts was compared to that of a synthetic surfactant (Tween 20) that is currently widely used in the food and beverage industry to formulate emulsion-based products. The influence of environmental stresses (pH, ionic strength, and temperature) and storage on the stability of the resulting emulsions against droplet growth and gravitational separation was evaluated and the in vitro digestion was also investigated to provide information on their gastrointestinal transformation and/or absorption.

2. Materials and Methods

2.1. Materials

The extracts from *Q. saponaria* (190 g/L saponin) and *C. quinoa* were provided from South extracts S.A. (Perquenco, Chile), canola oil was purchased from a local market and astaxanthin oleoresin (Supreme Asta oil 5.0%) from Atacama Bio Natural Products S.A. (Iquique, Chile). Distilled water used in this study had a conductivity 0.90 μS/cm.

Non-ionic surfactant polyoxyethylene (20) sorbitan monolaurate (Tween 20; P1379), sodium chloride (NaCl; 746398), sodium hydroxide (NaOH; S5881), dipotassium hydrogen phosphate (K_2HPO_4; P3786), mucin from porcine stomach Type II (M2378), pepsin from porcine gastric mucosa (P7012), bile extract porcine (B8631), and pancreatin from porcine pancreas (P1750) were purchased from Sigma Aldrich Co. (St. Louis, MO, USA). All other chemicals were of analytical grade.

The *Q. saponaria* and *C. quinoa* extracts were characterized according to the AOAC methods [29], measuring their dry weight, refractive index, and solids percent in a refractometer (Abbe Mark II plus, Reichert Inc., Depew NY, USA) and pH.

2.2. Emulsion Formation and Characterization

O/W emulsions were prepared using 99% aqueous phase and 1% oil phase. The aqueous phase was obtained, dispersing 1% emulsifier (*Q. saponaria*, *C. quinoa* or Tween 20) in distilled water, and the oil phase was prepared by mixing astaxanthin (2g/L) with canola oil (1:1). The emulsions were homogenized (5000 rpm, 10 min, Pro400DS benchtop homogenizer, Pro Scientific Inc., Oxford, CT, USA) and subsequently passed through the high-pressure homogenizer (4 cycles, 100 MPa, PandaPlus 2000, GEA Niro Soavi, Parma, Italy).

The average particle size and polydispersity index (PDI) of emulsions were determined by dynamic light scattering and the surface charge (zeta potential) by electrophoretic mobility in a Zetasizer (Nano-ZS90, Malvern Instruments, Worcestershire, UK). Measurements were performed on diluted (1:100 distilled water) emulsions.

The influence of emulsifier percentage on the mean particle size and zeta potential of emulsions was evaluated using 0.1, 0.5, 1, 2, and 5% of emulsifier.

2.3. Influence of Environmental Changes on Emulsion Physical Stability

The influence of different environmental conditions that might be encountered in the processing of food on the emulsion stability were evaluated. Emulsions were prepared using 1% oil phase and 1% emulsifier. The effect of pH on emulsion stability was evaluated by manually adjusting the pH of emulsions at 1 unit interval from 2 to 10, after dilution (1:100), emulsions were evaluated in a Zetasizer, measuring the mean particle size and zeta potential. The effect of temperature on emulsion stability was investigated by measuring the average droplet size and zeta potential using a step-wise protocol, in which the temperature was changed in steps of 5 °C from 20 to 50 °C. Temperature was stabilized with a Peltier temperature control of the Zetasizer equipment. The influence of ionic strength on emulsion stability was determined by adjusting the salt concentration to between 0 and 500 mM NaCl prior to dilution (1:100) and analysis with the Zetasizer at 25 °C. Representative photographs of the emulsions were taken after 24 h of incubation with different conditions of pH (2–10) and salinity (0–500 mM NaCl).

The emulsion stability index (ESI) was determined by monitoring the extent of gravitational phase separation during storage for 35 days at 25 °C in darkness according to previous reports [30].

2.4. In Vitro Digestion of Emulsions

An in vitro gastrointestinal tract (GIT) model was used to simulate mouth, gastric, and small intestine digestion according to our previous report [31]. Freshly prepared emulsions were mixed (1:1) with simulated saliva fluid (SSF) containing mucin (5 g/L). The pH of the mixture was adjusted to 6.8 prior to incubation at 37 °C for 10 min with continuous agitation at 100 rpm to simulate the mouth phase. For the gastric phase, simulated gastric fluid (SGF) was prepared by dissolving NaCl (2 g) and HCl (7 mL) in a liter of water adjusted to pH 1.2. The previously processed emulsion was mixed with SGF (1:1, *v*/*v*) and pH adjusted to 1.5 prior to incubation at 37 °C for 10 min with continuous agitation at 100 rpm. Pepsin (3 mg) was added to the mixture after 10 min and samples were incubated for 2 h with the previous conditions (37 °C, 100 rpm). For the intestinal phase, simulated intestinal fluid (SIF) containing K_2HPO_4 (6.8 g/L), 0.2 M NaOH (190 mL/L), and maintained at pH 7.5 was mixed with the samples from the gastric phase (1:3, *v*/*v*, for a total of 30 mL) and bile extract (0.15 g). This mixture was maintained at 37 °C after adjusting the pH to 7. The small intestinal phase was simulated with a pH-stat (Metrohm USA Inc., Riverview, FL, USA) to maintain constant pH (7) of the solution by adding 0.05 M NaOH solution. The volume of NaOH required to neutralize the free fatty acids (FFA) was recorded for 20 min. Once the equipment was prepared, freshly prepared pancreatin suspension (2.5 mL; 24 mg/mL) dissolved in phosphate buffer was added to the mixture, to initiate the reaction. The amount of free fatty acids released from lipid digestion was calculated as follows:

$$FFA(mM) = (V_{NaOH\ T} - V_{NaOH\ T0}) \times M_{NaOH} \times 1000$$

Here, V_{NaOHT} is the volume (L) of sodium hydroxide required to neutralize the FFA produced, $V_{NaOH\ T0}$ is the volume (L) of sodium hydroxide added at the beginning of the reaction, and M_{NaOH} is the molarity (M) of the sodium hydroxide solution used.

2.5. Statistical Analysis

All measurements were performed in triplicate and results were expressed as the mean and the standard deviation. All the results of this study were subjected to one-way analysis of variance (ANOVA). Significant differences ($p \leq 0.05$) between means were determined by Tukey's tests.

3. Results

3.1. Emulsion Formation and Characterization

Q. saponaria and *C. quinoa* extracts were characterized (Table 1). The extracts had similar characteristics, a solid concentration of ~21%, a refractive index of 1.36, and a humidity of ~80%. The only characteristic that was a little different between the extracts was the pH. *Q. saponaria* extracts had a pH of 3.94, and the *C. quinoa* extract had a slightly lower pH of 3.49.

Table 1. Characteristics of *Quillaja saponaria* and *Chenopodium quinoa* extracts.

Characteristic	*Quillaja saponaria*	*Chenopodium quinoa*
% Solids (°Brix-TC [1])	20.83 ± 0.060	20.70 ± 0.200
Refractive Index	1.36 ± 0.000	1.36 ± 0.000
pH	3.94 ± 0.010	3.49 ± 0.010
Humidity (%)	79.39 ± 0.005	81.22 ± 0.140

[1] TC: Temperature compensated.

Emulsions were effectively produced using saponins from *Q. saponaria* and *C. quinoa* (Table 2). Tween 20 emulsions were produced as a control for comparison. *Q. saponaria* emulsions had the smallest mean particle size, 189 nm, while *C. quinoa* emulsions had a mean particle size of 316 nm. Emulsions prepared under the same conditions using Tween 20 had a mean particle size of 205 nm. Polydispersity index was similar between the emulsions, 0.32, 0.33, and 0.35, for emulsions prepared with *Q. saponaria*, *C. quinoa*, or Tween 20 as emulsifier, respectively.

Table 2. Mean particle size, zeta potential and polydispersity index (PDI) of emulsions prepared with *Q. saponaria* or *C. quinoa* extracts as emulsifier.

Emulsifier	Mean Particle Size (nm)	PDI	Zeta Potential (mV)
Q. saponaria	189 ± 5 a	0.32 ± 0.005 a	−29.6 ± 0.3 a
C. quinoa	316 ± 8 b	0.33 ± 0.003 a	−27.7 ± 1.1 ab
Tween 20	205 ± 10 a	0.35 ± 0.014 a	−26.0 ± 1.2 b

Different lowercase letters in a column indicate significant differences ($p \leq 0.05$) among the different experimental groups (*Q. saponaria*, *C. quinoa*, Tween 20).

Regarding the zeta potential values, all emulsions presented highly negative surface charges. *Q. saponaria* emulsions had a zeta potential of −29.6 mV, *C. quinoa* emulsions had a less negative charge of −27.7 mV and Tween 20 emulsions had a zeta potential of −26.0 mV. Statistical analysis showed that emulsions prepared with *C. quinoa* saponins had significantly higher particle size.

The influence of emulsifier concentration was assessed (Figure 1), evaluating five emulsifier percentages (0.1, 0.5, 1, 2, and 5%). For *Q. saponaria* emulsions, the average particle size decreased from 329 nm to 189 nm when the emulsifier concentration increased from 0.1 to 1%; however, the average particle size increased from 189 nm to 348 nm when the emulsifier concentration increased from 1 to 5%. The smallest mean particle size (189 nm, PDI: 0.32) was obtained using 1% emulsifier and 1% oil phase. *C. quinoa* emulsions had a similar mean particle size (~330 nm) between 0.1 and 1% emulsifier, statistical analysis did not show significant differences within this range. However, a significant increase in particle size was observed using 2 and 5% emulsifier (Figure 1a). Using 2% *C. quinoa* saponins as emulsifier, an average particle size of 931 nm and a PDI of 0.61 were obtained, while using 5% emulsifier an average size of 3893 nm and a PDI of 1 were obtained. The smallest particle size (316 nm, PDI: 0.33) was obtained using a 1% concentration of *C. quinoa* emulsifier.

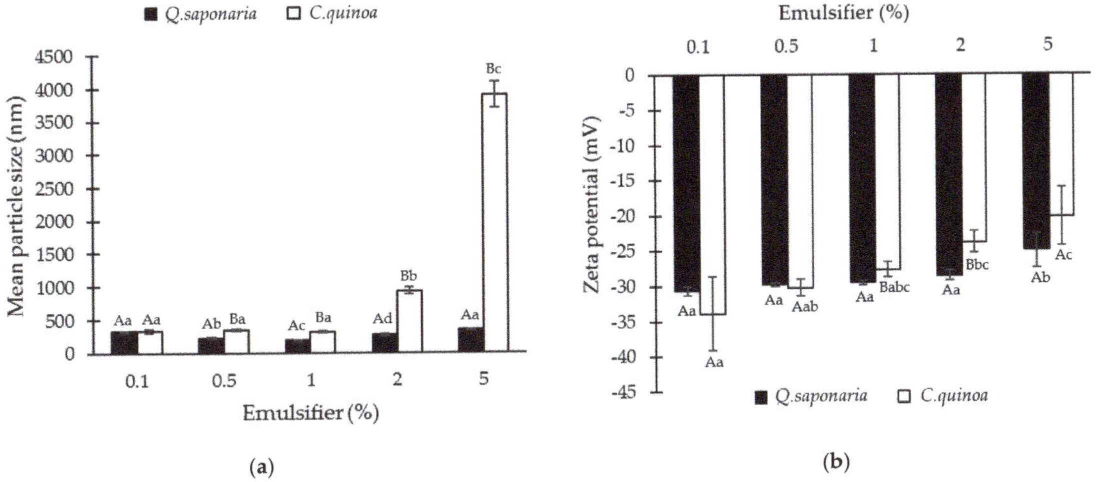

Figure 1. Influence of emulsifier concentration on emulsion (**a**) mean particle size and (**b**) zeta potential. Different capital letters indicate significant differences ($p \leq 0.05$) of the mean particle size or zeta potential among the different emulsion types (*C. quinoa*, *Q. saponaria*) within the same emulsifier concentration. Different lowercase letters indicate significant differences ($p \leq 0.05$) of the mean particle size or zeta potential among the different emulsifier concentrations (0.1, 0.5, 1, 2, 5%) within the same emulsion type.

Regarding the zeta potential, the effect of emulsifier concentration had the same tendency for emulsions generated with saponins of *Q. saponaria* and *C. quinoa*, the zeta potential increased as the emulsifier percent increased (Figure 1b), obtaining a less negative charge. Statistically significant differences between *Q. saponaria* and *C. quinoa* emulsions were obtained using 1 and 2% of emulsifier. The highest zeta potential value was −20.3 mV for *C. quinoa* emulsions with a 5% of emulsifier.

The 1% emulsifier concentration was used in the following evaluations, considering these results, small particle size, and zeta potential above ±20 mV.

3.2. Influence of Environmental Changes on Emulsion Physical Stability

Emulsions were physically stable over a wide range of pH values (4–10), the mean particle size had no significant changes between this range (Figure 2a), which can also be observed in Figure 3 where no phase separation, creaming, or other form of destabilization of the emulsion is observed. In addition, an increase in the negative charge of the particles was observed (Figure 2b) for all emulsions. For example, *Q. saponaria* zeta potential changed from −28.3 mV at pH 4 to −39.3 mV at pH 10. However, *Q. saponaria* and

Tween 20 emulsions were destabilized at pH 2, while *C. quinoa* emulsions were destabilized at pH 2 and 3 (Figure 2a).

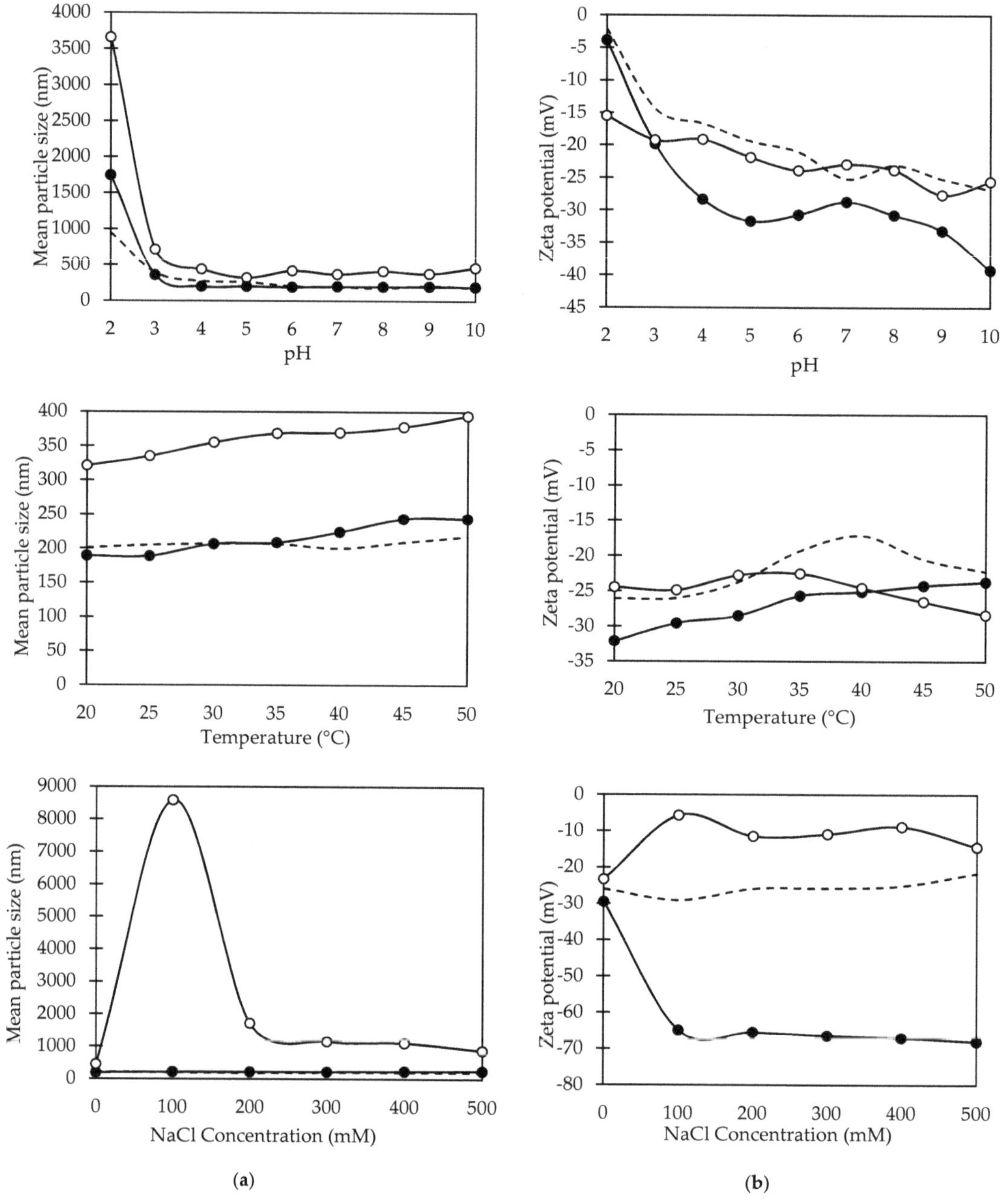

Figure 2. Influence of environmental changes: temperature (20–50 °C), pH (2–10) and ionic strength (0–500 mM) on emulsion stability expressed as changes in (**a**) mean particle size and (**b**) zeta potential for emulsions produced with: (•) *Quillaja saponaria*, (○) *Chenopodium quinoa* or (–) Tween 20 as emulsifier. Emulsions were prepared using 1% oil phase and 1% emulsifier.

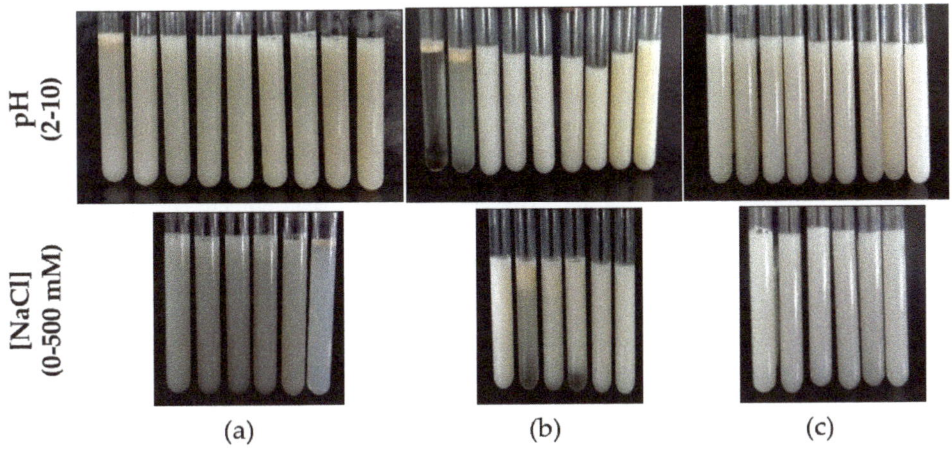

Figure 3. Photographs of emulsions prepared with (**a**) *Quillaja saponaria*, (**b**) *Chenopodium quinoa*, or (**c**) Tween 20 after incubation for 24 h with different environmental conditions: pH (2–10) or ionic strength (0–500 mM NaCl).

The mean particle size increased significantly at pH 2 for all emulsions, *C. quinoa* emulsions reached a mean particle size of 3656 nm, *Q. saponaria* 1741 nm and Tween 20,946 nm. Particle aggregation and emulsion destabilization can also be observed in Figure 3, a creaming line is produced for emulsions generated with *Q. saponaria* at pH 2 and for emulsions generated with *C. quinoa* at pH 2 and 3. Emulsion destabilization can be correlated to the reduction of the negative charge of the particles. The zeta potential of the particles increased significantly at acid pH. At pH 2, *Q. saponaria* emulsions had a zeta potential of −3.9 mV, *C. quinoa* emulsions showed a zeta potential of −15.5 mV and Tween 20 emulsions had a zeta potential of −2.1 mV.

The effect of temperature on emulsions prepared with different emulsifiers was evaluated by measuring the change in the mean particle size (PS) and zeta potential (ZP). The changes on temperature led to some fluctuations in the mean particle size and zeta potential. For emulsions prepared with *Q. saponaria* and *C. quinoa*, the particle size increased linearly with increasing temperature, but at a different rate for each emulsion (Figure 2). The equations for emulsions, *Q. saponaria* and *C. quinoa* were: $PS = 2.08 \cdot [T°] + 142.08$, $R^2 = 0.9473$; $PS = 2.28 \cdot [T°] + 280.44$, $R^2 = 0.9555$, respectively. For *Q. saponaria* emulsions the mean particle size increased significantly over 40 °C. For emulsions prepared with Tween 20, the size remained relatively stable, with a standard deviation between samples of only 6 nm, however, a linear regression did not provide a good fit ($R^2 < 0.5$), given the fluctuations in size.

The zeta potential of emulsions prepared with *Q. saponaria* extract as emulsifier increased linearly with increasing temperature ($ZP = 0.2812 \cdot [T°] - 36.827$, $R^2 = 0.9362$). For emulsions prepared with *C. quinoa* extract and Tween 20 as emulsifier, the correlation between zeta potential and temperature was not clear. For Tween 20 emulsions, the lowest negative charge (−17.2 mV) was observed at 40 °C, in the case of emulsions with *C. quinoa*, the lowest negative charge was observed at 35 °C, obtaining a zeta potential of −22.5 mV.

The presence of salts in the aqueous phase significantly affected the stability of the emulsions. A linear increase in the mean particle size was observed for *Q. saponaria* emulsions ($PS = 0.0911 \cdot [NaCl] + 193.84$, $R^2 = 0.893$), emulsion destabilization is observed at 500 mM NaCl (Figure 3a). *C. quinoa* emulsions proved to be more sensitive to the influence of salinity, observing destabilization at 100 mM NaCl, where a mean particle size of 8584 ± 672 nm was determined (Figure 2a) and the PDI reached the value of 1. Between 200 mM NaCl and 500 mM NaCl the mean particle size was 3 times higher than the emulsion at its normal state and the PDI values were above 0.59. Emulsion destabilization effect can be observed in Figure 3b. Tween 20 emulsions remained stable at different concentrations of NaCl (Figure 3c), a slight decrease in the mean particle size was observed, varying from 205 nm at 0 mM NaCl to 194 nm at 500 mM NaCl.

Regarding zeta potential, *Q. saponaria* emulsions showed a significant increase in the negative charge of the particles with increasing ionic strength. At 0 mM NaCl concentration, *Q. saponaria* emulsions had a zeta potential of −29.6 mV, while emulsions exposed to higher concentrations of NaCl (100–500 mM) had a zeta potential that varied between −64.9 and −68.1 mV. For *C. quinoa* emulsions, the zeta potential reacted in the opposite way, the negative charge was reduced, at 100 mM the highest zeta potential of −5.7 mV was determined, between 200 and 500 mM NaCl zeta potential remained close to −10 mV. In the case of Tween 20 emulsions, the negative charge remained stable, decreasing slightly as the NaCl concentration increased.

The emulsion stability index (ESI) was determined, monitoring the extent of gravitational phase separation for 35 days (Figure 4).

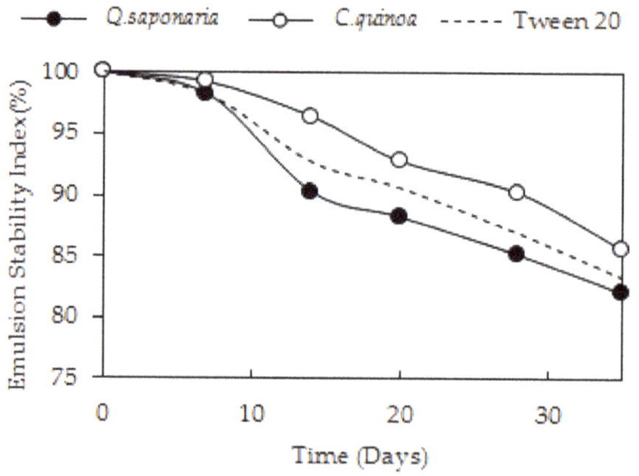

Figure 4. Emulsion stability index (ESI) after 35 days of storage at 25 °C for emulsions prepared with (•) *Quillaja saponaria*, (○) *Chenopodium quinoa* or (–) Tween 20 as emulsifier.

All emulsions were stable and homogeneous immediately after preparation, starting with an ESI of 100%. Phase separation increased with storage time, revealing the creaming of emulsions. ESI decreased with storage at slightly different rates depending on the emulsifier used, represented by the equations: ESI = −0.5363 · [time] + 99.935, R^2 = 0.9585, for *Q. saponaria*; ESI = −0.4192 · [time] + 101.33, R^2 = 0.9685, for *C. quinoa* and ESI = −0.4904 · [time] + 100.44 R^2 = 0.9885, for Tween 20. At day 35 the ESI values were 82.04 ± 1.13%, 85.64 ± 0.89%, and 83.27 ± 4.33% for emulsions prepared with *Q. saponaria*, *C. quinoa*, or Tween 20, respectively. No significant differences were found between emulsions at day 35 of storage, indicating that the emulsifier type did not affect their stability during storage time.

3.3. *In Vitro Digestion of Emulsions*

Emulsions remained stable during the mouth and stomach phases, and the release of fatty acids occurred at intestinal level where their absorption usually takes place. After approximately seven minutes of reaction, all fatty acids were released. The curves of the fatty acid release were very similar for the different types of emulsifier (Figure 5). *C. quinoa* and Tween 20 emulsions reached a similar concentration of fatty acids released, 87.6 mM and 90.6 mM, respectively. However, *Q. saponaria* emulsions achieved a slightly higher amount of fatty acid released, 101.7 mM. Statistical analysis indicates that there are no significant differences between the curves, which suggests that the emulsifier type does not play an important role in this assay.

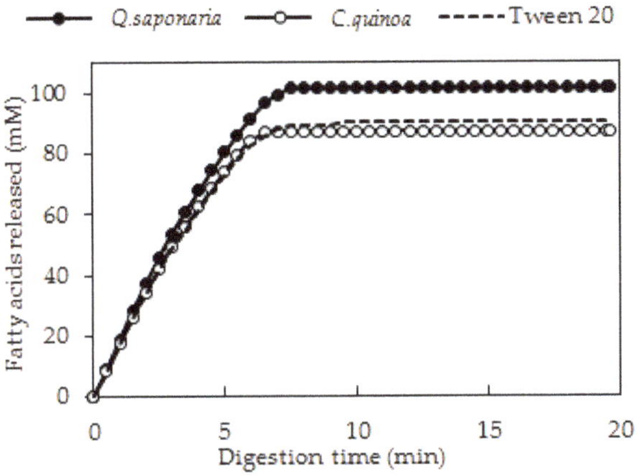

Figure 5. Fatty acids release curves generated during intestinal digestion for emulsions generated with (•) *Quillaja saponaria*, (○) *Chenopodium quinoa*, or (–) Tween 20 as emulsifier.

4. Discussion

Saponins from *Q. saponaria* and *C. quinoa* were effective for the formation of astaxanthin enriched oil-in-water emulsions. The smallest mean particle size was obtained using *Q. saponaria* saponins. According to previous studies, saponins from *Q. saponaria* are usually effective at forming small droplets, due to their relatively low molecular weight (~1.67 kDa) they tend to form thin interfacial layers [3]. However, for emulsions prepared with *C. quinoa* saponins, a significantly higher size was obtained. In general, a small particle size is sought since they can improve the system, having a better stability and bioaccessibility after ingestion [31]. All emulsions presented highly negative surface charges, ranging from −29.6 to −26.0 mV. According to previous studies [32], the magnitude of zeta potential gives an indication of the potential stability of the colloidal system and generally, values greater than ±20 mV produce systems that are stable over time.

After evaluating the effectiveness of the extracts to form emulsions, the effect of different proportions of the emulsifier was evaluated. Studies forming oil-in-water emulsions with saponins use emulsifier percentages ranging from 0.001% to 2% [8,10,11]. In our study, the range of 0.1 to 5% was evaluated.

Previous studies have reported a dependence between emulsifier concentration and particle size, where droplet size decreases with increasing emulsifier concentration [32]. For oil-in-water emulsions containing heavy crude, droplet size decreased with increasing Tween 20 concentrations from 0.1 to 2.1 wt% [33]. Similarly, in the production of food grade Pickering emulsions average particle size decreased from ~25 to ~0.15 µm as the concentration of Tween 20 increased from 1 to 4% [34]. Authors attributed this effect to (i) a higher surfactant concentration means that a larger surface area can be stabilized during homogenization; (ii) a higher surfactant concentration leads to faster coverage of the droplet surfaces by surfactant molecules, and therefore better protection against recoalescence [35]. Our study coincided with the tendency reported by other authors in the range between 0.1 to 1%, however, above 2% an increase in the obtained particle size was observed. For *Q. saponaria* emulsions, the average particle size increased 1.84-fold between 1% and 5% emulsifier concentration, and the PDI increased to 0.4. For *C. quinoa* emulsions, the effect was more significant, between 1 and 2% of emulsifier concentration, the average particle size increased 2.95-fold and the PDI increased to 0.6; between 1 and 5% of emulsifier concentration, the average particle size increased approximately 12-fold and the PDI reached the value of 1, indicating that the sample is highly polydisperse with multiple size populations. This destabilization of the emulsion could be an effect of the reduction of the negative charge of the particles by increasing the concentration of the emulsifier. When using 5% of *C. quinoa* saponins as emulsifier, the limit of the stable zone for colloidal particles is reached. For electrostatically-stabilized emulsions, the magnitude of the zeta potential should be greater than about 20 mV to produce systems that are stable during long-term storage [32]. It should be considered that the extracts from *Q. saponaria* and *C. quinoa* are compositionally complex materials that will contain a variety of surface active components with different surface activities, for *Q. saponaria* around 100 saponins have been reported, where the majority of these consist of quillaic acid substituted with oligosaccharides at C-3 and C-28 [36], while for *C. quinoa* the main sapogenins reported are oleanolic acid, hederagenin, and phytolaccagenin [17], consequently, it is possible that a portion of the components of these extracts may affect the stability of the emulsion.

Accordingly, the use of these extracts as an emulsifier should be kept in the range of 0.1 to 1% to ensure the formation and stability of the emulsion.

Emulsions may become unstable through a number of different instability mechanisms (e.g., flocculation, coalescence, Ostwald ripening, and gravitational separation), which depend on storage conditions such as pH, ionic strength, and temperature [37]. We therefore examined the influence of different environmental conditions that might be encountered in the processing of food on the stability of emulsions.

The acid pH had a strong effect on the stability of emulsions, zeta potential values increased significantly, and particle size increased generating a cream layer at pH 2 for *Q. saponaria* emulsions, and at pH 2 and 3 for *C. quinoa* emulsions. Previous studies have shown that changes in pH can have an immediate and significant effect on the zeta potential of emulsions [38,39]. An explication to this effect is that the high concentration of protons present in the aqueous phase at acidic pH neutralize the negative initial charges of the particles. Therefore, the increase in the mean particle size could be a secondary effect of lowering the pH, i.e., at lower pH, the charge of the particles decreases, and hence the electrostatic repulsion becomes insufficient, causing the phenomenon of aggregation observed. This correlation between the reduction in absolute zeta potential and the increase in particle size has been reported previously [39].

Previous studies report a similar trend regarding the zeta potential, where, as the pH increases, the zeta potential becomes more negative, and as the pH decreases, the negative charge is reduced, explaining the effect on the adsorption of hydrogen H^+ and hydroxyl OH^- ions [40]. Saponin emulsions, especially emulsions prepared with *C. quinoa* were more susceptible to the effect of pH, unlike the emulsions prepared with Tween 20 in which the creaming process was not observed. Furthermore, studies suggest that acidic pH values could be unfavorable for emulsions containing carotenoids, since the rate of degradation of the carotenoid at acidic pH is higher [41]. Consequently, the acid pH would not be favorable for emulsions containing carotenoids due to their degradation and destabilization.

The effect of temperature was also evaluated, observing an increase in particle size, especially in emulsions prepared with saponins. The increase on the mean particle size might be related to the Brownian motion of the particles. Considering that higher temperatures increase the movement of particles and hence increase the collisions between the particles, which could generate coalescence or aggregation phenomena [32]. The emulsions prepared with Tween 20 remained relatively stable when facing different temperatures, having a standard deviation between the samples of only 6 nm. Although linear regression did not provide a good fit ($R^2 = 0.4723$), it can be seen that there are no significant changes in the average particle size.

The ionic strength of emulsified foods may vary considerably depending on the nature of the food products in which the oil droplets are present. Consequently, the effect of ionic strength on the emulsion stability was evaluated. Studies suggest that high concentration of salts destabilize emulsions generating particle aggregation [37]. The phenomenon where high concentration of ions in the aqueous phase (produced by the dissociation of salts) invalidate the repulsive charges between the particles is known as "electrostatic screening" [42].

In the case of *Q. saponaria* emulsions, an adsorption of anions from the aqueous phase is presumed, which is observed in the significant decrease in the zeta potential, generating a destabilization of the system at high NaCl concentrations. *For C. quinoa,* an unusual effect was observed where the greatest destabilization was generated at a concentration of 100 mM NaCl, the negative charge was reduced, the particle size increased almost 20 times, and the polydispersity index reached the value of 1, generating a thick line of creaming, which can be observed in Figure 3b. Between 200 and 500 mM NaCl, an increase in the average particle size and emulsion destabilization were also observed, however, it was not as significant as previously observed. In the case of Tween 20 emulsions, significant changes were determined, however, they did not generate creaming by particle aggregation. The size-enhancing effect has been previously reported in n-alkane emulsions where an increase from 450 nm to 1300 nm in the presence of NaCl is described [38].

The effect of storage time on emulsion stability was evaluated. After 7 days, the gravitational separation of phases begins to be observed, the ESI values are reduced to ~85% at 35 days of evaluation. In emulsion with small droplets, this phenomenon is expected, given that they are thermodynamically unstable systems. In nanoemulsions, Brownian motion dominates the movement of the particles and destabilization is caused by Ostwald ripening [43].

Finally, the release of fatty acids was evaluated after in vitro digestion, release curves were very similar; however, *Q. saponaria* emulsions achieved a higher amount of fatty acids released. The slight differences between the curves might be given by the initial different mean particle sizes between the types of emulsions [31].

5. Conclusions

Our study demonstrates that oil-in-water astaxanthin enriched emulsions can be effectively produced using a 1% oil phase and 1% saponins from *Q. saponaria* or *C. quinoa* extract as emulsifier. The average particle size depended on the emulsifier used, the smallest particle size was obtained with saponins from *Q. saponaria* and the largest particle size with saponins from *C. quinoa.* Additionally, it was determined that the concentration of saponins as emulsifier significantly affected the particle size and the zeta potential obtained. Emulsions with a size smaller than 350 nm could be obtained using *Q. saponaria* saponin concentrations between 0.1 and 5%, and *C. quinoa* saponin concentrations between 0.1 and 1%. Concentrations of *C. quinoa* saponins between 2 and 5% generated destabilization of the emulsion.

The emulsions had slightly different responses to the effect of environmental conditions. Tween 20 emulsions were stable over a wide range of pH values (3–8), salt concentrations (0–500 mM NaCl), and temperatures (20–50 °C); *Q. saponaria* emulsions were unstable at low pH values (2), high NaCl concentrations, and high temperatures (over 40 °C) and *C. quinoa* emulsions were highly unstable to droplet aggregation and phase separation at low pH values (2–3) and moderate ionic strengths (>100 mM NaCl). Emulsions remained stable during in vitro digestion, releasing fatty acids at intestinal level. *C. quinoa* emulsions released fatty acids at the same level as Tween 20 emulsions; however, *Q. saponaria* emulsions achieved a slightly higher amount of released fatty acids.

The results of our study contribute to increase the knowledge about the use of saponins from different natural sources in the formation of oil-in-water emulsions, and suggest that saponins can be an effective alternative to synthetic emulsifiers and even superior in terms of releasing bioactive compounds.

Author Contributions: Conceptualization, D.S.-G. and M.R.; methodology, D.S.-G.; validation, D.S.-G.; formal analysis, D.S.-G. and E.M.; investigation, D.S.-G.; resources, M.R.; data curation, D.S.-G.; writing—original draft preparation, D.S.-G.; writing—review and editing, D.S.-G., E.M. and M.R.; visualization, D.S.-G.; supervision, M.R.; project administration, M.R.; funding acquisition, D.S.-G. and M.R. All authors have read and agreed to the published version of the manuscript.

Funding: This research was funded by ANID through Doctoral Scholarship n° 21150735 and through FONDECYT project n° 1160558.

Institutional Review Board Statement: Not applicable.

Informed Consent Statement: Not applicable.

Data Availability Statement: Not applicable.

Acknowledgments: The authors would also like to acknowledge the support of the Scientific and Technological Bioresource Nucleus (BIOREN) at the Universidad de La Frontera for granting access to their equipment. Special thanks to Reinaldo Briones from South Extracts company that supplied the *Quillaja saponaria* and *Chenopodium quinoa* extracts.

Conflicts of Interest: The authors declare no conflict of interest. The funders had no role in the design of the study; in the collection, analyses, or interpretation of data; in the writing of the manuscript; or in the decision to publish the results.

References

1. Hasenhuettl, G.L. Overview of Food Emulsifiers. In *Food Emulsifiers and Their Applications*, 2nd ed.; Hasenhuettl, G.L., Hartel, R.W., Eds.; Springer: New York, NY, USA, 2008; pp. 1–9. [CrossRef]
2. Marhamati, M.; Ranjibarm, G.; Rezaie, M. Effects of emulsifiers on the physicochemical stability of Oil-in-water Nanoemulsions: A critical review. *J. Mol. Liq.* **2021**, *340*, 117218. [CrossRef]
3. Ozturk, B.; McClements, D.J. Progress in natural emulsifiers for utilization in food Emulsions. *Curr. Opin. Food Sci.* **2016**, *7*, 1–6. [CrossRef]
4. Cheok, C.Y.; Salman, H.A.K.; Sulaiman, R. Extraction and quantification of saponins: A review. *Food Res. Int.* **2014**, *59*, 16–40. [CrossRef]
5. McClements, D.J.; Gumus, C.E. Natural emulsifiers — Biosurfactants, phospholipids, biopolymers, and colloidal particles: Molecular and physicochemical basis of functional performance. *Adv. Colloid Interface Sci.* **2016**, *234*, 3–26. [CrossRef]
6. Sparg, S.G.; Light, M.E.; van Staden, J. Biological activities and distribution of plant saponins. *J. Ethnopharmacol.* **2004**, *94*, 219–243. [CrossRef]
7. Tippel, J.; Gies, K.; Harbaum-Piayda, B.; Steffen-Heins, A.; Drusch, S. Composition of Quillaja saponin extract affects lipid oxidation in oil-in-water emulsions. *Food Chem.* **2016**, *221*, 386–394. [CrossRef] [PubMed]
8. Yang, Y.; Leser, M.E.; Sher, A.A.; McClements, D.J. Formation and stability of emulsions using a natural small molecule surfactant: Quillaja saponin (Q-Naturale). *Food Hydrocoll.* **2013**, *30*, 589–596. [CrossRef]
9. Zhu, Z.; Wen, Y.; Yi, J.; Cao, Y.; Liu, F.; McClements, D.J. Comparison of natural and synthetic surfactants at forming and stabilizing nanoemulsions: Tea saponin, Quillaja saponin, and Tween 80. *J. Colloid Interface Sci.* **2019**, *536*, 80–87. [CrossRef]
10. Zhang, J.; Bing, L.; Reineccius, G.A. Comparison of modified starch and Quillaja saponins in the formation and stabilization of flavor nanoemulsions. *Food Chem.* **2016**, *192*, 53–59. [CrossRef]
11. Ozturk, B.; Argin, S.; Ozilgen, M.; McClements, D.J. Formation and stabilization of nanoemulsion-based vitamin E delivery systems using natural surfactants: Quillaja saponin and lecithin. *J. Food Eng.* **2014**, *142*, 57–63. [CrossRef]
12. Reichert, C.L.; Salminen, H.; Badolato Bönisch, G.; Schäfer, C.; Weiss, J. Concentration effect of Quillaja saponin - Co-surfactant mixtures on emulsifying properties. *J. Colloid Interface Sci.* **2018**, *519*, 71–80. [CrossRef] [PubMed]
13. Riquelme, N.; Zuñiga, R.N.; Arancibia, C. Physical stability of nanoemulsions with emulsifier mixtures: Replacement of Tween 80 with quillaja saponin. *LWT - Food Sci. Technol.* **2019**, *111*, 760–766. [CrossRef]
14. de Faria, J.T.; de Oliveira, E.B.; Rodrigues Minim, V.P.; Minim, L.M. Performance of Quillaja bark saponin and β-lactoglobulin mixtures on emulsion formation and stability. *Food Hydrocoll.* **2017**, *67*, 178–188. [CrossRef]
15. Xu, X.; Sun, Q.; McClements, D.J. Enhancing the formation and stability of emulsions using mixed natural emulsifiers: Hydrolyzed rice glutelin and quillaja saponin. *Food Hydrocoll.* **2019**, *89*, 396–405. [CrossRef]
16. Nickel, J.; Spanier, L.P.; Botelho, F.T.; Gularte, M.A.; Helbig, E. Effect of different types of processing on the total phenolic compound content, antioxidant capacity, and saponin content of *Chenopodium quinoa* Willd grains. *Food Chem.* **2016**, *209*, 139–143. [CrossRef]
17. Gómez-Caravaca, A.M.; Iafelice, G.; Verardo, V.; Marconi, E.; Caboni, M.F. Influence of pearling process on phenolic and saponin content in quinoa (*Chenopodium quinoa Willd*). *Food Chem.* **2014**, *157*, 174–178. [CrossRef]
18. Lin, X.; Li, S.; Yin, J.; Chang, F.; Wang, C.; He, X.; Huang, Q.; Zhang, B. Anthocyanin-loaded double Pickering emulsion stabilized by octenylsuccinate quinoa starch: Preparation, stability and in vitro gastrointestinal digestion. *Int. J. Biol. Macromol.* **2020**, *152*, 1233–1241. [CrossRef]
19. Kierulf, A.; Whaley, J.; Liu, W.; Enayati, M.; Tan, C.; Perez-Herrera, M.; You, Z.; Abbaspourrad, A. Protein content of amaranth and quinoa starch plays a key role in their ability as Pickering emulsifiers. *Food Chem.* **2020**, *315*, 126246. [CrossRef]

20. Zhang, L.; Xiong, T.; Wang, X.F.; Chen, C.L.; He, X.D.; Zhang, C.; Wu, C.; Li, Q.; Ding, X.; Qiang, J.Y. Pickering emulsifiers based on enzymatically modified quinoa starches: Preparation, microstructures, hydrophilic property and emulsifying property. *Int. J. Biol. Macromol.* **2021**, *190*, 130–140. [CrossRef]
21. Lingiardi, N.; Galante, M.; de Sanctis, M.; Spelzini, D. Are quinoa proteins a promising alternative to be applied in plant-based emulsion gel formulation? Review. *Food Chem.* **2022**, *394*, 133485. [CrossRef]
22. Cen, K.; Yu, X.; Gao, C.; Feng, X.; Tang, X. Effects of different vegetable oils and ultrasonicated quinoa protein nanoparticles on the rheological properties of Pickering emulsion and freeze-thaw stability of emulsion gels. *J. cereal Sci.* **2021**, *102*, 103350. [CrossRef]
23. Zhang, R.; Cheng, L.; Luo, L.; Hemar, Y.; Yang, Z. Formation and characterisation of high-internal-phase emulsions stabilized by high-pressure homogenised quinoa protein isolate. *Colloids Surf. A Physicochem. Eng. Asp.* **2021**, *631*, 127688. [CrossRef]
24. Zuo, Z.; Zhang, X.; Li, T.; Zhou, J.; Yang, Y.; Bian, X.; Wang, L. High internal phase emulsions stabilized solely by sonicated quinoa protein isolate at various pH values and concentrations. *Food Chem.* **2022**, *378*, 132011. [CrossRef] [PubMed]
25. Graf, B.L.; Rojas-Silva, P.; Rojo, L.E.; Delatorre-Herrera, J.; Baldeon, M.E.; Raskin, I. Innovations in Health Value and Functional Food Development of Quinoa (*Chenopodium quinoa Willd.*). *Compr. Rev. Food Sci. Food Saf.* **2015**, *14*, 431–445. [CrossRef]
26. Zhang, R.; Zhai, Q.; Yu, Y.; Li, X.; Zhang, F.; Hou, Z.; Cao, Y.; Feng, J.; Xue, P. Safety assessment of crude saponins from *Chenopodium quinoa* willd. husks: 90-day oral toxicity and gut microbiota & metabonomics study in rats. *Food Chem.* **2022**, *375*, 131655. [CrossRef] [PubMed]
27. Fakhri, S.; Abbaszadeh, F.; Dargahi, L.; Jorjani, M. Astaxanthin: A mechanistic review on its biological activities and health benefits. *Pharmacol. Res.* **2018**, *136*, 1–20. [CrossRef]
28. Anderson, J.S.; Sunderland, R. Effect of extruder moisture and dryer processing temperature on vitamin C and E and astaxanthin stability. *Aquaculture* **2002**, *207*, 137–149. [CrossRef]
29. AOAC. *Official Methods of Analysis*, 16th ed.; Association of Official Analytical Chemists: Washington, DC, USA, 1999.
30. Karimi, N.; Mohammadifar, M.A. Role of water soluble and water swellable fractions of gum tragacanth on stability and characteristic of model oil in water emulsion. *Food Hydrocoll.* **2014**, *37*, 124–133. [CrossRef]
31. Sotomayor-Gerding, D.; Oomah, B.D.; Acevedo, F.; Morales, E.; Bustamante, M.; Shene, C.; Rubilar, M. High carotenoid bioaccessibility through linseed oil nanoemulsions with enhanced physical and oxidative stability. *Food Chem.* **2016**, *199*, 463–470. [CrossRef]
32. Piorkowski, D.T.; McClements, D.J. Beverage emulsions: Recent developments in formulation, production, and applications. *Food Hydrocoll.* **2014**, *42*, 5–41. [CrossRef]
33. Shen, W.; Koirala, N.; Mukherjee, D.; Lee, K.; Zhao, M.; Li, J. Tween 20 Stabilized Conventional Heavy Crude Oil-In-Water Emulsions Formed by Mechanical Homogenization. *Front. Environ. Sci.* **2022**, *10*, 873730. [CrossRef]
34. Pawlik, A.; Kukurukji, D.; Norton, I.; Spyropoulos, F. Food-grade Pickering emulsions stabilised with solid lipid particles. *Food Funct.* **2016**, *7*, 2712–2721. [CrossRef] [PubMed]
35. Jafari, S.M.; Assadpoor, E.; He, Y.H.; Bhandari, B. Re-coalescence of emulsion droplets during high-energy emulsification. *Food Hydrocoll.* **2008**, *22*, 1191–1202. [CrossRef]
36. Kite, G.C.; Howes, M.J.; Simmonds, M.S. Metabolomic analysis of saponins in crude extracts of *Quillaja saponaria* by liquid chromatography/mass spectrometry for product authentication. *Rapid Commun. Mass Spectrom.* **2004**, *18*, 2859–2870. [CrossRef]
37. McClements, D.J. *Food Emulsions: Principles, Practices and Techniques*, 2nd ed.; CRC Press: Boca Raton, FL, USA, 2004. [CrossRef]
38. Chibiwski, E.; Soltys, E.; Lazarz, M. Model studies on the n-alkane emulsions stability. In *Trends in Colloid and Interface Science XI*; Rosenholm, J.B., Lindman, B., Stenius, P., Eds.; Progress in Colloid & Polymer Science; Steinkopff: Dresden, Germany, 1997; Volume 105, pp. 260–267. [CrossRef]
39. Wiacek, A.E. Effect of ionic strength on electrokinetic properties of oil/water emulsions with dipalmitoylphosphatidylcholine. *Colloids Surf. A Physicochem. Eng. Asp.* **2007**, *302*, 141–149. [CrossRef]
40. Wiacek, A.E.; Chibowski, E.; Wilk, K. Studies of oil-in-water emulsion stability in the presence of new dicephalic saccharide-derived surfactants. *Colloids Surf. B* **2002**, *25*, 243–256. [CrossRef]
41. Boon, C.S.; McClements, D.J.; Weiss, J.; Decker, E.A. Factors influencing the chemical stability of carotenoids in foods. *Crit. Rev. Food Sci. Nutr.* **2010**, *50*, 515–532. [CrossRef]
42. Aoki, T.; Decker, E.A.; McClements, D.J. Influence of environmental stresses on stability of O/W emulsions containing droplets stabilized by multilayered membranes produced by a layer-by-layer electrostatic deposition technique. *Food Hydrocoll.* **2005**, *19*, 209–220. [CrossRef]
43. McClements, D.J. Edible nanoemulsions: Fabrication, properties, and functional performance. *Soft Matter* **2011**, *7*, 2297–2316. [CrossRef]

MDPI
St. Alban-Anlage 66
4052 Basel
Switzerland
www.mdpi.com

Colloids and Interfaces Editorial Office
E-mail: colloids@mdpi.com
www.mdpi.com/journal/colloids

www.ingramcontent.com/pod-product-compliance
Lightning Source LLC
LaVergne TN
LVHW070143100526
838202LV00015B/1881
9783036589503